INFINITE-DIMENSIONAL TOPOLOGY

North-Holland Mathematical Library

VOLUME 43

NORTH-HOLLAND
AMSTERDAM · NEW YORK · OXFORD · TOKYO

Infinite-Dimensional Topology

Prerequisites and Introduction

J. van Mill

Vrije Universiteit
Amsterdam, The Netherlands

1989

NORTH-HOLLAND
AMSTERDAM · NEW YORK · OXFORD · TOKYO

ISBN: 0 444 87134 9

Publishers:

ELSEVIER SCIENCE PUBLISHERS B.V.
P.O. BOX 103
1000 AC AMSTERDAM
THE NETHERLANDS

Sole distributors for the U.S.A. and Canada:

ELSEVIER SCIENCE PUBLISHING COMPANY, INC.
655 Avenue of the Americas
NEW YORK, NY 10010
U.S.A.

Library of Congress Cataloging-in-Publication Data

Mill, J. van.
Infinite Dimensional Topology. Prerequisites and
Introduction/J. van Mill.
p. cm. – (North-Holland mathematical library; v. 43)
Bibliography: p.
Includes index.
ISBN 0 444 87133 0. ISBN 0 444 87134 9 (pbk.)
1. Topology. 2. Dimension theory (Topology) 3. Infinite
dimensional manifolds. I. Title. II. Series.
QA611.M54 1988
514-dc 19

88-31299
CIP

PRINTED IN THE NETHERLANDS

Aan mijn kinderen Josine en Marije.

Aan de topologen van vakgroep AMT.

Preface

The first part of this book is intended as a text for graduate courses in topology. In chapters 1 through 5 part of the basic material of plane topology, combinatorial topology, dimension theory and **ANR** theory is presented. For a student who will go on in geometric or algebraic topology this material is a prerequisite for later work. Chapter 6 is an introduction to infinite-dimensional topology; it uses for the most part geometric methods, and gets to spectacular results fairly quickly. The second part of this book, chapters 7 and 8, is part of geometric topology and is meant for the more advanced mathematician with interest in manifolds.

The text is self-contained for readers with a modest knowledge of general topology and linear algebra; the necessary background material is collected in chapter 1, or developed as needed.

One can look upon this book as a complete and self-contained proof of Toruńczyk's Hilbert cube manifold characterization theorem: *a compact* **ANR** X *is a manifold modeled on the Hilbert cube if and only if* X *satisfies the disjoint-cells property.* In the process of proving this result we make several interesting detours. Most of the results presented however are small steps in the proof of Toruńczyk's Theorem, or are needed for one of the applications.

I am very much indebted to J. Baars, J. van der Bijl, F. van Engelen, J. de Groot, K.P. Hart, K. Sakai and M. van de Vel for their critical reading of the manuscript and their many valuable suggestions for improvements.

Sleeuwijk, December, 1988 J. van Mill

How To Use This Book

This book can be used as a text for the following one semester courses:

I. Elementary Dimension Theory,
II. Elementary **ANR** Theory,
III. An Introduction to Infinite-Dimensional Topology,
IV. Infinite-Dimensional Topology.

The first three courses are elementary but the fourth one is not.

Chapter 1 consists of introductory material. Before the formulation of a result we have indicated there in square brackets where it will be used: so depending on the reader's plan he can decide to skip or to postpone it.

I. Elementary Dimension Theory.

The aim of this course is to present basic results from the classical dimension theory of separable metrizable spaces, as well as some recent developments concerning infinite-dimensional spaces. The main results are that the topological dimension of the euclidean spaces $\mathbb{R}^n$ is equal to n and that there exists an example of an infinite-dimensional compactum all whose subspaces are either zero-dimensional or infinite-dimensional.

Cover §§3.1 - 3.5 and chapter 4 (if time permits, start by covering chapter 2).

*II. Elementary **ANR** Theory.*

In this course we present basic results from the theory of Absolute (Neighborhood) Retracts. The main results are that each compact **AR** has the fixed-point property, that the hyperspace of a Peano continuum is an **AR** and that certain appropriate adjunction spaces are **ANR**'s.

Cover §§1.5 and 1.6, chapters 3 and 5 and parts of chapter 4 (as needed).

III. An introduction to Infinite-Dimensional Topology.

In this course basic homeomorphism theory in the Hilbert cube Q is presented. The main results are that Q is homogeneous, that the product of a letter T and Q is homeomorphic to Q, and that Hilbert space l^2 is homeomorphic to the countable infinite product of lines $\mathbb{R}^\infty$.

There are virtually no prerequisites for this course. One only needs to have some basic understanding of topology and elementary plane geometry.

Cover chapter 6.

IV. Infinite-Dimensional Topology.

In this course we present a proof of Toruńczyk's Characterization Theorem for Q-manifolds: a compact **ANR** is a Q-manifold if and only if it satisfies the disjoint-cells property. As an application we show that the hyperspace 2^X of a Peano continuum X is homeomorphic to Q (this is the so-called Curtis-Schori-West Hyperspace Theorem) and that Q is (topologically) the only compact **AR** which is homeomorphic to its own cone.

Prerequisites: chapters 1,3,4,5 and 6.

Cover chapters 7 and 8.

Table of Contents

1. Extension Theorems

Suppose that X,Y and Z are topological spaces with Y a subspace of X. Let f: Y $\to$ Z be continuous. In topology it is often of interest to detect whether f is the restriction to Y of a continuous function $\bar{f}$: X $\to$ Z. Easy examples show that this need not be the case. If f is the restriction to Y of a continuous function $\bar{f}$: X $\to$ Z then we say that f is *continuously extendable over* X and that $\bar{f}$ is a continuous extension of f; continuous extensions need not be unique. In this chapter we shall present examples of spaces Z having the property that if Y is closed in X, then every continuous function f: Y $\to$ Z is continuously extendable over X (respectively, over some neighborhood of Y). Some basic results are presented, which will be used frequently throughout the remaining part of this monograph.

Prerequisites:

We assume that the reader understands the material of basic courses in general topology and linear algebra. Concepts such as topological space, metric space, complete metric space, compact space, product space, continuous function, etc., should be familiar. For all undefined notions see Engelking [59].

1.1. Topological Spaces

By a *space* we mean a *separable metrizable topological space* (equivalently, a second countable regular T_1-space). Observe that each space has a countable base and consequently is Lindelöf. We let I and J denote the intervals [0,1] and [-1,1], respectively. The set of all real numbers is denoted by $\mathbb{R}$ and $\mathbb{N}$ denotes the set of all natural numbers.

The *Hilbert cube* Q is the countable infinite product

$$\prod_{i=1}^{\infty} J_i,$$

where each J_i is a copy of J. A countable infinite product of the same space X shall sometimes for convenience be denoted by X^∞. The i-th coordinate of a point x in a product space shall be denoted by x_i. A natural metric for Q that generates its topology is

$$d(x,y) = \sum_{i=1}^{\infty} 2^{-i}|x_i - y_i|.$$

Geometrically one should think of Q as an infinite-dimensional brick the sides of which get shorter and shorter. This can be demonstrated in the following way. Let $x(n) \in Q$ be the point having all coordinates 0 except for the n-th coordinate which equals 1. So x(n) is the "endpoint" of the n-th axis in Q. In addition, let y be the "origin" of Q, i.e. the point all coordinates of which are 0. Intuitively, each x(n) has distance 1 from y and hence x(n) and y are far apart. However, the appearance of the factor 2^{-i} in the definition of d implies that

$$d(x(n),y) = 2^{-n},$$

whence the sequence $(x(n))_n$ converges to y in Q.

It is well-known that every space is homeomorphic to a subspace of Q (see theorem 1.4.18). Consequently, we can think of spaces as being subspaces of Q. For certain considerations however this obscures the argumentation and it is better to deal with abstract spaces.

For any space X we let d denote an *admissible* metric on X, i.e. a metric that generates the topology. Many of the spaces we deal with are concrete objects and then d will be defined explicitly.

If $A \subseteq X$ then diam(A) denotes the diameter of A. Similarly, $\overline{A}$ denotes the closure of A and Int(A) denotes its interior. As usual, Bd(A) is the boundary of A, i.e. $Bd(A) = \overline{A} \setminus Int(A)$. Finally, $B(A,\varepsilon) = \{y \in X: d(A,y) < \varepsilon\}$ and $D(A,\varepsilon) = \{y \in X: d(A,y) \le \varepsilon\}$ denote the open and closed balls about A with radius ε, respectively. If $A = \{x\}$ for some x then we write $B(x,\varepsilon)$ instead of $B(\{x\},\varepsilon)$ and $D(x,\varepsilon)$ instead of $D(\{x\},\varepsilon)$, respectively.

An admissible metric d for X is called *bounded* if $diam(X) < \infty$. Each admissible metric can be replaced by one which in addition is bounded, see exercise 1.1.7.

Unless stated to the contrary, all functions between spaces are assumed to be continuous.

The symbol "$X \approx Y$" means that X and Y are homeomorphic spaces. If $\mathcal{U}$ is a familie of sets and X is a set then $\mathcal{U} \cap X$ denotes the collection $\{U \cap X : U \in \mathcal{U}\}$. If $f: X \to Y$ is a function and $A \subseteq X$ then $f \mid A$ denotes the restriction of f to A. Finally, if $f: X \to \mathbb{R}$ is a function and $t \in \mathbb{R}$ then $f \equiv t$ means that f is the constant function with value t.

It will be convenient to introduce the following notation. If $\mathcal{U}$ and $\mathcal{V}$ are open covers of a space X then $\mathcal{U} < \mathcal{V}$ means that $\mathcal{U}$ is a *refinement* of $\mathcal{V}$, i.e. for every $U \in \mathcal{U}$ there exists $V \in \mathcal{V}$ such that $U \subseteq V$.

Let X be a space, let $A \subseteq X$ and let $\mathcal{U}$ be an open cover of X. The *star of* A *with respect to* $\mathcal{U}$ is the set

$$St(A,\mathcal{U}) = \bigcup\{U \in \mathcal{U} : A \cap U \neq \emptyset\}.$$

The cover $\{St(U,\mathcal{U}) : U \in \mathcal{U}\}$ is denoted by $St(\mathcal{U})$. We say that an open cover $\mathcal{V}$ of X is a *star-refinement* of $\mathcal{U}$ if $St(\mathcal{V}) < \mathcal{U}$, i.e. if for every $V \in \mathcal{V}$ there exists $U \in \mathcal{U}$ such that

$$St(V,\mathcal{V}) \subseteq U;$$

notation: $\mathcal{V} \overset{*}{<} \mathcal{U}$.

We finish this section with the following simple but important lemma.

1.1.1. LEMMA [[chapters 1,3,4,6,7,8]]: *Let* X *be a compact subspace of a space* Y *and let* $\mathcal{U}$ *be a collection of open subsets of* Y *which covers* X. *Then there exists* $\delta > 0$ *with the property that every* $A \subseteq Y$ *with* $\operatorname{diam}(A) < \delta$ *and which moreover intersects* X, *is contained in an element* $U \in \mathcal{U}$.

PROOF: Suppose, to the contrary, that such δ does not exist. Then for every $n \in \mathbb{N}$ we can find a subset A_n of Y such that

(1) $\operatorname{diam}(A_n) < 1/n$,
(2) A_n intersects X, say $x_n \in A_n \cap X$,
(3) A_n is not contained in any element of $\mathcal{U}$.

Since X is compact, every sequence in X has a convergent subsequence, so without loss of generality we may assume that $x = \lim_{n\to\infty} x_n$ exists and belongs to X of course. There exists $U \in \mathcal{U}$ such that $x \in U$. Since U is open, there exists $\varepsilon > 0$ such that $B(x,\varepsilon) \subseteq U$. In addition, there exists $N \in \mathbb{N}$ such that $x_m \in B(x,\varepsilon/2)$ for every $m \geq N$. Now choose $m \geq N$ so large that $1/m \leq$

$\varepsilon/2$. Since the diameter of A_m is less than $1/m \le \varepsilon/2$, it now follows easily that $A_m \subseteq B(x,\varepsilon) \subseteq U$, which is a contradiction. □

The number δ in the above lemma is called a *Lebesgue number* for $\mathcal{U}$.

Exercises for §1.1.

1. For each $i \in \mathbb{N}$ let X_i be a space. Show that a sequence $(x(n))_n$ in $\prod_{i=1}^{\infty} X_i$ converges to a point $x \in \prod_{i=1}^{\infty} X_i$ if and only if the sequence $(x(n)_i)_n$ converges to x_i for every $i \in \mathbb{N}$.

2. Show that Q^n is homeomorphic to Q for every $n \in \mathbb{N} \cup \{\infty\}$. Give an example of a space X such that X^n and X are homeomorphic for every $n \in \mathbb{N}$ while X^∞ is not homeomorphic to X.

3. Let X and Y be compact spaces and let $f\colon X \to Y$ be continuous. Prove that f is a *closed map*, i.e. that $f(A)$ is closed in Y for every closed subset A of X.

4. Let X be compact and let $f\colon X \to Y$ be one-to-one. Prove that if $f(X)$ is dense in Y then f is a homeomorphism.

5. Let X be a space and let K be a compact subset of X. Prove that there exist $x,y \in K$ such that $d(x,y) = \operatorname{diam}(K)$.

A topological space X is called *discrete* if every subset of X is open, or, equivalently, closed. For example, the set of natural numbers $\mathbb{N}$ with the subspace topology inherited from $\mathbb{R}$ is an example of an infinite discrete space.

6. Let X be a space and assume that X is not compact. Prove that X contains an infinite closed discrete subspace.

7. Let X be a space and let d be an admissible metric for X. Prove that the function $\rho\colon X \times X \to [0,\infty)$ defined by

$$\rho(x,y) = \min\{1, d(x,y)\}$$

is an admissible bounded metric for X.

8. Show that all nondegenerate open intervals in $\mathbb{R}$ are homeomorphic to $\mathbb{R}$.

9. Show that all nondegenerate bounded closed intervals in $\mathbb{R}$ are homeomorphic.

10. Let $f\colon X \to Y$ be a closed map, let $A \subseteq Y$ and let U be an open neighborhood of $f^{-1}(A)$ in X. Prove that there is an open neighborhood V of A in Y such that $f^{-1}(V) \subseteq U$.

1.2. Linear Spaces

A *linear space* is a real vector space carrying a (separable metrizable) topology with the properties that the algebraic operations

$$(x,y) \to x+y$$
$$(t,x) \to t\cdot x \qquad (t \in \mathbb{R})$$

are continuous (warning: a *vector* space is an algebraic structure which may or may not carry a topology while a *linear* space is automatically a topological space). A subset A of a linear space L is called *convex* if for all $x,y \in A$ and $\alpha \in I$ we have $\alpha x + (1-\alpha)y \in A$.

A linear space L is called *locally convex* if the zero of L (which we shall always denote by $\underline{0}$) has arbitrarily small convex neighborhoods. Obviously, $\mathbb{R}^n$ ($n \in \mathbb{N}$) and $\mathbb{R}^\infty$ with their usual product topologies are locally convex linear spaces under coordinatewise defined addition and scalar multiplication.

Let L be a vector space. A *norm* on L is a function $\|\cdot\|: L \to [0,\infty)$ having the following properties:

(1) $\|x+y\| \le \|x\| + \|y\|$ for all $x,y \in L$ (triangle inequality),

(2) $\|tx\| = |t|\cdot\|x\|$ for all $t \in \mathbb{R}$ and $x \in L$,

(3) $\|x\| = 0$ if and only if $x = 0$.

If $\|\cdot\|$ is a norm on L then the function

(4) $d(x,y) = \|x - y\|$

defines a metric on L; this metric is called the *metric derived from the norm* $\|\cdot\|$. We call a linear space L *normable* provided that there exists a norm on L such that the metric derived from this norm generates the topology on L; such a norm is called *admissible*. Observe that each normable linear space is locally convex. A *normed linear space* is a pair $(L,\|\cdot\|)$, where L is a vector space and $\|\cdot\|$ is a norm on L; we shall always endow the underlying vector space of a normed linear space with the topology derived from its norm (since by convention we only deal with separable spaces, we restrict ourselves to normed linear spaces the underlying topological spaces of which are separable; this may seem to be unnecessarily restrictive or artificial, but in practice this will not be likely to cause confusion since the linear spaces we deal with are mostly concrete objects such as Hilbert space and the countable infinite product of lines and these are separable). So we make a formal distinction between *normed linear space* and *normable linear space:* a normable linear space may possess many different norms that generate its topology, see exercise 1.2.2, whereas in a normed linear space the norm is fixed.

A *Banach space* is a normable space for which there exists an admissible norm such that the metric derived from this norm is complete.

EXAMPLES: (1) The standard norm making $\mathbb{R}^n$ into a normed linear space is defined by

$$\|x\| = \sqrt{\sum_{i=1}^{n} x_i^2}\ .$$

So each $\mathbb{R}^n$ is a normable linear space and since $\mathbb{R}^\infty$ is in many respects the "limit" of the spaces $\mathbb{R}^n$, the question naturally arises whether $\mathbb{R}^\infty$ admits a norm. Define $\Omega = \{x \in \mathbb{R}^\infty : x_n = 0$ for all but finitely many $n \in \mathbb{N}\}$. Our question is answered by the following

1.2.1. LEMMA [[chapters 1,4]]: *If* L *is a linear subspace of* $\mathbb{R}^\infty$ *with* $\Omega \subseteq L$ *then* L *with the subspace topology is not a normable linear space.*

PROOF: Assume, to the contrary, that $\|\cdot\|$ is an admissible norm on L. Then

$$U = \{x \in L : \|x\| < 1\}$$

is an open neighborhood of the zero of L. By definition of the product topology on $\mathbb{R}^\infty$ there are an open neighborhood V of 0 in $\mathbb{R}$ and an $n \in \mathbb{N}$ such that

(a) $$(\Pi_{i=1}^{n} V_i \times \Pi_{i=n+1}^{\infty} \mathbb{R}_i) \cap L \subseteq U,$$

where $V_i = V$ for $i \leq n$ and $\mathbb{R}_i = \mathbb{R}$ for $i > n$. Let $y \in \mathbb{R}^\infty$ be defined by $y_i = 0$ if $i \neq n+1$ and $y_{n+1} = 1$. Since $y \in \Omega \subseteq L$ and $y \neq 0$ it follows that $\varepsilon = \|y\| > 0$. By (a), $ty \in U$ for every $t \in \mathbb{R}$. In particular, $\|y/\varepsilon\| < 1$; but also $\|y/\varepsilon\| = \|y\|/\varepsilon = \varepsilon/\varepsilon = 1$, which is a contradiction.□

From the proof of lemma 1.2.1 it is clear that the linear structure of $\mathbb{R}^\infty$ together with its topology, prevent it from being normable. (The question naturally arises whether every vector space can be endowed with a norm which is compatible with its linear structure. The answer to this question is in the affirmative, see exercise 1.2.3.) Consequently, although $\mathbb{R}^\infty$ seems a natural "limit" of the spaces $\mathbb{R}^n$, it is notably different from any of its finite-dimensional analogues $\mathbb{R}^n$. For this reason we seek another "natural" limit of the spaces $\mathbb{R}^n$ not having this defect.

A space X is called *topologically complete* if there exists an admissible metric on X which is complete. Observe that each Banach space is by definition topologically complete. In addition, $\mathbb{R}^\infty$ is topologically complete; its standard complete metric is the following one:

$$d(x,y) = \sum_{n=1}^{\infty} 2^{-n} \frac{|x_n - y_n|}{1 + |x_n - y_n|}.$$

(2) *The spaces* C(X). Let X be a compact nonempty space. Let C(X) denote the set of all continuous real valued functions on X. Obviously, C(X) is a vector space; addition of functions and scalar multiplication are defined pointwise. If $f \in C(X)$ then define the norm, $\|f\|$, of f by

$$\|f\| = \sup\{|f(x)|: x \in X\}.$$

It is easily seen that $\|\cdot\|: C(X) \to [0,\infty)$ is indeed a norm; it is called the *sup-norm* on C(X). Consequently, the function

$$(*) \qquad d(f,g) = \|f - g\|$$

defines a metric on C(X) and therefore generates a topology. From now on we shall endow C(X) with this topology; we shall see in exercise 1.2.4 and proposition 1.3.3 that C(X) is separable. Observe that C(X) is normable since its topology is defined via a norm. In corollary 1.3.5 we shall prove that the metric d defined in (*) is complete.

Let L be a linear space. If $A \subseteq L$ then conv(A) denotes the smallest convex subset of L containing A; this set is sometimes called the *convex hull* of A. A *convex combination* of elements of A is a vector of the form $\Sigma_{i=1}^n \lambda_i a_i$ with $a_1,\cdots,a_n \in A$, $\lambda_1,\cdots,\lambda_n \in I$ and $\Sigma_{i=1}^n \lambda_i = 1$.

For each $n \in \mathbb{N}$, define $\text{conv}_n(A)$ by

$$x \in \text{conv}_n(A) \Leftrightarrow \exists\, a_1,\cdots,a_n \in A\ \exists\, \lambda_1,\cdots,\lambda_n \in I \text{ with } \sum_{i=1}^n \lambda_i = 1 \text{ such that } x = \sum_{i=1}^n \lambda_i a_i.$$

Finally, put

$$\text{conv}_\infty(A) = \bigcup_{n=1}^{\infty} \text{conv}_n(A);$$

observe that $\text{conv}_\infty(A)$ is the set of all convex combinations of elements of A.

1.2.2. LEMMA [[chapters 1,3,4]]:
(1) $\text{conv}(A) = \text{conv}_\infty(A)$, *and*

(2) *if* A *is finite then* conv(A) *is compact.*

PROOF: It is clear that $\mathrm{conv}_\infty(A)$ is convex. Consequently, $\mathrm{conv}(A) \subseteq \mathrm{conv}_\infty(A)$. By induction on $n \in \mathbb{N}$ we shall prove that $\mathrm{conv}_n(A) \subseteq \mathrm{conv}(A)$. This is clearly true for n = 1. Assume that $\mathrm{conv}_n(A) \subseteq \mathrm{conv}(A)$ and take an arbitrary point $x \in \mathrm{conv}_{n+1}(A)$. By definition there exist $a_1,\cdots,a_{n+1} \in A$ and $\lambda_1,\cdots,\lambda_{n+1} \in I$ with $\Sigma_{i=1}^{n+1}\lambda_i = 1$ such that $x = \Sigma_{i=1}^{n+1}\lambda_i a_i$. Without loss of generality we may assume that $\lambda_i \neq 0$ for every $i \leq n+1$. Define $\mu = 1 / \Sigma_{i=1}^{n}\lambda_i$. Clearly $\mu\cdot\lambda_i \in I$ for every $i \leq n$ and $\Sigma_{i=1}^{n}\mu\lambda_i = 1$.

From this we conclude that $y = \Sigma_{i=1}^{n}\mu\lambda_i a_i \in \mathrm{conv}_n(A)$ and therefore, by our inductive assumption, $y \in \mathrm{conv}(A)$. Consequently, y, $a_{n+1} \in \mathrm{conv}(A)$ and since conv(A) is convex,

$$x = (\sum_{i=1}^{n}\lambda_i)\cdot y + \lambda_{n+1}a_{n+1} \in \mathrm{conv}(A),$$

which is as required.

Now let $A = \{a_1,\cdots,a_n\}$ be finite. It follows from (1) that $\mathrm{conv}(A) = \mathrm{conv}_n(A)$. Put $\Delta_n = \{\lambda \in I^n: \Sigma_{i=1}^{n}\lambda_i = 1\}$. It is easily seen that Δ_n is a closed and bounded subset of $\mathbb{R}^n$ and therefore is compact. Define $f: \Delta_n \times A^n \to \mathrm{conv}(A)$ by $f((\lambda_1,\cdots,\lambda_n),(a_1,\cdots,a_n)) = \Sigma_{i=1}^{n}\lambda_i a_i$. Then f is well-defined and surjective. By the continuity of the algebraic operations on L it follows that f is continuous. We conclude that conv(A) is the continuous image of a compact space and is therefore compact itself. □

The function space C(X) has the following property:

1.2.3. LEMMA [[chapters 1,5]]: *For every compact space* (X,d) *there exists an isometric imbedding* $i: X \to C(X)$ *such that*

(1) *for every subset* $Y \subseteq X$ *the image set* i(Y) *is closed in* conv(i(Y)), *and*

(2) $i(X) \subseteq \{f \in C(X): \|f\| \leq \mathrm{diam}\, X\}$.

PROOF: Define $i: X \to C(X)$ by $i(x)(y) = d(x,y)$. Then i is an isometry. Take arbitrary x_1, $x_2 \in X$. Then

$$d(i(x_1),i(x_2)) \geq |i(x_1)(x_2) - i(x_2)(x_2)| = |d(x_1,x_2)| = d(x_1,x_2).$$

In addition, for every $y \in X$ we have

$$|i(x_1)(y) - i(x_2)(y)| = |d(x_1,y) - d(x_2,y)| \leq d(x_1,x_2),$$

from which it follows that $d(i(x_1),i(x_2)) \leq d(x_1,x_2)$. Consequently, $d(i(x_1),i(x_2)) = d(x_1,x_2)$, which is as required.

To prove (1), let $Y \subseteq X$ be an arbitrary subset. Assume that $f \in \mathrm{conv}(i(Y))$ and that for some sequence $(y_n)_n$ in Y, $f = \lim_{n\to\infty} i(y_n)$. We shall prove that $f \in i(Y)$. Since $f \in \mathrm{conv}(i(Y))$ there exist distinct $a_1,\cdots, a_m \in Y$ and elements $t_1,\cdots,t_m \in I$ such that

$$f = \Sigma_{j=1}^m t_j i(a_j) \text{ and } \Sigma_{j=1}^m t_j = 1$$

(lemma 1.2.2). Without loss of generality we may assume that $t_1 \neq 0$. Then for every $n \in \mathbb{N}$,

$$d(f,i(y_n)) \geq |f(y_n) - i(y_n)(y_n)| = |f(y_n)| \geq t_1 i(a_1)(y_n) = t_1 d(a_1,y_n).$$

Since the sequence $(i(y_n))_n$ converges to f it therefore follows that $\lim_{n\to\infty} y_n = a_1$. This obviously implies that $f = i(a_1)$.

(2) is a triviality of course. □

Observe that by the above lemma it follows that *every* space X can be imbedded as a closed subspace of a convex subset of a normed linear space. Since every space can be thought of as a subspace of the compact Hilbert cube (theorem 1.4.18), this is immediate.

Let L be a vector space. An *inner product* on L is a function $\langle\cdot,\cdot\rangle: L \times L \to \mathbb{R}$ satisfying the following axioms:

(1) $\langle x,y\rangle = \langle y,x\rangle$ for all $x,y \in L$,

(2) $\langle tx,y\rangle = t\langle x,y\rangle$ for all $t \in \mathbb{R}$, $x,y \in L$,

(3) $\langle x+y,z\rangle = \langle x,z\rangle + \langle y,z\rangle$ for all $x,y,z \in L$,

and

(4) if $x \in L$ then $\begin{cases} \langle x,x\rangle \geq 0, \\ \langle x,x\rangle = 0 \text{ if and only if } x = 0. \end{cases}$

If $\langle\cdot,\cdot\rangle$ is an inner product on L then the following statements hold:

(5) (Schwarz's inequality) $\langle x,y\rangle^2 \leq \langle x,x\rangle\cdot\langle y,y\rangle$ for all $x,y \in L$,

(6) the function $\|\cdot\|$ defined by $\|x\| = \sqrt{\langle x,x\rangle}$ is a norm on L,

(7) (parallelogram law) $\|x+y\|^2 + \|x-y\|^2 = 2\|x\|^2 + 2\|y\|^2$ for all $x,y \in L$.

The norm in (6) is called the *norm derived from the inner product* $\langle\cdot,\cdot\rangle$.

For proofs of (5), (6) and (7) see Rudin [121, p. 75-76] or consult any textbook on linear algebra.

An *inner product space* is a pair $(L,\langle\cdot,\cdot\rangle)$, where L is a vector space and $\langle\cdot,\cdot\rangle$ is an inner product on L.

Let $(L,\langle\cdot,\cdot\rangle)$ be an inner product space. By (6), the formula

$$\|x\| = \sqrt{\langle x,x\rangle}$$

defines a natural norm on L and consequently the function

$$d(x,y) = \|x - y\| = \sqrt{\langle x-y,x-y\rangle}$$

is a metric on L which therefore generates a topology. We shall always endow L with this topology (as in the normed case, we restrict our attention to those inner product spaces whose underlying topological spaces are separable).

Let $(L,\|\cdot\|)$ be a normed linear space. Then L is said to be a *pre-Hilbert space* provided that there exists an inner product on L such that the norm given by this inner product, coincides with the given norm on L.

EXAMPLES: (1) The usual inner product on $\mathbb{R}^n$ is given by

$$\langle x,y\rangle = \sum_{i=1}^{n} x_i y_i.$$

Observe that the norm derived from this inner product is the usual norm on $\mathbb{R}^n$. Consequently, $\mathbb{R}^n$ is a pre-Hilbert space. Observe that $\mathbb{R}^\infty$ is not a pre-Hilbert space since it is not even normable (lemma 1.2.1).

(2) Let X be a compact space. If X contains only one point then it is easily seen that C(X) is both topologically and algebraically isomorphic to $\mathbb{R}$. So in this case, $(C(X),\|\cdot\|)$, where $\|\cdot\|$ denotes the sup-norm of course, is a pre-Hilbert space (for trivial reasons). If X contains more than one point then this is not the case.

1.2.4. LEMMA: *Let* X *be a compact space containing more than one point. Then* $(C(X),\|\cdot\|)$ *is not a pre-Hilbert space.*

PROOF: Let x and y be different points in X. Define $f\colon X \to \mathbb{R}$ by $f(z) = d(x,z)$. Since $f(y) =$

$d(x,y) > 0$, we have $f \neq 0$ and consequently, $\|f\| \neq 0$. Define functions $g,h : X \to \mathbb{R}$ by

$$g = \frac{f}{\|f\|} \text{ and } h = 1 - g.$$

Observe that $\|g\| = 1$ and that the range of g is contained in I. From this it follows that the range of h is contained in I and since $h(x) = 1$ we get $\|h\| = 1$. Since $g + h = 1$ and $g - h = 2g - 1$ it follows that $\|g + h\| = 1$ and $\|g - h\| \leq 1$. Consequently, $2\|g\|^2 + 2\|h\|^2 = 4$ and $\|g + h\|^2 + \|g - h\|^2 \leq 2$, i.e. if $(C(X),\|\cdot\|)$ were a pre-Hilbert space it would violate the parallelogram law.□

EXAMPLE: *Hilbert space l^2*. We saw that the linear structure on $\mathbb{R}^\infty$ is very different from the linear structure on any of its finite-dimensional analogues $\mathbb{R}^n$. We shall now construct another natural "limit" of the spaces $\mathbb{R}^n$ which behaves better (in this respect).

Consider the usual inner product on $\mathbb{R}^n$ given by

$$\langle x,y\rangle = \sum_{i=1}^{n} x_i y_i.$$

If we try to generalize this inner product to the case of $\mathbb{R}^\infty$ then we have to deal with infinite series and it is therefore quite natural to restrict our attention to the following subset of $\mathbb{R}^\infty$:

$$l^2 = \{x \in \mathbb{R}^\infty : \sum_{i=1}^{\infty} x_i^2 < \infty\}.$$

This subset of $\mathbb{R}^\infty$ is called *Hilbert's set*. We shall first prove that l^2 is a vector subspace of $\mathbb{R}^\infty$. For every $x \in l^2$ we write

$$p(x) = \sqrt{\sum_{i=1}^{\infty} x_i^2}\,.$$

If $x,y \in l^2$, then Schwarz's inequality applied to $\mathbb{R}^n$ shows that

$$\left|\sum_{i=1}^{n} x_i y_i\right| \leq p(x)\cdot p(y).$$

From this it follows that

$$|\sum_{i=1}^{\infty} x_i y_i| \leq p(x)\cdot p(y) < \infty,$$

so

$$\sum_{i=1}^{\infty}(x_i + y_i)^2 = \sum_{i=1}^{\infty} x_i^2 + \sum_{i=1}^{\infty} y_i^2 + 2\sum_{i=1}^{\infty} x_i y_i < \infty$$

since all infinite series considered are convergent. We conclude that for every $x,y \in l^2$ we have $x+y \in l^2$. If $x \in l^2$ and $t \in \mathbb{R}$ then trivially $tx \in l^2$. Consequently, l^2 is a vector subspace of $\mathbb{R}^\infty$.

Since

$$\sum_{i=1}^{\infty} x_i y_i < \infty$$

for all $x,y \in l^2$ we have a well-defined function $<\cdot,\cdot>: l^2 \times l^2 \to \mathbb{R}$,

$$<x,y> = \sum_{i=1}^{\infty} x_i y_i,$$

which is easily seen to be an inner product. Consequently, $\|x\| = p(x)$ defines a norm on l^2 and the metric derived from this norm is:

$$d(x,y) = \sqrt{\sum_{i=1}^{\infty}(x_i - y_i)^2}\ .$$

We endow l^2 with the topology generated by this metric and refer to l^2 with this topology as *Hilbert space*.

Let $\Omega \subseteq \mathbb{R}^\infty$ be the such as in lemma 1.2.1. Clearly $\Omega \subseteq l^2$. Consequently, lemma 1.2.1 shows that the topology that l^2 inherits from $\mathbb{R}^\infty$ is different from the topology on l^2 which we just defined. We will comment on the precise relation between these topologies later.

1.2.5. LEMMA [[chapter 8]]: *The metric* d *on* l^2 *defined above is complete.*

PROOF: Let $(x(n))_n$ be a d-Cauchy sequence in l^2. For every $i \in \mathbb{N}$ the sequence $(x(n)_i)_n$ is clearly Cauchy (in $\mathbb{R}$). Put

$$x_i = \lim_{n\to\infty} x(n)_i \qquad (i \in \mathbb{N})$$

and let $x = (x_1,x_2,\cdots) \in \mathbb{R}^\infty$. We claim that $x \in l^2$ and that $\lim_{n\to\infty} x(n) = x$ (in l^2).

Let $\varepsilon > 0$. There is $N \in \mathbb{N}$ such that for all $n,m \geq N$ we have $d(x(m),x(n)) < \varepsilon$. Choose $k \in \mathbb{N}$ arbitrarily. Then for all $n,m \geq N$,

$$\sum_{i=1}^{k}(x(n)_i - x(m)_i)^2 \leq \sum_{i=1}^{\infty}(x(n)_i - x(m)_i)^2 < \varepsilon^2.$$

Consequently, for all $m \geq N$ we have

$$\sum_{i=1}^{k}(x_i - x(m)_i)^2 = \lim_{n\to\infty} \sum_{i=1}^{k}(x(n)_i - x(m)_i)^2 \leq \varepsilon^2.$$

This proves that the series

$$\sum_{i=1}^{\infty}(x_i - x(m)_i)^2$$

is convergent for all $m \geq N$ and also that

$$\sum_{i=1}^{\infty}(x_i - x(m)_i)^2 \leq \varepsilon^2 \qquad (m \geq N).$$

We conclude that $x - x(N) \in l^2$ and since $x(N) \in l^2$ and l^2 is a vector subspace of $\mathbb{R}^\infty$ this proves that $x \in l^2$.

The last inequality also implies that

$$d(x,x(m)) = \sqrt{\sum_{i=1}^{\infty}(x_i - x(m)_i)^2} \leq \varepsilon$$

for all $m \geq N$. Since ε was chosen arbitrarily it now follows that $\lim_{n\to\infty}x(n) = x$ (in l^2). □

The topology on $\mathbb{R}^\infty$ is the topology of "coordinatewise convergence", see exercise 1.1.1. Topologists usually find such product topologies easier to handle than topologies derived from a norm. However, convergence in l^2 can be handled with the same ease, as is shown in the next result.

1.2.6. LEMMA [[chapters 1,4,6]]: *Suppose that* $x(n) \in l^2$, $n \in \mathbb{N}$, *and* $x \in l^2$. *The following statements are equivalent*

(a) $\lim_{n\to\infty} x(n) = x$ *(in* l^2*), and*

(b) $\lim_{n\to\infty} \|x(n)\| = \|x\|$ *and for every* $i \in \mathbb{N}$, $\lim_{n\to\infty} x(n)_i = x_i$.

PROOF: *We prove (a)* $\Rightarrow$ *(b).* The triangle inequality for $\|\cdot\|$ directly implies that

$$\forall x,y \in l^2 : |\, \|x\| - \|y\| \,| \leq \|x - y\|.$$

Therefore, for every $n \in \mathbb{N}$,

$$0 \leq |\, \|x\| - \|x(n)\| \,| \leq \|x - x(n)\| = d(x,x(n)).$$

Since $\lim_{n\to\infty} d(x,x(n)) = 0$ this implies that $\lim_{n\to\infty} \|x(n)\| = \|x\|$.

That $\lim_{n\to\infty} x(n)_i = x_i$ for every $i \in \mathbb{N}$ is a triviality.

We prove (b) $\Rightarrow$ *(a).* Let $\varepsilon > 0$. As $x \in l^2$ there is a $p \in \mathbb{N}$ such that

$$(1) \qquad \sum_{i=p}^{\infty} x_i^2 < \varepsilon.$$

Now $\|x(n)\| \to \|x\|$ as $n \to \infty$ and hence

$$(2) \qquad \exists m_0 : n \geq m_0 \Rightarrow |\, \|x(n)\|^2 - \|x\|^2 \,| < \varepsilon.$$

Finally, $x(n)_i \to x_i$ for $i = 1,\cdots, p-1$ as $n \to \infty$, and hence

$$(3) \qquad \exists m_1 : n \geq m_1 \Rightarrow \sum_{i=1}^{p-1} (x(n)_i - x_i)^2 < \varepsilon, \text{ and}$$

$$(4) \qquad \exists m_2 : n \geq m_2 \Rightarrow \sum_{i=1}^{p-1} |x(n)_i^2 - x_i^2| < \varepsilon.$$

Then for $n \geq \max(m_0,m_1,m_2)$ we have by (2) and (4),

$$\left|\sum_{i=p}^{\infty} x(n)_i^2 - \sum_{i=p}^{\infty} x_i^2\right| = \left|\, \|x(n)\|^2 - \sum_{i=1}^{p-1} x(n)_i^2 + \sum_{i=1}^{p-1} x_i^2 - \|x\|^2 \right|$$

$$\leq |\|x(n)\|^2 - \|x\|^2| + |\sum_{i=1}^{p-1}(x(n)_i^2 - x_i^2)|$$

$$< \quad \varepsilon \quad + \quad \varepsilon$$

$$= 2\varepsilon,$$

from which it follows by (1) that

$$\sum_{i=p}^{\infty} x(n)_i^2 < 3\varepsilon. \tag{5}$$

Since $(a - b)^2 \leq 2a^2 + 2b^2$ for all $a,b \in \mathbb{R}$,

$$\begin{aligned} \sum_{i=1}^{\infty}(x(n)_i - x_i)^2 &= \sum_{i=1}^{p-1}(x(n)_i - x_i)^2 + \sum_{i=p}^{\infty}(x(n)_i - x_i)^2 \\ &\leq \sum_{i=1}^{p-1}(x(n)_i - x_i)^2 + 2\sum_{i=p}^{\infty}x(n)_i^2 + 2\sum_{i=p}^{\infty}x_i^2. \end{aligned} \tag{6}$$

Consequently, for $n \geq \max(m_0,m_1,m_2)$ we obtain by (3), (5) and (1) that

$$\sum_{i=1}^{\infty}(x(n)_i - x_i)^2 < \varepsilon + 6\varepsilon + 2\varepsilon = 9\varepsilon.$$

We conclude that $\lim_{n\to\infty} x(n) = x$ (in l^2).☐

From lemma 1.2.6 we immediately derive that the topology on l^2 is finer than the topology that l^2 inherits from $\mathbb{R}^\infty$. However, more can be concluded. For example, consider the unit sphere $S = \{x \in l^2: \|x\| = 1\}$. Since all points in S have the same norm, the topology that S inherits from l^2 is precisely the same as the topology that S inherits from $\mathbb{R}^\infty$, i.e. the topology of "coordinatewise convergence". This remark will play an important role in the proof of theorem 6.6.11.

We conclude this section by giving some remarks. We introduced linear spaces, locally convex linear spaces, normable linear spaces and inner product spaces. Obviously, the "underlying" linear space of an inner product space is normable and each normable linear space is locally convex. Since $\mathbb{R}^\infty$ is clearly locally convex, the spaces $(C(I),\|\cdot\|)$ and $\mathbb{R}^\infty$ show that these "implications" are strict. In exercise 5 of this section a linear space is defined that is not locally convex.

The question naturally arises whether the linear space C(I) admits an equivalent norm $|||\cdot|||$ such that $(C(I),|||\cdot|||)$ is a pre-Hilbert space. The answer to this question is in the negative. The proof of this fact requires techniques that are outside the scope of this book. The interested reader may consult for example [132, p.195].

Hilbert space l^2 and the countable infinite product of lines $\mathbb{R}^\infty$ are both topologically complete, locally convex linear spaces. In addition, they are both natural "limits" (or generalizations) of the euclidean spaces $\mathbb{R}^n$. Since l^2 is normable and $\mathbb{R}^\infty$ is not, it follows that there does not exist a homeomorphism $h: l^2 \to \mathbb{R}^\infty$ which is *linear*, i.e. has the property that

$$h(\lambda x + \mu y) = \lambda h(x) + \mu h(y)$$

for all $x,y \in l^2$ and $\lambda,\mu \in \mathbb{R}$. The question naturally arises whether l^2 and $\mathbb{R}^\infty$ are (topologically) homeomorphic. This question was raised by Fréchet [66] in 1928 and also by Banach [16] in 1932. In 1966 Anderson [7] answered Fréchet's question in the affirmative. We shall present a complete and elementary proof of Anderson's Theorem in chapter 6.

Exercises for §1.2.

1. Let L be a linear space and let $x \in L$. Show that x has arbitrarily small neighborhoods of the form $x + W$, where W is a neighborhood of the zero of L, and $x + W = \{x+w: w \in W\}$.

2. Define a function $|||\cdot|||: \mathbb{R}^n \to \mathbb{R}$ by $|||x||| = \max\{|x_i|: 1\le i\le n\}$. Prove that $|||\cdot|||$ is a norm on $\mathbb{R}^n$ and that it generates the euclidean topology.

Let L be a linear space. A *maximal* linearly independent subset B of L (i.e. a subset of L which is maximal with respect to the property of being linearly independent) is called a *Hamel basis* for L. The Kuratowski-Zorn Lemma easily implies that every linear space has a Hamel basis. It is well-known, and easy to prove, that if B is a Hamel basis for L then each $x \in L\setminus\{0\}$ can be written uniquely in the form

$$x = \alpha_1 x_1 + \alpha_2 x_2 + \cdots + \alpha_n x_n,$$

with $x_i \in B$ and $\alpha_i \in \mathbb{R}\setminus\{0\}$ for every $i \le n$ (for details, consult any textbook on Linear Algebra).

3. Let L be a linear space and let B be a Hamel basis for L. If $x = \alpha_1 x_1 + \alpha_2 x_2 + \cdots + \alpha_n x_n$, with $x_i \in B$ and $\alpha_i \in \mathbb{R}$ for every $i \le n$, then put $\|x\| = |\alpha_1| + |\alpha_2| + \cdots + |\alpha_n|$. Prove that $\|\cdot\|$ defines a norm on L.

Let $f \in C(I)$. It is known that for every $\varepsilon > 0$ there exists a polynomial

$$p(x) = \sum_{i=0}^{n} a_i x^i (1-x)^{n-i} \qquad (a_i \in \mathbb{R},\ 0 \leq i \leq n, \text{ and } x \in I),$$

such that $\|f - p(x)\| < \varepsilon$ (for details, see [126, §7.3])

Let **C** denote the subspace of I consisting of all points that have a tryadic expansion in which the digit 1 does not occur. This set is called the *Cantor middle-third set* and will be studied in detail in §4.2.

4. Let K be a compact subset of $\mathbb{R}$. Prove that C(K) is separable. Let $\mathbf{C} \subseteq I$ be the Cantor middle third set and let X be any compact space. It is known that there exists a surjection $f: \mathbf{C} \to X$ (theorem 4.2.9). Use this fact to prove that C(X) is separable (see also proposition 1.3.3 for a different proof that C(X) is separable).

5. Let $0 < p < 1$ and put $l^p = \{x \in \mathbb{R}^\infty : \sum_{n=1}^{\infty} |x_n|^p < \infty\}$. Prove that l^p is a vector subspace of $\mathbb{R}^\infty$. Define the following metric on l^p: $d(x,y) = \sum_{n=1}^{\infty} |x_n - y_n|^p$. Prove that l^p with the topology derived from this metric is a linear space which is not locally convex (Remark: so the formula $\|x\| = (\sum_{n=1}^{\infty} |x_n|^p)^{1/p}$ does not define a norm on l^p; it is known that for $p > 1$ this formula does define a norm on l^p).

Let X be a space. For compact subset K in X and open U in $\mathbb{R}$ define $[K,U] = \{f \in C(X): f(K) \subseteq U\}$. Topologize C(X) by taking the collection $\{[K,U]: K \subseteq X \text{ is compact and } U \subseteq \mathbb{R} \text{ is open}\}$ as an open subbase. This topology is called the *compact-open topology* on C(X).

6. Let X be a compact space. Prove that the compact-open topology on C(X) coincides with the topology derived from the sup-norm $\|\cdot\|$.

7. Prove that l^2 and $\mathbb{R}^\infty$ are separable.

8. Prove that l^2 and $l^2 \times \mathbb{R}$ are linearly homeomorphic (self-explanatory).

9. Let $X = \{0\} \cup \{1/n: n \in \mathbb{N}\}$ and $Y = \{0\} \cup \{1/n: n \in \mathbb{N}\} \cup \{2\} \cup \{2+1/n: n \in \mathbb{N}\}$. Prove that C(X) and C(Y) are linearly homeomorphic.

10. Let $X = \{x \in l^2: (\forall n \in \mathbb{N})(|x_n| \leq 1/n)\}$. Prove that X and Q are homeomorphic.

11. Define a function $\langle \cdot,\cdot \rangle$ on $C(I) \times C(I)$ by $\langle f,g \rangle = \int_0^1 f(t)g(t)dt$. Prove that $\langle \cdot,\cdot \rangle$ is an inner product on C(I).

A linear space L is called *finite-dimensional* if it has a finite Hamel basis. Otherwise it is called *infinite-*

dimensional.

12. Let $|||\cdot|||$ be a norm on $\mathbb{R}^n$. Prove that the topology derived from this norm is the euclidean topology. In particular, for every $x \in \mathbb{R}^n$ and $\varepsilon > 0$ we have that the set $D(x,\varepsilon) = \{y \in \mathbb{R}^n: |||x-y||| \le \varepsilon\}$ is compact.

13. Let V be a normed linear space and let W be a finite-dimensional linear subspace. Prove that W is closed in V.

14. Let $||\cdot||$ be a norm on $\mathbb{R}^n$ and let $f: \mathbb{R}^n \to \mathbb{R}$ be linear. Prove that there exists $t > 0$ such that

    (1) $f^{-1}(t) \cap B(\underline{0},1) = \emptyset$,
    (2) $f^{-1}(t) \cap D(\underline{0},1) \neq \emptyset$.

15. Let V be an infinite-dimensional normed linear space. Prove that there is a sequence $(e_n)_n$ in V such that

    (1) $\{e_n: n \in \mathbb{N}\}$ is linearly independent,
    (2) $||e_n|| = 1$ $(n \in \mathbb{N})$,
    (3) $||e_m - e_n|| \ge 1$ $(m,n \in \mathbb{N},\ m \neq n)$.

16. Let V be an infinite-dimensional normed linear space. Prove that the unit sphere $S = \{x \in V: ||x|| = 1\}$ is not compact.

1.3. Function Spaces

In this section we shall present some elementary results on function spaces that will be important in the remaining part of this monograph. *All spaces in this section are assumed to be nonempty.*

To begin with, let us fix some terminology. If X and Y are spaces then $C(X,Y)$ denotes the set of all continuous functions from X to Y. It will be convenient to topologize $C(X,Y)$ and interesting subsets of it with a natural topology. Unfortunately, there are many ways to do that; we shall discuss one of them.

For all $f,g \in C(X,Y)$ put

$$D(f,g) = \sup\{d(f(x),g(x)): x \in X\}.$$

Observe that $D(f,g) \in [0,\infty]$.

Let d be an admissible metric for Y. For any space X define

$$C(X,Y;d) = \{f \in C(X,Y): \mathrm{diam}(f(X)) < \infty\}.$$

It is sometimes convenient to refer to elements of C(X,Y;d) as *bounded* functions.

1.3.1. LEMMA: *Let* X *and* Y *be spaces. Then*

(1) *for all* $f,g \in C(X,Y;d)$ *we have* $D(f,g) < \infty$,

(2) *the function* $D: C(X,Y;d) \times C(X,Y;d) \to [0,\infty)$ *is a metric.*

PROOF: For (1), take an arbitrary point $z \in X$ and observe that for all $f,g \in C(X,Y;d)$ the following holds:

$$D(f,g) \leq \mathrm{diam}(f(X)) + d(f(z),g(z)) + \mathrm{diam}(g(X)) < \infty.$$

The proof of (2) is routine and is left as an exercise to the reader. □

From now on we shall endow C(X,Y;d) with the topology induced by D (there is a problem since C(X,Y;d) need not be separable; we find it convenient to ignore this difficulty). By abuse of notation we shall also write d(f,g) instead of D(f,g) for all $f,g \in C(X,Y)$.

1.3.2. LEMMA: *Let* X *and* Y *be spaces with* X *compact. In addition, let* d_1 *and* d_2 *be admissible metrics for* Y. *Then*

(1) $C(X,Y;d_1) = C(X,Y;d_2) = C(X,Y)$, *and*

(2) *the topologies on* C(X,Y) *induced by* d_1 *and* d_2 *are the same.*

PROOF: (1) is trivial.

For each $\varepsilon > 0$, $y \in Y$ and $i \in \{1,2\}$ we put

$$B_i(y,\varepsilon) = \{z \in Y: d_i(y,z) < \varepsilon\}.$$

For (2), take $f \in C(X,Y;d_1)$ and $\varepsilon > 0$, arbitrarily. Since f(X) is compact, the open cover

$$\mathcal{U} = \{B_1(f(x),\varepsilon/4): x \in X\}$$

has a d_2-Lebesgue number, say δ (lemma 1.1.1). Without loss of generality, $\delta < \varepsilon/4$. Now take $g \in C(X,Y;d_2)$ such that $d_2(f,g) < \delta$. For each $x \in X$ we have $d_2(f(x),g(x)) < \delta$, so there exists

$p_x \in X$ such that $\{f(x),g(x)\} \subseteq B_1(f(p_x),\varepsilon/4)$. Consequently, for each $x \in X$ we have

$$d_1(f(x),g(x)) < \varepsilon/2,$$

from which it follows that $d_1(f,g) \leq \varepsilon/2 < \varepsilon$. We conclude that

$$\{g \in C(X,Y;d_2): d_2(f,g) < \delta\} \subseteq \{g \in C(X,Y;d_1): d_1(f,g) < \varepsilon\}$$

and hence that the topology on C(X,Y) induced by d_2 is finer than the topology on C(X,Y) induced by d_1. By interchanging the roles of d_1 and d_2 in the above argument we obtain the desired result. □

Let X and Y be spaces with X compact. From the above lemma we conclude that all the topologies we defined on the set C(X,Y) coincide. For that reason, if X is compact then C(X,Y;d) shall be denoted simply by C(X,Y) from now on. Observe that the norm topology which we defined on C(X) in §1.2 is the same as the topology for C(X,ℝ) just defined: thus, we will denote C(X,ℝ) by C(X).

We shall now prove that C(X,Y) for compact X is separable. The following notation will be useful. If $\mathcal{U}$ is a collection of subsets of a space X then $\overline{\mathcal{U}}$ denotes the collection $\{\overline{U}: U \in \mathcal{U}\}$.

1.3.3. PROPOSITION [[chapters 1,4,7]]: *Let* X *and* Y *be spaces with* X *compact. Then* C(X,Y) *is separable.*

PROOF: Let $\mathcal{B}$ and $\mathcal{E}$ be countable open bases for X and Y, respectively, which are closed under finite unions. For $B \in \mathcal{B}$ and $E \in \mathcal{E}$ put

$$A(B,E) = \{f \in C(X,Y): f(\overline{B}) \subseteq E\}.$$

Let $\mathcal{A}$ be the collection of all A(B,E)'s. Observe that $\mathcal{A}$ is countable. We claim that the family $\mathcal{A}^*$ of all finite intersections of elements of $\mathcal{A}$ is an open basis for C(X,Y). As a consequence it will follow that C(X,Y) is separable.

CLAIM 1: Each $A(B,E) \in \mathcal{A}$ is open in C(X,Y).

Take an arbitrary $f \in A(B,E)$. Since $f(\overline{B})$ is compact and E is open, there exists $\varepsilon > 0$ such that $B(f(\overline{B}),\varepsilon) \subseteq E$. Now suppose that $g \in C(X,Y)$ is such that $d(f,g) < \varepsilon$. For each $x \in X$,

$d(f(x),g(x)) < \varepsilon$, from which it follows that $g(\overline{B}) \subseteq B(f(\overline{B}),\varepsilon) \subseteq E$. We conclude that $g \in A(B,E)$. Consequently, the ball (in $C(X,Y)$) about f of radius ε is contained in $A(B,E)$.

CLAIM 2: $\mathcal{A}^*$ is an open basis for $C(X,Y)$.

Let $f \in C(X,Y)$ and $\varepsilon > 0$. We shall prove that there exists an element $F \in \mathcal{A}^*$ such that $f \in F \subseteq \{g \in C(X,Y): d(f,g) < \varepsilon\}$. Since $f(X)$ is compact, there are finitely many elements of $\mathcal{E}$, say $E_1,E_2,\cdots,E_n$, such that

(1) $f(X) \subseteq \bigcup_{i=1}^n E_i$, and
(2) for every $i \leq n$, $\mathrm{diam}(E_i) < \varepsilon$.

Let $\mathcal{U} = \{f^{-1}(E_i) : i \leq n\}$ and let $\mathcal{V}$ be a finite cover of X consisting of elements of $\mathcal{B}$ such that $\overline{\mathcal{V}} < \mathcal{U}$. For each $i \leq n$ let W_i be the union of the elements of $\mathcal{V}$ the closures of which are contained in $f^{-1}(E_i)$. Since $\mathcal{B}$ is closed under finite unions, $\mathcal{W} = \{W_i: i \leq n\}$ is a subcollection of $\mathcal{B}$, $\mathcal{W}$ covers X, and $\mathcal{W}$ has the property that the closure of each W_i is contained in $f^{-1}(E_i)$. Now put

$$F = \bigcap_{i=1}^n A(\overline{W}_i,E_i).$$

It is clear that $f \in F$. In addition, F is open by claim 1. We claim that $F \subseteq \{g \in C(X,Y): d(f,g) < \varepsilon\}$. To this end, take $g \in F$ and $x \in X$. There exists $i \leq n$ with $x \in W_i$. Since $f,g \in F$, $f(W_i) \cup g(W_i) \subseteq E_i$. Consequently, both $f(x)$ and $g(x)$ belong to E_i from which it follows by (2) that $d(f(x),g(x)) < \varepsilon$. □

Let X and (Y,d) be spaces. The function d on $C(X,Y)$ need not be a metric; it is easy to see that $d(f,g)$ can be ∞ in certain cases. Although d need not be a metric, we find it convenient to adopt some of the terminology of metrics and to treat d as some sort of generalized metric. For example, we call a sequence $(f_n)_n$ in $C(X,Y)$ *Cauchy* if for each $\varepsilon > 0$ there exists an $N \in \mathbb{N}$ such that $d(f_n,f_m) < \varepsilon$ for all $n,m \geq N$. This is of course nothing but the ordinary definition of a Cauchy sequence in a metric space. However, since $d(f,g) \in [0,\infty]$, the definition also makes sense in our situation.

We now turn to completeness properties of $C(X,Y)$ and $C(X,Y;d)$.

1.3.4. PROPOSITION [[chapters 1,3,4,5,6,7]]: *Let* X *and* (Y,d) *be spaces. Let* $(f_n)_n$ *be a* d-*Cauchy sequence in* $C(X,Y)$ *such that for every* $x \in X$, $\lim_{n\to\infty} f_n(x)$ *exists. Then the function* $f: X \to Y$ *defined by* $f(x) = \lim_{n\to\infty} f_n(x)$ *is continuous. In addition, if* $f_n \in C(X,Y;d)$ *for every* n *then* $f \in C(X,Y;d)$ *and* $f = \lim_{n\to\infty} f_n$ *(in* $C(X,Y;d)$*).*

PROOF: To begin with, let us establish the following

CLAIM: $\forall \varepsilon > 0\ \exists\ N \in \mathbb{N}$ such that for every $x \in X$ and $m \geq N$, $d(f(x),f_m(x)) < \varepsilon$.

Take $N \in \mathbb{N}$ such that $d(f_n,f_m) < \varepsilon/2$ for all $n,m \geq N$. We claim that N is as required. Let $x \in X$. Since $d(f_n(x),f_m(x)) < \varepsilon/2$ for all $n,m \geq N$ and since $f(x)$ is equal to $\lim_{n\to\infty} f_n(x)$, we obtain that for every $m \geq N$, $d(f(x),f_m(x)) \leq \varepsilon/2 < \varepsilon$.

We conclude that the sequence $(f_n)_n$ converges uniformly to f on X.

We shall now prove that f is continuous. Take $x \in X$ and $\varepsilon > 0$ arbitrarily. By the above, there exists $N \in \mathbb{N}$ such that $d(f(x),f_m(x)) < \varepsilon/3$ for all $m \geq N$. Since f_N is continuous at x, there exists $\delta > 0$ such that if $d(x,z) < \delta$ then $d(f_N(x),f_N(z)) < \varepsilon/3$. Now take $z \in X$ with $d(x,z) < \delta$. Then

$$\begin{aligned} d(f(x),f(z)) &\leq d(f(x),f_N(x)) + d(f_N(x),f_N(z)) + d(f_N(z),f(z)) \\ &< \quad \varepsilon/3 \quad + \quad \varepsilon/3 \quad + \quad \varepsilon/3 \\ &= \varepsilon. \end{aligned}$$

We conclude that f is continuous at x.

Now suppose that $f_n \in C(X,Y;d)$ for every n. We shall prove that $\mathrm{diam}(f(X)) < \infty$. Again by the above, there exists an $M \in \mathbb{N}$ such that for every $x \in X$, $d(f(x),f_M(x)) < 1$. Take arbitrary $x,z \in X$. Then

$$\begin{aligned} d(f(x),f(z)) &\leq d(f(x),f_M(x)) + d(f_M(x),f_M(z)) + d(f_M(z),f(z)) \\ &< 2 + \mathrm{diam}(f_M(X)), \end{aligned}$$

so $\mathrm{diam}(f(X)) < \infty$.

It remains to prove that $f = \lim_{n\to\infty} f_n$ (in $C(X,Y;d)$). However, this follows easily from the claim. □

1.3.5. COROLLARY [[chapters 1,4,6,7]]: *Let* X *and* (Y,d) *be spaces. Then* (Y,d) *is com-*

plete if and only if $(C(X,Y;d),d)$ *is complete.*

PROOF: Suppose that (Y,d) is complete and let $(f_n)_n$ be a d-Cauchy sequence in $C(X,Y;d)$. Fix $z \in X$ arbitrarily and let $\varepsilon > 0$. There exists $N \in \mathbb{N}$ such that $d(f_n,f_m) < \varepsilon$ for all $n,m \geq N$. Since for all n and m,

$$d(f_n(z),f_m(z)) \leq d(f_n,f_m),$$

we conclude that $(f_n(z))_n$ is Cauchy in (Y,d). The completeness of (Y,d) and proposition 1.3.4 now yield that the sequence $(f_n)_n$ converges.

Now assume that $(C(X,Y;d),d)$ is complete. Let $(y_n)_n$ be a d-Cauchy sequence in Y: For each n let $f_n: X \to Y$ be the constant function with value y_n. It is easy to see that $(f_n)_n$ is a d-Cauchy sequence in $C(X,Y;d)$. By assumption, $f = \lim_{n\to\infty} f_n$ exists and belongs to $C(X,Y;d)$. It is left as an exercise to the reader to prove that f is constant and that the sequence $(y_n)_n$ converges to the unique point in the range of f. □

A space X is called a *Baire space* if the intersection of every countable family of dense open subsets of X is dense in X or, equivalently, the union of every countable family of nowhere dense subsets of X is again nowhere dense. It is well-known, and easy to prove, that every topologically complete space is a Baire space. From this result and the completeness of $(\mathbb{R},|\cdot|)$ we obtain the following:

1.3.6. COROLLARY: *If* X *is a compact space then the standard metric* $d(f,g) = \|f - g\|$ *on* $C(X)$ *is complete. Consequently,* $C(X)$ *is a Baire space.* □

Now let X and Y be spaces and define

$$\mathcal{S}(X,Y) = \{f \in C(X,Y): f \text{ is surjective}\}.$$

There are spaces X and Y for which $\mathcal{S}(X,Y)$ is empty, see exercise 1.3.4.

1.3.7. PROPOSITION [[chapters 4,6,7]]: *Let* X *and* Y *be compact spaces. Then* $\mathcal{S}(X,Y)$ *is closed in* $C(X,Y)$.

PROOF: Assume that $f \notin \mathcal{S}(X,Y)$, i.e. there exists a point $y \in Y\setminus f(X)$. By compactness, $\varepsilon = d(y,f(X)) > 0$. It is a triviality to verify that $B(f,\varepsilon) \cap \mathcal{S}(X,Y) = \emptyset$. We conclude that $C(X,Y)\setminus\mathcal{S}(X,Y)$ is open in $C(X,Y)$. □

Let X and Y be spaces with X compact, and let $\varepsilon > 0$. A function $f \in C(X,Y)$ is called an *ε-map* if for every $y \in Y$,

$$\operatorname{diam}(f^{-1}(y)) < \varepsilon.$$

Put $C_\varepsilon(X,Y) = \{f \in C(X,Y): f \text{ is an } \varepsilon\text{-map}\}$ and $\mathcal{S}_\varepsilon(X,Y) = C_\varepsilon(X,Y) \cap \mathcal{S}(X,Y)$, respectively. In addition, let $\mathcal{G}_\varepsilon(X,Y) = \mathcal{S}(X,Y) \setminus \mathcal{S}_\varepsilon(X,Y)$. □

1.3.8. LEMMA: *Let* X *and* Y *be spaces with* X *compact and let* $\varepsilon > 0$. *Then* $C_\varepsilon(X,Y)$ *is an open subspace of* $C(X,Y)$. *Consequently,* $\mathcal{G}_\varepsilon(X,Y)$ *is closed in* $C(X,Y)$.

PROOF: Take $f \in C_\varepsilon(X,Y)$. Since X is compact, $f: X \to f(X)$ is a closed map (exercise 1.1.3). Consequently there exists for every $y \in f(X)$ an open neighborhood U_y (in $f(X)$) such that

$$\operatorname{diam}(f^{-1}(U_y)) < \varepsilon$$

(exercise 1.1.10). Let $\delta > 0$ be a Lebesgue number for the open covering $\{U_y: y \in f(X)\}$ of $f(X)$ (lemma 1.1.1). Let $g \in C(X,Y)$ be such that $d(g,f) < \delta/2$. We claim that $g \in C_\varepsilon(X,Y)$. To this end, take an arbitrary $y \in Y$. Since $d(f,g) < \delta/2$ it follows easily that $\operatorname{diam}(fg^{-1}(y)) < \delta$. Consequently there exists a point $z \in f(X)$ such that $fg^{-1}(y) \subseteq U_z$ which implies that $\operatorname{diam}(f^{-1}fg^{-1}(y) < \varepsilon$. Since $g^{-1}(y) \subseteq f^{-1}fg^{-1}(y)$, we conclude that $\operatorname{diam}(g^{-1}(y)) < \varepsilon$, i.e. g is an ε-map. □

Let X and Y be spaces and let $\mathcal{H}(X,Y)$ denote the set of all homeomorphisms from X onto Y considered as a subspace of $C(X,Y)$. If $X = Y$ then for $\mathcal{H}(X,X)$ we shall simply write $\mathcal{H}(X)$. As usual, $\mathcal{H}(X)$ is called the *autohomeomorphism group* of X.

1.3.9. LEMMA [[chapters 1,6]]: *Let* X *and* Y *be compact spaces. Then* $\mathcal{H}(X,Y) = \bigcap_{n=1}^{\infty} \mathcal{S}_{1/n}(X,Y)$. *As a consequence,* $\mathcal{H}(X,Y)$ *is a* G_δ*-subset of* $\mathcal{S}(X,Y)$ *and hence of* $C(X,Y)$.

PROOF: That $\mathcal{H}(X,Y) \subseteq \bigcap_{n=1}^{\infty} \mathcal{S}_{1/n}(X,Y)$ is a triviality. Pick $f \in \bigcap_{n=1}^{\infty} \mathcal{S}_{1/n}(X,Y)$. Then f is a $1/n$-map for every n, hence f is one-to-one. Since f is onto, the compactness of X implies that f is a homeomorphism (exercise 1.1.4). □

Let X be a compact space. The above lemma implies that $\mathcal{H}(X)$ is a G_δ-subset of $C(X,X)$, hence $\mathcal{H}(X)$ is completely metrizable since $C(X,X)$ is (corollary 1.3.5 and theorem 4.7.4). We shall now explicitly describe a complete metric for $\mathcal{H}(X)$ which generates its topology.

1.3.10. PROPOSITION [[chapter 1]]: *Let* X *be a compact space. For* f,g $\in \mathcal{H}(X)$ *define* $\sigma(f,g) = d(f,g) + d(f^{-1},g^{-1})$. *Then* σ *is a complete metric on* $\mathcal{H}(X)$ *that generates its topology.*

PROOF: That σ is a metric is routine and is left as an exercise to the reader. We shall first prove that d and σ generate the same topology. Since $d(f,g) \leq \sigma(f,g)$ for all f,g $\in \mathcal{H}(X)$, the only thing to verify is that

$$\forall \varepsilon > 0\ \forall f \in \mathcal{H}(X)\ \exists\, \delta > 0 \text{ such that if } g \in \mathcal{H}(X) \text{ and } d(f,g) < \delta \text{ then } \sigma(f,g) < \varepsilon.$$

Choose arbitrary $\varepsilon > 0$ and $f \in \mathcal{H}(X)$. By compactness, f^{-1} is uniformly continuous and consequently there exists $\gamma > 0$ such that for all $x,y \in X$ with $d(x,y) < \gamma$ we have $d(f^{-1}(x),f^{-1}(y)) < \varepsilon/4$. Let $\delta = \min\{\gamma,\varepsilon/4\}$. Take $g \in \mathcal{H}(X)$ such that $d(f,g) < \delta$. Pick an arbitrary $x \in X$ and put $z = g^{-1}(x)$. Since $d(f,g) < \delta$, it follows that $d(f(z),g(z)) = d(fg^{-1}(x),x) < \gamma$ (recall that $\delta \leq \gamma$). Consequently,

$$d(f^{-1}fg^{-1}(x),f^{-1}(x)) < \varepsilon/4,$$

so that $d(g^{-1}(x),f^{-1}(x)) < \varepsilon/4$. We conclude that $d(g^{-1},f^{-1}) \leq \varepsilon/4$ (in fact by compactness of X, even $d(g^{-1},f^{-1}) < \varepsilon/4$). Therefore, $\sigma(f,g) = d(f,g) + d(f^{-1},g^{-1}) \leq \varepsilon/4 + \varepsilon/4 = \varepsilon/2 < \varepsilon$.

Now let $(f_n)_n$ be a σ-Cauchy sequence in $\mathcal{H}(X)$. Then $(f_n)_n$ is a d-Cauchy sequence in C(X,X) and therefore the limit $f = \lim_{n\to\infty} f_n$ exists and belongs to C(X,X) (corollary 1.3.5). Similarly, the limit $g = \lim_{n\to\infty} f_n^{-1}$ exists and belongs to C(X,X). It is easily seen that $f \circ g = 1_X = g \circ f$ from which it follows that $f \in \mathcal{H}(X)$.□

Lemma 1.3.9 and proposition 1.3.10 both imply the following

1.3.11. COROLLARY: *If* X *is compact then* $\mathcal{H}(X)$ *is a Baire space.*□

Exercises for §1.3.

1. Let X be a compact space. Prove that the function $\xi: \mathcal{H}(X) \times \mathcal{H}(X) \to \mathcal{H}(X)$ defined by $\xi(f,g) = f \circ g^{-1}$ is continuous (i.e. $\mathcal{H}(X)$ is a topological group).

2. Prove that the function $f: I \to I$ defined by

$$f(x) = \begin{cases} 2x & (0 \le x \le 1/4), \\ \frac{1}{2} & (1/4 \le x \le 3/4), \\ 2x-1 & (3/4 \le x \le 1), \end{cases}$$

belongs to the closure of $\mathcal{H}(I)$ in $C(I,I)$.

3. Prove that $\mathcal{H}(I)$ has exactly two components.

4. Give an example of two compact connected spaces X and Y such that $\mathcal{S}(X,Y) = \emptyset = \mathcal{S}(Y,X)$.

5. Let $\mathbb{N}$ denote the discrete space of natural numbers. Prove that $C(\mathbb{N},\mathbb{R};|\cdot|)$ is not separable.

6. Give an example of a Baire space which is not topologically complete.

7. Let X, Y and Z be compact spaces, let $f \in C(Z,X)$ and let $g,h \in C(X,Y)$. Prove that $d(g \circ f, h \circ f) \le d(g,h)$. In addition, prove that if f is surjective then $d(g \circ f, h \circ f) = d(g,h)$.

1.4. The Michael Selection Theorem and Applications

In this section we shall prove that certain set-valued functions admit a continuous selection (for definitions, see below) and present applications of this result.

We shall first present a few elementary results that shall be important later in this section.

Let $\mathcal{A}$ be a collection of subsets of a space X. We say that $\mathcal{A}$ is *locally finite* provided that for every $x \in X$ there is a neighborhood U_x of x such that the set

$$\{A \in \mathcal{A} : A \cap U_x \neq \emptyset\}$$

is finite. Observe that $\mathcal{A}$ is countable. This can be seen as follows: every point in X has a neighborhood meeting only finitely many elements of $\mathcal{A}$ and countably many of these neighborhoods cover X.

A space X is called *paracompact* if for every open cover $\mathcal{U}$ of X there exists an open refinement $\mathcal{V}$ of $\mathcal{U}$ such that $\mathcal{V}$ is locally finite.

1.4.1. LEMMA [[chapters 1,4,5,7]]: *Every space is paracompact.*

PROOF: Let X be a space and let $\mathcal{U}$ be an open cover of X. For each $x \in X$ let $E(x)$ and $F(x)$ be open neighborhoods of x such that $E(x) \subseteq \overline{E(x)} \subseteq F(x) \in \mathcal{U}$. Since X is Lindelöf, the cover

$\{E(x): x \in X\}$ of X has a countable subcover, say $\mathcal{E} = \{E(x_n): n \in \mathbb{N}\}$. For every $n \in \mathbb{N}$ define

$$V_n = F(x_n) \backslash \bigcup_{m<n} \overline{E(x_m)}.$$

We claim that $\mathcal{V} = \{V_n: n \in \mathbb{N}\}$ is the required locally finite open refinement of $\mathcal{U}$. Clearly $\mathcal{V}$ consists of open sets. For each $x \in X$ let $n(x)$ be the smallest integer with $x \in F(x_{n(x)})$. For this number $n(x)$ we clearly have $x \in V_{n(x)}$. Consequently, $\mathcal{V}$ is an open cover of X which obviously refines $\mathcal{U}$. We shall prove that $\mathcal{V}$ is locally finite. Take an arbitrary $x \in X$. Since $\mathcal{E}$ covers X there is an $n \in \mathbb{N}$ with $x \in E(x_n)$. Clearly $E(x_n) \cap V_m = \emptyset$ for all $m > n$. Consequently, $E(x_n)$ is a neighborhood of x which intersects at most n members from $\mathcal{V}$. □

Let X be a space. A family $\mathcal{F}$ of continuous functions from X to $I = [0,1]$ is called a *partition of unity on* X if for each $x \in X$ there exist a neighborhood U_x of x and a *finite* subset $\mathcal{F}(x)$ of $\mathcal{F}$ such that

(1) $\sum_{f \in \mathcal{F}(x)} f(y) = 1$ for each $y \in U_x$, and

(2) if $f \in \mathcal{F} \backslash \mathcal{F}(x)$ and $y \in U_x$ then $f(y) = 0$.

Observe that each partition of unity on X is countable. If $\mathcal{F}$ is a partition of unity on X then we define $\mathcal{U}(\mathcal{F}) = \{f^{-1}((0,1]): f \in \mathcal{F}\}$.

1.4.2. LEMMA: *Let* X *be a space and let* $\mathcal{F}$ *be a partition of unity on* X. *Then* $\mathcal{U}(\mathcal{F})$ *is a locally finite open cover of* X.

PROOF: That $\mathcal{U}(\mathcal{F})$ consists of open sets is clear. Take $x \in X$ arbitrarily and let U_x and $\mathcal{F}(x)$ be as in the above definition. By (1) there exists an $f \in \mathcal{F}(x)$ such that $f(x) \neq 0$. We conclude that $\mathcal{U}(\mathcal{F})$ covers X. That $\mathcal{U}(\mathcal{F})$ is locally finite follows immediately from (2). □

Let $\mathcal{U}$ be a locally finite open cover of a space (X,d). We shall associate to $\mathcal{U}$ a certain family of continuous functions which will be useful here as well as in chapters 2, 3 and 4, as follows. For every $U \in \mathcal{U}$ define $\kappa_U: X \to \mathbb{R}$ by

$$(*) \qquad \kappa_U(x) = \frac{d(x, X \backslash U)}{\sum_{V \in \mathcal{U}} d(x, X \backslash V)}.$$

These functions are called the κ-*functions with respect to the cover* $\mathcal{U}$. Observe that the sum in the denominator of (*) contains at least one but at most finitely many non-zero terms, so that κ_U is well-defined. Also observe that $\kappa_U(x) \geq 0$.

We next claim that each κ_U is continuous. This is easy. Take an arbitrary $x \in X$. There is an open neighborhood W of x such that the set $\mathcal{F} = \{U \in \mathcal{U}: U \cap W \neq \emptyset\}$ is finite. Let ρ_U denote the restriction $\kappa_U \mid W$. Then for every $y \in W$ we have

$$\rho_U(x) = \frac{d(x,X\setminus U)}{\Sigma_{V\in\mathcal{F}}\, d(x,X\setminus V)}.$$

Since $\mathcal{F}$ is finite, ρ_U is clearly a continuous function on W. Since W is a neighborhood of x this implies that κ_U is continuous at x.

Finally, observe that $\Sigma_{U\in\mathcal{U}}\, \kappa_U(x) = 1$ for every $x \in X$.

We say that a partition of unity $\mathcal{F}$ on a space X is *subordinated to a cover* $\mathcal{V}$ of X if the cover $\mathcal{U}(\mathcal{F})$ refines $\mathcal{V}$.

1.4.3. THEOREM [[chapter 1,4]]: *Let* X *be a space and let* $\mathcal{U}$ *be an open cover of* X. *Then there exists a partition of unity* $\mathcal{K}$ *on* X *which is subordinated to* $\mathcal{U}$.

PROOF: By lemma 1.4.1, without loss of generality we may assume that $\mathcal{U}$ is locally finite. Let $\mathcal{K} = \{\kappa_U: U \in \mathcal{U}\}$ be the set of κ-functions with respect to $\mathcal{U}$. Since $\mathcal{U}$ is locally finite, from the above it is clear that $\mathcal{K}$ is a partition of unity on X which is subordinated to $\mathcal{U}$. □

If X is a set then $\mathcal{P}(X)$ denotes the *power set* of X, i.e. the set consisting of all subsets of X. Let X and Y be sets. A *set-valued function* F *from* X *to* Y is defined to be a function from X to $\mathcal{P}(Y)\setminus\{\emptyset\}$, i.e. $F: X \to \mathcal{P}(Y)\setminus\{\emptyset\}$. By the symbol $F: X \Rightarrow Y$ we shall mean that F is a set-valued function from X to Y.

Let X and Y be topological spaces and let $F: X \Rightarrow Y$. For every $V \subseteq Y$ we put

$$F^{\Leftarrow}(V) = \{x \in X: F(x) \cap V \neq \emptyset\}.$$

We say that F is *lower semi-continuous* (abbreviated **LSC**) provided that for every open subset U of Y, $F^{\Leftarrow}(U)$ is open in X. Observe that if $\mathcal{U}$ is a covering of Y then

$$F^{\Leftarrow}(\mathcal{U}) = \{F^{\Leftarrow}(U): U \in \mathcal{U}\}$$

covers X since for every $x \in X$, $F(x) \neq \emptyset$. A basic example of an **LSC** set-valued function is the following one: let $f: X \to Y$ be an open surjection and define $F: Y \to \mathcal{P}(X)$ by $F(y) = f^{-1}(y)$. Then F is **LSC** since for every open $U \subseteq Y$ we have

$$F^{\Leftarrow}(U) = \{y \in Y: F(y) \cap U \neq \emptyset\} = \{y \in Y: f^{-1}(y) \cap U \neq \emptyset\} = f(U).$$

Other examples of **LSC** set-valued mappings will be presented later.

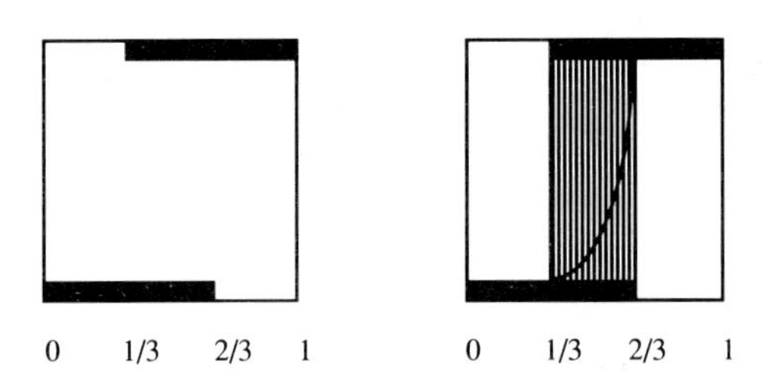

Figure 1.4.1.

Let X and Y be sets and let $F: X \Rightarrow Y$. A function $f: X \to Y$ is called a *selection* for F if for all $x \in X$, $f(x) \in F(x)$. Since for all $x \in X$ the set $F(x)$ is nonempty, by the Axiom of Choice a selection for F always exists. The question naturally arises whether it is possible to find a *continuous* selection in case X and Y are topological spaces. This question is natural but rather naive of course. Simple examples show that the answer in general is in the negative.

1.4.4. EXAMPLE: Define $F: I \Rightarrow I$ by

$$F(x) = \begin{cases} \{0\} & (0 \leq x \leq \frac{1}{3}), \\ \{0,1\} & (\frac{1}{3} < x < \frac{2}{3}), \\ \{1\} & (\frac{2}{3} \leq x \leq 1). \end{cases}$$

From figure 1.4.1 it is immediately clear that there does not exist a continuous function $f: I \to I$ the graph of which is contained in the "graph" of F. It is left as an exercise to the reader to prove that F is **LSC**.

The values of F in the above example are too "small" for F to admit a *continuous* selection. If we enlarge these values by for example to require that $F(x) = I$ for all $\frac{1}{3} < x < \frac{2}{3}$ then F *does* admit a continuous selection.

The following result shows that the concept of lower semi-continuity is the right one if one wants to deal with *continuous* selections.

1.4.5. PROPOSITION: *Let* X *and* Y *be spaces and let* $F: X \Rightarrow Y$ *be a set-valued function with the following property:*

$$\forall x \in X\ \forall y \in F(x)\ \exists \textit{continuous selection } f \textit{ for } F \textit{ with } f(x) = y.$$

Then F *is* **LSC**.

PROOF: Let $U \subseteq Y$ be open and take $x \in F^{\Leftarrow}(U)$. Pick $y \in F(x) \cap U$. By assumption there exists a continuous selection $f: X \to Y$ for F such that $f(x) = y$. Put $V = f^{-1}(U)$. By continuity of f, V is a neighborhood of x. In addition, $V \subseteq F^{\Leftarrow}(U)$ since if $x' \in V$ then

$$f(x') \in F(x') \cap U.$$

We conclude that $F^{\Leftarrow}(U)$ is open. □

One of the main results in this section is that for **LSC** set-valued functions continuous selections exist provided that the range of the set-valued function under consideration is a normed linear space and its values are, among other things, convex, cf. figure 1.4.1. The following three technical lemmas are needed in the proof of this result.

1.4.6. LEMMA [[chapter 1]]: *Let* $F: X \Rightarrow Y$ *be* **LSC**. *Then*

(a) *the function* $F_c: X \Rightarrow Y$ *defined by* $F_c(x) = \overline{F(x)}$ *is also* **LSC**,

(b) *if* $f: X \to Y$ *is continuous and* d *is an admissible metric for* Y *and* $r > 0$ *is such that*

$$\forall x \in X: d(f(x),F(x)) < r,$$

then the function $G: X \Rightarrow Y$ *defined by* $G(x) = \overline{F(x) \cap B(f(x),r)}$ *is* **LSC**.

PROOF: For (a), simply observe that for all $x \in X$ and open $U \subseteq Y$ we have $F(x) \cap U \neq \emptyset$ if and only if $\overline{F(x)} \subseteq U \neq \emptyset$. Consequently, for every open $U \subseteq Y$ the equality $F^{\Leftarrow}(U) = F_c^{\Leftarrow}(U)$ holds.

For (b) we use (a) to conclude that it suffices to prove that the set-valued mapping $\vec{G}: X \Rightarrow Y$ defined by

$$\vec{G}(x) = F(x) \cap B(f(x),r)$$

is **LSC**. Let $V \subseteq Y$ be open and take a point $x \in \vec{G}^{\Leftarrow}(V)$. There exists $y \in Y$ such that

$$y \in (F(x) \cap B(f(x),r)) \cap V.$$

Let $\varepsilon = r - d(y,f(x))$ and choose $\delta > 0$ such that $\delta < \varepsilon$ and

$$B(y,\delta) \subseteq B(f(x),r) \cap V.$$

Since $F(x) \cap B(y,\delta/2) \neq \emptyset$ and F is **LSC,** $U_0 = F^{\Leftarrow}(B(y,\delta/2))$ is a neighborhood of x. In addition, since f is continuous, $U_1 = f^{-1}(B(f(x),\delta/2))$ is also a neighborhood of x. Put $U = U_0 \cap U_1$. We claim that $U \subseteq \vec{G}^{\Leftarrow}(V)$.

To this end, take an arbitrary $x' \in U$. Since $x' \in U_0$, there exists a point $y' \in F(x') \cap B(y,\delta/2)$. In addition, $f(x') \in B(f(x),\delta/2)$. Consequently,

$$\begin{aligned} d(y',f(x')) &\leq d(y',y) + d(y,f(x)) + d(f(x),f(x')) \\ &< \delta/2 + r - \varepsilon + \delta/2 \\ &< \delta/2 + r - \delta + \delta/2 \\ &= r \end{aligned}$$

and therefore $y' \in F(x') \cap B(f(x'),r) \cap V = \vec{G}(x') \cap V$. We conclude that $x' \in \vec{G}^{\Leftarrow}(V)$.□

1.4.7. LEMMA [[chapter 7]]: *Let* L *be a normed linear space, let* X *be a space and let* $F: X \Rightarrow L$ *be* **LSC** *such that* $F(x)$ *is convex for every* $x \in X$. *Then for every* $r > 0$ *there exists a continuous function* $f: X \to L$ *such that for every* $x \in X$, $d(f(x),F(x)) < r$.

PROOF: Put $\mathcal{B} = \{B(y,r): y \in L\}$. By theorem 1.4.3 there exists a partition of unity $\mathcal{P}$ on X which is subordinated to $F^{\Leftarrow}(\mathcal{B})$. Consequently, for each $p \in \mathcal{P}$ there exists $b_p \in L$ such that

$$p^{-1}(0,1] \subseteq F^{\Leftarrow}(B(b_p,r)).$$

Define $f: X \to L$ by

$$f(x) = \Sigma_{p \in \mathcal{P}}\, p(x) \cdot b_p.$$

For each $x \in X$ there exists a neighborhood U_x and a finite subset $G(x)$ of $\mathcal{P}$ such that $U_x \cap p^{-1}(0,1] \neq \emptyset$ if and only if $p \in G(x)$.

Now fix $x \in X$ arbitrarily. Observe that the restriction of f to U_x is given by

$$f(y) = \Sigma_{p\in G(x)}\, p(y)\cdot b_p,$$

which is a continuous expression in y since $G(x)$ is finite. From this we conclude that f is well-defined and continuous at x.

Put $\mathcal{G}(x) = \{p \in G(x): x \in p^{-1}(0,1]\}$. For each $p \in \mathcal{G}(x)$ we have $x \in p^{-1}(0,1] \subseteq F^{\Leftarrow}(B(b_p,r))$ and consequently there exists $y_p \in F(x) \cap B(b_p,r)$. Observe that

$$f(x) = \Sigma_{p\in \mathcal{G}(x)}\, p(x)\cdot b_p \text{ and } \Sigma_{p\in \mathcal{G}(x)}\, p(x) = 1.$$

This implies that

$$\|f(x) - \Sigma_{p\in \mathcal{G}(x)}\, p(x)\cdot y_p\| \leq \Sigma_{p\in \mathcal{G}(x)}\, p(x)\cdot\|b_p - y_p\| < \Sigma_{p\in \mathcal{G}(x)}\, p(x)\cdot r = r.$$

Since $F(x)$ is convex and $\Sigma_{p\in \mathcal{G}(x)}\, p(x) = 1$ we have $\Sigma_{p\in \mathcal{G}(x)}\, p(x)\cdot y_p \in F(x)$ (lemma 1.2.2), and consequently, $d(f(x),F(x)) < r$.□

We need one more technical lemma.

1.4.8. LEMMA [[chapter 1]]: *Let* X *and* Y *be spaces and let* $F: X \Rightarrow Y$ *be* **LSC**. *Suppose that* $A \subseteq X$ *is closed and that* $f: A \to Y$ *is a continuous selection for the function* $F \mid A: A \Rightarrow Y$. *Define* $G: X \Rightarrow Y$ *by*

$$G(x) = \begin{cases} \{f(x)\} & (x \in A), \\ F(x) & (x \in X \setminus A). \end{cases}$$

Then G *is* **LSC**.

PROOF: Let $U \subseteq Y$ be open. We first claim that $f^{-1}(U) \subseteq F^{\Leftarrow}(U)$. Take an arbitrary $x \in f^{-1}(U)$. Then $f(x) \in F(x) \cap U$ which implies that $x \in F^{\Leftarrow}(U)$. Now, since $f^{-1}(U)$ is open in A, there exists an open $V \subseteq X$ such that $V \cap A = f^{-1}(U)$. Since $f^{-1}(U) \subseteq F^{\Leftarrow}(U)$ and $F^{\Leftarrow}(U)$ is open, without loss of generality, $V \subseteq F^{\Leftarrow}(U)$. Consequently,

$$G^{\Leftarrow}(U) = V \cup (F^{\Leftarrow}(U)\setminus A)$$

is open. □

Let $(L,\|\cdot\|)$ be a normed linear space. A subset A of L is called *complete with respect to* $\|\cdot\|$ if the restriction of the metric $d(x,y) = \|x - y\|$ to A is complete. Observe that such an A is automatically closed in L and also that every *compact* subset of L is complete with respect to $\|\cdot\|$.

We now come to one of the main results in this section which can be viewed as an extension theorem, since it says that a partially defined selection on a closed set can be extended to a selection over the entire space.

1.4.9. THEOREM ("The Michael Selection Theorem") [[chapter 1]]: *Let* $(L,\|\cdot\|)$ *be a normed linear space, let* X *be a space and let* $F: X \Rightarrow L$ *be* **LSC** *such that each* F(x) *is convex in* L, *and complete with respect to* $\|\cdot\|$. *Then for every closed subset* A *of* X *and every continuous selection* $f: A \to L$ *for the function* $F \mid A: A \Rightarrow L$, *there exists a continuous selection* $g: X \to L$ *for* F *which extends* f.

PROOF: We shall first prove the theorem in the special case $A = \emptyset$.

By induction on n we shall construct a sequence $(f_n)_n$ in C(X,L) such that

$$(1)\ d(f_n,f_{n+1}) < 2^{-(n-1)}, \text{ and}$$
$$(2)\ d(f_n(x),F(x)) < 2^{-n} \text{ for every } x \in X.$$

For $r = 2^{-1}$ apply lemma 1.4.7 to find $f_1: X \to L$ with $d(f_1(x),F(x)) < 2^{-1}$ for every $x \in X$. Suppose that f_n has been defined. Define $F_n: X \Rightarrow L$ by

$$F_n(x) = \overline{F(x) \cap B(f_n(x),2^{-n})}\ .$$

Then F_n is **LSC** by lemma 1.4.6. By another appeal to lemma 1.4.7 we can find $h: X \to L$ such that $d(h(x),F_n(x)) < 2^{-(n+1)}$ for every $x \in X$. It is easy to see that $f_{n+1} = h$ is as required.

CLAIM: For every $x \in X$, $\lim_{n\to\infty} f_n(x)$ exists and belongs to F(x).

Take an arbitrary $x \in X$. By (2) there exists for every $n \in \mathbb{N}$ a point $a_n \in F(x)$ such that $d(f_n(x),a_n) < 2^{-n}$. Consequently, by (1) we have

$$d(a_n,a_{n+1}) \le d(a_n,f_n(x)) + d(f_n(x),f_{n+1}(x)) + d(f_{n+1}(x),a_{n+1})$$
$$< \quad 2^{-n} \quad + \quad 2^{-(n-1)} \quad + \quad 2^{-(n+1)}$$

$< 2^{-(n-2)}$.

We conclude that the sequence $(a_n)_n$ is Cauchy, and by assumption, $a = \lim_{n\to\infty} a_n$ exists. Since $d(f_n(x),a_n) < 2^{-n}$ for every n, this implies that $\lim_{n\to\infty} f_n(x) = f(x)$ also exists and is equal to a. Since $a \in F(x)$, we are done.

By proposition 1.3.4 we now conclude that the function f is continuous. We conclude that f is as required.

Now if $A \neq \emptyset$, the above special case and lemma 1.4.8 yield the desired result. □

We shall now present several applications of the Michael Selection Theorem. We need the following:

1.4.10. THEOREM ("The Open Mapping Theorem") [[chapter 1]]: *Let* T *be a continuous linear mapping of a Banach space* E *onto a Banach space* F. *Then* T *is open.*

PROOF: As announced, $\underline{0}$ denotes the zero of F. The proof is in three steps.

CLAIM 1: There exists $\alpha > 0$ such that such that $\{y \in F: \|y\| \le 1\} \subseteq \overline{T(\{x \in E: \|x\| \le \alpha\})}$.

For each $\alpha > 0$ put $B_\alpha = \{x \in E: \|x\| \le \alpha\}$. Since

$$F = \bigcup_{n=1}^{\infty} \overline{T(B_n)},$$

and F is a Baire space there exist $m \in \mathbb{N}$ such that $\overline{T(B_m)}$ has nonempty interior. Since T is linear, it follows easily that $T(B_m)$ is convex, and by the continuity of the algebraic operations on F, so is $\overline{T(B_m)}$. In addition, B_m is symmetric, i.e. $-B_m = B_m$. Again since T is linear, it follows that $T(B_m)$ is symmetric, from which it follows easily that $\overline{T(B_m)}$ is symmetric as well. Now choose $y \in F$ and $\beta > 0$ such that $D(y,\beta) \subseteq \overline{T(B_m)}$. Let $z \in F$ with $\|z\| \le \beta$. Then $\|(z+y)-y\| \le \beta$, hence $z+y \in \overline{T(B_m)}$. Similarly, $\|(y-z)-y\| \le \beta$, from which it follows that $(y-z) \in \overline{T(B_m)}$ and by the fact that $\overline{T(B_m)}$ is symmetric, that $(z-y) \in \overline{T(B_m)}$. Now observe that since $\underline{0} \in \overline{T(B_m)}$, by the convexity of $\overline{T(B_m)}$ we get

$$z = (z+y)/2 + (z-y)/2 \in \overline{T(B_m)}.$$

This proves that $D(\underline{0},\beta) \subseteq \overline{T(B_m)}$ from which it follows easily that $D(\underline{0},1) \subseteq \overline{T(B_\alpha)}$, where $\alpha = m/\beta$.

In the remaining part of the proof we adopt the notation introduced in claim 1.

CLAIM 2: $\{y \in F: \|y\| \le 1\} \subseteq T(B_{2\alpha})$.

Let $y \in F$ with $\|y\| \le 1$. We shall define inductively a sequence $(y_n)_n$ in $T(B_\alpha)$ such that for all n,

$$(*) \qquad \|y - \sum_{k=1}^{n} 2^{-(k-1)} y_k\| \le 2^{-n}.$$

By claim 1, there exists $y_1 \in T(B_\alpha)$ such that $\|y - y_1\| \le \frac{1}{2}$. Suppose that $y_1, \cdots, y_n$ are chosen properly. Then

$$\|2^n(y - \sum_{k=1}^{n} 2^{-(k-1)} y_k)\| \le 1$$

and therefore, again by claim 1, there exists $y_{n+1} \in T(B_\alpha)$ such that

$$\|2^n(y - \sum_{k=1}^{n} 2^{-(k-1)} y_k) - y_{n+1}\| \le \tfrac{1}{2}.$$

It is clear that y_{n+1} is as required.

From (*) it easily follows that

$$y = \lim_{n\to\infty} \sum_{k=1}^{n} 2^{-(k-1)} y_k.$$

For every n choose a point $x_n \in B_\alpha$ with $T(x_n) = y_n$. Since $\|x_n\| \le \alpha$ for every n, it is easily seen that

$$x = \lim_{n\to\infty} \sum_{k=1}^{n} 2^{-(k-1)} x_k$$

exists. Since $\|\cdot\|: E \to \mathbb{R}$ is continuous,

$$\|x\| \le \sum_{k=1}^{\infty} 2^{-(k-1)} \|x_k\| \le \sum_{k=1}^{\infty} 2^{-(k-1)} \cdot \alpha = 2\alpha.$$

By observing that T is continuous and linear we obtain

$$T(x) = \lim_{n\to\infty} \sum_{k=1}^{n} 2^{-(k-1)}T(x_k) = y,$$

from which we conclude that $y \in T(B_{2\alpha})$.

CLAIM 3: T is open.

Let U be a nonempty open subset of E. Let $x \in U$ and choose $\varepsilon > 0$ such that $D(x,\varepsilon) \subseteq U$. Consequently, $B_\varepsilon \subseteq U - x$. It follows by step 2 that $\{y \in F: \|y\| \leq \varepsilon/2\alpha\} \subseteq T(B_\varepsilon)$. Consequently, since $T(B_\varepsilon) \subseteq T(U - x) = T(U) - T(x)$, we have

$$\{y \in F: \|y - T(x)\| \leq \tfrac{\varepsilon}{2\alpha}\} \subseteq T(U).$$

So T(U) is a neighborhood of T(x) and hence it follows that T(U) is open. □

This result and the Michael Selection Theorem now imply the following

1.4.11. COROLLARY: *Let* T *be a continuous linear mapping of a Banach space* E *onto a Banach space* F *and let* ker T *denote the kernel of* T. *Then there exists a continuous function* $f: F \to E$ *such that* $T \circ f = 1_F$ *and the function* $h: E \to \ker T \times F$ *defined by*

$$h(x) = (x - f(T(x)), T(x))$$

is a homeomorphism.

PROOF: Define $H: F \Rightarrow E$ by $H(y) = T^{-1}(y)$. Then for each open $U \subseteq E$, $H^{\Leftarrow}(U) = T(U)$, so by theorem 1.4.10, F is **LSC**. Since the fibers of T are convex and closed by linearity and continuity of T, respectively, the existence of f follows directly from theorem 1.4.9.
The easy proof that h is a homeomorphism is left as an exercise to the reader. □

We shall now present another important application of theorem 1.4.9. To begin with, let us first formulate and prove the following technical:

1.4.12. LEMMA [chapters 1,4,5,6]]: *Let* X *be a space and let* A *be a closed subset of* X. *Then there exist a locally finite (countable) open cover* $\mathcal{U}$ *of* X\A *and a sequence of points*

$\{a_U: U \in \mathcal{U}\}$ *in* A *such that*

(1) *for all* $U \in \mathcal{U}$ *and* $x \in U$, $d(x,a_U) \le 2d(x,A)$, *and*

(2) *if* $U_n \in \mathcal{U}$ *for every* n *and* $\lim_{n\to\infty} d(U_n,A) = 0$ *then* $\lim_{n\to\infty} \mathrm{diam}(U_n) = 0$.

PROOF: Let

$$\mathcal{V} = \{B(x,\tfrac{1}{4}d(x,A)): x \in X\backslash A\}.$$

Since A is closed, $\mathcal{V}$ is an open cover of $X\backslash A$. By lemma 1.4.1 there exists a locally finite open cover $\mathcal{U}$ of $X\backslash A$ that refines $\mathcal{V}$. Observe that $\mathcal{U}$, being locally finite, is countable (see the remark following the definition of locally finite collection at the beginning of this section). Since $\mathcal{U} < \mathcal{V}$, for each $U \in \mathcal{U}$ there exists $x_U \in X\backslash A$ with

$$U \subseteq B(x_U,\tfrac{1}{4}d(x_U,A)).$$

In addition, for each $U \in \mathcal{U}$ there exists $a_U \in A$ with

$$d(x_U,a_U) \le \tfrac{5}{4}d(x_U,A).$$

We claim that the U's and the a_U's are as required.

CLAIM: For every $U \in \mathcal{U}$ and $x \in U$ the following inequalities hold:

$$\tfrac{3}{4}d(x_U,A) \le d(x,A) \text{ and } d(x,a_U) \le \tfrac{3}{2}d(x_U,A).$$

The second part of the claim is easy since

$$d(x,a_U) \le d(x,x_U) + d(x_U,a_U) \le \tfrac{1}{4}d(x_U,A) + \tfrac{5}{4}d(x_U,A) = \tfrac{3}{2}d(x_U,A).$$

Also,

$$d(x_U,A) \le d(x_U,x) + d(x,A) \le \tfrac{1}{4}d(x_U,A) + d(x,A),$$

from which it follows that

$$\tfrac{3}{4}d(x_U,A) \le d(x,A),$$

as required.

We now conclude that for every $U \in \mathcal{U}$ and $x \in U$,

$$d(x,a_U) \le \tfrac{3}{2}d(x_U,A) \le \tfrac{3}{2}\cdot\tfrac{4}{3}d(x,A) = 2d(x,A),$$

which proves (1).

Now assume that $U_n \in \mathcal{U}$ for all n and that $\lim_{n\to\infty} d(U_n,A) = 0$. For each n pick $p_n \in U_n$ such that $\lim_{n\to\infty} d(p_n,A) = 0$. By the first part of the claim we obtain

$$\lim_{n\to\infty} d(x_{U_n},A) \le \lim_{n\to\infty} \tfrac{4}{3}d(p_n,A) = 0.$$

Since for every n, $U_n \subseteq B(x_n,\tfrac{1}{4}d(x_{U_n},A))$ this implies that $\lim_{n\to\infty} \operatorname{diam}(U_n) = 0$. □

We now come to the announced important application.

1.4.13. THEOREM ("The Dugundji Extension Theorem") [[chapters 1,2]]: *Let* L *be a normable linear space. For every space* X, *for every closed* $A \subseteq X$ *and for every continuous function* $f\colon A \to L$ *there exists a continuous extension* $\bar{f}\colon X \to L$ *such that* $\bar{f}(X) \subseteq \operatorname{conv}(f(A))$.

PROOF: Let X be a space, $A \subseteq X$ be closed, and $f\colon A \to L$ be continuous. By lemma 1.4.12 there exists a locally finite open cover $\mathcal{U}$ of $X\setminus A$ and a sequence of points $(a_U)_{U\in\mathcal{U}}$ in A such that for every $x \in U \in \mathcal{U}$ we have $d(x,a_U) \le 2d(x,A)$. For each $x \in X\setminus A$, let $\mathcal{E}(x) = \{U \in \mathcal{U}: x \in U\}$. Define a function $F\colon X \Rightarrow L$ by

$$F(x) = \begin{cases} \{f(x)\} & (x \in A), \\ \operatorname{conv}(\{f(a_U): U \in \mathcal{E}(x)\}) & (x \in X\setminus A). \end{cases}$$

We claim that F is **LSC**. To this end, let $W \subseteq L$ be open and take an arbitrary $x \in F^{\Leftarrow}(W)$.

CASE 1: $x \notin A$.

Put $V = \bigcap\mathcal{E}(x)$. Then V is an open neighborhood of x and if $y \in V$ then $\mathcal{E}(x) \subseteq \mathcal{E}(y)$, hence $F(x) \subseteq F(y)$, which implies that $F(y) \cap W \ne \emptyset$. We conclude that $V \subseteq F^{\Leftarrow}(W)$.

CASE 2: $x \in A$.

By continuity of f, $f^{-1}(W)$ is open in A. Let $\varepsilon > 0$ be such that $B(x,\varepsilon) \cap A \subseteq f^{-1}(W) = F^{\Leftarrow}(W) \cap A$. We claim that $B(x,\varepsilon/3) \subseteq F^{\Leftarrow}(W)$. Clearly, $B(x,\varepsilon/3) \cap A \subseteq F^{\Leftarrow}(W)$. Take $y \in B(x,\varepsilon/3)\backslash A$. There exists $U \in \mathcal{U}$ with $y \in U$. Then

$$d(y,a_U) \leq 2d(y,A) < 2\varepsilon/3,$$

from which it follows that

$$d(x,a_U) \leq d(x,y) + d(y,a_U) < \varepsilon/3 + 2\varepsilon/3 = \varepsilon,$$

hence $a_U \in f^{-1}(W)$. Since $y \in U$, by the definition of F we obtain $f(a_U) \in F(y) \cap W$. We conclude that $y \in F^{\Leftarrow}(W)$.

Since the convex hull of any finite subset of L is compact, lemma 1.2.2, we are in a position to apply theorem 1.4.9 from which the desired result now follows immediately.□

We shall see that theorem 1.4.13 is also true for *locally convex* linear spaces, see exercise 1.4.3. One of the main unsolved problems in infinite-dimensional topology and **ANR** theory is whether the local convexity assumption in this result can be dropped.

1.4.14. COROLLARY ("The Tietze Extension Theorem") [[chapters 2,3,6,7]]: *Every continuous function from a closed subspace* A *of a space* X *to* $\mathbb{R}$ *or* I *is continuously extendable over* X.

PROOF: Observe that the convex hull of a convex subset A of $\mathbb{R}$ is equal to A, and observe that I is convex. Now apply theorem 1.4.13.□

1.4.15. COROLLARY ("Urysohn's Lemma") [[chapters 1,3,4,5,6,7]]: *For every pair of disjoint closed subsets* A *and* B *of a space* X *there exists a continuous function* $\alpha\colon X \to I$ *such that* $\alpha \mid A \equiv 0$ *and* $\alpha \mid B \equiv 1$.□

Functions α such as in corollary 1.4.15 are called *Urysohn functions.*

1.4.16. COROLLARY [[chapters 4,5,7]]: *Let* X *be a space. Then for every pair of disjoint closed subsets* $A,B \subseteq X$ *there exist open subsets* $U,V \subseteq X$ *such that* $A \subseteq U$, $B \subseteq V$ *and* $\overline{U} \cap \overline{V} = \emptyset$.□

In certain situations we shall need a slightly stronger result than corollary 1.4.15 of which we now present a direct elementary proof.

1.4.17. THEOREM [[chapters 2,3,4]]: *For every pair of disjoint closed subsets* A *and* B *of a space* X *there exists a continuous function* $\alpha\colon X \to I$ *with* $\alpha^{-1}(0) = A$ *and* $\alpha^{-1}(1) = B$.

PROOF: If $A = \emptyset$ or $B = \emptyset$, then the result is obvious (take $\alpha \equiv 0$ or $\alpha \equiv 1$), so assume $A \neq \emptyset$ and $B \neq \emptyset$. Define $\alpha\colon X \to \mathbb{R}$ by

$$\alpha(x) = \frac{d(x,A)}{d(x,A) + d(x,B)} .$$

Since $A \cap B = \emptyset$, for every x we have $d(x,A) + d(x,B) \neq 0$ which implies that α is well-defined and continuous. The range of α is clearly contained in I. Finally it is a triviality that $\alpha(x) = 0$ if and only if $d(x,A) = 0$ and that $\alpha(x) = 1$ if and only if $d(x,B) = 0$. We conclude that α is as required. □

We finish this section by deriving two additional applications.

1.4.18. THEOREM [[chapters 1,3,4,5,6,7]]: *Every space* X *is homeomorphic to a subspace of the Hilbert cube* Q.

PROOF: Let $\mathcal{U}$ be a countable open basis for X. For every pair (U,V) of elements of $\mathcal{U}$ with $\overline{U} \subseteq V \neq Q$, pick a Urysohn function $f\colon X \to I$ such that $f \mid \overline{U} \equiv 0$ and $f \mid X \setminus V \equiv 1$ (corollary 1.4.15). Let $\mathcal{F} = \{f_n : n \in \mathbb{N}\}$ be the collection of functions obtained in this way. Define a function $i\colon X \to Q$ by

$$i(x)_n = f_n(x) \qquad (n \in \mathbb{N}).$$

An easy check shows that i is an imbedding. □

1.4.19. THEOREM [[chapter 7]]: *Let* X *be a space and let* $\mathcal{E}$ *be an open cover of* X. *Then there exists an admissible metric* d *on* X *such that the family of all open* d*-balls of radius* 1 *form a refinement of* $\mathcal{E}$.

PROOF: Let $\mathcal{K} = \{g_n : n \in \mathbb{N}\}$ be a partition of unity on X which is subordinated to $\mathcal{E}$ (theorem 1.4.3). For each n, put

$$U_n = g_n^{-1}((0,1]),\ V_n = g_n^{-1}((2^{-(n+1)},1])\ \text{and}\ W_n = g_n^{-1}([2^{-(n+1)},1]),$$

respectively. Observe that U_n and V_n is open, that W_n is closed ($n \in \mathbb{N}$), and that the cover $\mathcal{U} = \{U_n : n \in \mathbb{N}\}$ refines $\mathcal{E}$.

CLAIM 1: The collection $\{V_n : n \in \mathbb{N}\}$ covers X.

To the contrary, assume that this is not the case. Then there exists $x \in X$ such that for every n, $g_n(x) \leq 2^{-(n+1)}$. Consequently,

$$\Sigma_{n=1}^{\infty} g_n(x) \leq \Sigma_{n=1}^{\infty} 2^{-(n+1)} = \tfrac{1}{2} < 1,$$

which contradicts the fact that $\mathcal{K}$ is a partition of unity.

For every n, let $\alpha_n : X \to I$ be a Urysohn function such that

$$\alpha_n \mid W_n \equiv 1\ \text{and}\ \alpha_n \mid X \backslash U_n \equiv 0$$

(corollary 1.4.15). We may assume that X is a subspace of the Hilbert cube Q (theorem 1.4.18). By lemma 1.2.3 we may therefore assume that X is a subspace of the unit ball (i.e. the set of vectors with norm not exceeding 1) of a normed linear space $(L, \|\cdot\|)$. For each $n \in \mathbb{N}$ define $d_n : X \times X \to \mathbb{R}$ by

$$d_n(x,y) = \|\alpha_n(x)\cdot x - \alpha_n(y)\cdot y\| + |\alpha_n(x) - \alpha_n(y)|.$$

It is easily seen that d_n is continuous and that d_n is a pseudometric on X (i.e. it satisfies all axioms for a metric, except the implication "$d(x,y) = 0 \Rightarrow x = y$"). Define $d : X \times X \to \mathbb{R}$ by

$$d(x,y) = \sum_{n=1}^{\infty} d_n(x,y).$$

Since $\mathcal{U}$ is locally finite, the set $\{d_n(x,y) : n \in \mathbb{N}\}$ is finite for every $x,y \in X$, from which it follows easily that d is a continuous pseudometric on X.

CLAIM 2: d is an admissible metric on X.

Take $x \in X$, and assume that $d(x,y) = 0$ for certain $y \in X$. There exists $n \in \mathbb{N}$ such that

$x \in V_n$. Then $d_n(x,y) = 0$ and $\alpha_n(x) = 1$ so that

$$0 = d_n(x,y) \geq |\alpha_n(x) - \alpha_n(y)| = |1 - \alpha_n(y)|,$$

from which we conclude that $\alpha_n(y) = 1$. Consequently,

$$0 = d_n(x,y) = \|x - y\|,$$

i.e. $x = y$. This shows that d is a metric.

If $x,y \in V_n$ for certain n then $\alpha_n(x) = \alpha_n(y) = 1$, hence

$$d(x,y) \geq d_n(x,y) = \|1\cdot x - 1\cdot y\| + |1 - 1| = \|x - y\|.$$

Since d is continuous and the collection $\{V_n: n \in \mathbb{N}\}$ is an open cover of X (claim 1), this easily implies that d is admissible.

CLAIM 3: The family of all open d-balls of radius 1 form a refinement of $\mathcal{E}$.

Take $x \in X$ and $n \in \mathbb{N}$ such that $x \in V_n$. Now take an arbitrary $y \in X$ with $d(x,y) < 1$. Then $\alpha_n(x) = 1$ and since

$$1 > d(x,y) \geq d_n(x,y) \geq |\alpha_n(x) - \alpha_n(y)| = 1 - \alpha_n(y),$$

we find that $\alpha_n(y) > 0$, i.e. $y \in U_n$. Since $\mathcal{U} < \mathcal{E}$, we are done. □

Exercises for §1.4.

Let X be a space. A (single-valued) function (not necessarily continuous) $f: X \to \mathbb{R}$ is called *lower semi-continuous* (abbreviated **lsc**) if $f^{-1}(t,\infty)$ is open in X for every $t \in \mathbb{R}$; observe that even if we regard f to be a set-valued function then this concept of lower semi-continuity does not agree with the concept of lower semi-continuity defined in this section. Similarly, f is called *upper semi-continuous* (abbreviated **usc**) if $f^{-1}(-\infty,t)$ is open in X for every $t \in \mathbb{R}$.

1. Let X be a space and let $f,g: X \to \mathbb{R}$ with f **lsc** and g **usc.** Prove that if $g \leq f$, i.e. $g(x) \leq f(x)$ for every $x \in X$, then the function $F: X \Rightarrow \mathbb{R}$ defined by $F(x) = [g(x),f(x)]$ is **LSC.** Use this to conclude that there exists a *continuous* $h: X \to \mathbb{R}$ such that $g \leq h \leq f$.

2. Let L be a locally convex linear space, let X be a space and let F: X ⇒ L be **LSC** such that each F(x) is convex and compact in L. Prove that for every closed subset A of X and every continuous selection f: A → L for the function F | A: A ⇒ L, there exists a continuous selection g: X → L for F which extends f.

3. Let L be a locally convex linear space. Prove that for every space X, for every closed $A \subseteq X$ and for every continuous function f: A → L there exists a continuous extension g: X → L such that $g(X) \subseteq \mathrm{conv}(f(A))$.

Let L be a locally convex linear space. In addition, let $A \subseteq X$ be closed and let f: A → L be continuous. As in the proof of the Dugundji Extension Theorem, let $\mathcal{U}$ be a locally finite open cover of X\A for which there exists a sequence $(a_U)_{U\in\mathcal{U}}$ in A such that for every $x \in U$ we have $d(x,a_U) \leq 2d(x,A)$. Let $\{\kappa_U: U \in \mathcal{U}\}$ be the set of κ-functions with respect to $\mathcal{U}$. Define L(f): X → L by

$$L(f)(x) = \begin{cases} f(x) & (x \in A), \\ \Sigma_{U\in\mathcal{U}}\kappa_U(x)\cdot f(a_U) & (x \in X\backslash A). \end{cases}$$

4. Prove that L(f) defines a continuous extension of f such that $L(f)(X) \subseteq \mathrm{conv}(f(A))$.

5. Let X be a compact space and let $A \subseteq X$ be closed. For every $f \in C(A)$ define $L(f) \in C(X)$ as above. Prove that the function L: C(A) → C(X) is a linear imbedding (i.e. is linear and L: C(A) → L(C(A)) is a homeomorphism).

6. Let A_1 and A_2 be closed subsets of the locally convex linear spaces L_1 and L_2, respectively. Prove that for each homeomorphism $h: A_1 \to A_2$ there exists a homeomorphism $H: L_1 \times L_2 \to L_1 \times L_2$ such that $H(a,0) = (0,h(a))$ for every $a \in A$ (Hint: Prove that there exist continuous functions $f_1: L_1 \to L_2$ and $f_2: L_2 \to L_1$ such that $f_1 \mid A_1 = h$ and $f_2 \mid A_2 = h^{-1}$. Define $H_1,H_2: L_1 \times L_2 \to L_1 \times L_2$ by $H_1(x_1,x_2) = (x_1,x_2+f_1(x_1))$ and $H_2(x_1,x_2) = (x_1-f_2(x_2),x_2)$. Show that H_1 and H_2 are homeomorphisms of $L_1 \times L_2$ and that $H = H_2 \circ H_1$ is as required).

7. Let A be closed subspace of a space X and let f: A → I be continuous. Show that the function g: X → I defined by

$$g(x) = \begin{cases} \inf\{f(a) + \dfrac{d(x,a)}{d(x,A)} - 1 : a \in A\} & (x \in X\backslash A), \\ f(x) & (x \in A) \end{cases}$$

is a continuous extension of f.

8. Let X be a space and let $A \subseteq X$. Prove that if the identity function $1_A: A \to A$ can be extended to a continuous

function $r: X \to A$ then A is closed in X.

9. Prove that every space is homeomorphic to a subspace of $\mathbb{R}^\infty$.

10. Let X be a noncompact space. Prove that there exists a continuous function $\lambda: X \to (0,1]$ such that $\inf \lambda(X) = 0$ (Hint: Use exercise 1.1.6).

11. Let X be a space, and for each $n \in \mathbb{N}$ let $f_n: X \to \mathbb{R}$ be **lsc**. Suppose that for each x, $\sup_n f_n(x)$ exists. Prove that the function $f(x) = \sup_n f_n(x)$ is **lsc**. Assume moreover that each f_n is *continuous*. Is f continuous ?

12. Let $C^1(I)$ be the subspace of $C(I)$ consisting of all continuously differentiable functions. For each $f \in C^1(I)$, define

$$L(f) = \int_0^1 \sqrt{1 + \left(\frac{df}{dx}\right)^2}\, dx.$$

Prove that L is **lsc**.

13. Let X be a space and let $\mathcal{F}$ be a locally finite collection of closed subsets of X. Prove that $\bigcup \mathcal{F}$ is closed in X.

1.5. AR's and ANR's

Let X be a space and let $A \subseteq X$ be closed. We say that A is a *retract* of X provided that there is a continuous function $r: X \to A$ such that r restricted to A is the identity. Such a function r is called a *retraction*. We say that A is a *neighborhood retract* of X provided that there exists a neighborhood U of A in X such that A is a retract of U.

Retractions are very interesting functions since they preserve many topological properties. For example, if $r: X \to A$ is a retraction and X has the fixed-point property then A has the fixed-point property, etc.

A space X is called an *Absolute Retract* (abbreviated **AR**) provided that X is a retract of every space Y containing X as a closed subspace. If X is an **AR** and $f: X \to Y$ is a homeomorphism then Y is clearly also an **AR**. Consequently, X is an **AR** if and only if for every space Y containing a closed subspace Z which is homeomorphic to X, there exists a retraction $r: Y \to Z$. Theorem 1.5.2 below easily implies that a retract of an **AR** is an **AR**.

A space X is called an *Absolute Neighborhood Retract* (abbreviated **ANR**) provided that X is a neighborhood retract of every space Y containing X as a closed subspace. The space $X = \{0,1\}$

is easily seen to be an **ANR** but is not an **AR** since retractions preserve connectivity, from which it follows that there does not exist a retraction $r: I \to X$. Notice that every **AR** is an **ANR**.

As above, theorem 1.5.2 below easily implies that X is an **ANR** if and only if for every space Y containing a closed subspace Z which is homeomorphic to X, Z is a neighborhood retract of Y. Also, every neighborhood retract of an **ANR** is again an **ANR**, cf. exercise 1.5.1.

The Dugundji Extension Theorem gives us a rich supply of **AR**'s.

1.5.1. THEOREM [[chapters 1,3,4,5,8]]: *Let* C *be a convex subset of a normable linear space. Then* C *is an* **AR**.

PROOF: Assume that C is a closed subspace of a space X. By theorem 1.4.13, the identity function $1: C \to C$ can be extended to a continuous function $f: X \to \mathrm{conv}(C) = C$. It is clear that f is a retraction. □

We shall now present a characterization of **A(N)R**'s.

1.5.2. THEOREM [[chapter 1, and at other places without explicit reference]]: *Let* X *be a space. The following statements are equivalent:*

(a) X *is an* **A(N)R**,

(b) *for every space* Y *and for every closed subspace* A *of* Y, *every continuous function* f: $A \to X$ *can be extended over* Y *(over a neighborhood (depending on* f*) of* A *in* Y*).*

PROOF: The implication (b) $\Rightarrow$ (a) is trivial.

For (a) $\Rightarrow$ (b), let us assume that X is an **ANR.** The proof for **AR**'s is entirely similar, and shall therefore be omitted. Let A,Y and f be given such as in (b). Since X is homeomorphic to a subspace of the compact space Q (theorem 1.4.18), by lemma 1.2.3 we may assume that X is a closed subspace of a convex set C of a normed linear space L. Since such C is an **AR** (theorem 1.5.1), we can extend f to a continuous function $\bar{f}: Y \to C$. Since X is an **ANR**, there is a neighborhood U of X in C and a retraction $r: U \to X$. Put $V = \bar{f}^{-1}(U)$. Since $f(A) = \bar{f}(A) \subseteq X$, V is clearly a neighborhood of A in Y. Let h denote the restriction of $r \circ \bar{f}$ to V (observe that h is well-defined since $\bar{f}(V) \subseteq U$). Then h is continuous and extends f since for every $x \in A$ we have $h(x) = r(\bar{f}(x)) = f(x)$. □

1.5.3. *Remark:* Theorem 1.5.2 is of fundamental importance. Define a space X to be an **A(N)E** (this is an abbreviation for *Absolute (Neighborhood) Extensor*) whenever it has the property mentioned in theorem 1.5.2(b). In this terminology we have proved that $X \in$ **A(N)R** $\Leftrightarrow X \in$ **A(N)E**. In the sequel we shall not always conscientiously refer to theorem 1.5.2 when dealing with **A(N)R**'s. The reader should keep this in mind.

Proving that a given space is an **AR** or an **ANR** is usually a difficult task. We will come back to this in chapter 5. For the moment we shall prove a few elementary results about **AR**'s and **ANR**'s only.

1.5.4. PROPOSITION [[chapters 1,6,8]]: *A countable product of nonempty spaces is an* **AR** *iff all factors are.*

PROOF: For each $n \in \mathbb{N}$, let X_n be a space. The projection from $X = \prod_{n=1}^{\infty} X_n$ onto its n-th factor is denoted by π_n. Then π_n induces a retraction of X onto a homeomorphic copy of X_n in X. Consequently, if X is an **AR** then so is X_n for all $n \in \mathbb{N}$. Now let each X_n be an **AR.** Let Y be a space, $A \subseteq Y$ be closed and $f: A \to X$ be continuous. For each $n \in \mathbb{N}$ define $g_n: A \to X_n$ by $g_n = \pi_n \circ f$. By theorem 1.5.2 we can extend g_n to a continuous function $\bar{g}_n: Y \to X_n$. Now define $F: Y \to X$ by

$$\bar{f}(y)_n = \bar{g}_n(y) \qquad (n \in \mathbb{N}).$$

It is clear that $\bar{f}$ is continuous and extends f. □

1.5.5. COROLLARY [[chapters 4,6,7]]: $\mathbb{R}^n$, I^n, Q *and* $\mathbb{R}^\infty$ *are* **AR's.**

PROOF: Apply proposition 1.5.4 and theorem 1.5.1. □

For each $n \geq 0$ let $S^n \subseteq \mathbb{R}^{n+1}$ denote the n-*sphere*, i.e.

$$S^n = \{x \in \mathbb{R}^{n+1}: \|x\| = 1\}.$$

1.5.6. COROLLARY [[chapters 2,3]]: *For each* $n \geq 0$, S^n *is an* **ANR**.

PROOF: Let $U = \mathbb{R}^{n+1} \setminus \{(0,0,\cdots,0)\}$. The function $r: U \to S^n$ defined by

$$r(x) = \frac{x}{\|x\|}$$

is clearly a retraction. Consequently, S^n is a neighborhood retract of an **ANR** and is therefore an **ANR** itself. □

In chapter 3 we shall prove that no S^n is an **AR**.

1.5.7. PROPOSITION [[chapter 1]]: *The product of finitely many* **ANR**'s *is an* **ANR.**

PROOF: The simple proof is left as an exercise to the reader. □

A product of a countable infinite number of **ANR**'s need not be an **ANR** (in contrast to proposition 1.5.4 on **AR**'s). The following result gives more information on this:

1.5.8. THEOREM [[chapters 2,5,8]]: *Let* X_n *be a nonempty space for every* $n \in \mathbb{N}$ *and let* $X = \prod_{n=1}^{\infty} X_n$. *The following statements are equivalent:*

(a) X *is an* **ANR**,

(b) *each* X_n *is an* **ANR** *and there is an* $n \in \mathbb{N}$ *such that* X_m *is an* **AR** *for every* $m \geq n$.

PROOF: Since X is homeomorphic to $\prod_{i=1}^{n} X_i \times \prod_{i=n+1}^{\infty} X_i$ for every $n \in \mathbb{N}$, the implication (b) $\Rightarrow$ (a) immediately follows from propositions 1.5.4 and 1.5.7.

For (a) $\Rightarrow$ (b), first observe that each X_n can be viewed as a retract of X (cf. the proof of proposition 1.5.4), hence X_n is an **ANR** for every n. Now take a sequence of points $x_n \in X_n$, $n \in \mathbb{N}$ and let C_n be a convex subset of a certain normed linear space L_n which contains X_n as a closed subset (lemma 1.2.3 and theorem 1.4.18). Since X is an **ANR** and is closed in $C = \prod_{n=1}^{\infty} C_n$, there is a neighborhood U of X in C for which there exists a retraction $r\colon U \to X$. Put $x = (x_1, x_2, \cdots)$. Since U is a neighborhood of x there are an $n \in \mathbb{N}$ and neighborhoods V_i of x_i in C_i for every $i \leq n-1$ such that

$$x \in W = V_1 \times \cdots \times V_{n-1} \times C_n \times C_{n+1} \times \cdots \subseteq U.$$

Observe that by theorem 1.5.1 and proposition 1.5.4, the product $\prod_{m=n}^{\infty} C_m$ is an **AR**. Now define a function

$$s\colon \{x_1\} \times \cdots \times \{x_{n-1}\} \times C_n \times C_{n+1} \times \cdots \to \{x_1\} \times \cdots \times \{x_{n-1}\} \times X_n \times X_{n+1} \times \cdots$$

by

$$s(x) = (x_1, \cdots, x_{n-1}, r(x)_n, r(x)_{n+1}, \cdots).$$

An easy check shows that s is a retraction from which it follows that $\prod_{m=n}^{\infty} X_m$ is an **AR**. Now apply proposition 1.5.4. □

The following result enables us to create many new **A(N)R**'s from old ones in yet another way.

1.5.9. THEOREM [[chapters 1,3,7]]: *Let* $X = X_1 \cup X_2$, *where* X_1 *and* X_2 *are closed, and let* $X_0 = X_1 \cap X_2$. *Then*

(1) *If* X_0, X_1 *and* X_2 *are* **A(N)R**'s *then* X *is an* **A(N)R**.

(2) *If* X *and* X_0 *are* **A(N)R**'s *then* X_1 *and* X_2 *are* **A(N)R**'s *as well.*

PROOF: For (1), assume that X_0, X_1 and X_2 are **ANR**'s. We shall prove that X is an **ANR**. The proof of (1) for **AR**'s is similar.

Assume that X is a closed subset of a space Z. Our task is to construct a neighborhood U of X in Z and a retraction $r: U \to X$. To this end, define

$$Z_0 = \{z \in Z: d(z,X_1) = d(z,X_2)\},$$
$$Z_1 = \{z \in Z: d(z,X_1) \le d(z,X_2)\}, \text{ and}$$
$$Z_2 = \{z \in Z: d(z,X_1) \ge d(z,X_2)\},$$

respectively. It is clear that $Z_i \cap X = X_i$ for $i \in \{0,1,2\}$, that $Z_1 \cap Z_2 = Z_0$ and finally that $Z_1 \cup Z_2 = Z$. Also observe that the Z_i are closed in Z.

Since X_0 is closed in Z_0 there are a closed neighborhood W_0 of X_0 in Z_0 and a retraction r_0: $W_0 \to X_0$. Now for $i \in \{1,2\}$ define r_i: $W_0 \cup X_i \to X_i$ by

$$r_i(z) = \begin{cases} r_0(z) & (z \in W_0), \\ z & (z \in X_i). \end{cases}$$

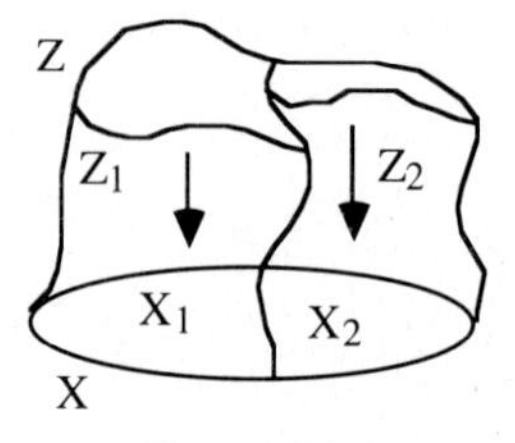

Figure 1.5.1.

Observe that r_i is a retraction. By theorem 1.5.2 there exists a closed neighborhood V_i of $W_0 \cup X_i$ in Z_i such that r_i can be extended to a continuous function

$$\bar{r}_i: V_i \to X_i \qquad (i \in \{1,2\}).$$

It is clear that there exists a closed neighborhood U_i of X_i in Z_i such that $U_i \subseteq V_i$ and $U_i \cap Z_0 \subseteq W_0$. Then

$$U_1 \cap U_2 \subseteq W_0$$

from which it follows that the function $\bar{r}\colon U_1 \cup U_2 \to X$ defined by

$$\bar{r}(z) = \begin{cases} \bar{r}_1(z) & (z \in U_1), \\ \bar{r}_2(z) & (z \in U_2), \end{cases}$$

is a well-defined retraction. Since $U_1 \cup U_2$ is a neighborhood of X in the space Z, we are done.

For (2), let X_0 and X be **ANR**'s. Again, the proof for **AR**'s is similar. Since X_0 is an **ANR**, there are a neighborhood U_0 of X_0 in X and a retraction $r\colon U_0 \to X_0$. Define $\bar{r}\colon X_1 \cup U_0 \to X_1$ by

$$\bar{r}(x) = \begin{cases} x & (x \in X_1), \\ r(x) & (x \in U_0 \cap X_2). \end{cases}$$

An easy check shows that $\bar{r}$ is a retraction. Since $X_1 \cup U_0$ is a neighborhood of X_1 and X is an **ANR**, it now follows that X_1 is an **ANR.** The proof for X_2 is the same. □

Let $X \subseteq \mathbb{R}^2$ be defined by

$$X = \{(x,y) \in \mathbb{R}^2\colon \max\{|x|,|y|\} \le 1\} \cup \{(x,0)\colon 1 \le x \le 2\}.$$

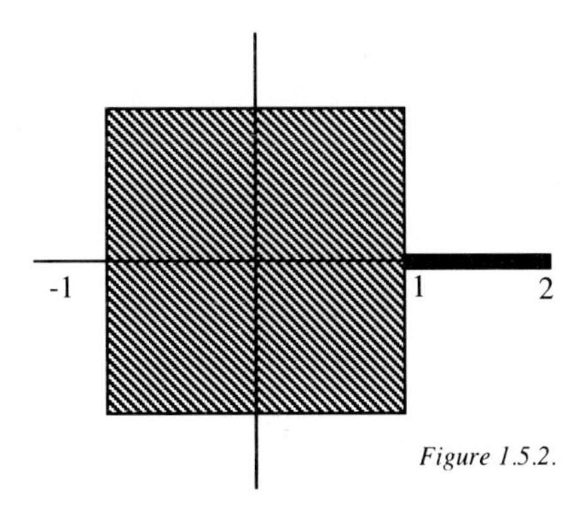

Figure 1.5.2.

Put $X_1 = \{(x,y) \in \mathbb{R}^2\colon \max\{|x|,|y|\} \le 1\}$ and $X_2 = \{(x,0)\colon 1 \le x \le 2\}$. Then X_1 is a product of two intervals and is consequently an **AR** by theorem 1.5.1 and proposition 1.5.4. Also, clearly X_2 is an **AR**. Since $X_1 \cap X_2 = \{(1,0)\}$ is a one-point space and hence is an **AR** for trivial reasons, by theorem 1.5.9(1) it follows that X is an **AR**. It is illustrative to go through the proof of theorem 1.5.9(1) with X_0, X_1, X_2 and X as above and $Z = \{(x,y): -1 \leq x \leq 2 \text{ and } -1 \leq y \leq 1\}$ in order to obtain a more or less explicitly defined retraction $r: Z \to X$.

Exercises for §1.5.

1. Prove that a neighborhood retract of an **ANR** is an **ANR**.

2. Prove that a subset A of the real line $\mathbb{R}$ is an **AR** if and only if A is an interval (i.e. order convex). Give an example of a subspace X of $\mathbb{R}$, such that X is an **ANR**, but $X \cup \{p\}$ is not an **ANR** for some $p \in \mathbb{R}$.

3. A space X is called *path-connected* if for all $x,y \in X$ there exists a continuous function $f: I \to X$ with $f(0) = x$ and $f(1) = y$. Prove that each **AR** is path-connected. A space X is called *locally path-connected* if every point $x \in X$ has arbitrarily small path-connected neighborhoods. Prove that every **ANR** is locally path-connected. Observe that, in particular, every **ANR** is locally connected.

4. Let X be the following subset of the plane: $X = \{(x,\sin 1/x): 0 < x \leq 1/\pi\} \cup \{(0,y): -1 \leq y \leq 1\}$. Prove that X is connected and that X is not an **ANR**.

A space X has the *fixed-point property* if every continuous function $f: X \to X$ has a fixed-point, i.e. a point $x \in X$ with $f(x) = x$.

5. (1) Prove that I has the fixed-point property.
   (2) Let Y be a retract of X. Prove that if X has the fixed-point property then Y has the fixed-point property.
   (3) Prove that no (nontrivial) linear space has the fixed-point property.
   (4) Prove that the space from exercise 1.5.4 has the fixed-point property.

1.6. The Borsuk Homotopy Extension Theorem

Let X and Y be spaces. A *homotopy* from X to Y is a continuous function $H: X \times I \to Y$. If $t \in I$ then the function $H_t: X \to Y$ defined by $H_t(x) = H(x,t)$ is called the t-*level* of H. One should think of a homotopy as a continuous family of functions connecting the functions H_0 and H_1. Two continuous functions $f,g: X \to Y$ are called *homotopic*, in symbols $f \simeq g$, if there is a homotopy $H: X \times I \to Y$ such that $H_0 = f$ and $H_1 = g$. A continuous function $f: X \to Y$ is called *nullhomotopic* if there exists a constant function g from X to Y such that f and g are homotopic.

1.6.1. LEMMA [[chapters 1,2]]: *Let* X *and* Y *be spaces and let* $f,g,h: X \to Y$ *be continuous. Then*

(a) $f \simeq f$,

(b) *if* $f \simeq g$ *then* $g \simeq f$,

(c) *if* $f \simeq g$ *and* $g \simeq h$ *then* $f \simeq h$.

PROOF: For (a), define $H: X \times I \to X$ by $H(x,t) = f(x)$. For (b), let $H: X \times I \to Y$ be a homotopy such that $H_0 = f$ and $H_1 = g$. Define $F: X \times I \to Y$ by

$$F(x,t) = H(x,1-t).$$

Then F is a homotopy with $F_0 = g$ and $F_1 = f$.

For (c), let $H,F: X \times I \to Y$ be homotopies with $H_0 = f$, $H_1 = g$, $F_0 = g$ and $F_1 = h$. Define $S: X \times I \to Y$ by

$$S(x,t) = \begin{cases} H(x,2t) & (t \le \frac{1}{2}), \\ F(x,2t-1) & (t \ge \frac{1}{2}). \end{cases}$$

Then S is a homotopy with $S_0 = f$ and $S_1 = h$.□

From the above lemma we conclude that the homotopy relation "$\simeq$" is an equivalence relation in $C(X,Y)$. If X and Y are spaces and if $f: X \to Y$ is continuous then $[f]$ denotes its $\simeq$-equivalence class and $[X,Y]$ denotes $\{[f]: f \in C(X,Y)\}$. We say that $[X,Y]$ is *trivial* if it contains only one point (i.e. all continuous functions from X to Y are homotopic).

As we will see, homotopic functions have much in common. The aim of the present section is to prove that a continuous function f from a closed subspace A of a space X into an **ANR** Z is extendable over X if and only if f is homotopic to an extendable function $g: A \to Z$.

We shall need the following simple lemma:

1.6.2. LEMMA [[chapters 1,5]]: *Let* A *be a closed subset of a space* X. *Then for every neighborhood* V *of* $B = (X \times \{0\}) \cup (A \times I)$ *in* $X \times I$ *there is a continuous function* $\alpha: X \times I \to V$ *which is the identity on* B.

PROOF: Take an arbitrary $x \in A$. Since $S(x) = \{x\} \times I$ is compact and V is a neighborhood of

S(x), it follows easily that there exists a neighborhood U_x of x in X such that $U_x \times I \subseteq V$. Put $U = \bigcup_{x \in A} U_x$. Then U is a neighborhood of A in X such that $U \times I \subseteq V$.

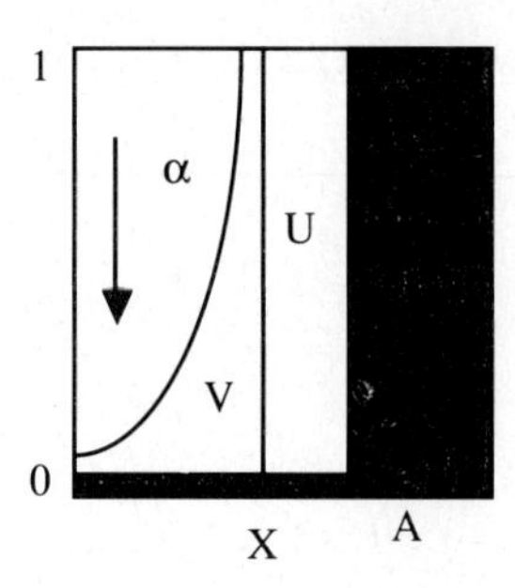

Figure 1.6.1.

Now let $\beta: X \to I$ be a Urysohn function (corollary 1.4.15) such that $\beta \mid A \equiv 1$ and $\beta \mid X \setminus U \equiv 0$. Define $\alpha: X \times I \to X \times I$ by

$$\alpha(x,t) = (x, \beta(x) \cdot t).$$

Then α is clearly continuous. It is easily seen that α restricts to the identity on B. We shall prove that $\alpha(X \times I) \subseteq V$, thereby showing that α is as required. To this end, take $(x,t) \in X \times I$ arbitrarily. If $x \notin U$ then $\beta(x) = 0$ and consequently, $\alpha(x,t) = (x,0) \in B \subseteq V$. On the other hand, if $x \in U$ then $\alpha(x,t) = (x, \beta(x) \cdot t) \in U \times I \subseteq V$.□

We now come to the main result in this section.

1.6.3. THEOREM ("Borsuk Homotopy Extension Theorem") [[chapters 1,5]]: *Let* A *be a closed subspace of a space* X, *let* Z *be an* **ANR** *and let* $H: A \times I \to Z$ *be a homotopy such that* H_0 *is extendable to a function* $f: X \to Z$. *Then there is a homotopy* $F: X \times I \to Z$ *such that*

(1) $F_0 = f$,

(2) *for every* $t \in I$, $F_t \mid A = H_t$.

PROOF: Using the notation as in lemma 1.6.2, define a function $\xi: B \to Z$ by

$$\begin{cases} \xi(x,0) = f(x) & (x \in X), \\ \xi(x,t) = H(x,t) & (x \in A, t \in I). \end{cases}$$

Since $X \times \{0\}$ and $A \times I$ are closed in B, it follows easily that ξ is continuous. As Z is an **ANR** and as B is closed in $X \times I$, we can find a neighborhood V of B in $X \times I$ such that ξ can be extended to a continuous function ξ': $V \to Z$. Let α: $X \times I \to V$ be as in lemma 1.6.2. Define F: $X \times I \to Z$ by

$$F(x,t) = \xi'(\alpha(x,t)).$$

It is easily seen that F is as required. □

Theorem 1.6.3 has some corollaries.

1.6.4. COROLLARY [[chapters 1,3]]: *Let* A *be a closed subset of a space* X. *Let* Z *be an* **ANR**, *and let* f: $A \to Z$ *be continuous. The following statements are equivalent:*

(a) f *can be extended over* X,

(b) f *is homotopic to an extendable function* g: $A \to Z$.

PROOF: That (a) ⇒ (b) is clear since $f \simeq f$ by lemma 1.6.1(a).

For (b) ⇒ (a), let H: $A \times I \to Z$ be a homotopy such that $H_0 = g$ and $H_1 = f$ (since $f \simeq g$, by lemma 1.6.1(b) it follows that $g \simeq f$). Since g is extendable, by theorem 1.6.3 f is extendable as well. □

A space X is called *contractible* provided that there exists a homotopy H: $X \times I \to X$ such that H_0 is the identity and H_1 is a constant function; the homotopy H is called a *contraction* of X. It is clear that every contractible space is path-connected from which it follows that $S^0 = \{-1,1\}$ is not contractible. We shall see later that no S^n is contractible (corollary 3.5.6). The proof of this fact is difficult.

We shall now present a simple but useful characterization of contractible spaces.

1.6.5. PROPOSITION [[chapter 3]]: *Let* X *be a path-connected space. The following statements are equivalent:*

(a) X *is contractible,*

(b) *if* Y *is any path-connected space then* [X,Y] *is trivial,*

(c) *if* Y *is any space then* [Y,X] *is trivial.*

PROOF: For (a) ⇒ (b), let F: $X \times I \to X$ be a contraction.If f: $X \to Y$ is continuous then $f \circ F$ is a homotopy from X to Y that connects f with a constant function. It therefore suffices to prove that any two constant functions from X to Y are homotopic. Let c_p, c_q: $X \to Y$ be the constant functions with values p and q, respectively. Since Y is path-connected, there is a path α: $I \to Y$

with $\alpha(0) = p$ and $\alpha(1) = q$. Now define the required homotopy $H: X \times I \to X$ by $H(x,t) = \alpha(t)$.

For (b) ⇒ (c), first observe that X is contractible since [X,X] is trivial. Let $F: X \times I \to X$ be a contraction such that F_1 is the constant function with value c. If Y is any space and $f: Y \to X$ is continuous then $H = F \circ (f \times 1_I): Y \times I \to X$ is a homotopy that connects f with the constant function with value c.

Since (c) ⇒ (a) is a triviality, we are done. □

Despite the fact that no S^n is contractible, the class of contractible spaces is quite large.

1.6.6. THEOREM [[chapter 3]]: *Every retract of a contractible space is contractible.*

PROOF: Let Y be a contractible space, let X be a subspace of Y, and let $r: Y \to X$ be a retraction. By definition, there exists a contraction $H: Y \times I \to Y$. Now define $F: X \times I \to X$ by

$$F(x,t) = r(H(x,t)).$$

The trivial proof that F is a contraction of X is left to the reader. □

Corollary 1.6.4 and theorem 1.6.6 now yield the following characterization of **AR**'s.

1.6.7. COROLLARY [[chapters 1,3,5]]: *Let* X *be a space. The following statements are equivalent:*

(a) X *is an* **AR**,

(b) X *is a contractible* **ANR**.

PROOF: For the implication (a) ⇒ (b) we only need to prove that X is contractible. By lemma 1.2.3 and theorem 1.4.18 we may assume that X is a closed subset of a convex subset C of a normed linear space L. Since X is a retract of C, it suffices to prove that C is contractible (theorem 1.6.6). To this end, pick an arbitrary point $c \in C$ and define $H: C \times I \to C$ by

$$H(x,t) = (1-t)x + tc.$$

Then H is well-defined, H_0 is the identity and H_1 is the constant function with value c.

For (b) ⇒ (a), let $H: X \times I \to X$ be a homotopy such that $H_0 = 1$ and H_1 is constant, say with constant value c. In addition, let Y be a space, $A \subseteq Y$ be closed and $f: A \to X$ be continuous. Define $F: A \times I \to X$ by

$$F(a,t) = H(f(a),t).$$

Then F is clearly a homotopy with $F_0 = f$ and F_1 a constant function. Consequently, F_1 can be extended to a continuous function from Y to X. By corollary 1.6.4 it follows that $F_0 = f$ can be extended over Y. We conclude that X is an **AR.** □

For a generalization of corollary 1.6.7 see theorem 5.2.15.

Exercises for §1.6.

1. For each $n \in \mathbb{N}$, let X_n be contractible. Prove that the product $\prod_{n=1}^{\infty} X_n$ is contractible.

A space X is called *locally contractible* (abbreviated **LC**) at $x \in X$ if for every neighborhood U of x in X there is a neighborhood V of x and a homotopy $H: V \times I \to U$ such that H_1 is the identity and H_0 is constant. In addition, X is called *locally contractible* if X is locally contractible at every point.

2. Prove that every **ANR** is locally contractible.

3. For each $n \in \mathbb{N}$, let X_n be a space. Prove that the following statements are equivalent:

   (a) $\prod_{n=1}^{\infty} X_n$ is **LC**.

   (b) Each X_n is **LC** and there is $N \in \mathbb{N}$ such that X_m is contractible for every $m \geq N$.

ᴵ et X be a space and let ∞ be a point not in $X \times [0,1)$. Topologize $\Delta(X) = (X \times [0,1)) \cup \{\infty\}$ as follows: points of the form (x,t) have their usual product neighborhoods and a basic neighborhood of ∞ has the form

$$(X \times (s,1)) \cup \{\infty\},$$

where $0 < s < 1$. We call $\Delta(X)$ the *cone over* X (see figure 1.6.2). If X is compact then it is easy to see that $\Delta(X)$ is the one-point compactification of $X \times [0,1)$ and also that $\Delta(X)$ is homeomorphic to the quotient space, obtained by identifying the subset $X \times \{1\}$ of $X \times [0,1]$ to one point (for the definition of quotient space see the remarks following exercise 4.7.8). Observe that $\Delta(X)$ is a separable metric space.

4. Let X be a space. Prove that $\Delta(X)$ is contractible.

5. Let $\mathbf{C} \subseteq \mathbb{R}$ be the Cantor middle-third set, cf. exercise 1.2.4 or §4.2. Prove that $\Delta(\mathbf{C})$ is an example of a

contractible space which is not an **ANR.**

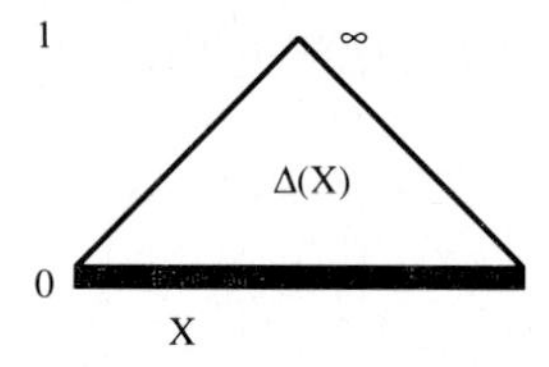

Figure 1.6.2.

For each $n \in \mathbb{N}$ let $B^n = \{x \in \mathbb{R}^n: \|x\| \leq 1\}$.

6. Let $A \subseteq S^{n-1}$. Prove that $B^n \setminus A$ is an **AR** (Hint: Let X be any space and let $Y \subseteq X$ be closed, and let $f: Y \to B^n \setminus A$ be continuous. Use the fact that B^n is an **AR** to find a continuous extension $g: X \to B^n$ of f. Now define $h: X \to B^n \setminus A$ by $h(x) = (1 - d(x,Y)) \cdot g(x)$. Prove that h is a well-defined continuous extension of f.).

7. Let A be a closed subspace of a space X and let T be the subspace $(X \times \{0\}) \cup (A \times I)$ of $X \times I$. Prove that for every space Y, if a continuous function $f: T \to Y$ can be extended over $(X \times \{0\}) \cup U$, where U is a neighborhood of $A \times I$ in $X \times I$, then f can be extended over $X \times I$.

Notes and comments for chapter 1.

§2.

Lemma 1.2.3 is due to Kuratowski [88] and Wojdysławski [151].

§4.

Theorem 1.4.9 is due to Michael [105]. A more general result than theorem 1.4.10 is due to Banach; the proof presented here was taken from Brown and Page [37, p.316-317]. A more general result than corollary 1.4.11 is due to Bartle and Graves [18]. Lemma 1.4.12 and theorem 1.4.13 are due to Dugundji [54]; the proof of theorem 1.4.13 presented here is due to M. van de Vel. The Tietze Extension Theorem and Urysohn's Lemma are classical results. For references and historical comments see Engelking [59, p. 101]. Exercise 1.4.5 is due to Michael [104]. Exercise 1.4.6 is Bessaga and Pełczyński [21, chapter II proposition 3.2] and exercise 1.4.7 is due to Hausdorff [70]. Finally, exercise 1.4.9 is known as Urysohn's Metrization Theorem; for details see Engelking [59, 4.2.10].

§5.

AR's and **ANR**'s were introduced by Borsuk [24], [25]. The material presented in §5 is standard. For more information, see Borsuk [29], Hu [74] and chapter 5 of this monograph.

§6.

Theorem 1.6.3 and its applications are due to Borsuk [27].

2. Elementary Plane Topology

The aim of this chapter is to present a few elementary results concerning the topology of the plane $\mathbb{R}^2$. Some of the results will motivate the later developments.

2.1. The Brouwer Fixed Point Theorem and Applications

By exercise 1.5.5(1), the unit interval I has the fixed-point property. We shall generalize this result and show that I^2 has the fixed-point property. In chapter 3 this will again be generalized to the result that I^n has the fixed-point property for every $n \in \mathbb{N}$.

We shall sometimes identify $\mathbb{R}^2$ and the complex plane $\mathbb{C}$. In addition, we assume that the reader knows about addition and multiplication of complex numbers and also about the exponential function $z \to e^z$, $z \in \mathbb{C}$. Finally, if $z = a + bi$ is a complex number then

$$|z| = \sqrt{a^2 + b^2} \text{ and } \bar{z} = a - bi$$

denote its *absolute value* and its *complex conjugate*, respectively.

Let X be a space. A continuous function $f: X \to S^1 = \{x \in \mathbb{R}^2: \|x\| = 1\}$ is called *inessential* if f is homotopic to a constant function from X to S^1. A continuous function that is not inessential is called *essential*. This divides the class of continuous functions from X to S^1 into two classes. A second, more technical concept is needed. We say that a function $f \in C(X,S^1)$ has a *continuous logarithm* provided that there is a continuous function $\phi: X \to \mathbb{R}$ such that $f = e^{i\phi}$, i.e. $f(x) = e^{i\phi(x)}$ for every $x \in X$. We shall prove that for compact X a function $f \in C(X,S^1)$ is inessential if and only if it has a continuous logarithm.

If X is a space and f,g: $X \to \mathbb{C}$ are continuous then $|f - g|$ denotes $\sup\{|f(x) - g(x)|: x \in X\}$ whenever this supremum exists, cf. §1.3.

2.1.1. LEMMA: *Let* X *be a space and let* f: $X \to S^1$ *be continuous such that* $f(X) \neq S^1$. *Then* f *has a continuous logarithm.*

PROOF: Choose $q \in \mathbb{R}$ such that $e^{iq} \notin f(X)$. The function $t \to e^{it}$ is a homeomorphism from the interval $(q,q+2\pi)$ onto $S^1\setminus\{e^{iq}\}$. Let L: $S^1\setminus\{e^{iq}\} \to (q,q+2\pi)$ be the inverse of this mapping. Define $\phi: X \to \mathbb{R}$ by $\phi(x) = L(f(x))$. Then ϕ is clearly continuous and

$$e^{i\phi(x)} = e^{iL(f(x))} = f(x),$$

for every $x \in X$, since if $z \in S^1\setminus\{e^{iq}\}$ then $z = e^{iL(z)}$. □

2.1.2. LEMMA: *Let* X *be a space and let* f_1,f_2: $X \to S^1$ *be continuous functions such that*

$$|f_1 - f_2| < 1.$$

Then f_1 *has a continuous logarithm if and only if* f_2 *has a continuous logarithm.*

PROOF: Define h: $X \to S^1$ by $h = f_1/f_2$. Observe that h is well-defined. If $x \in X$ then

$$|h(x) - 1| = \left|\frac{(f_1 - f_2)(x)}{f_2(x)}\right| = \frac{|(f_1 - f_2)(x)|}{|f_2(x)|} = |(f_1 - f_2)(x)| < 1,$$

by assumption (the "1" in the expression "$|h(x) - 1|$" is the complex number $1 + 0\cdot i$). It follows that the image of h is contained in the half plane

$$\{z = a + bi: a > 0\},$$

and consequently h is not surjective. According to lemma 2.1.1 there is a continuous function ϕ: $X \to \mathbb{R}$ such that $h = e^{i\phi}$. Since $f_1 = f_2\cdot h$, the proof is complete. □

2.1.3. LEMMA: *Let* X *be a compact space and let* H: $X \times I \to S^1$ *be a homotopy. Then* H_0 *has a continuous logarithm if and only if* H_1 *has a continuous logarithm.*

PROOF: Since H is uniformly continuous there is an $n \in \mathbb{N}$ such that if $s,t \in I$ and $|s - t| \leq \frac{1}{n}$ then $|H(x,t) - H(x,s)| < 1$ for every $x \in X$ (in fact, much more is true). Put $f_j = H_{j/n}$ for every $0 \leq j \leq n$. Then

$$|f_{j+1}(x) - f_j(x)| = |H_{(j+1)/n}(x) - H_{j/n}(x)| < 1,$$

for every $0 \leq j \leq n-1$ and $x \in X$. By applying lemma 2.1.2 successively to the level functions f_j $(0 \leq j \leq n)$, we arrive at the desired result. □

We now come to the following result.

2.1.4. THEOREM: *Let* X *be compact and let* $f: X \to S^1$ *be continuous. The following statements are equivalent:*

(a) f *is inessential,*

(b) f *has a continuous logarithm.*

PROOF: For (a) ⇒ (b), let $H: X \times I \to S^1$ be a homotopy such that $H_0 = f$ and H_1 is constant. Since H_1 has clearly a continuous logarithm we find that $f = H_0$ has one by lemma 2.1.3.

For (b) ⇒ (a), suppose that $f = e^{i\phi}$ for a certain continuous function $\phi: X \to \mathbb{R}$. It is easy to see that the function $H: X \times I \to S^1$ defined by

$$H(x,t) = e^{ti\phi(x)}$$

is a homotopy such that H_0 is constant with value 1 and $H_1 = f$. □

This characterization of inessential maps enables us to conclude quite easily that there are essential maps.

2.1.5. COROLLARY: *Let* $n \in \mathbb{Z}\backslash\{0\}$. *Then the mapping* $\psi_n: S^1 \to S^1$ *defined by* $\psi_n(z) = z^n$ *is essential.*

PROOF: To the contrary, assume that ψ_n is inessential. By theorem 2.1.4 there exists a continuous function $\phi: S^1 \to \mathbb{R}$ such that

$$\psi_n(z) = z^n = e^{2\pi i\phi(z)}$$

for all $z \in S^1$ (the constant "2π" is added for technical reasons). Fix $\theta \in \mathbb{R}$. Then for $z = e^{2\pi i\theta}$ we have

$$(e^{2\pi i\theta})^n = e^{2\pi i\phi(z)}$$

i.e.

$$e^{2\pi i(\phi(z)-n\theta)} = 1.$$

From this we conclude that the continuous function f: $\mathbb{R} \to \mathbb{R}$ defined by

$$f(\theta) = \phi(e^{2\pi i\theta}) - n\theta$$

takes its values in $\mathbb{Z}$ only. By connectivity of $\mathbb{R}$ it therefore follows that f is constant. So there exists $N \in \mathbb{Z}$ such that

$$\phi(e^{2\pi i\theta}) = n\theta + N \qquad (\forall\theta \in \mathbb{R}).$$

From this we conclude that

$$n\cdot 1 + N = \phi(e^{2\pi i.1}) = \phi(e^{2\pi i.0}) = n\cdot 0 + N,$$

so that n = 0, which is a contradiction. □

2.1.6. COROLLARY: *The identity mapping* 1: $S^1 \to S^1$ *is essential, i.e* S^1 *is not contractible.* □

We now come to the main result in this section.

2.1.7. THEOREM ("Brouwer Fixed-Point Theorem"): *Let* $D = \{z \in \mathbb{C}: |z| \leq 1\}$. *Then for every continuous function* f: $D \to \mathbb{C}$ *with* $f(S^1) \subseteq D$ *there is a point* $x \in D$ *with* f(x) = x.

PROOF: To the contrary, assume that $f(x) \neq x$ for every $x \in D$. Define r: $\mathbb{C}\backslash\{0\} \to S^1$ and g: $S^1 \to S^1$ by $r(z) = \frac{z}{|z|}$ and $g(u) = r(u-f(u))$, respectively. For every $t \in I$ and $u \in S^1$ we have

$$H(u,t) = u - tf(u) \neq 0.$$

For t = 1 this is clear and for t < 1 observe that

$$|tf(u)| = t|f(u)| \leq t < 1 = |u|.$$

Consequently, r ∘ H is a homotopy connecting the identity function on S^1 and the function g. Observe that for every $t \in I$ and $u \in S^1$ we have

$$(2) \qquad G(u,t) = tu - f(tu) \neq 0.$$

We conclude that $r \circ G$ is a homotopy from S^1 to S^1 connecting the constant function with value $r(-f(0))$ and the function g.

By transitivity of the homotopy relation (lemma 1.6.1(c)) it therefore follows that the identity on S^1 is homotopic to a constant function, violating corollary 2.1.6.□

2.1.8. COROLLARY: I^2 *has the fixed-point property.*

PROOF: It is geometrically obvious that I^2 and D are homeomorphic, cf. exercise 2.1.1. The result therefore follows directly from theorem 2.1.7.□

We finish this section by giving four applications of the techniques developed in this section.

Application 1: The No-Retraction Theorem (part 1).

2.1.9. THEOREM: *Let* $f: D \to S^1$ *be continuous. Then for every* $\lambda \in S^1$ *there exists a point* $u \in S^1$ *(depending on* λ*) such that* $f(u) = \lambda u$.

PROOF: Put $F = \lambda^{-1} \cdot f$. By theorem 2.1.7 there exists $u \in D$ with $F(u) = u$. So $f(u) = \lambda u$. Notice that since $f(u) \in S^1$ and $\lambda \in S^1$ we have $u \in S^1$.□

2.1.10. COROLLARY: S^1 *is not a retract of* D.□

2.1.11. *Remark:* The "Brouwer Fixed-Point Theorem" and the "No Retraction Theorem" are equivalent statements, in the sense that they are easily deduced from one another, see exercise 3.5.1.

2.1.12. COROLLARY: (a) S^1 *is an* **ANR** *but not an* **AR**.
(b) *The "infinite-dimensional torus"* $(S^1)^\infty$ *is not an* **ANR**.

PROOF: (a) By corollary 1.5.6, S^1 is an **ANR**. That S^1 is not an **AR** is a direct consequence of corollary 2.1.10.
(b) This follows immediately from (a) and theorem 1.5.8.□

Application 2: The Theorem on Partitions (part 1).

Let X be a space and let A and B be disjoint closed subsets of X. A closed subset C of X is called a *partition between* A *and* B if there exist open subsets $U,V \subseteq X$ such that

$$A \subseteq U, B \subseteq V, U \cap V = \emptyset \text{ and } X\backslash C = U \cup V.$$

EXAMPLES:

(a) Let $X = \{0,1\}$. Then $\emptyset$ is a partition between $\{0\}$ and $\{1\}$.

(b) Let $X = [0,1]$. Then $\{\frac{1}{3},\frac{2}{3}\}$ is a partition between $\{\frac{1}{2}\}$ and $\{0,1\}$; let $U = (\frac{1}{3},\frac{2}{3})$ and $V = [0,\frac{1}{3}) \cup (\frac{2}{3},1]$.

(c) Let $X = \mathbb{R}^n$. Then $\{x \in \mathbb{R}^n: \|x\| = 1\}$ is a partition between $\{(0,\cdots,0)\}$ and the set $\{x \in \mathbb{R}^n: \|x\| \geq 2\}$; let $U = \{x \in \mathbb{R}^n: \|x\| < 1\}$ and $V = \{x \in \mathbb{R}^n: \|x\| > 1\}$.

(d) Let X be a space, let $A,B \subseteq X$ be closed and disjoint and let U be an open neighborhood of A such that $\overline{U} \cap B = \emptyset$. Then Bd(U) is a partition between A and B.

Now consider J^2 (recall that $J = [-1,1]$) and its faces

$$A_1 = \{-1\} \times J, B_1 = \{1\} \times J, A_2 = J \times \{-1\} \text{ and } B_2 = J \times \{1\}.$$

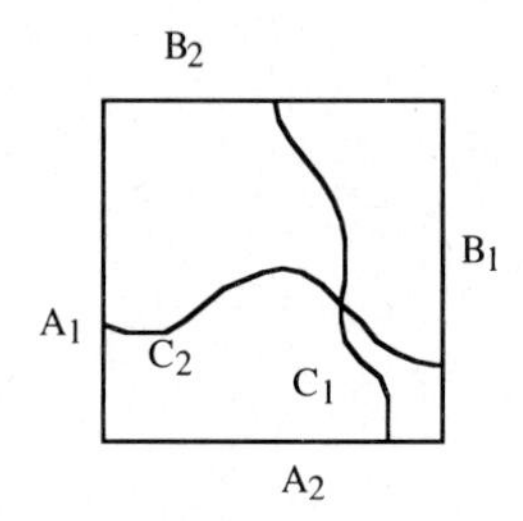

Figure 2.1.1.

It is geometrically obvious that if C_i is a partition between A_i and B_i for $i = 1,2$ then $C_1 \cap C_2 \neq \emptyset$. However, to present a rigorous proof of this fact is not easy at all. Fortunately, the Brouwer Fixed-Point Theorem does the job for us.

2.1.13. THEOREM: *If* C_i *is a partition between* A_i *and* B_i *for* $i = 1,2$ *then* $C_1 \cap C_2 \neq \emptyset$.

PROOF: To the contrary, assume that C_i is a partition between A_i and B_i for $i = 1,2$ such that $C_1 \cap C_2 = \emptyset$. There exist closed subsets $E,F \subseteq J^2$ such that

$$A_1 \subseteq E, B_1 \subseteq F, E \cup F = J^2 \text{ and } E \cap F = C_1.$$

By theorem 1.4.17 there exist continuous functions $\alpha: E \to I$ and $\beta: F \to I$ such that

$$\alpha^{-1}(0) = C_1, \alpha^{-1}(1) = A_1, \beta^{-1}(0) = C_1 \text{ and } \beta^{-1}(1) = B_1.$$

Define $\xi: J^2 \to J$ by

$$\xi(z) = \begin{cases} \alpha(z) & (z \in E), \\ -\beta(z) & (z \in F). \end{cases}$$

Observe that ξ is continuous and that

(1) $\xi(A_1) = \{1\}$, $\xi(B_1) = \{-1\}$ and $\xi^{-1}(0) = C_1$.

Similarly, construct a continuous function $\eta: J^2 \to J$ such that

(2) $\eta(A_2) = \{1\}$, $\eta(B_2) = \{-1\}$ and $\eta^{-1}(0) = C_2$.

Now define $f: J^2 \to J^2$ by $f(z) = (\xi(z),\eta(z))$. Since $C_1 \cap C_2 = \emptyset$, by (1) and (2), the range of f misses (0,0). For every $z \in J^2 \setminus \{(0,0)\}$ the ray from (0,0) through z intersects the "boundary" B $= A_1 \cup A_2 \cup B_1 \cup B_2$ of J^2 in precisely one point, say r(z).

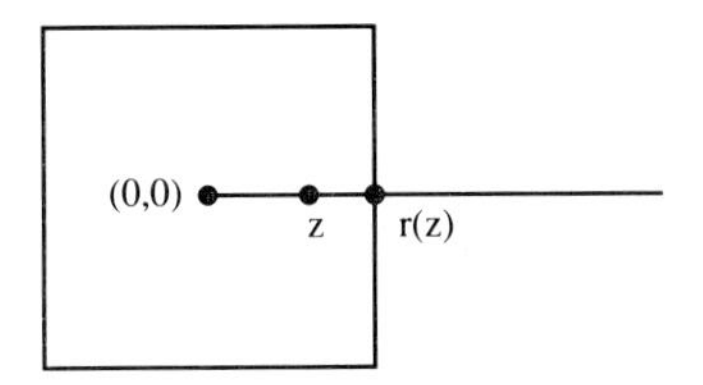

Figure 2.1.2.

The function $r: J^2 \setminus \{(0,0)\} \to B$ is easily seen to be continuous. Consequently, the function $g = r \circ f: J^2 \to B$ has the following properties:

$$g((-1,1)^2) \cap (-1,1)^2 = \emptyset,\ g(A_1) \subseteq B_1,\ g(B_1) \subseteq A_1,\ g(A_2) \subseteq B_2, \text{and } g(B_2) \subseteq A_2.$$

Therefore, g has no fixed-point, which contradicts corollary 2.1.8.□

Application 3. The Non-Homogeneity Theorem (part 1).

A space X is called *homogeneous* (or, *topologically homogeneous*) provided that for all x,y $\in X$ there is a homeomorphism $h: X \to X$ with $h(x) = y$. So, loosely speaking, a space X is homogeneous if from the topological standpoint all points in X behave the same way.

It is trivial that the space $J = [-1,1]$ is not homogeneous. Simply observe that $J\setminus\{-1\}$ is connected while $J\setminus\{0\}$ is not and consequently, no homeomorphism $h: J \to J$ takes -1 onto 0. Let us now consider the space J^2. Intuitively it is clear that this space is not homogeneous since it has an identifiable "boundary" and an identifiable "interior". How to make this precise is not obvious at all. Recall that $D = \{z \in \mathbb{C}: |z| \leq 1\}$ and that D is homeomorphic to J^2.

2.1.14. THEOREM: (1) *If* $A \subseteq S^1$ *then the space* $D \setminus A$ *is contractible.*
(2) *If* $A \subseteq D \setminus S^1$ *is nonempty then* $D \setminus A$ *is not contractible.*

PROOF: (1) Define $H: (D\setminus A) \times I \to D \setminus A$ by $H(x,t) = (1-t)x$. It is easily seen that H is a contraction (cf. exercise 1.6.6).
(2) Without loss of generality, $(0,0) \in A$. Assume, to the contrary, that $H: (D\setminus A) \times I \to D\setminus A$ is a contraction. Define $F: S^1 \times I \to S^1$ by

$$F(x,t) = \frac{H(x,t)}{|H(x,t)|}.$$

Then F contracts S^1 to a point and therefore contradicts corollary 2.1.6.□

2.1.15. COROLLARY: J^2 *is not homogeneous.*□

Application 4: The Fundamental Theorem of Algebra.

The following result is known as the "Fundamental Theorem of Algebra".

2.1.16. THEOREM: *Every polynomial with complex coefficients of degree at least* 1 *has a root in* $\mathbb{C}$.

PROOF: Assume, to the contrary, that $f(z) = \Sigma_{i=0}^{n} a_i z^i$ ($a_i \in \mathbb{C}$, $0 \leq i \leq n$, $n \geq 1$ and $a_n \neq 0$), has no roots in $\mathbb{C}$. Observe that then $a_0 \neq 0$. Define $F: \mathbb{C} \to \mathbb{C}$ by

$$F(z) = \Sigma_{i=0}^{n} a_i z^{n-i}.$$

Then $F(0) = a_n \neq 0$ and for $z \neq 0$, $F(z) = z^n \cdot f(\frac{1}{z})$, so F has no roots, too. Let $\eta: \mathbb{C}\setminus\{0\} \to S^1$ be the retraction $z \to \frac{z}{|z|}$. Define $\xi: S^1 \times I \to S^1$ by

$$\xi(z,t) = \eta\big(f(tz) \,/\, F(\tfrac{t}{z})\big).$$

Then ξ is a well-defined homotopy connecting the constant function

$$z \to \eta(\frac{f(0)}{F(0)}) = \eta(\frac{a_0}{a_n}).$$

with the function $z \to z^n$ on S^1. Since $n \geq 1$ this contradicts corollary 2.1.5.□

Exercises for §2.1.

1. Prove that all compact convex subsets of $\mathbb{R}^2$ with nonempty interiors are homeomorphic (Hint: Let C be such a set. First observe that without loss of generality $(0,0) \in C$. Prove that every ray starting from (0,0) meets the boundary of C in precisely one point. Use this to define a homeomorphism $h: C \to D$).

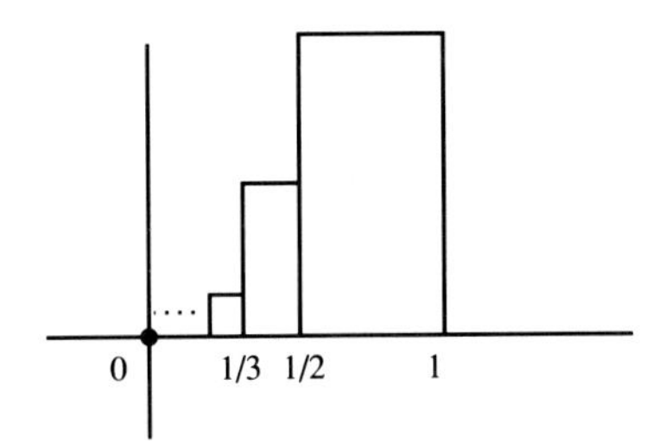

Figure 2.1.3.

2. Let $h: D \to D$ be a homeomorphism. Prove that $h(S^1) = S^1$.

Let X,Y and Z be spaces. We say that a continuous function $f: X \to Y$ *factorizes through* Z provided that there exist continuous functions $g: X \to Z$ and $h: Z \to Y$ such that $f = h \circ g$.

3. Let X be a space. Prove that every continuous function $f: X \to S^1$ that factorizes through a contractible space is inessential. Prove that the space X in exercise 1.5.4 is not contractible while yet every continuous function $f: X \to S^1$ is inessential.

4. Let X be the subset of the plane pictured in figure 2.1.3, i.e.

$$X = (I \times \{0\}) \cup \bigcup \{Bd([1/n+1,1/n] \times [0,1/n]) : n \in \mathbb{N}\}.$$

Prove that X is compact and that X is not an **ANR.**

5. Let $f: D \to D$ be continuous such that $f \mid S^1$ is the identity. Prove that f is surjective.

6. Let X be a compact space and let $f: X \to X$ be a continuous function without fixed-point. Prove that there exists $\varepsilon > 0$ such that $d(x,f(x)) > \varepsilon$ for every $x \in X$.

2.2. The Borsuk-Ulam Theorem

Let $n,m \in \mathbb{N}$. It is intuitively clear that if $n \neq m$ then $\mathbb{R}^n$ and $\mathbb{R}^m$ are not homeomorphic. These intuitively simple facts sometimes have very complicated proofs. If $n = 1$ then there is no problem. Each point of $\mathbb{R}$ disconnects $\mathbb{R}$ while $\mathbb{R}^m\backslash\{pt\}$ is always connected for $m \geq 2$. In this section we shall prove that if $m \neq 2$ then $\mathbb{R}^2$ and $\mathbb{R}^m$ are not homeomorphic. Our proof involves the so-called *Borsuk-Ulam Theorem,* a result of independent interest.

We start with the following simple result.

2.2.1. THEOREM: *Let* $f: S^1 \to \mathbb{R}$ *be continuous. Then there is a* $p \in S^1$ *such that* $f(p) = f(-p)$.

PROOF: Define $F: J \to \mathbb{R}$ by

$$F(t) = f(t,\sqrt{1-t^2}) - f(-t,-\sqrt{1-t^2}).$$

Observe that $F(1) = -F(-1)$. From this and the connectivity of J it follows easily that there exists $t \in J$ such that $F(t) = 0$. Clearly, $p = (t,\sqrt{1-t^2})$ is as required.□

We want to generalize theorem 2.2.1 for the case of functions $f: S^2 \to \mathbb{R}^2$. The same strategy as in the proof of theorem 2.2.1 works, but a more delicate "connectivity argument" is needed.

2.2.2. LEMMA: *Let* $\omega = e^{2\pi i/n}$, *for certain* $n \in \mathbb{N}$, $n > 1$. *In addition, let* $\phi: S^1 \to S^1$ *be a continuous function such that* $\phi(\omega u) = \omega\phi(u)$ *for all* u. *Then* ϕ *has no* n*-th root, i.e. there is no continuous function* $g: S^1 \to S^1$ *such that* $\phi(u) = (g(u))^n$ *for every* $u \in S^1$. *As a consequence,* ϕ *is essential.*

PROOF: To the contrary, assume that $g: S^1 \to S^1$ is a continuous function such that $\phi(u) = (g(u))^n$ for every $u \in S^1$. For $u \in S^1$ we have

$$(g(\omega u))^n = \phi(\omega u) = \omega\phi(u) = \omega(g(u))^n.$$

Consequently,

$$\left(\frac{g(\omega u)}{g(u)}\right)^n = \omega \qquad (\forall u \in S^1).$$

We conclude that the function $\xi: S^1 \to S^1$ defined by $\xi(u) = g(\omega u)\cdot(g(u))^{-1}$ has the property that $\xi(S^1)$ is contained in the finite set

$$A = \{z \in \mathbb{C}: z^n = \omega\}.$$

Since ξ is obviously continuous and S^1 is connected, there exists a point $\nu \in A$ such that

$$\frac{g(\omega u)}{g(u)} = \nu \qquad (\forall u \in S^1).$$

Since $\omega = \omega^{n+1}$, we therefore conclude that

$$1 = \frac{g(\omega)}{g(\omega)} = \frac{g(\omega^{n+1})}{g(\omega)} = \prod_{j=1}^{n} \frac{g(\omega^{j+1})}{g(\omega^j)} = \nu^n = \omega,$$

which is a contradiction.

Now if $f: S^1 \to \mathbb{R}$ is a continuous function such that $\phi = e^{if}$, then

$$\phi(u) = (e^{if(u)/n})^n \qquad (\forall u \in S^1).$$

Consequently, by the above, such a function does not exist, i.e. ϕ is essential (theorem 2.1.4).□

Recall that a continuous function $\phi: S^1 \to S^1$ is *odd* if $\phi(u) = -\phi(-u)$ for all $u \in S^1$. As $-1 = e^{2\pi i/2}$ we have:

2.2.3. COROLLARY: *If* $\phi: S^1 \to S^1$ *is odd then* ϕ *is essential.*□

We are now in a position to prove the main result in this section.

2.2.4. THEOREM ("The Borsuk-Ulam Theorem"): *Let* $f: S^2 \to \mathbb{C}$ *be continuous. Then there exists a* $p \in S^2$ *such that* $f(p) = f(-p)$.

PROOF: For convenience, we shall identify $\mathbb{R}^3$ and $\mathbb{C} \times \mathbb{R}$. A point in S^2 is then represented by a pair (z,t), where $z \in \mathbb{C}$, $t \in \mathbb{R}$ and $|z|^2 + t^2 = 1$. Define $F: D \to \mathbb{C}$ by

$$F(z) = f(z, \sqrt{1-|z|^2}) - f(-z, -\sqrt{1-|z|^2}).$$

Assume that $F(z) \neq 0$ for every $z \in D$. Define $\tilde{F}: D \to S^1$ by

$$\tilde{F}(z) = \frac{F(z)}{|F(z)|}.$$

Observe that if $z \in S^1$ then $1-|z|^2 = 0$ from which it follows that

$$F(z) = f(z,0) - f(-z,0) = -F(-z),$$

and consequently,

$$\tilde{F}(z) = \frac{F(z)}{|F(z)|} = \frac{-F(-z)}{|-F(-z)|} = -\tilde{F}(-z).$$

We conclude that the restriction from $\tilde{F}$ to S^1 is odd, and hence is essential by corollary 2.2.3. On the other hand, D is contractible, from which it follows that $\tilde{F}$ is inessential (exercise 2.1.3). Contradiction.

We conclude that there is a $z \in D$ with $F(z) = 0$. Put $p = (z, \sqrt{1-|z|^2})$. Then $f(p) = f(-p)$.□

We finish this section with the following applications.

Application 1: The Classification Theorem (part 1).

2.2.5. THEOREM: *Let* $n,m \in \mathbb{N}$. *If* $n \neq m$ *and* $n \in \{1,2\}$ *then* $\mathbb{R}^n$ *and* $\mathbb{R}^m$ *are not homeomorphic.*

PROOF: As remarked in the introduction to this section, the case $n = 1$ is a triviality. Therefore, let $n = 2$ and assume that $m > 2$. Then $\mathbb{R}^m$ contains a copy of $\mathbb{R}^3$ which contains S^2. Consequently, $\mathbb{R}^m$ contains a copy of S^2. By theorem 2.2.4, no continuous function $f: S^2 \to \mathbb{R}^2$ is one-to-one. This implies that $\mathbb{R}^m$ and $\mathbb{R}^2$ are not homeomorphic.□

Application 2: Involutions.

Let X be a space. An *involution* on X is a homeomorphism $h: X \to X$ such that $h \circ h = 1_X$. If L is a linear space then a typical example of an involution is the map $x \to -x$.

2.2.6. THEOREM: *Let* $n \in \{1,2\}$ *and let* $h: \mathbb{R}^n \to \mathbb{R}^n$ *be an involution. Then* h *has a fixed-point.*

PROOF: Define a function f_0: $S^0 \to \mathbb{R}^n$ by $f_0(1) = 0$ and $f_0(-1) = h(0)$. For $n \in \{1,2\}$ let

$$S^n_+ = \{x \in S^n: x_{n+1} \geq 0\},$$

i.e. the northern hemisphere of S^n. By theorem 1.4.13 we can extend f_0 to a continuous function

$$g: S^1_+ \to \mathbb{R}^n.$$

Now define f_1: $S^1 \to \mathbb{R}^n$ by

$$f_1(x) = \begin{cases} g(x) & (x \in S^1_+), \\ h(g(-x)) & (x \notin S^1_+). \end{cases}$$

Then f is clearly well-defined and continuous. Observe that for every $x \in S^1$,

$$(*) \qquad f_1(-x) = h(f_1(x)).$$

If n =1 then stop the procedure. If n = 2 then proceed as above, i.e. first extend f_1 over the northern hemisphere of S^2 and then over all of S^2 precisely such as in the definition of f_1. This yields a map f_2: $S^2 \to \mathbb{R}^2$ having the property that for every $x \in S^2$,

$$(**) \qquad f_2(-x) = h(f_2(x)).$$

Now by applying theorems 2.2.1 and 2.2.4 we find that there exists $x \in S^n$ such that

$$f_n(x) = f_n(-x) \qquad (n = 1,2).$$

By (*) and (**) this gives us the desired fixed-point of h.□

Exercises for §2.2.

1. Let f: $D \to S^1$ be continuous. Prove that there exists a point $u \in S^1$ such that $f(u) = f(-u)$ (Hint: Assume the contrary. Define H: $S^1 \times I \to \mathbb{C}$ by $H(u,t) = (2-t)f(u) - tf(-u)$).

2. Let $\omega = e^{2\pi i/n}$, for certain $n \in \mathbb{N}$, $n > 1$. Let f: $S^1 \to S^1$ be a continuous function such that $f(\omega u) = \bar{\omega}f(u)$ for all $u \in S^1$. Prove that f is essential.

3. Prove that there does not exist a space X such that $X \times X$ and $\mathbb{R}$ are homeomorphic.

4. Prove that for any covering $\{M_1, M_2\}$ of S^1 consisting of closed sets, at least one M_i must contain a pair of antipodal points.

5. Prove that for any covering $\{M_1, M_2, M_3\}$ of S^2 consisting of closed sets, at least one M_i must contain a pair of antipodal points (this is a special case of the so-called Lusternik-Schnirelman-Borsuk Theorem).

Let $\alpha: S^1 \to S^1$ be the antipodal map $\alpha(x) = -x$. We call a continuous function $f: S^1 \to S^1$ *antipodal-preserving* if $f \circ \alpha = \alpha \circ f$.

6. Let $f: S^1 \to S^1$ be antipodal-preserving. Prove that f is not nullhomotopic (this is a special case of the Borsuk Antipodal Theorem).

7. Let $f_1, f_2 : S^2 \to \mathbb{R}$ be continuous. Prove that there exists $p \in S^2$ such that $f_1(p) = f_2(-p)$.

8. Let $f: S^1 \to S^1$ be a continuous function such that $f(x) \neq -x$ for every x. Prove that f is homotopic to the identity on S^1.

9. Let $f: S^1 \to S^1$ be a continuous function such that $f(x) \neq x$ for every x. Prove that f is homotopic to the antipodal mapping α.

10. Let $f: S^1 \to S^1$ be nullhomotopic. Prove that f has a fixed-point, and that it must send some point to its antipode.

2.3. The Poincaré Theorem

Let $\langle \cdot , \cdot \rangle$ be the usual inner product on $\mathbb{R}^3$, i.e.

$$\langle x,y \rangle = x_1y_1 + x_2y_2 + x_3y_3.$$

A *continuous tangent vectorfield* V on S^2 is a continuous function $V: S^2 \to \mathbb{R}^3$ such that x and V(x) are *orthogonal* for every $x \in S^2$, i.e. $\langle x, V(x) \rangle = 0$.

Intuitively, V is the wind that blows on the surface of the earth at a given time. In this section we shall prove that if V is a continuous tangent vectorfield on S^2 then there is a point $x \in S^2$ with $V(x) = (0,0,0)$. Consequently, at any time there is a place on earth where there is no wind.

As usual, if $z = a + bi$ is a complex number then we shall write $\operatorname{Re} z = a$ and $\operatorname{Im} z = b$.

Define $S_+ = \{x \in S^2: x_3 \geq 0\}$ and $S_- = \{x \in S^2: x_3 \leq 0\}$, respectively.

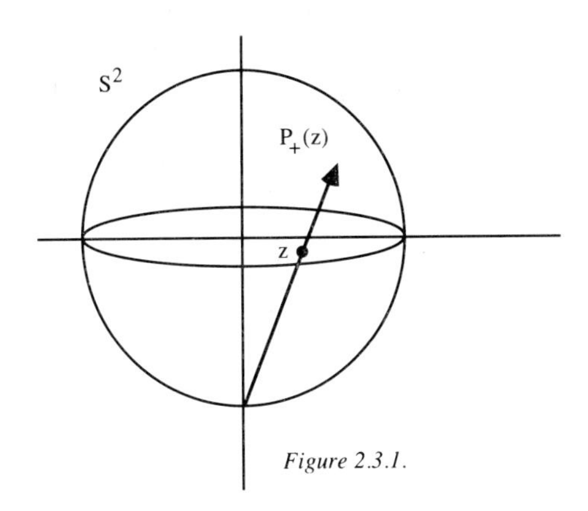

Figure 2.3.1.

In addition, define $P_+: D \to S_+$ and $P_-: D \to S_-$ by

$$P_+(z) = (\frac{2\text{Re } z}{1+|z|^2}, \frac{2\text{Im } z}{1+|z|^2}, \frac{1-|z|^2}{1+|z|^2}),$$

and

$$P_-(z) = (\frac{2\text{Re } z}{|z|^2+1}, \frac{2\text{Im } z}{|z|^2+1}, \frac{|z|^2-1}{|z|^2+1}),$$

respectively. Observe that P_+ and P_- are stereographic projections. Also, it is easy to see that they are homeomorphisms (see figure 2.3.1).

As before, we shall identify $\mathbb{R}^3$ and $\mathbb{C} \times \mathbb{R}$.

Now let us assume that V is a continuous tangent vectorfield on S^2 such that $V(x) \neq (0,0,0)$ for every $x \in S^2$. By proving a series of technical lemmas, which unfortunately have no clear geometric interpretation, we shall derive a contradiction.

Let us define $V_c: S^2 \to \mathbb{C}$ and $V_r: S^2 \to \mathbb{R}$ by

$$V_c(x) = (V(x)_1, V(x)_2), \text{ and } V_r(x) = V(x)_3,$$

i.e. V_c and V_r are the compositions of V and the projections on the $\mathbb{C}$-axis and the $\mathbb{R}$-axis of $\mathbb{R}^3 = \mathbb{C} \times \mathbb{R}$, respectively. In addition, define $W_+, W_-: D \to \mathbb{C}$ by

$$W_+(z) = V_r(P_+(z))z - V_c(P_+(z)),$$

and

$$W_-(z) = V_r(P_-(z))z + V_c(P_-(z)).$$

2.3.1. LEMMA: W_+ *and* W_- *are continuous functions and* $W_+(D) \cup W_-(D) \subseteq \mathbb{C}\backslash\{0\}$.

PROOF: That the functions $W_\pm$ are continuous is trivial.

Now assume that e.g. $W_+(z) = 0$ for certain $z \in D$. Put $y = P_+(z)$. Then $V_r(y)z = V_c(y)$ from which it follows that

$$0 = \langle y,V(y)\rangle = \Sigma_{i=1}^{3} y_i V(y)_i = \langle (y_1,y_2),V_c(y)\rangle + y_3 \cdot V_r(y)$$

$$= V_r(y)\cdot\langle (y_1,y_2),(\text{Re } z,\text{Im } z)\rangle + y_3 \cdot V_r(y)$$

$$= V_r(y)(\langle (y_1,y_2),(\text{Re } z,\text{Im } z)\rangle + y_3).$$

Since

$$\begin{aligned} \langle (y_1,y_2),(\text{Re } z,\text{Im } z)\rangle + y_3 &= \langle (\frac{2\text{Re } z}{1+|z|^2}, \frac{2\text{Im } z}{1+|z|^2}),(\text{Re } z,\text{Im } z)\rangle + \frac{1-|z|^2}{1+|z|^2} \\ &= \frac{1}{1+|z|^2}(2(\text{Re } z)^2 + 2(\text{Im } z)^2 + 1 - |z|^2) \\ &= 1, \end{aligned}$$

we conclude that $V_r(y) = 0$. Consequently, $V_c(y) = (0,0)$, and we obtain $V(y) = V(P_+(z)) = (0,0,0)$, which is a contradiction. □

2.3.2. LEMMA: *If* $u \in S^1$ *then* $\bar{u}W_+(u) = u\overline{W_-(u)}$; *in particular,* $|W_+(u)| = |\overline{W_-(u)}|$.

PROOF: Let $u \in S^1$ and put $\mathbf{u} = (u,0)$. Since $\langle \mathbf{u},V(\mathbf{u})\rangle = 0$ and $\mathbf{u}_3 = 0$ we obtain $\langle u,V_c(\mathbf{u})\rangle = 0$. Observe that for every $a,b \in \mathbb{C}$ we have

$$\langle a,b\rangle = \text{Re}(\bar{a}\cdot b) \text{ and } \bar{a}\cdot b + a\cdot\bar{b} = 2\text{Re}(\bar{a}\cdot b).$$

Therefore, we obtain

$$-\bar{u}\cdot V_c(\mathbf{u}) = u\cdot\overline{V_c(\mathbf{u})}.$$

Consequently,

$$-\bar{u}\cdot V_c(\mathbf{u}) + V_r(\mathbf{u})\bar{u}\cdot u = u\cdot\overline{V_c(\mathbf{u})} + V_r(\mathbf{u})\bar{u}\cdot u,$$

from which it follows easily that

$$\bar{u}(V_r(\mathbf{u})u - V_c(\mathbf{u})) = u(\overline{V_r(\mathbf{u})u + V_c(\mathbf{u})}).$$

Since $P_+(u) = \mathbf{u} = P_-(u)$ we obtain

$$\bar{u}(V_r(P_+(u))u - V_c(P_+(u))) = u(\overline{V_r(P_-(u))u + V_c(P_-(u))}),$$

which is as required. □

We now come to the main result in this section.

2.3.3. THEOREM ("The Poincaré Theorem"): *If* V *is a continuous tangent vectorfield on* S^2 *then there exists* $x \in S^2$ *such that* $V(x) = (0,0,0)$.

PROOF: Naturally, we adopt the notation introduced above. Define $f: D \to \mathbb{C}\backslash\{0\}$ by

$$f(u) = W_+(u) \cdot \overline{W_-(u)}^{-1}.$$

Observe that f is well-defined and continuous (lemma 2.3.1). In addition, define $\bar{f}: D \to S^1$ by $\bar{f} = f/|f|$. Then $\bar{f}$ is inessential since D is contractible (theorem 2.1.14). On the other hand, for $u \in S^1$ we have by lemma 2.3.2,

$$\bar{f}(u) = f(u) = \frac{u}{\bar{u}} = u^2.$$

Consequently, the restriction of $\bar{f}$ to S^1 is essential by corollary 2.1.5. Contradiction. □

We now present a nice application of the Poincaré Theorem.

Application: A fixed-point theorem for S^2.

It is clear that S^2 does not have the fixed-point property because the antipodal map $x \to -x$ has no fixed-points. We shall prove that any fixed-point free map on S^2 sends at least one point to its antipode and therefore has someting in common with the antipodal mapping.

Observe that for S^1 this does not hold. Consider e.g. the rotation

$$f(e^{i\phi}) = e^{i(\phi + \frac{\pi}{4})}.$$

2.3.4. THEOREM: *Let* f: $S^2 \to S^2$ *be continuous. Then* f *has a fixed-point or there exists* $x \in S^2$ *such that* $f(x) = -x$.

PROOF: Define V: $S^2 \to \mathbb{R}^3$ by $V(p) = p \times f(p)$, the vector product of p and f(p), i.e.

$$V(p) = (p_2f(p)_3 - p_3f(p)_2, p_3f(p)_1 - p_1f(p)_3, p_1f(p)_2 - p_2f(p)_1).$$

It is clear that V is continuous. A straightforward verification shows that for all $p \in S^2$, $\langle p,V(p)\rangle = 0$, i.e. V is a continuous tangent vectorfield. By theorem 2.3.3 there exists $x \in S^2$ such that $V(x) = 0$. We shall prove that $f(x) = \pm x$.

Since $f(x) \in S^2$, without loss of generality, $f(x)_1 \neq 0$. Since $V(x)_2 = V(x)_3 = 0$, this implies that

$$(*) \qquad x_3 = \frac{x_1f(x)_3}{f(x)_1} \text{ and } x_2 = \frac{x_1f(x)_2}{f(x)_1}.$$

Thus, since $x \in S^2$, $x_1 \neq 0$. Consequently,

$$1 = x_1^2 + x_2^2 + x_3^2 = x_1^2\left(1 + \left[\frac{f(x)_2}{f(x)_1}\right]^2 + \left[\frac{f(x)_3}{f(x)_1}\right]^2\right) = \left[\frac{x_1}{f(x)_1}\right]^2.$$

CASE 1: $x_1 = f(x)_1$.

By (*) this immediately implies $x_3 = f(x)_3$ and $x_2 = f(x)_2$, i.e. x is a fixed-point of f.

CASE 2: $x_1 = -f(x)_1$.

Again by (*) we get $x_3 = -f(x)_3$ and $x_2 = -f(x)_2$, i.e. $f(x) = -x$. □

Exercises for §2.3.

1. Prove that there exists a continuous function V: $S^1 \to \mathbb{R}^2\backslash\{(0,0)\}$ such that $\langle x,V(x)\rangle = 0$ for every $x \in S^1$.

2. Let f: $D \to \mathbb{C}$ be continuous. Let $x_0 \in D$ be a point such that $f(x_0) = y_0 \notin f(S^1)$. Suppose that the function f_0: $S^1 \to S^1$ defined by $f_0(u) = (f(u)-y_0)/|f(u)-y_0|$ is essential. Prove that y_0 belongs to the interior of f(D).

3. Let f: $D \to \mathbb{C}$ be continuous and one-to-one. Prove that f(0) belongs to the interior of f(D).

4. Let U be an open subset of $\mathbb{C}$ and let f: U $\to$ $\mathbb{C}$ be continuous and one-to-one. Prove that f(U) is open in $\mathbb{C}$ and that f: U $\to$ f(U) is a homeomorphism (this is called the Brouwer Invariance of Domain Theorem; for more information see theorem 4.6.7).

2.4. The Jordan Curve Theorem

The aim of this section is to prove one of the most classical theorems in topology. A *simple closed curve* is a space homeomorphic to the 1-sphere S^1 and an *arc* is a space homeomorphic to the closed unit interval I. The Jordan Curve Theorem states that the complement in the plane of any simple closed curve $C \subseteq \mathbb{R}^2$ has precisely two components, each of which has C as its boundary.

Before we give the actual proof of the main result in this section we derive a few elementary results.

2.4.1. LEMMA: *Let* $n \in \mathbb{N}$, $n > 1$, *and let* $K \subseteq \mathbb{R}^n$ *be compact. Then* $\mathbb{R}^n \backslash K$ *has a component* U *such that* $\mathbb{R}^n \backslash U$ *is bounded. Consequently,* U *is the only unbounded component of* $\mathbb{R}^n \backslash K$.

PROOF: Choose $r > 0$ such that for $E = \{x \in \mathbb{R}^n : \|x\| > r\}$ we have $K \cap E = \emptyset$. As $n > 1$ it follows that E is connected. Since $K \cap E = \emptyset$, there is a component U of $\mathbb{R}^n \backslash K$ such that $E \subseteq U$. If U' is a different component then $U' \cap U = \emptyset$ and consequently, $U' \subseteq \{x \in \mathbb{R}^n : \|x\| \le r\}$, so U' is bounded. □

2.4.2. LEMMA: *Every connected, locally path-connected space is path-connected.*

PROOF: Choose an arbitrary $x \in X$ and put $U = \{y \in X$: there exists a path $\alpha: I \to X$ with $\alpha(0) = x$ and $\alpha(1) = y\}$. Choose an arbitrary $y \in \overline{U}$ and let V be a path-connected neighborhood of y. We claim that $V \subseteq U$. Let $w \in V$. Since $y \in \overline{U}$ and V is a neighborhood of y, there is a point $v \in V \cap U$. Since $v, w \in V$ and V is path-connected, there is a path $\beta: I \to V$ such that

$$\beta(0) = v \text{ and } \beta(1) = w.$$

In addition, since $v \subset U$ there is a path $\alpha: I \to X$ such that

$$\alpha(0) = x \text{ and } \alpha(1) = v.$$

Now define $\gamma: I \to X$ by

$$\gamma(t) = \begin{cases} \alpha(2t) & (0 \leq t \leq \frac{1}{2}), \\ \beta(2t-1) & (\frac{1}{2} \leq t \leq 1). \end{cases}$$

Then γ is a path such that $\gamma(0) = x$ and $\gamma(1) = w$. We conclude that $w \in U$.

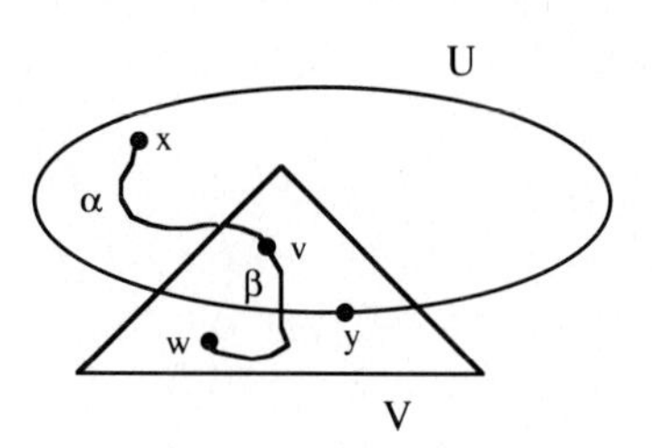

Figure 2.4.1.

Since $y \in \overline{U}$ was chosen arbitrarily, it now follows that U is both open and closed. Since clearly $x \in U$, the connectivity of X implies that $U = X$.

Now choose arbitrary $a,b \in X$. There is a path $\xi: I \to X$ such that $\xi(0) = x$ and $\xi(1) = a$. Similarly, there is a path $\eta: I \to X$ with $\eta(0) = x$ and $\eta(1) = b$. Define $\delta: I \to X$ by

$$\delta(t) = \begin{cases} \xi(1-2t) & (0 \leq t \leq \frac{1}{2}), \\ \eta(2t-1) & (\frac{1}{2} \leq t \leq 1). \end{cases}$$

Then δ is a path such that $\delta(0) = a$ and $\delta(1) = b$. □

2.4.3. COROLLARY: *A connected open subset of* $\mathbb{R}^n$, $n \in \mathbb{N}$, *is path-connected.*

PROOF: Since each open subset of $\mathbb{R}^n$ is clearly locally path-connected, this follows from lemma 2.4.2. □

We now start with the proof of the Jordan Curve Theorem. Our first result is much in the spirit of theorem 2.1.13.

2.4.4. LEMMA: *Let* $X = [a,b] \times [c,d]$, *where* $a,b,c,d \in \mathbb{R}$, $a < b$ *and* $c < d$. *In addition, let* $g,h: J \to X$ *be continuous functions such that*

$$g(-1)_1 = a,\ g(1)_1 = b,\ h(-1)_2 = c \textit{ and } h(1)_2 = d.$$

Then $g(J) \cap h(J) \neq \emptyset$.

Remark: It is clear from the picture that $h(J)$ need not be a partition between the faces $\{a\} \times [c,d]$ and $\{b\} \times [c,d]$. For that reason, theorem 2.1.13 is not applicable and we have to find another argument which will turn out to depend also on the Brouwer Fixed-Point Theorem.

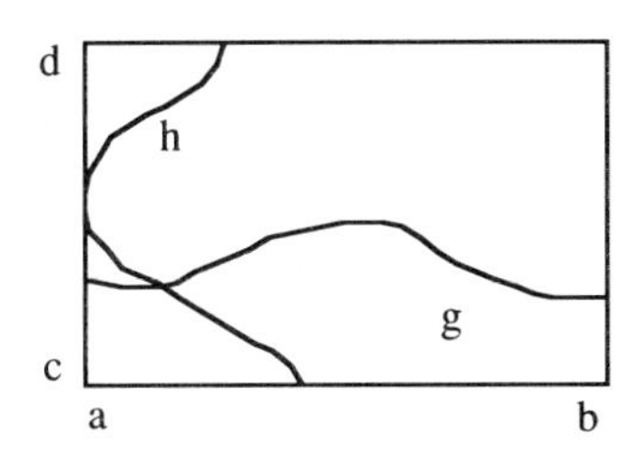

Figure 2.4.2.

PROOF: To the contrary, assume that $g(J) \cap h(J) = \emptyset$. For $s,t \in J$ define

$$N(s,t) = \max\{|g(s)_1 - h(t)_1|, |g(s)_2 - h(t)_2|\}.$$

Clearly, $N(s,t) \neq 0$ for all s and t. Define $F: J^2 \to J^2$ by

$$F(s,t) = \left(\frac{h(t)_1 - g(s)_1}{N(s,t)}, \frac{g(s)_2 - h(t)_2}{N(s,t)}\right).$$

Then F is continuous and

$$(1) \qquad F(J^2) \cap (-1,1)^2 = \emptyset.$$

By the fixed-point property of J^2 (corollary 2.1.8) there exists $(s,t) \in J^2$ such that $F(s,t) = (s,t)$. By (1), $|s| = 1$ or $|t| = 1$. Suppose that, for example, $s = 1$. Then

$$1 = \frac{h(t)_1 - b}{N(1,t)} \leq 0,$$

which is a contradiction. Similarly, the other possibilities cannot occur, concluding the proof of the lemma. □

2.4.5. LEMMA: *Let* $K \subseteq \mathbb{R}^2$ *be compact and let* U *be a bounded component of* $\mathbb{R}^2 \setminus K$. *Then* Bd U $\subseteq$ K *and if* F *is an arbitrary closed subset of* K *which contains* Bd U *then* F *is not an* **AR**.

PROOF: Since $\mathbb{R}^2$ is locally connected, every component of $\mathbb{R}^2\backslash K$ is open. From this it follows easily that Bd U $\subseteq$ K. Now let F be a closed subset of K which contains Bd U and, to the contrary, assume that F is an **AR**. Let $r > 0$ be such that

$$K \subseteq \{x \in \mathbb{R}^2: \|x\| < r\} \subseteq X = \{x \in \mathbb{R}^2: \|x\| \leq r\}$$

and put $S = \{x \in \mathbb{R}^2: \|x\| = r\}$. It is easy to see that $U \cap S = \emptyset$ and that $U \subseteq X$, cf. the proof of lemma 2.4.1. Since F is an **AR**, there is a retraction $p: X \to F$. Define a function $q: X \to X$ by

$$q(z) = \begin{cases} p(z) & (z \in \overline{U}), \\ z & (z \in X\backslash U). \end{cases}$$

Since p is a retraction and $\overline{U} \cap (X\backslash U) = \text{Bd } U \subseteq F$, q is well-defined and continuous. In addition, since $F \cap U = \emptyset$ we have $q(X) \cap U = \emptyset$ and since $S \cap U = \emptyset$, the function q is the identity on S. Pick an arbitrary $x \in U$.

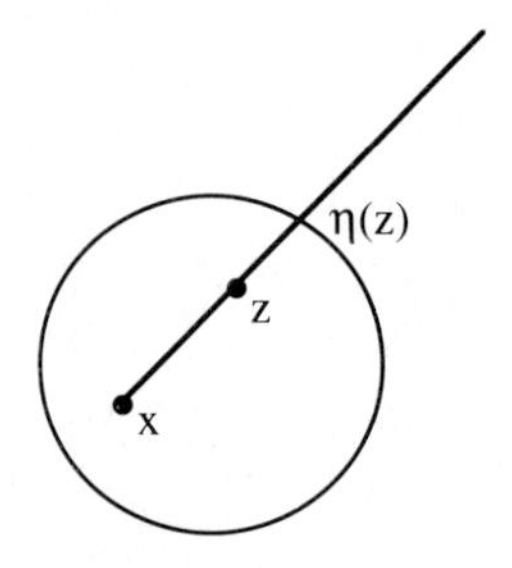

Figure 2.4.3.

For every $z \in X\backslash\{x\}$ the ray from x through z intersects S in precisely one point, say $\eta(z)$, and the function $\eta: X\backslash\{x\} \to S$ is easily seen to be a retraction. Then $\xi = \eta \circ q: X \to S$ is a retraction, which contradicts corollary 2.1.10.□

2.4.6. COROLLARY: *Let* $C \subseteq \mathbb{R}^2$ *be a simple closed curve and let* U *be a bounded component of* $\mathbb{R}^2\backslash C$. *Then* Bd U = C.

PROOF: By lemma 2.4.5, Bd U $\subseteq$ C. If Bd U $\neq$ C then by the special nature of C there exists an arc $A \subseteq C$ which contains Bd U. Since A is **AR** by the Tietze Extension Theorem (corollary 1.4.14) this is impossible by lemma 2.4.5.□

2.4.7. THEOREM ("The Jordan Curve Theorem"): *The complement in the plane of any simple closed curve* C *has precisely two components, each of which has* C *as its boundary.*

PROOF: For all $x,y \in \mathbb{R}^2$ let xy denote the straight line segment between x and y.

Let $C \subseteq \mathbb{R}^2$ be a simple closed curve. Since C is compact, there exist points $x,y \in C$ such that $d(x,y) = \text{diam}(C)$ (exercise 1.1.5). Let L be the straight line through x and y and, in addition, let P and Q denote the straight lines through x and y, respectively, which are orthogonal to L (see figure 2.4.4). We claim that $P \cap C = \{x\}$, and similarly, that $Q \cap C = \{y\}$. If e.g. $p \in P \cap C$ and $p \neq x$ then by the Theorem of Pythagoras, $d(y,p) > d(y,x)$, which contradicts the special choice of x and y.

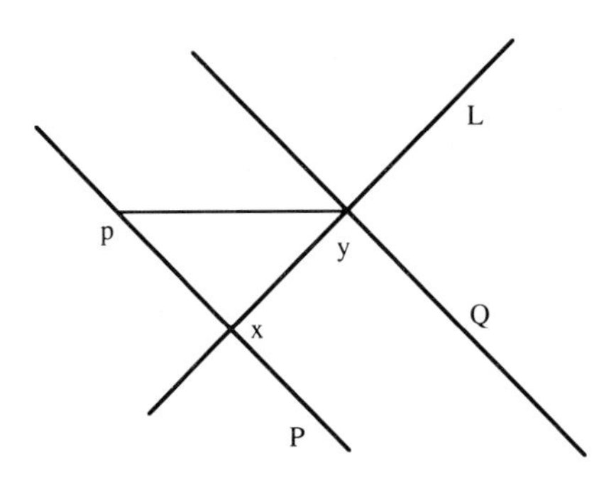

Figure 2.4.4.

Since C is bounded, there exist two straight lines parallel to L, one "above" L and the other "below" L, which do not intersect C. Consequently, by performing a suitable rotation of the plane, we may assume, without loss of generality, that there exist points $a,b,c,d \in \mathbb{R}$ with $a < b$ and $c < d$ such that if $X = [a,b] \times [c,d]$ then

(1) $C \subseteq X$,
(2) $C \cap (\{a\} \times [c,d]) = \{x\}$, $C \cap (\{b\} \times [c,d]) = \{y\}$,
(3) $([a,b] \times \{c,d\}) \cap C = \emptyset$.

Put

$$u = (\tfrac{a+b}{2}, d) \text{ and } l = (\tfrac{a+b}{2}, c),$$

respectively. If $e,f \in ul$ then we say that $e \preccurlyeq f$ if and only if $e_2 \preccurlyeq f_2$. By connectivity of C, the segment ul intersects C. Put $u^- = \max(ul \cap C)$ (with respect to the order " $\preccurlyeq$ "). This maximum exists by compactness. The points x and y divide C into two arcs; we denote the one containing u^- by C(u) and the other one by C(l). Let $m^+ = \min(ul \cap C(u))$ (possibly, $u^- = m^+$).

CLAIM 1: The segment m^+l intersects C(l).

Let $I(u^-,m^+)$ denote the subarc of $C(u)$ between u^- and m^+. The set $uu^- \cup I(u^-,m^+) \cup m^+l$ clearly constitutes a path from u to l. Consequently, by lemma 2.4.4,

$$(uu^- \cup I(u^-,m^+) \cup m^+l) \cap C(l) \neq \emptyset$$

from which it follows that $m^+l \cap C(l) \neq \emptyset$ since by construction

$$(uu^- \cup I(u^-,m^+)) \cap C(l) = \emptyset.$$

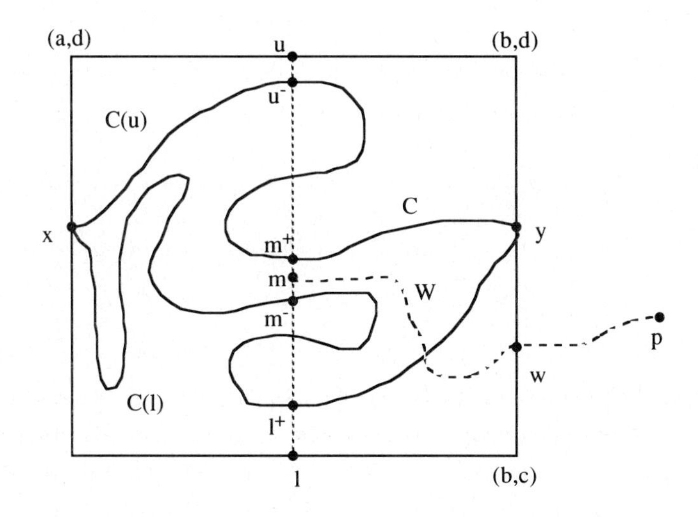

Figure 2.4.5.

Now put $m^- = \max(m^+l \cap C(l))$ and $l^+ = \min(m^+l \cap C(l))$. In addition, let m be the midpoint of the interval m^+m^-. Observe that $m \notin C$.

CLAIM 2: The component U of $\mathbb{R}^2 \setminus C$ which contains m is bounded.

To the contrary, assume that U is unbounded. Pick a point $p \in U \setminus X$. By corollary 2.4.3, U is path-connected. Consequently, there is a path $\beta: I \to U$ connecting m and p. Obviously, $\beta(I)$ intersects Bd X. Let $t \in I$ be the first point for which $\beta(t) \in \text{Bd } X$; put $w = \beta(t)$ and $W = \beta([0,t])$. Then W is a path connecting m and w which moreover is contained in $X \cap U$.

The points x and y divide Bd X in an upper half and a lower half. Suppose first that w is in the lower half. By (1), (2) and (3) there exists an arc $A \subseteq \text{Bd } X$ connecting l and w such that $A \cap C = \emptyset$. Now $uu^- \cup I(u^-,m^+) \cup m^+m \cup W \cup A$ is a path from u to l which misses $C(l)$, contradicting lemma 2.4.4.

If w is in the upper half one can argue similarly.

CLAIM 3: If V is a bounded component of $\mathbb{R}^2\backslash C$ then V = U.

To the contrary, let V be a bounded component of $\mathbb{R}^2\backslash C$ such that $V \neq U$. Then $V \cap U = \emptyset$ and since the complement of X is connected and unbounded it follows that $V \subseteq X$. We let $I(m^-,l^+)$ denote the subarc of C(l) between m^- and l^+. Put

$$H = uu^- \cup I(u^-,m^+) \cup m^+m^- \cup I(m^-,l^+) \cup l^+l.$$

Then H constitutes a path from u to l and we shall prove that $H \cap V = \emptyset$. Since u and l are clearly in the unbounded component of $\mathbb{R}^2\backslash C$, we have $(uu^- \cup l^+l) \cap V = \emptyset$. In addition, $m \in U$ and hence $m^+m^- \cap V = \emptyset$. Finally, since $I(u^-,m^+) \cup I(m^-,l^+) \subseteq C$ and $C \cap V = \emptyset$, we get what we want. Since $x,y \notin H$ and H is compact, there are convex neighborhoods V_x and V_y of x and y, respectively, such that

$$(V_x \cup V_y) \cap H = \emptyset.$$

By corollary 2.4.6, $C \subseteq \overline{V}$. Consequently, there exist points $x_1 \in V_x \cap V$ and $y_1 \in V_y \cap V$. Since V_x is convex and $x_1 \in V \subseteq X$, the straight line segment xx_1 between x and x_1 is contained in $X \cap V_x$. Similarly, $yy_1 \subseteq X \cap V_y$. Since V is path-connected by corollary 2.4.3, there is a path E in V connecting x_1 and y_1. Now $xx_1 \cup E \cup y_1y$ is a path in X connecting x and y which misses H. This (again) contradicts lemma 2.4.4.

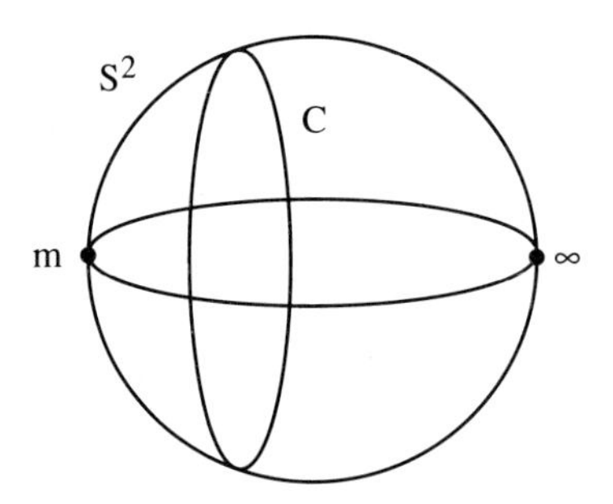

Figure 2.4.6.

By claims 2 and 3, $\mathbb{R}^2\backslash C$ has precisely one bounded component. Also, by lemma 2.4.1, $\mathbb{R}^2\backslash C$ has precisely one unbounded component. Consequently, $\mathbb{R}^2\backslash C$ has exactly two components, which is as required.

Let U and V denote the bounded and the unbounded component of $\mathbb{R}^2\backslash C$, respectively. By corollary 2.4.6 we know that the bounded component U satisfies Bd U = C. We have to prove the same thing for the unbounded component. This is achieved by the following trick on "exchanging" points at infinity. Recall that $m \in U$ (any other point of U will do equally well). The one point compactification $\mathbb{R}^2 \cup \{\infty\}$ of $\mathbb{R}^2$ is homeomorphic to S^2. We now take m as the point at infinity: the space $Y = (\mathbb{R}^2 \cup \{\infty\})\backslash\{m\}$ is homeomorphic to $\mathbb{R}^2$. The "bounded" component of $Y\backslash C$ in Y has to be $V \cup \{\infty\}$ since the closure of $U\backslash\{m\}$ in Y is not compact. By corollary 2.4.6, the boundary of $V \cup \{\infty\}$ in Y equals C. This directly gives that the boundary of V in $\mathbb{R}^2$ is also equal to C.□

We finish this section with an application.

Application: The Non-imbedding Theorem.

Choose points $x_0,x_1,x_2,x_3,x_4 \in \mathbb{R}^3$ such that no four of them lie in the same plane and let X be the union of all the straight line segments connecting two of the x_i's. If the reader tries to draw pictures of X in the plane then he will notice that there are always lines that intersect. However, no elementary proofs confirming this are known.

2.4.8. THEOREM: X *cannot be imbedded in the plane.*

PROOF: For all $x,y \in \mathbb{R}^2$ let xy denote the straight line segment between x and y. Let us assume that $f: X \to \mathbb{R}^2$ is an imbedding and let us try to derive a contradiction. Consider the simple closed curve $C_1 = f(x_1x_3 \cup x_3x_2 \cup x_2x_1)$. Since $f(x_0x_4)$ is an arc connecting $f(x_0)$ and $f(x_4)$, which moreover is disjoint from C_1, it follows that $f(x_0)$ and $f(x_4)$ belong to the same component U of $\mathbb{R}^2\backslash C_1$. Since $\mathbb{R}^2\backslash C_1$ has two components (theorem 2.4.7) by a similar trick as used at the end of the proof of theorem 2.4.7, we can assume that U is the bounded component of $\mathbb{R}^2\backslash C_1$. By applying theorem 2.4.7 a few times it follows that $U\backslash f(x_1x_4 \cup x_2x_4 \cup x_3x_4)$ is the union of three disjoint connected open sets U_1, U_2 and U_3 such that $f(x_i) \notin \overline{U_i}$ for i = 1,2,3 (this requires some work). Without loss of generality $f(x_0) \in U_1$. The set $f(x_1x_0)$ is an arc joining the points $f(x_1)$ and $f(x_0)$ and it misses $C_2 = f(x_2x_3 \cup x_3x_4 \cup x_4x_2)$. Since $f(x_0)$ and $f(x_1)$ are contained in distinct components of $\mathbb{R}^2\backslash C_2$, we have obtained the desired contradiction.□

Exercises for §2.4.

1. Let $X \subseteq \mathbb{R}^2$ be a compact **AR**. Prove that $\mathbb{R}^2\backslash X$ is connected.

2. Let $X \subseteq \mathbb{R}^2$ be homeomorphic to a figure eight. Prove that $\mathbb{R}^2 \backslash X$ has three components.

3. Let U be a nonempty open bounded subset of $\mathbb{R}^2$. Prove that $\mathbb{R}^2 \backslash U$ is not a retract of $\mathbb{R}^2$.

4. Let $X \subseteq \mathbb{R}^2$ be a compact **ANR**. Prove that $\mathbb{R}^2 \backslash X$ has only finitely many components (Hint: use exercise 3).

Notes and comments for chapter 2.

§1.

The Brouwer fixed-point Theorem is due to Brouwer [33]. The approach in this section is due to Burckel [40]. The theorem on partitions is implicit in Eilenberg and Otto [58]. The proof of theorem 2.1.16 presented here is a simplification (due to E. Verheul) of Burckel's proof.

§2.

The Borsuk-Ulam Theorem was conjectured by Ulam and proved by Borsuk [26]. The proof presented here is taken from Burckel [40]. The Lusternik-Schnirelman-Borsuk Theorem is due to Lusternik and Schnirelman [96] and Borsuk [26]. The Borsuk Antipodal Theorem is also due to Borsuk [26].

§3.

The Poincaré Theorem and its applications are due to Poincaré [115]. The approach in this section is again due to Burckel [40]. Exercise 4 is due to Brouwer [34].

§4.

The Jordan Curve Theorem is due to Veblen [141]. The elegant proof presented here is taken from Maehara [97].

For more information on plane topology, see e.g. Kuratowski [90], Moise [109] and Wall [142].

3. Elementary Combinatorial Techniques

In this chapter we present some elementary combinatorial techniques and apply these to get nontrivial information about the topology of the euclidean spaces $\mathbb{R}^n$, $n \in \mathbb{N}$. Our main result is the Brouwer Fixed-Point Theorem.

3.1. Affine Notions

Let V be a vector space and let F be a finite subset of V, say $F = \{v_1,\cdots,v_n\}$.

(1) A *linear combination* of $v_1,\cdots,v_n$ is a vector v of the form $\Sigma_{i=1}^n t_i v_i$ with $t_1,\cdots,t_n \in \mathbb{R}$. Such a v is called *linearly dependent* on F.

(2) An *affine combination* of $v_1,\cdots,v_n$ is a vector v that can be written as $\Sigma_{i=1}^n t_i v_i$ with $t_1,\cdots,t_n \in \mathbb{R}$ and $\Sigma_{i=1}^n t_i = 1$. Such a v is called *geometrically dependent* on F.

A *linear (affine) subspace* of V is a subset of V closed under the formation of linear (affine) combinations.

3.1.1. THEOREM: *Let* A *be a subset of* V *and let* $a \in A$. *Then* A *is an affine subspace if and only if* $A - a$ *is a linear subspace.*

PROOF: Suppose that A is an affine subspace, and take arbitrary elements $a_1 - a,\cdots,a_n - a \in A-a$. For $t_1,\cdots,t_n \in \mathbb{R}$ we have

(1) $$\sum_{i=1}^{n} t_i(a_i - a) = \sum_{i=1}^{n} t_i a_i + (1 - \sum_{i=1}^{n} t_i)a - a.$$

Observe that the right-hand side of (1) is of the form $p - a$, where p is an affine combination of the $a_1, \cdots, a_n, a$. This implies that the left-hand side of (1) belongs to $A - a$.

Now assume that $A - a$ is a linear subspace. Let $\Sigma_{i=1}^{n} t_i a_i$ be an affine combination of elements of A. Since $A - a$ is a linear subspace, we have

$$p = \sum_{i=1}^{n} t_i(a_i - a) \in A - a$$

and since $\Sigma_{i=1}^{n} t_i = 1$ this implies that

$$\sum_{i=1}^{n} t_i a_i = \sum_{i=1}^{n} t_i(a_i - a) + a = p + a \in A. \ \square$$

If $S \subseteq V$, then the intersection of all affine (linear) subspaces of V containing S is the smallest affine (linear) subspace of V containing S. This subset is called the *affine hull (linear hull)* of S and is denoted by aff(S) (lin(S)).

3.1.2. THEOREM: aff(S) *is the set of all affine combinations of elements of* S. *Moreover, for every* $a \in S$ *the following equality holds:*

$$\text{aff}(S) - a = \text{lin}(S - a).$$

PROOF: The first part of the theorem follows easily by arguments similar to the ones in the proof of lemma 1.2.2.

We shall now prove the second part. Since aff(S) is an affine subspace, by theorem 3.1.1, aff(S) - a is a linear subspace. Since $S - a \subseteq \text{aff}(S) - a$ this implies that $\text{lin}(S - a) \subseteq \text{aff}(S) - a$. Conversely, since lin(S - a) is a linear subspace, again by theorem 3.1.1 it follows that lin(S - a) + a is an affine subspace. Since $S \subseteq \text{lin}(S - a) + a$ this implies $\text{aff}(S) \subseteq \text{lin}(S - a) + a$. $\square$

We see that if $S \subseteq V$ then aff(S) is a *hyperplane*, i.e. a translated linear subspace of V, since for every $a \in S$ the equality

$$\text{aff}(S) = a + \text{lin}(S - a)$$

holds. We say that S *spans* the hyperplane aff(S).

Let $v_1,\cdots,v_n \in V$. Then

(1) the vectors $v_1,\cdots,v_n$ are said to be *linearly independent* if for all $t_1,\cdots,t_n \in \mathbb{R}$ with $\Sigma_{i=1}^n t_i v_i = \underline{0}$ we have $t_1 = \cdots = t_n = 0$;

(2) the vectors $v_1,\cdots,v_n$ are said to be *geometrically independent* if for all elements $t_1,\cdots,t_n \in \mathbb{R}$ with $\Sigma_{i=1}^n t_i = 0$ and $\Sigma_{i=1}^n t_i v_i = \underline{0}$ we have $t_1 = \cdots = t_n = 0$;

(3) a subset $S \subseteq V$ is *linearly (geometrically) independent* if and only if every finite subset is.

3.1.3. THEOREM: *Let* $S \subseteq V$. *The following statements are equivalent:*

(a) S *is geometrically independent,*

(b) *no* $x \in S$ *is geometrically dependent on a finite subset* $F \subseteq S\backslash\{x\}$,

(c) *for every* $s \in S$, $\{x - s: x \in S, x \neq s\}$ *is linearly independent,*

(d) *for some* $s \in S$, $\{x - s: x \in S, x \neq s\}$ *is linearly independent.*

PROOF: For (a) $\Rightarrow$ (b), suppose that there exist $x \in S$, $v_1,\cdots,v_n \in S\backslash\{x\}$ and $t_1,\cdots,t_n \in \mathbb{R}$ such that $\Sigma_{i=1}^n t_i = 1$ and $x = \Sigma_{i=1}^n t_i v_i$. Then $1\cdot x + \Sigma_{i=1}^n(-t_i)v_i = \underline{0}$ and $1 + \Sigma_{i=1}^n(-t_i) = 0$. This contradicts (a).

For (b) $\Rightarrow$ (c), take pairwise distinct $v_1,\cdots,v_n \in S\backslash\{s\}$ and $t_1,\cdots,t_n \in \mathbb{R}$ such that $\Sigma_{i=1}^n t_i(v_i - s) = \underline{0}$. Assume that e.g. $t_1 \neq 0$. By dividing the equality by t_1 we find that without loss of generality we may assume $t_1 = 1$. Then

$$v_1 = s + \Sigma_{i=2}^n(-t_i)(v_i - s) = (1 + \Sigma_{i=2}^n t_i)s + \Sigma_{i=2}^n(-t_i)v_i.$$

This contradicts (b).

Since (c) $\Rightarrow$ (d) is a triviality, it suffices to establish (d) $\Rightarrow$ (a). To this end, take pairwise distinct $x_1,\cdots,x_n \in S\backslash\{s\}$, elements $t_1,\cdots,t_n,t \in \mathbb{R}$ with $\Sigma_{i=1}^n t_i + t = 0$ and assume that $\Sigma_{i=1}^n t_i x_i + ts = \underline{0}$. Then $\Sigma_{i=1}^n t_i(x_i - s) = \underline{0}$ so that by (d), $t_1 = \cdots = t_n = 0$. Since $\Sigma_{i=1}^n t_i + t = 0$ we also get $t = 0$. □

3.1.4. PROPOSITION: *Let* $S \subseteq V$ *be geometrically independent. If* $A,B \subseteq S$ *then*

$$\mathrm{aff}(A) \cap \mathrm{aff}(B) = \mathrm{aff}(A \cap B).$$

PROOF: Clearly, $\mathrm{aff}(A \cap B) \subseteq \mathrm{aff}(A) \cap \mathrm{aff}(B)$. Take an arbitrary $x \in \mathrm{aff}(A) \cap \mathrm{aff}(B)$. There exist (distinct) $a_1,\cdots,a_n \in A$, $t_1,\cdots,t_n \in \mathbb{R}$, (distinct) $b_1,\cdots,b_m \in B$, $s_1,\cdots,s_m \in \mathbb{R}$ such that $\Sigma_{i=1}^n t_i = \Sigma_{j=1}^m s_j = 1$ and

$$x = \sum_{i=1}^n t_i a_i = \sum_{j=1}^m s_j b_j.$$

Then $\Sigma_{i=1}^{n} t_i a_i + \Sigma_{j=1}^{m}(-s_j) b_j = \underline{0}$ and $\Sigma_{i=1}^{n} t_i + \Sigma_{j=1}^{m} -s_j = 1 - 1 = 0$. Consequently, since S is geometrically independent, all the t_i and s_j are equal to 0, unless a_i equals b_j in which case $t_i = s_j$. This clearly implies that $x \in \mathrm{aff}(A \cap B)$.□

3.1.5. THEOREM: *Let* $a_0, a_1, \cdots, a_n$ *be elements of* V *such that for every* $i < n$,

$$a_{i+1} \notin \mathrm{aff}(\{a_0, \cdots, a_i\}).$$

Then $\{a_0, \cdots, a_n\}$ *is geometrically independent.*

PROOF: For $0 \le i \le n$ put $A_i = \{a_0, \cdots, a_i\}$. By induction on i we shall prove that A_i is geometrically independent. Clearly this is true for $i = 0$. Therefore suppose that A_i is geometrically independent for certain $i \ge 0$. Assume that $s_{i+1} a_{i+1} + s_i a_i + \cdots + s_0 a_0 = \underline{0}$ with $\Sigma_{j=1}^{i+1} s_j = 0$. If $s_{i+1} = 0$ then since A_i is geometrically independent, we have $s_i = \cdots = s_0 = 0$. So assume that $s_{i+1} \neq 0$. Then since

$$\sum_{j=0}^{i} \frac{-s_j}{s_{i+1}} = \frac{-\sum_{j=0}^{i} s_j}{s_{i+1}} = \frac{s_{i+1}}{s_{i+1}} = 1$$

we have

$$\frac{-s_i}{s_{i+1}} a_i + \cdots + \frac{-s_0}{s_{i+1}} a_0 = a_{i+1} \in \mathrm{aff}(\{a_0, \cdots, a_i\}).$$

This is a contradiction.□

A function between affine subspaces of linear spaces is called *affine* if it preserves affine combinations. Images and preimages of affine sets under affine functions are again affine. The following triviality is left as an exercise to the reader.

3.1.6. THEOREM: *Let* V_1 *and* V_2 *be linear spaces, let* $A_1 \subseteq V_1$ *and* $A_2 \subseteq V_2$ *be affine subspaces and let* $f: A_1 \to A_2$ *be a function. Then the following statements are equivalent:*

(1) f *is affine,*

(2) *the composition*

$$A_1 - a_1 \xrightarrow{+a_1} A_1 \xrightarrow{f} A_2 \xrightarrow{-a_2} A_2 - a_2 \qquad (a_1 \in A_1 \text{ and } a_2 = f(a_1)),$$

i.e. the function $\xi: A_1 - a_1 \to A_2 - a_2$ *defined by* $\xi(x) = f(x + a_1) - a_2$, *is linear.*□

From theorem 3.1.6 it follows that an affine function, the domain and range of which are both contained in a finite-dimensional normed linear space, is continuous. This can be seen as follows. First observe that each linear function f: $\mathbb{R} \to \mathbb{R}$ is continuous (the verification of this is left as an exercise to the reader). By induction, this gives us that each linear function f: $\mathbb{R}^n \to \mathbb{R}$ is continuous. Consequently, each linear function f: $\mathbb{R}^n \to \mathbb{R}^m$ is continuous. By exercise 1.2.12 it follows that each finite-dimensional normed linear space is topologically isomorphic to some $\mathbb{R}^n$. We conclude that a linear function between finite-dimensional normed linear spaces is continuous.

Exercises for §3.1.

1. Let V be a linear space. Prove that if C,D are convex subsets of V then $\text{conv}(C \cup D) = \cup\{[c,d]: c \in C, d \in D\}$. Here [c,d] denotes the segment $\{t \cdot c + (1-t) \cdot d: t \in I\}$ connecting c and d.

2. Let $x(0),\cdots,x(p)$ be p+1 points in $\mathbb{R}^n$, where $p \le n$. Let A be the (p+1,n+1)-matrix the i-th row $(0 \le i \le p)$ of which is equal to $(x(i)_1,\cdots,x(i)_n,1)$. Prove that $x(0),\cdots,x(p)$ is geometrically independent if and only if A has rank p+1.

3.2. Simplexes

Let V be a linear space. An *n-simplex* in V is a geometrically independent subset of V having precisely n+1 points. Simplexes are denoted by Greek letters, i.e. σ,μ,τ, etc. If σ and τ are simplexes and $\sigma \subseteq \tau$ then σ is called a *face* of τ; to indicate that σ is a face of τ we shall also use the notation: $\sigma \preccurlyeq \tau$. An n-simplex in V is sometimes also called an *n-dimensional simplex.*

3.2.1. THEOREM: *Let* σ *be a simplex in* V *and let* $A = \text{aff}(\sigma)$. *Then every element* $x \in A$ *can be written uniquely as an affine combination* $\Sigma_{v \in \sigma} t_v \cdot v$ *of* σ. *In addition, the functions* $\alpha_v: A \to \mathbb{R}$ *defined by* $\alpha_v(x) = t_v$ *are affine.*

PROOF: Let $\sigma = \{v_0,\cdots,v_n\}$ and take an arbitrary $x \in A$. Theorem 3.1.2 implies that x can be written in the form

$$x = \sum_{i=0}^{n} t_i v_i \text{ with } \sum_{i=0}^{n} t_i = 1.$$

Assume that x can also written in the form

$$x = \sum_{i=0}^{n} s_i v_i \text{ with } \sum_{i=0}^{n} s_i = 1.$$

Then $\Sigma_{i=0}^{n}(t_i - s_i)v_i = \underline{0}$ and $\Sigma_{i=0}^{n}(t_i - s_i) = 0$. Since σ is geometrically independent this implies that for every i, $t_i = s_i$.

To prove that α_{v_i} is affine, take $a_1,a_2 \in A$, $t \in \mathbb{R}$, and put $a = t\,a_1 + (1-t)a_2$. Then

$$a = \sum_{i=0}^{n}(t\,\alpha_{v_i}(a_1) + (1-t)\alpha_{v_i}(a_2))v_i,$$

and

$$\sum_{i=0}^{n} t\,\alpha_{v_i}(a_1) + \sum_{i=0}^{n}(1-t)\alpha_{v_i}(a_2) = t + (1-t) = 1.$$

Consequently, $\alpha_{v_i}(a) = t\,\alpha_{v_i}(a_1) + (1-t)\alpha_{v_i}(a_2)$ for every i. □

The real numbers $\alpha_v(x)$ for $v \in \sigma$ are called the *affine coordinates* of x with respect to σ. We call the α_v the coordinate functions of aff(σ). This notation will remain in force throughout the chapter.

A *geometric simplex* is the convex hull of a simplex. We use $|\sigma|$ as an abbreviation for conv(σ) and sometimes say that $|\sigma|$ is the geometric simplex *spanned* by σ. Observe that the affine coordinates of $x \in |\sigma|$ are non-negative (lemma 1.2.2). The elements of σ are called the *vertices* of $|\sigma|$. If τ is a face of σ then $|\tau|$ is also called a *face* of $|\sigma|$. The union of all proper faces of $|\sigma|$ is called the *(geometric) boundary* $\partial|\sigma|$ of $|\sigma|$ and its complement is the *(geometric) interior* of $|\sigma|$.

Since a geometric simplex $|\sigma|$ is the convex hull of the finite set σ it follows that $|\sigma|$ is compact (lemma 1.2.2(2)); this remark will be used without explicit reference in the remaining part of this book.

3.2.2. THEOREM: *If σ is a simplex and $\sigma_1,\sigma_2 \subseteq \sigma$ then $|\sigma_1| \cap |\sigma_2| = |\sigma_1 \cap \sigma_2|$.*

PROOF: This follows immediately by observing that affine coordinates are unique (theorem 3.2.1) which implies that for i = 1,2,

$$|\sigma_i| = \{x \in |\sigma|: (\forall v \in \sigma\backslash\sigma_i)(\alpha_v(x) = 0)\}. \square$$

3.2.3. THEOREM: *Let* $(V,\|\cdot\|)$ *be a normed linear space, and let* $\sigma \subseteq V$ *be a simplex. Then* diam(σ) = diam($|\sigma|$).

PROOF: We prove that diam($|\sigma|$) $\le$ diam(σ). Take $p,q \in |\sigma|$. Then

$$\|p - q\| = \|\Sigma_{v\in\sigma}\alpha_v(p)\cdot v - \Sigma_{v\in\sigma}\alpha_v(p)\cdot q\| = \|\Sigma_{v\in\sigma}\alpha_v(p)(v - q)\|$$

$$\leq \Sigma_{v\in\sigma}\alpha_v(p)\|v - q\| \leq \Sigma_{v\in\sigma}\alpha_v(p)\cdot\max_{v\in\sigma}\|v - q\|$$

$$= \max_{v\in\sigma}\|v - q\|.$$

As a consequence we obtain for every $v \in \sigma$ that $\|q - v\| \leq \max_{w\in\sigma}\|w - v\|$. Consequently,

$$\|p - q\| \leq \max_{v\in\sigma}\|v - q\| \leq \max_{v,w\in\sigma}\|v - w\|,$$

as required. □

For each $0 \leq i \leq k$ let $d_i \in \mathbb{R}^{k+1}$ be the vector all coordinates of which are 0, except the (i+1)-th coordinate which equals 1. The vectors $d_0,\cdots,d_k$ are clearly linearly independent (hence geometrically independent) and for $x \in \text{aff}(\{d_0,\cdots,d_k\})$ we have that the affine coordinates of $x = (x_0,\cdots,x_k)$ with respect to $d_0,\cdots,d_k$ are equal to $x_0,\cdots,x_k$, respectively. Put $\tau_k = \{d_0,\cdots,d_k\}$

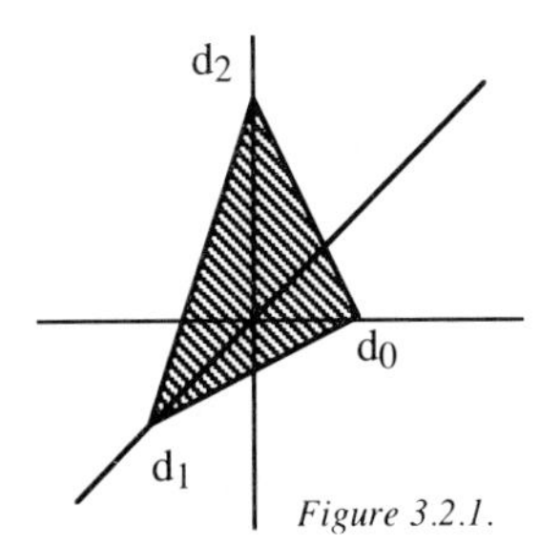

Figure 3.2.1.

Observe that $|\tau_k| = \{x \in I^{k+1}: \Sigma_{i=0}^{k}x_i = 1\}$ is a closed and bounded subset of $\mathbb{R}^{k+1}$ and hence is compact (alternatively, use that $|\tau_k|$ is a geometric simplex).

3.2.4. THEOREM: *Let* $(V,\|\cdot\|)$ *be a normed linear space, and let* $\sigma = \{v_0,v_1,\cdots,v_k\} \subseteq V$ *be a* k-*simplex. Then the function* $f: |\sigma| \to |\tau_k|$ *defined by*

$$f(a) = (\alpha_{v_0}(a),\cdots,\alpha_{v_k}(a))$$

is a homeomorphism.

PROOF: First observe that the functions α_{v_i} are affine (theorem 3.2.1). Consequently, f is continuous by the remark following theorem 3.1.6, and one-to-one by unicity of affine coordinates (theorem 3.2.1). Moreover, f is clearly surjective since if $y \in |\tau_k|$ then $f(x) = y$, where $x = \Sigma_{i=0}^{k} y_i \cdot v_i$. By compactness of $|\sigma|$ it therefore follows that f is a homeomorphism (exercise 1.1.4).□

The *barycenter* of a simplex σ is the point in $|\sigma|$ whose affine coordinates with respect to σ are all equal: if $\sigma = \{v_0, \cdots, v_n\}$ then the barycenter b_σ of σ is given by

$$b_\sigma = \sum_{i=0}^{n} \frac{1}{n+1} \cdot v_i.$$

Observe that b_σ belongs to the geometric interior of $|\sigma|$.

3.2.5. LEMMA: *Let $\tau_1 \subset \tau_2 \subset \cdots \subset \tau_k$ be a strictly increasing collection of faces of a simplex σ. Then the barycenters $b_{\tau_1}, b_{\tau_2}, \cdots, b_{\tau_k}$ are geometrically independent.*

PROOF: By theorem 3.1.5 it suffices to show that for $i < k$, $b_{\tau_{i+1}} \notin \mathrm{aff}(\{b_{\tau_1}, b_{\tau_2}, \cdots, b_{\tau_i}\})$. To this end, take an arbitrary vertex $v \in \tau_{i+1} \setminus \tau_i$. Then

$$\alpha_v(b_{\tau_{i+1}}) = \frac{1}{\text{the cardinality of } \tau_{i+1}} \neq 0,$$

while for every $x \in \mathrm{aff}(\{b_{\tau_1}, b_{\tau_2}, \cdots, b_{\tau_i}\})$ we have $\alpha_v(x) = 0$ because $\{b_{\tau_1}, b_{\tau_2}, \cdots, b_{\tau_i}\} \subseteq \mathrm{aff}(\tau_i)$, so $\mathrm{aff}(\{b_{\tau_1}, b_{\tau_2}, \cdots, b_{\tau_i}\}) \subseteq \mathrm{aff}(\tau_i)$.□

Let σ be a simplex and let $\mathcal{K} = \{\tau_1, \tau_2, \cdots, \tau_k\}$ be a chain of faces of σ (by a chain we mean of course a chain with respect to inclusion). For every $v \in \bigcup\mathcal{K}$ define its height ht(v) to be the first $i \leq k$ with $v \in \tau_i$.

Define a quasi-order $\preccurlyeq$ on $\bigcup\mathcal{K}$ by putting: $v \preccurlyeq w \Leftrightarrow \mathrm{ht}(v) \leq \mathrm{ht}(w)$.

3.2.6. LEMMA: *Let σ be an n-simplex and let $\mathcal{K}$ be a maximal chain of faces of σ. Then*

(1) *$\mathcal{K}$ has cardinality* n+1 *and the corresponding quasi-order $\preccurlyeq$ is a total order on $\sigma = \bigcup\mathcal{K}$;*

(2) *The geometric simplex $|\beta|$ of all barycenters of elements of $\mathcal{K}$ is equal to the set*

$$\{x \in |\sigma| : \forall v, w \in \sigma : v \preccurlyeq w \Rightarrow \alpha_v(x) \geq \alpha_w(x)\};$$

(3) *If $v \preccurlyeq w$ and for certain $x \in |\beta|$ we have $\alpha_v(x) = \alpha_w(x)$ then for every $\tau \in \mathcal{K}$ with $v \in \tau$ and $w \notin \tau$ we have that the affine coordinate of x in $|\beta|$ with respect to b_τ is equal to* 0.

PROOF: If α is the direct successor of β in $\mathcal{K}$ then $\alpha\backslash\beta$ is a singleton: if it contains two points v and w then

$$\cdots\beta, \beta \cup \{v\}, \alpha\cdots$$

would be a chain extending $\mathcal{K}$; likewise one sees that min $\mathcal{K}$ is a singleton. Thus $\mathcal{K} = \{\tau_0,\tau_1,\cdots,\tau_n\}$ where τ_i has i+1 points. Now write $\sigma = \{v_0,v_1,\cdots,v_n\}$ such that for every $0 \le i \le n$, $\tau_i = \{v_0,v_1,\cdots,v_i\}$. Then $v_i \preccurlyeq v_j$ if and only if $i \le j$. This proves (1).

An arbitrary element $x \in |\beta|$ can be written as

$$x = \sum_{i=0}^{n} t_i \cdot b\tau_i \text{ where } t_i \ge 0 \text{ for every i and } \sum_{i=0}^{n} t_i = 1$$

(lemma 1.2.2). For every i define

$$s_i = \frac{t_i}{i+1} + \cdots + \frac{t_n}{n+1}.$$

Then

$$x = \sum_{i=0}^{n} s_i \cdot v_i \text{ and } \sum_{i=0}^{n} s_i = 1.$$

From this we conclude that the s_i are the affine coordinates of x (theorem 3.2.1) and satisfy $s_0 \ge s_1 \ge \cdots \ge s_n$.

Conversely, let p be a point in $|\sigma|$ whose affine coordinates $s_0, \cdots, s_n$ have the property that $s_0 \ge s_1 \ge \cdots \ge s_n$. Then the numbers

$$t_i = (i+1)(s_i - s_{i+1}) \qquad (0 \le i < n),$$
$$t_n = (n+1)s_n$$

are non-negative and have the property that $p = \Sigma_{i=0}^{n} t_i \cdot b\tau_i$ and $\Sigma_{i=0}^{n} t_i = \Sigma_{i=0}^{n} s_i = 1$. The t_i are consequently the affine coordinates of p in $|\beta|$ (theorem 3.2.1). This proves (2).

For (3), observe that if for p above we have that if $s_i = s_j$ for certain $i < j \le n$ then all t_k for $i < k \le j$ are equal to 0. □

3.2.7. LEMMA: *Let σ be a simplex and let $\mathcal{K}_1$ and $\mathcal{K}_2$ be chains of faces of σ. Let β_1 and β_2 denote the sets of barycenters of elements of $\mathcal{K}_1$ and $\mathcal{K}_2$, respectively. Then*

$$|\beta_1| \cap |\beta_2| = |\beta_1 \cap \beta_2|.$$

PROOF: Since $\beta_1 \cap \beta_2 \subseteq |\beta_1| \cap |\beta_2|$ and $|\beta_1 \cap \beta_2| = \mathrm{conv}(\beta_1 \cap \beta_2)$ it is clear that

$$|\beta_1 \cap \beta_2| \subseteq |\beta_1| \cap |\beta_2|.$$

For the reverse inclusion, take an arbitrary $x \in |\beta_1| \cap |\beta_2|$. Since $(\mathcal{K}_1 \setminus \mathcal{K}_2) \cap (\mathcal{K}_2 \setminus \mathcal{K}_1) = \emptyset$, without loss of generality we may assume that $\sigma \notin \mathcal{K}_1 \setminus \mathcal{K}_2$. It is clear that there exists a total order $\preccurlyeq_1$ on σ such that every element of $\mathcal{K}_1$ is an initial $\preccurlyeq_1$-segment. This order corresponds to a maximal chain of faces of σ. Let $\tau \in \mathcal{K}_1 \setminus \mathcal{K}_2$; observe that $\tau \neq \sigma$. The collection $\mathcal{K}_2$ can easily be extended to a maximal chain $\mathcal{L}$ of simplexes such that $\tau \notin \mathcal{L}$. Consequently, τ is not an initial segment with respect to the corresponding total order $\preccurlyeq_2$ (lemma 3.2.6(1)), i.e. there exist distinct $v,w \in \sigma$ such that $v \in \tau$, $w \notin \tau$ and $w \preccurlyeq_2 v$; consequently, $\alpha_w(x) \geq \alpha_v(x)$ (lemma 3.2.6(2)). Observe that $v \preccurlyeq_1 w$ from which it follows that $\alpha_v(x) \geq \alpha_w(x)$ (lemma 3.2.6(2)). By lemma 3.2.6(3) we therefore conclude that the affine coordinate of x in $|\beta_1|$ with respect to b_τ is equal to 0. □

Exercises for §3.2.

1. Prove that if $|\sigma|$ is a geometric n-simplex in $\mathbb{R}^n$ then $\mathrm{Int}(|\sigma|)$ is equal to the geometric interior of $|\sigma|$.

2. Let σ be a simplex and let $p \in |\sigma|$. Prove that $|\sigma| \setminus \{p\}$ is convex if and only if $p \in \sigma$.

3. Let σ and τ be simplexes. Prove that $|\sigma| = |\tau|$ if and only if $\sigma = \tau$.

4. Let σ be a simplex and let C be a convex set which is contained in $\partial|\sigma|$. Prove that C is contained in a proper face of $|\sigma|$.

5. Let $\sigma = \{v_0,\cdots,v_k\}$ be a simplex. Prove that the geometric interior of $|\sigma|$ equals $\{x \in |\sigma|: (\forall i \leq n)(\alpha_{v_i}(x) > 0)\}$.

6. Let $\sigma = \{v_0,\cdots,v_k\}$ be a simplex. Prove that the geometric boundary of $|\sigma|$ equals $\{x \in |\sigma|: (\exists i \leq n)(\alpha_{v_i}(x) = 0)\}$.

3.3. Triangulation

A *simplicial complex* is a *countable* collection $\mathcal{S}$ of non-empty, finite sets such that:

(SC)$_{\text{abstract}}$ if $\sigma \in \mathcal{S}$ and $\emptyset \neq \tau \subseteq \sigma$ then $\tau \in \mathcal{S}$,

The elements of the set $S = \bigcup \mathcal{S}$ are called the *vertices* of the simplicial complex.

Although S need not be a subset of a normed linear space, without loss of generality we may assume that this is the case. This can be seen as follows. For every $n \in \mathbb{N}$ let $x_n \in l^2$ be the vector all whose coordinates are 0 except for the n-th coordinate which equals 1. Observe that the set $D = \{x_n: n \in \mathbb{N}\}$ is linearly independent. Let $f: S \to D$ be a bijection and identify every $\sigma \in \mathcal{S}$ with $f(\sigma) \subseteq l^2$. Observe that S, now regarded to be a subset of l^2, is geometrically independent. By theorem 3.2.2 it therefore follows that for every $\sigma_1, \sigma_2 \in \mathcal{S}$,

$$(*) \qquad |\sigma_1| \cap |\sigma_2| = |\sigma_1 \cap \sigma_2|.$$

In the sequel, when dealing with a simplicial complex $\mathcal{S}$, we shall always require that $S = \cup\mathcal{S}$ is contained in a normed linear space L, and that (*) holds. The set $|\mathcal{S}| = \cup\{|\sigma|: \sigma \in \mathcal{S}\}$ is called the *geometric realization* of $\mathcal{S}$ and $|\mathcal{S}|$ is said to be *triangulated* by $\mathcal{S}$ (or $\mathcal{S}$ is a *triangulation* of $|\mathcal{S}|$). Observe that the collection $\mathcal{S}_{||} = \{|\sigma|: \sigma \in \mathcal{S}\}$ consists of geometric simplexes in L and has the following properties:

$(SC)_{geometric}$ (1) if $|\sigma| \in \mathcal{S}_{||}$ and $|\tau|$ is a face of $|\sigma|$ then $|\tau| \in \mathcal{S}_{||}$,

(2) for all $|\sigma_1|, |\sigma_2| \in \mathcal{S}_{||}$ with $|\sigma_1| \cap |\sigma_2| \neq \emptyset$, $|\sigma_1| \cap |\sigma_2|$ is a face of $|\sigma_1|$ as well as $|\sigma_2|$.

Let $|\sigma|$ be a geometric simplex in L. The collection of all faces of $|\sigma|$ is clearly a simplicial complex which triangulates $|\sigma|$. It is called the *standard triangulation* of $|\sigma|$ and is denoted by $\mathcal{F}(\sigma)$.

3.3.1. LEMMA: *Let $\mathcal{S}$ be a simplicial complex. Then:*

if $\sigma, \sigma' \in \mathcal{S}$ and $b_\sigma \in |\sigma'|$ then σ is a face of σ'.

PROOF: By (*),

$$|\sigma| \cap |\sigma'| = |\sigma \cap \sigma'|$$

is a face of $|\sigma|$. This face contains b_σ so it must be $|\sigma|$ itself. □

Let X be a subset of L which is triangulated by $\mathcal{S}$. A triangulation $\mathcal{T}$ of X is called a *subdivision* of $\mathcal{S}$ if for every simplex $\sigma \in \mathcal{S}$ we have that the collection

$$\mathcal{T}(\sigma) = \{\tau \in \mathcal{T}: \tau \subseteq |\sigma|\}$$

is finite and is a triangulation of $|\sigma|$.

3.3.2. THEOREM: *Let $\mathcal{S}$ and $\mathcal{T}$ be triangulations of* X. *If $\mathcal{T}$ is a subdivision of $\mathcal{S}$ then for every $\tau \in \mathcal{T}$ there exists $\sigma \in \mathcal{S}$ such that $\tau \subseteq |\sigma|$.*

PROOF: There is a simplex $\sigma \in \mathcal{S}$ such that $b_\tau \in |\sigma|$. Take an element $\tau' \in \mathcal{T}(\sigma)$ such that $b_\tau \in |\tau'|$. Then by lemma 3.3.1 it follows that $\tau \subseteq \tau'$ and consequently, $\tau \subseteq |\sigma|$. □

3.3.3. THEOREM: *Let $\mathcal{S}$, $\mathcal{T}$ and $\mathcal{R}$ be simplicial complexes. If $\mathcal{S}$ is a subdivision of $\mathcal{T}$ and if $\mathcal{T}$ is a subdivision of $\mathcal{R}$, then $\mathcal{S}$ is a subdivision of $\mathcal{R}$.*

PROOF: We have to prove that $\mathcal{S}(\rho)$ is a finite triangulation of $|\rho|$ for every $\rho \in \mathcal{R}$. That $\mathcal{S}(\rho)$ is a simplicial complex (condition $(SC)_{abstract}$) is clear, and condition (*) holds for all simplexes in $\mathcal{S}$. Finally, $\mathcal{S}(\rho)$ is finite and $\bigcup_{\sigma\in\mathcal{S}(\rho)}|\sigma|$ is equal to $|\rho|$; this follows easily from theorem 3.3.2 and the fact that $\bigcup_{\tau\in\mathcal{T}(\rho)}|\tau| = |\rho|$ and $\bigcup_{\sigma\in\mathcal{S}(\tau)}|\sigma| = |\tau|$ for $\tau \in \mathcal{T}(\mathcal{S})$. □

Let $|\sigma|$ be a geometric simplex. We shall now define a special triangulation of $|\sigma|$, the so-called *barycentric triangulation*. Let $B(\sigma)$ denote the set of all barycenters of faces of σ. We shall define a simplicial complex $\mathcal{B}(\sigma)$ consisting of subsets of $B(\sigma)$ as follows: a finite nonempty subset β of $B(\sigma)$ belongs to $\mathcal{B}(\sigma)$ if and only if the faces of σ the barycenters of which belong to β form a chain. Observe that by lemma 3.2.5 the elements of $\mathcal{B}(\sigma)$ are geometrically independent. $\mathcal{B}(\sigma)$ clearly satisfies condition $(SC)_{abstract}$. In addition, by lemma 3.2.7 it follows that the collection $|\mathcal{B}(\sigma)| = \{|\beta|: \beta \in \mathcal{B}(\sigma)\}$ satisfies condition (*). It remains to prove that

$$\bigcup\{|\beta|: \beta \in \mathcal{B}(\sigma)\} = |\sigma|.$$

Let $x \in |\sigma|$. Without loss of generality we have $\sigma = \{v_0,\cdots,v_n\}$ and

$$\alpha_{v_n}(x) \le \alpha_{v_{n-1}}(x) \le \cdots \le \alpha_{v_0}(x).$$

This ordering corresponds to a (maximal) chain of faces in σ, which in turn corresponds to a (maximal) simplex β in $\mathcal{B}(\sigma)$. By lemma 3.2.6(2), $x \in |\beta|$.

Let X be a set which is triangulated by the simplicial complex $\mathcal{S}$. As above, for every $\sigma \in \mathcal{S}$ let $\mathcal{B}(\sigma)$ be the barycentric triangulation of $|\sigma|$. The collection

$$\mathcal{B}(\mathcal{S}) = \bigcup\{\mathcal{B}(\sigma): \sigma \in \mathcal{S}\}$$

is called the *barycentric subdivision of $\mathcal{S}$*. That $\mathcal{B}(\mathcal{S})$ satisfies condition $(SC)_{abstract}$ is clear. We claim that it is a subdivision of $\mathcal{S}$. For that we only need to verify condition (*) because for every

$\sigma \in \mathcal{S}$, $\{\tau \in \mathcal{B}(\mathcal{S}): \tau \subseteq |\sigma|\} = \mathcal{B}(\sigma)$ and $\mathcal{B}(\sigma)$ is a finite triangulation of $|\sigma|$. Let $\beta_i \in \mathcal{B}(\sigma_i)$ for $\sigma_i \in \mathcal{S}$, $i = 1,2$. If $\sigma_1 \cap \sigma_2 = \emptyset$ then by condition (*), $|\sigma_1| \cap |\sigma_2| = \emptyset$ so then there is nothing to prove. If $\sigma_1 \cap \sigma_2 \neq \emptyset$ then for $\sigma = \sigma_1 \cap \sigma_2$, $|\sigma_1| \cap |\sigma_2| = |\sigma|$ and $|\sigma|$ is a face of $|\sigma_1|$ as well as $|\sigma_2|$. Now put $\gamma_i = \beta_i \cap |\sigma|$ for $i = 1,2$. Then γ_1 and γ_2 belong to $\mathcal{B}(\sigma)$ and we claim that

$$|\gamma_i| = |\beta_i| \cap |\sigma| \qquad (i = 1,2).$$

That $|\gamma_i| \subseteq |\beta_i| \cap |\sigma|$ is clear. Take an arbitrary $x \in |\beta_i| \cap |\sigma|$. There exists an element $\beta \in \mathcal{B}(\sigma)$ such that $x \in |\beta|$. Observe that $\beta \in \mathcal{B}(\sigma_1) \cap \mathcal{B}(\sigma_2)$. Consequently,

$$x \in |\beta_i| \cap |\beta| = |\beta_i \cap \beta| \subseteq |\beta_i \cap |\sigma|| = |\gamma_i|.$$

By lemma 3.2.7 we therefore conclude that

$$|\beta_1| \cap |\beta_2| = |\gamma_1| \cap |\gamma_2| = |\gamma_1 \cap \gamma_2| = |\beta_1 \cap \beta_2|,$$

which is as required.

Sometimes it will be convenient to denote $\mathcal{S}$ by $\mathrm{sd}^{(0)}\mathcal{S}$ and $\mathcal{B}(\mathcal{S})$ by $\mathrm{sd}^{(1)}\mathcal{S}$. We define the *second* barycentric subdivision $\mathrm{sd}^{(2)}\mathcal{S}$ of $\mathcal{S}$ by $\mathcal{B}(\mathcal{B}(\mathcal{S}))$. Similarly, one defines $\mathrm{sd}^{(n)}\mathcal{S}$, the n-th barycentric subdivision of $\mathcal{S}$, for every $n \in \mathbb{N}$. Notice that by theorem 3.3.3, $\mathrm{sd}^{(n)}\mathcal{S}$ is a subdivision of $\mathcal{S}$ for every $n \in \mathbb{N}$.

Let $(L,\|\cdot\|)$ be a normed linear space and let $\mathcal{S}$ be a triangulation of a subset of L. The *mesh*, $\mathrm{mesh}(\mathcal{S})$, of $\mathcal{S}$ is defined to be the number

$$\sup_{\sigma \in \mathcal{S}} \mathrm{diam}(\sigma).$$

We allow $\mathrm{mesh}(\mathcal{S})$ to be equal to ∞.

3.3.4. THEOREM: *Let σ be an n-simplex in the normed linear space $(L,\|\cdot\|)$ and let $\mathcal{B}(\sigma)$ be the barycentric triangulation of $|\sigma|$. Then*

$$\mathrm{mesh}(\mathcal{B}(\sigma)) \le \frac{n}{n+1} \cdot \mathrm{diam}(\sigma).$$

PROOF: Let $\beta \in \mathcal{B}(\sigma)$ and consider two barycenters $c,d \in \beta$ of faces ξ and δ of σ, respectively. Without loss of generality, assume that $\xi \subseteq \delta$, that ξ consists of k+1 points and that δ consists of m+1 points. Choose an enumeration $\{v_0,v_1,\cdots,v_n\}$ of σ such that $\xi = \{v_0,v_1,\cdots,v_k\}$ and $\delta = \{v_0,v_1,\cdots,v_m\}$. Then

$$\begin{aligned}
\|d - c\| &= \|\sum_{i=0}^{m} \frac{v_i}{m+1} - \sum_{j=0}^{k} \frac{v_j}{k+1}\| \\
&= \frac{1}{k+1} \|\sum_{i=0}^{m} \frac{k+1}{m+1} v_i - \sum_{j=0}^{k} v_j\| \\
&= \frac{1}{k+1} \|\sum_{j=0}^{k} (\frac{1}{m+1} \sum_{i=0}^{m} v_i - v_j)\| \\
&\le \frac{1}{k+1} \sum_{j=0}^{k} \|\frac{1}{m+1} \sum_{i=0}^{m} v_i - v_j\| \\
&\le \max_{0\le j\le k} \|\frac{1}{m+1} \sum_{i=0}^{m} v_i - v_j\|.
\end{aligned}$$

Moreover,

$$\begin{aligned}
\|\frac{1}{m+1} \sum_{i=0}^{m} v_i - v_j\| &= \frac{1}{m+1} \|\sum_{i=0}^{m} (v_i - v_j)\| \\
&\le \frac{1}{m+1} \sum_{i=0}^{m} \|v_i - v_j\| \\
&\le \frac{m}{m+1} \cdot \mathrm{diam}(\sigma).
\end{aligned}$$

In the last inequality we used that one of the terms (if $i = j$) is equal to 0. We conclude that

$$\mathrm{diam}(\beta) \le \frac{m}{m+1} \cdot \mathrm{diam}(\sigma).$$

Finally observe that $m \le n$. □

3.3.5. COROLLARY: *Let $\mathcal{S}$ be a triangulation of a set* X *in a normed linear space* $(L,\|\cdot\|)$ *such that every $\sigma \in \mathcal{S}$ consists of at most* $n+1$ *points. If $\mathcal{B}$ is the barycentric subdivision of $\mathcal{S}$ then*

$$\mathrm{mesh}(\mathcal{B}) \le \frac{n}{n+1} \cdot \mathrm{mesh}(\mathcal{S}).\ \square$$

3.3.6. COROLLARY: *Let $|\tau|$ be a geometric simplex in a normed linear space* $(L,\|\cdot\|)$. *Then for every* $\varepsilon > 0$ *there is an* $m \in \mathbb{N}$ *such that* $\mathrm{mesh}(\mathrm{sd}^{(m)}\mathcal{F}(\tau)) < \varepsilon$.

PROOF: Let τ be an n-simplex. By corollary 3.3.5 it follows that for every $m \in \mathbb{N}$ we have

$$\mathrm{mesh}(\mathrm{sd}^{(m)}\mathcal{F}(\tau)) \le \left(\frac{n}{n+1}\right)^{m} \cdot \mathrm{diam}(\tau).$$

We conclude that for a sufficiently large m, $\mathrm{mesh}(\mathrm{sd}^{(m)}\mathcal{F}(\tau)) < \varepsilon$. □

Exercises for §3.3.

1. Let X be a set triangulated by the simplicial complex $\mathcal{S}$. Prove that the collection of all geometric interiors of elements of $|\mathcal{S}|$ is a partition of X.

2. Let $\mathcal{U}$ be a countable collection of sets. Prove that the collection $\{\mathcal{F}\colon \mathcal{F} \subseteq \mathcal{U}$ is finite and $\bigcap\mathcal{F} \neq \emptyset\}$ is a simplicial complex.

3.4. Simplexes in $\mathbb{R}^n$

In this section we shall formulate and prove some fundamental properties of simplexes in $\mathbb{R}^n$.

3.4.1. LEMMA: *Let σ^1 and σ^2 be two n-simplexes in $\mathbb{R}^n$ whose corresponding geometric simplexes intersect in a common (n-1)-face $|\tau|$. Then $|\sigma^1| \cup |\sigma^2|$ is a neighborhood of the barycenter of τ.*

PROOF: Observe that the lemma is trivial if $n = 1$; therefore assume that $n > 1$. Let $\sigma^1 = \{a_0,a_1,\cdots,a_{n-1},a_n^1\}$ and $\sigma^2 = \{a_0,a_1,\cdots,a_{n-1},a_n^2\}$, where

$$|\sigma^1| \cap |\sigma^2| = |\{a_0,a_1,\cdots,a_{n-1}\}| = |\tau|.$$

The points $a_0,a_1,\cdots,a_{n-1},a_n^1$ are geometrically independent, from which it follows by theorem 3.1.3 that the points $a_1-a_0, a_2-a_0,\cdots, a_n^1-a_0$ are linearly independent. Now consider the standard basis $e_1,e_2,\cdots,e_n$ in $\mathbb{R}^n$ and a linear isomorphism $g^1\colon \mathbb{R}^n \to \mathbb{R}^n$ with the properties:

$$\begin{cases} g^1(e_i) = a_i - a_0 & (i < n), \\ g^1(e_n) = a_n^1 - a_0. \end{cases}$$

This isomorphism followed by the translation $x \to x+a_0$ is an affine isomorphism (theorem 3.1.6) which we denote by f^1. Observe that f^1 has the following properties:

$$f^1(\underline{0}) = a_0,\ f^1(e_i) = a_i\ (1 \le i \le n-1),\ f^1(e_n) = a_n^1.$$

In the standard basis we now change e_n into $-e_n$. By a similar argumentation as above we obtain an affine isomorphism $f^2\colon \mathbb{R}^n \to \mathbb{R}^n$ such that

$f^2(\underline{0}) = a_0$, $f^2(e_i) = a_i$ $(1 \le i \le n-1)$, $f^2(-e_n) = a_n^2$.

Observe that f^1 and f^2 agree on the plane $P = \{x \in \mathbb{R}^n\colon x_n = 0\}$. Also observe that $f^1 \mid P = f^2 \mid P$ is an affine isomorphism between P and the hyperplane spanned by the elements of τ. Now define a homeomorphism $f\colon \mathbb{R}^n \to \mathbb{R}^n$ as follows

$$f(x) = \begin{cases} f^1(x) & (x_n \ge 0), \\ f^2(x) & (x_n \le 0). \end{cases}$$

Let b be the barycenter of τ and let $c = (\frac{1}{n}, \frac{1}{n}, \cdots, \frac{1}{n}, 0)$. Then $f(c) = b$ and the set

$$B(c, \tfrac{1}{n^2})$$

is a neighborhood of c which is mapped by f into $|\sigma^1| \cup |\sigma^2|$. We conclude that $|\sigma^1| \cup |\sigma^2|$ is a neighborhood of b. □

The next result is the key in the proof of the Brouwer Fixed-Point Theorem.

3.4.2. THEOREM: *Let τ be an n-simplex in $\mathbb{R}^n$, and let $\mathcal{S}$ be a finite triangulation of $|\tau|$ which subdivides its standard triangulation. Finally, let $\mu \in \mathcal{S}$ be an (n-1)-simplex and let b be the barycenter of $|\mu|$.*

(1) *If $b \in \partial|\tau|$ then $|\mu| \subseteq \partial|\tau|$ and μ is a face of precisely one n-simplex in $\mathcal{S}$.*

(2) *If $b \notin \partial|\tau|$ then μ is a face of precisely two n-simplexes in $\mathcal{S}$.*

PROOF: Assume that $b \in \partial|\tau|$. We claim that this implies that $|\mu| \subseteq \partial|\tau|$. Since $\mathcal{S}$ subdivides the standard triangulation of $|\tau|$ there is a simplex $\sigma \in \mathcal{S}$ with $b \in |\sigma|$ such that σ is contained in a proper face of τ, i.e. $|\sigma| \subseteq \partial|\tau|$. By lemma 3.3.1 we obtain that μ is a face of σ so that $|\mu| \subseteq \partial|\tau|$.

Again assume that $b \in \partial|\tau|$. If there exist two n-simplexes in $\mathcal{S}$ having μ as a common face then the union of these two simplexes is a neighborhood of b (lemma 3.4.1). But this contradicts the fact that $b \in \partial|\tau|$ which is equal to the *topological* boundary of $|\tau|$ (exercise 3.2.1).

Next assume that $b \notin \partial|\tau|$. Assume that there exist distinct n-simplexes σ_1, σ_2 and σ_3 in $\mathcal{S}$ having μ as a common face. Since $\mathcal{S}$ is a simplicial complex,

$$(|\sigma_1| \cup |\sigma_2|) \cap (|\sigma_2| \cup |\sigma_3|) \cap (|\sigma_3| \cup |\sigma_1|)$$

$$= |\sigma_1 \cap \sigma_2| \cup |\sigma_2 \cap \sigma_3| \cup |\sigma_3 \cap \sigma_1|$$

$$= |\mu|.$$

However, by lemma 3.4.1 this intersection is also a neighborhood of b in $\mathbb{R}^n$. Consequently, $|\mu|$ is a neighborhood of b in $\mathbb{R}^n$ which is impossible because $|\mu|$ is contained in an (n-1)-dimensional hyperplane.

For the remaining part of the proof define

$$U = |\tau| \setminus \bigcup\{|\sigma|: \sigma \in \mathcal{S};\ b \notin |\sigma|\}.$$

Observe that U is a nonempty relatively open subset of $|\tau|$.

Assume that there does not exist an n-simplex in $\mathcal{S}$ having μ as a face. According to lemma 3.3.1 this implies that $|\mu|$ is the only geometric simplex in $\mathcal{S}$ containing b which implies that

$$U \subseteq |\mu|.$$

Now if $b \in \partial|\tau|$ then we arrived at a contradiction because then $U \subseteq |\mu| \subseteq \partial|\tau|$ and $\partial|\tau|$ has empty interior in $|\tau|$. If $b \notin \partial|\tau|$ then $U \cap \mathrm{Int}(|\tau|)$ is a neighborhood of b in $\mathbb{R}^n$ which is contained in $|\mu|$. But this contradicts the fact that $|\mu|$ is contained in an (n-1)-dimensional hyperplane of $\mathbb{R}^n$.

Finally, assume that $b \notin \partial|\tau|$ and that there is precisely one n-simplex $\sigma \in \mathcal{S}$ having μ as a face. Then $U \subseteq |\sigma|$ which implies that $\mathrm{Int}(|\tau|) \cap U$ is a neighborhood of b in $\mathbb{R}^n$ which is contained in $|\sigma|$. Since μ is a proper face of σ we also have $b \in \partial|\sigma|$. This is a contradiction. □

Let $\tau = \{x_0,\cdots,x_k\}$ be an arbitrary k-simplex in $\mathbb{R}^n$. An m-*Sperner map* for τ is a function $h: \bigcup \mathrm{sd}^{(m)}\mathcal{F}(\tau) \to \{0,\cdots,k\}$ such that

$$\text{if } \{i(0),\cdots,i(l)\} \subseteq \{0,\cdots,k\} \text{ and } v \in |\{x_{i(0)},\cdots,x_{i(l)}\}| \text{ then } h(v) \in \{i(0),\cdots,i(l)\}.$$

We call a k-dimensional simplex σ in $\mathrm{sd}^{(m)}\mathcal{F}(\tau)$ *full* if $h(\sigma) = \{0,\cdots,k\}$.

There are always full simplexes.

3.4.3. THEOREM ("Sperner's Lemma"): *Let* τ *be a* k*-simplex in* $\mathbb{R}^k$ *and let* h *be an* m*-Sperner map for* τ*. The number of full simplexes in* $\mathrm{sd}^{(m)}\mathcal{F}(\tau)$ *is odd and hence non-zero.*

PROOF: Let $\tau = \{x_0,\cdots,x_k\}$. We shall prove the theorem by induction on k. If k = 0 then $\tau = \{x_0\}$, and there is one full simplex. Assume that the theorem is true for all (k-1)-simplexes, $k-1 \geq 0$. We shall prove the theorem for τ. Put

$$\tau^* = \{x_0,\cdots,x_{k-1}\},$$
$$\mathcal{P}_1 = \{\sigma \in \mathrm{sd}^{(m)}\mathcal{F}(\tau): \sigma \text{ is (k-1)-dimensional and } h(\sigma) = \{0,\cdots,k-1\}\},$$

$\mathcal{P}_2 = \{\sigma \in \mathrm{sd}^{(m)}\mathcal{F}(\tau)$: σ is k-dimensional and $\{0,\cdots,k-1\} \subseteq h(\sigma)\}$,

$\mathcal{P}_1(0) = \{\sigma \in \mathcal{P}_1: \sigma \subseteq \tau^*\}$,

$\mathcal{P}_1(1) = \mathcal{P}_1 \backslash \mathcal{P}_1(0)$,

$\mathcal{P}_2(0) = \{\sigma \in \mathcal{P}_2: h(\sigma) = \{0,\cdots,k\}\}$, and

$\mathcal{P}_2(1) = \mathcal{P}_2 \backslash \mathcal{P}_2(0)$,

respectively. Observe that $\mathrm{sd}^{(m)}\mathcal{F}(\tau^*) \subseteq \mathrm{sd}^{(m)}\mathcal{F}(\tau)$. By the definition of an m-Sperner map it follows therefore that the restriction of h to $\bigcup \mathrm{sd}^{(m)}\mathcal{F}(\tau^*)$ is an m-Sperner map for τ^*. Consequently, $\mathcal{P}_1(0)$ is the set of full simplexes in the m-th barycentric subdivision of τ^*, so $|\mathcal{P}_1(0)|$ is odd by our inductive assumptions. Also, $\mathcal{P}_2(0)$ is the set of full simplexes in the m-th barycentric subdivision of τ; so we have to prove that $|\mathcal{P}_2(0)|$ is an odd number.

Consider $\mathcal{R} = \{(\kappa,\mu) \in \mathcal{P}_1 \times \mathcal{P}_2: \kappa$ is a face of $\mu\}$. We compute the cardinality of $\mathcal{R}$ twice. For each $\kappa \in \mathcal{P}_1$ put $\mathcal{R}[\kappa] = \{\mu \in \mathcal{P}_2: (\kappa,\mu) \in \mathcal{R}\}$.

CLAIM 1: $|\mathcal{R}| = |\mathcal{P}_1(0)| + 2 \cdot |\mathcal{P}_1(1)|$.

Clearly $|\mathcal{R}| = \sum_{\kappa \in \mathcal{P}_1} |\mathcal{R}[\kappa]|$. If $\kappa \in \mathcal{P}_1(0)$ then $|\mathcal{R}[\kappa]| = 1$. For such κ is a face of precisely one k-dimensional simplex in $\mathrm{sd}^{(m)}\mathcal{F}(\tau)$ by theorem 3.4.2, and this simplex clearly belongs to $\mathcal{P}_2$. If $\kappa \in \mathcal{P}_1(1)$ then $|\mathcal{R}[\kappa]| = 2$. If $\kappa \subseteq \{x_{i(0)},\cdots,x_{i(k-1)}\}$ then $h(\kappa) \subseteq \{i(0),\cdots,i(k-1)\}$ which implies that $\{i(0),\cdots,i(k-1)\} = \{0,\cdots,k-1\}$, contradiction. So such a κ is not contained in a (k-1)-dimensional face of τ, and therefore, again by theorem 3.4.2, is a k-dimensional face of precisely two k-simplexes in $\mathrm{sd}^{(m)}\mathcal{F}(\tau)$, and these simplexes clearly belong to $\mathcal{P}_2$.

For each $\mu \in \mathcal{P}_2$ put $\mathcal{R}^{-1}[\mu] = \{\kappa \in \mathcal{P}_1: (\kappa,\mu) \in \mathcal{R}\}$.

CLAIM 2: $|\mathcal{R}| = |\mathcal{P}_2(0)| + 2 \cdot |\mathcal{P}_2(1)|$.

Clearly $|\mathcal{R}| = \sum_{\mu \in \mathcal{P}_2} |\mathcal{R}^{-1}[\mu]|$. Take $\mu \in \mathcal{P}_2(0)$. We claim that $|\mathcal{R}^{-1}[\mu]| = 1$. Indeed, $h(\mu) = \{0,\cdots,k\}$, i.e h is one-to-one on μ. So μ has exactly one (k-1)-dimensional face κ such that $\{0,\cdots,k-1\} = h(\kappa)$ and this face is clearly the only element of $|\mathcal{R}^{-1}[\mu]|$. If $\mu \in \mathcal{P}_2(1)$ then $h(\mu) = \{0,\cdots,k-1\}$. Since $V(\mu)$ has cardinality k+1, there exist precisely two subsets E and F of $V(\mu)$ of cardinality k with the property that $h(E) = \{0,\cdots,k-1\} = h(F)$. Consequently, there exist precisely two k-dimensional faces κ_1 and κ_2 of μ with $h(\kappa_1) = \{0,\cdots,k-1\} = h(\kappa_2)$. These simplexes are obviously the only elements of $|\mathcal{R}^{-1}[\mu]|$.

We find that $|\mathcal{P}_2(0)| - |\mathcal{P}_1(0)| = 2(|\mathcal{P}_2(1)| - |\mathcal{P}_1(1)|)$ is even. As was remarked at the beginning of the proof, $|\mathcal{P}_1(0)|$ is odd. Consequently, so is $|\mathcal{P}_2(0)|$. □

3.5. The Brouwer Fixed-Point Theorem

In this section we shall present a proof of the Brouwer Fixed-Point Theorem and some of its applications.

We begin with a lemma.

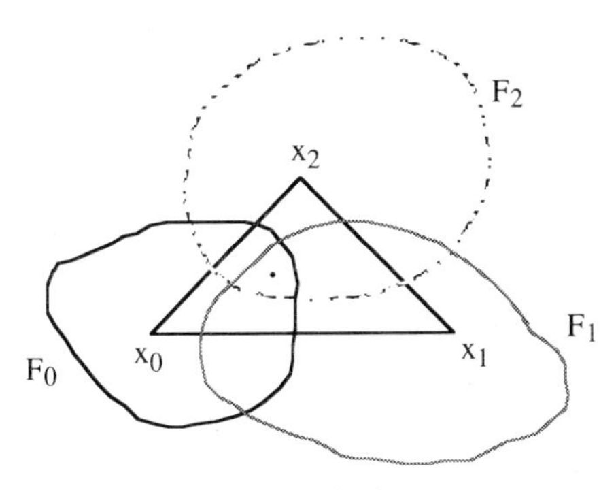

Figure 3.5.1.

3.5.1. LEMMA: *Let* $\tau = \{x_0,\cdots,x_k\}$ *be a* k*-simplex in* $\mathbb{R}^k$. Let $\{F_i\colon 0 \le i \le k\}$ *be a collection of closed subsets of* $\mathbb{R}^n$ *such that whenever* $\{i(0),\cdots,i(l)\} \subseteq \{0,\cdots,k\}$ *we have* $|\{x_{i(0)},\cdots,x_{i(l)}\}| \subseteq F_{i(0)} \cup \cdots \cup F_{i(l)}$. *Then* $\bigcap_{i=0}^{k} F_i \neq \emptyset$.

PROOF: If not, then $|\tau| = |\tau|\backslash F_0 \cup \cdots \cup |\tau|\backslash F_k$. Let ε be a Lebesgue number for this covering (use that τ is compact and apply lemma 1.1.1). By corollary 3.3.6 we can choose m so big that $\text{mesh}(\text{sd}^{(m)}\mathcal{F}(\tau)) < \varepsilon$. Define $V = \bigcup \text{sd}^{(m)}\mathcal{F}(\tau)$. Let $v \in V$ and put $I_v = \{0 \le i \le k\colon \alpha_i(v) > 0\}$ (here $\alpha_0(v),\cdots,\alpha_k(v)$ are the affine coordinates of v with respect to $x_0,\cdots,x_k$ of course). Let $I_v = \{i(0),\cdots,i(l)\}$. Then

$$v \in |\{x_{i(0)},\cdots,x_{i(l)}\}| \subseteq F_{i(0)} \cup \cdots \cup F_{i(l)}.$$

So for $v \in V$ we can pick $h(v) \in \{0,\cdots,k\}$ such that

$$\alpha_{h(v)}(v) > 0 \text{ and } v \in F_{h(v)}.$$

Then h is an m-Sperner map for τ since if $v \in |\{x_{j(0)},\cdots,x_{j(m)}\}|$ then for $i \notin \{j(0),\cdots,j(m)\}$ we have $\alpha_i(v) = 0$ so $i \neq h(v)$ and consequently, $h(v) \in \{j(0),\cdots,j(m)\}$. By theorem 3.4.3 there exists a full simplex $\sigma \in \text{sd}^{(m)}\mathcal{F}(\tau)$ for h, say $\sigma = \{v_0,\cdots,v_k\}$ with $h(v_i) = i$. Then $v_i \in \sigma \cap F_i$ for every $0 \le i \le k$. But $\text{diam}(\sigma) < \varepsilon$ so there exists i with $\sigma \subseteq |\tau|\backslash F_i$, i.e. $\sigma \cap F_i = \emptyset$. This is a contradiction.□

We now come to the main result of this section.

3.5.2. THEOREM ("Brouwer Fixed-Point Theorem"): *Let* $f\colon I^n \to I^n$ *be continuous. Then* f *has a fixed-point.*

PROOF: Let $\tau = |\{x_0,\cdots,x_n\}|$ be a geometric n-simplex in $\mathbb{R}^n$. It is geometrically obvious that τ is homeomorphic to I^n (for a proof see exercise 3.5.8(d)). So it suffices to prove the theorem for τ. To this end, let $f\colon \tau \to \tau$ be continuous. For $i = 0,1,\cdots,n$ let

$$F_i = \{x \in \tau\colon \alpha_i(f(x)) \leq \alpha_i(x)\}$$

(here $\alpha_0(x),\cdots,\alpha_n(x)$ are the affine coordinates of x with respect to $x_0,\cdots,x_n$ of course).

Let $|\{x_{i(0)},\cdots,x_{i(l)}\}|$ be a face of τ and $x \in |\{x_{i(0)},\cdots,x_{i(l)}\}|$. Then for $i \notin \{i(0),\cdots,i(l)\}$,

$$\alpha_i(x) = 0$$

so that

$$\alpha_{i(0)}(x) + \cdots + \alpha_{i(l)}(x) = 1 \geq \alpha_{i(0)}(f(x)) + \cdots + \alpha_{i(l)}(f(x))$$

and hence there must be $0 \leq j \leq l$ with

$$\alpha_{i(j)}(x) \geq \alpha_{i(j)}(f(x)).$$

Consequently, $x \in F_{i(j)}$. We conclude that $|\{x_{i(0)},\cdots,x_{i(l)}\}| \subseteq F_{i(0)} \cup \cdots \cup F_{i(l)}$.

Each F_i is closed by continuity of f and the functions α_i (theorems 3.1.6 and 3.2.1). So by lemma 3.5.1 there exists $x \in \bigcap_{i=0}^{n} F_i$. Then for every $0 \leq i \leq n$,

$$0 \leq \alpha_i(f(x)) \leq \alpha_i(x),$$

and

$$\sum_{i=0}^{n} \alpha_i(x) = 1 = \sum_{i=0}^{n} \alpha_i(f(x)).$$

This implies $\alpha_i(x) = \alpha_i(f(x))$ for every $0 \leq i \leq n$ and consequently, $x = f(x)$. □

3.5.3. COROLLARY: *The Hilbert cube* Q *has the fixed-point property.*

PROOF: Let $f\colon Q \to Q$ be continuous. For every $n \in \mathbb{N}$ define

$$K_n = \{x \in Q\colon (x_1,\cdots,x_n) = (f(x)_1,\cdots,f(x)_n)\}.$$

It is clear that for every n the set K_n is closed in Q and that $K_{n+1} \subseteq K_n$. Fix $n \in \mathbb{N}$, let $p_n: Q \to J^n$ denote the projection and define a continuous function $f_n: J^n \to J^n$ by

$$f_n(x_1,\cdots,x_n) = (p_n \circ f)(x_1,\cdots,x_n,0,0,\cdots).$$

By theorem 3.5.2, f_n has a fixed-point, say $(x_1,\cdots,x_n)$, from which it follows that

$$(x_1,\cdots,x_n,0,0,\cdots) \in K_n.$$

We conclude that the K_n's form a decreasing collection of nonempty closed subsets of Q so that by compactness of Q,

$$K = \bigcap_{n=1}^{\infty} K_n \neq \emptyset.$$

It is clear that every point in K is a fixed-point of f.□

3.5.4. COROLLARY: *Let X be a compact* **ANR**. *If f: $X \to X$ is a continuous function which is homotopic to a constant function, then f has a fixed-point. In particular, every compact* **AR** *has the fixed-point property.*

PROOF: Since every space is homeomorphic to a subspace of Q (theorem 1.4.18) we may assume that X is a subspace of Q. Since every constant function $X \to X$ is clearly extendable to a constant function $Q \to X$, by corollary 1.6.4 it follows that f can be extended to a continuous function $\bar{f}: Q \to X \subseteq Q$. Since Q has the fixed-point property, corollary 3.5.3, $\bar{f}$ has a fixed-point, say x. Since $\bar{f}(Q) \subseteq X$, x belongs to X, and since $\bar{f}$ extends f, we conclude that x is a fixed-point of f.

Since by theorem 1.6.6 every compact **AR** is contractible, the second part of the corollary immediately follows from the first part and proposition 1.6.5.□

We finish this section by giving three applications of the techniques developed in this section.

Application 1: The No-Retraction Theorem (part 2).

For each $n \in \mathbb{N}$ let B^n be the unit ball in $\mathbb{R}^n$, i.e.

$$B^n = \{x \in \mathbb{R}^n: \|x\| \leq 1\}.$$

It is geometrically obvious that B^n is contained in J^n and that the function $r: J^n \to B^n$ defined by

$$r(x) = \begin{cases} \frac{x}{\|x\|} & (\|x\| \geq 1), \\ x & (\|x\| \leq 1), \end{cases}$$

is a retraction. Consequently, B^n has the fixed-point property by theorem 3.5.2 and exercise 1.5.5(2) (also, it is geometrically obvious that I^n and B^n are homeomorphic, cf. exercise 2.1.1).

We generalize corollary 2.1.10 as follows:

3.5.5. THEOREM ("No-Retraction Theorem"): *For every* $n \in \mathbb{N}$, S^{n-1} *is not a retract of* B^n.

PROOF: To the contrary, suppose that S^{n-1} is a retract of B^n. As was remarked above, B^n has the fixed-point property. Consequently, S^{n-1} has the fixed-point property by exercise 1.5.5(2). However, the antipodal mapping on S^{n-1} clearly demonstrates that S^{n-1} does not have the fixed-point property. Contradiction. □

As was announced in §1.6 we now obtain

3.5.6. COROLLARY: *For every* $n \in \mathbb{N}$, S^{n-1} *is not contractible.*

PROOF: Suppose that S^{n-1} is contractible. Then S^{n-1} is an **AR** by corollaries 1.5.6 and 1.6.7. Consequently, there exists a retraction $r: B^n \to S^{n-1}$, which contradicts theorem 3.5.5. □

Application 2: The Theorem on Partitions (part 2).

Consider J^n and for $i \leq n$ its faces

$$A_i = \{x \in J^n : x_i = 1\} \text{ and } B_i = \{x \in J^n : x_i = -1\}.$$

We now generalize theorem 2.1.13 as follows:

3.5.7. THEOREM: *If* C_i *is a partition between* A_i *and* B_i *for every* i *then* $\bigcap_{i=1}^n C_i \neq \emptyset$.

PROOF: The proof is precisely the same as the proof of theorem 2.1.13, so we will be brief.

To the contrary, assume that C_i is a partition between A_i and B_i for $i \leq n$ such that $\bigcap_{i=1}^n C_i = \emptyset$. Precisely such as in the proof of theorem 2.1.13 we can find for each $i \leq n$ a continuous function $\xi_i: J^n \to J$ such that

$$\xi_i(A_i) = \{-1\}, \xi_i(B_i) = \{1\} \text{ and } \xi_i^{-1}(0) = C_i.$$

Define f: $J^n \to J^n$ by $f(x) = (\xi_1(x),\cdots,\xi_n(x))$. Then f is continuous and does not take on the value $(0,\cdots,0)$. For every $x \in J^n\backslash\{(0,\cdots,0)\}$ the ray from $(0,\cdots,0)$ through x intersects the "boundary" $B = \bigcup_{i=1}^n A_i \cup \bigcup_{i=1}^n B_i$ of J^n in precisely one point, say r(x). The function r: $J^n\backslash\{(0,\cdots,0)\} \to B$ is easily seen to be continuous. The function $g = r \circ f$: $J^n \to B$ has the following properties:

$$g((-1,1)^n) \cap (-1,1)^n = \emptyset, \text{ and for every } i \le n,\ g(A_i) \subseteq B_i \text{ and } g(B_i) \subseteq A_i.$$

Therefore, g has no fixed-point, which contradicts theorem 3.5.2.□

Now consider the Hilbert cube Q and its faces

$$W_i(\theta) = \{x \in Q: x_i = \theta\} \qquad (i \in \mathbb{N},\ \theta \in \{-1,1\}).$$

3.5.8. COROLLARY: *If* C_i *is a partition between* $W_i(-1)$ *and* $W_i(1)$ *for every* i, *then* $\bigcap_{i=1}^\infty C_i \neq \emptyset$.

PROOF: For every i let C_i be a partition between $W_i(-1)$ and $W_i(1)$. For every m, Define f_m: $J^m \to Q$ by

$$f_m(x_1,\cdots,x_m) = (x_1,\cdots,x_m,0,0,\cdots).$$

Then f_m is clearly an imbedding. It is easily seen that $f_m^{-1}(C_i)$ is a partition between A_i and B_i for every $i \le m$. Consequently, by theorem 3.5.7, $\bigcap_{i=1}^m f_m^{-1}(C_i) \neq \emptyset$. By compactness of Q we therefore obtain

$$\bigcap_{i=1}^\infty C_i \neq \emptyset,$$

as desired.□

Application 3. The Non-Homogeneity Theorem (part 2).

We generalize theorem 2.1.14 as follows:

3.5.9. THEOREM: *Let* $n \in \mathbb{N}$. *Then*

(1) *if* $A \subseteq S^{n-1}$ *then* $B^n\backslash A$ *is contractible.*

(2) *if* $A \subseteq B^n\backslash S^{n-1}$ *is nonempty then* $B^n\backslash A$ *is not contractible.*

PROOF: (1) Define H: $(B^n\backslash A)\times I\to B^n\backslash A$ by $H(x,t) = (1-t)x$. It is easily seen that H is a contraction.

(2) Without loss of generality, $(0,\cdots,0)\in A$. Assume, to the contrary, that H: $(B^n\backslash A)\times I\to B^n\backslash A$ is a contraction. Define F: $S^{n-1}\times I\to S^{n-1}$ by

$$F(x,t) = \frac{H(x,t)}{|H(x,t)|}.$$

Then F contracts S^{n-1} to a point and therefore contradicts corollary 3.5.6.□

Since J^n and B^n are homeomorphic, cf. exercise 3.5.8, this yields

3.5.10. COROLLARY: *Let* $n\in\mathbb{N}$. *Then* J^n *is not homogeneous.*□

In our second application we saw that the Hilbert cube Q very much behaves like its finite-dimensional analogues, the spaces J^n. It is natural to wonder about corollary 3.5.10 in this respect. We shall see in chapter 6 that Q is homogeneous, thereby demonstrating a striking difference between the finite-dimensional and the infinite-dimensional "world".

Exercises for §3.5.

1. Prove that the "No-Retraction Theorem", the "Brouwer Fixed-Point Theorem" and the fact that no S^n is contractible are equivalent statements, in the sense that they are easily deduced from each other.

2. Let $B(Q) = \bigcup_{n=1}^{\infty} W_n(-1)\cup\bigcup_{n=1}^{\infty} W_n(1)$ be the "boundary" of Q and let $s = Q\backslash B(Q)$ be its "interior". Prove that both B(Q) and s are dense in Q and that s is topologically complete. In addition, prove that B(Q) and s are not homeomorphic.

3. Prove that Q is not a topological group (warning: as remarked above, Q is homogeneous).

4. Prove that each compact convex subset of a normed linear space has the fixed-point property.

5. Let C be a convex subset of a normed linear space L. Prove that for every $\varepsilon > 0$ there exists a map f_ε: $C\to C$ such that (1) $d(f_\varepsilon,1) < \varepsilon$, and (2) $f_\varepsilon(C)$ is contained in a finite-dimensional linear subspace of L.

6. Let X be a space and let f: $X\to X$ be a map with the following properties: (1) the closure of f(X) is compact, and (2) for each $\varepsilon > 0$ there exists $x\in X$ with $d(x,f(x)) < \varepsilon$. Prove that f has a fixed-point.

7. Let C be a convex (not assumed to be closed) subset of a normed linear space L. Let f: C → C be continuous such that f(C) has compact closure in C. Prove that f has a fixed-point (this is called the Schauder Fixed-Point Theorem)(Hint: use exercises 5 and 6 of this section).

8.(a) Let τ be an n-dimensional simplex, $n \geq 1$, in $\mathbb{R}^m$. Prove that τ is homeomorphic to an n-dimensional simplex σ in $\mathbb{R}^n$.

(b) Prove that each n-dimensional simplex, $n \geq 1$, in $\mathbb{R}^n$ has nonempty interior in $\mathbb{R}^n$.

(c) Prove that each compact convex subset C of $\mathbb{R}^n$ having nonempty interior in $\mathbb{R}^n$ is homeomorphic to the n-ball B^n by a homeomorphism taking the boundary Bd(C) onto S^{n-1} (Hint: Take a point $x \in \text{Int}(C)$ and prove that each ray starting from x meets Bd(C) in exactly one point, cf. exercise 2.1.1).

(d) Let L be a linear space and let τ be an n-simplex in L. Prove that τ is homeomorphic to I^n.

3.6. Topologizing a Simplicial Complex

From this moment on, we will no longer distinguish between an abstract simplicial complex and its geometric realization. So a simplicial complex in a normed linear space L is a collection $\mathcal{S}$ of (geometric) simplexes in L such that for all $\sigma_1, \sigma_2 \in \mathcal{S}$ we have

(SC) (1) if $\sigma \in \mathcal{S}$ and τ is a face of σ then $\tau \in \mathcal{S}$,

(2) $\sigma_1 \cap \sigma_2 \neq \emptyset \Rightarrow \sigma_1 \cap \sigma_2$ is a face of σ_1 as well as σ_2.

We (again) use $|\mathcal{S}|$ as an abbreviation for $\bigcup \mathcal{S}$. A subcollection $\mathcal{T}$ of $\mathcal{S}$ which is also a simplicial complex is called a *subcomplex* of $\mathcal{S}$.

Let $\mathcal{T}$ be a simplicial complex in L. For each $m \geq 0$ define

$$\mathcal{T}^{(m)} = \{\sigma \in \mathcal{T} : \sigma \text{ is at most m-dimensional}\}.$$

So $\mathcal{T}^{(0)}$ is the set of vertices of all simplexes in $\mathcal{T}$, $\mathcal{T}^{(1)}$ is the collection of all at most 1-dimensional faces of all simplexes in $\mathcal{T}$, etc. The elements of $\mathcal{T}^{(0)}$ are called the *vertices* of $\mathcal{T}$. It is a triviality that $\mathcal{T}^{(m)}$ is a subcomplex of $\mathcal{T}$; we call it the m-*skeleton* of $\mathcal{T}$.

Let L be a normed linear space and let $\mathcal{T}$ be a simplicial complex in L. We are interested in useful topologies on the set $X = |\mathcal{T}|$. There are several natural candidates for such a topology. First, X is a subset of L and therefore carries the subspace topology inherited from L. It turns out that this is usually not an interesting topology. We shall now describe a different, more useful topology. To this end, let us first agree that there cannot be ambiguity about the topology that each simplex of $\mathcal{T}$ should carry. By theorem 3.2.4, each k-dimensional simplex in L with its subspace topology is naturally homeomorphic to a standard k-dimensional simplex in $\mathbb{R}^{k+1}$; *each simplex in* L *from this moment on is endowed with its subspace topology.* It would be very unnatural if the useful topology we want to define on X should induce a different topology on one of the simplexes of $\mathcal{T}$. So there is a natural candidate for a topology on X, namely, the largest topology that induces the "right" topology on every simplex of $\mathcal{T}$, i.e.

$$U \subseteq X \text{ is open iff for every simplex } \tau \in \mathcal{T},\ U \cap \tau \text{ is open in } \tau.$$

This is called the *Whitehead topology* on X and X with this topology is usually denoted by $|\mathcal{T}|$.

3.6.1. LEMMA: *Let $\mathcal{T}$ be a simplicial complex in* L *and let* $X = |\mathcal{T}|$. *Then*

(1) *The Whitehead topology on* X *is a topology and is finer than the topology that* X *inherits from* L, *i.e. the identity* $|\mathcal{T}| \to X \subseteq L$ *is continuous,*

(2) *On every simplex of $\mathcal{T}$ the Whitehead topology and the topology on* L *induce the same subspace topology.*

PROOF: The proof of (1) is a triviality the verification of which we leave to the reader.

For (2), let τ be a simplex of $\mathcal{T}$. It suffices to prove that for every relatively open subset $U \subseteq \tau$ there exists a Whitehead-open $V \subseteq |\mathcal{T}|$ such that $V \cap \tau = U$. Let $V = U \cup (|\mathcal{T}| \backslash \tau)$. Then $V \cap \tau = U$, so it remains to prove that V is Whitehead-open. To this end, let $\sigma \in \mathcal{T}$. Put $V' = V \cap (\tau \cup \sigma)$. Then V' is open in $\tau \cup \sigma$ since its complement is equal to $\tau \backslash U$ which is compact and hence closed in $\tau \cup \sigma$. Consequently, $V' \cap \sigma = V \cap \sigma$ is open in σ, as required. □

Observe that a subset A of $|\mathcal{T}|$ is closed if and only if $A \cap \tau$ is closed for every $\tau \in \mathcal{T}$. This immediately implies that $|\mathcal{T}|$ is a T_1- space, i.e. a space in which every singleton is closed (this follows also from lemma 3.6.1(1) of course).

Due to our self-chosen (sometimes unpleasant) restriction to deal with separable metric spaces exclusively, there is a small problem with the Whitehead topology since it need not be metrizable. However, there is a simple condition on $\mathcal{T}$ that ensures that $|\mathcal{T}|$ is separable and metrizable; call a simplicial complex $\mathcal{T}$ *locally finite* if every vertex of $\mathcal{T}$ is contained in at most finitely many simplexes of $\mathcal{T}$. In turns out that for our considerations it always suffices to consider locally finite simplicial complexes and these are precisely the simplicial complexes that are separable metric when given the Whitehead topology, see proposition 3.6.8 below. For the moment, we shall not worry about the metrizability of $|\mathcal{T}|$ but we shall prove a few easy but important lemmas from which metrizability in the locally finite case will follow rather easily later.

3.6.2. LEMMA: *Let $\mathcal{T}$ be a simplicial complex and let $\mathcal{S}$ be a subcollection of $\mathcal{T}$. Then*

(1) $\bigcup\mathcal{S}$ *is a closed subspace of* $|\mathcal{T}|$, *and*

(2) *if $\mathcal{S}$ is a subcomplex then the topology that $\bigcup\mathcal{S}$ inherits from $|\mathcal{T}|$ coincides with the Whitehead topology on $\bigcup\mathcal{S}$.*

PROOF: For (1), for convenience, put $Y = \bigcup\mathcal{S}$. Since each simplex of $\mathcal{T}$ has only finitely many faces, and $\mathcal{T}$ is a simplicial complex, it is clear that for every $\tau \in \mathcal{T}$, $Y \cap \tau$ is closed in τ. By definition of the Whitehead topology, we therefore conclude that Y is closed in $|\mathcal{T}|$.

For (2) observe that since $\mathcal{S} \subseteq \mathcal{T}$, the topology that $\bigcup\mathcal{S}$ inherits from $|\mathcal{T}|$ is coarser than its Whitehead topology. Now let $A \subseteq |\mathcal{S}|$ be closed. We shall prove that A is a closed subset of $|\mathcal{T}|$. To this end, let τ be an arbitrary simplex in $\mathcal{T}$. Since τ has only finitely many faces, there is a finite subcollection $\mathcal{F}$ of $\mathcal{S}$ such that

$$\bigcup\mathcal{F} \cap \tau = \bigcup\mathcal{S} \cap \tau.$$

Consequently, $A \cap \tau = (A \cap \bigcup\mathcal{S}) \cap \tau = (A \cap \bigcup\mathcal{F}) \cap \tau$. Now since $A \subseteq |\mathcal{S}|$ is closed, for every $\sigma \in \mathcal{F}$, $A \cap \sigma$ is closed in σ, and consequently, $A \cap \sigma \cap \tau$ is closed in τ. Since $\mathcal{F}$ is finite, it therefore follows that $(A \cap \bigcup\mathcal{F}) \cap \tau$ is closed in τ. We conclude that $A \cap \tau$ is closed in τ. □

From the above lemma it follows that if $\mathcal{T}$ is a simplicial complex and if $\mathcal{S}$ is a subcomplex of $\mathcal{T}$ then we can identify $|\mathcal{S}|$ and the subspace $\bigcup\mathcal{S}$ of $|\mathcal{T}|$. It will be convenient to do that.

3.6.3. COROLLARY: *Let $\mathcal{T}$ be a simplicial complex. Then $|\mathcal{T}^{(0)}|$ carries the discrete topology.*

PROOF: Since $\mathcal{T}^{(0)}$ is a subcollection of $\mathcal{T}$, by lemma 3.6.2(1) each subset of $\mathcal{T}^{(0)}$ is closed in $|\mathcal{T}|$. The corollary now follows from lemma 3.6.2(2). □

For each $x \in |\mathcal{T}|$ define the *star*, St x, and the *carrier*, car x, *of* x by

$$\text{St } x = |\mathcal{T}| \setminus \bigcup\{\tau \in \mathcal{T} : x \notin \tau\}, \text{ and}$$

$$\text{car } x = \bigcap\{\tau \in \mathcal{T} : x \in \tau\},$$

respectively. We shall prove that car $x \in \mathcal{T}$ and it is therefore the smallest simplex of $\mathcal{T}$ that contains x.

3.6.4. COROLLARY: *Let $\mathcal{T}$ be a simplicial complex. Then for every $x \in |\mathcal{T}|$,*

(1) St x *is open in* $|\mathcal{T}|$, *and*

(2) car x *belongs to* $\mathcal{T}$.

In addition, the family $\{\mathrm{St}\ x: x \in |\mathcal{T}^{(0)}|\}$ *covers* $|\mathcal{T}|$.

PROOF: For (1), let $\mathcal{S} = \{\tau \in \mathcal{T}: x \notin \tau\}$. From lemma 3.6.2 we obtain that $\cup\mathcal{S}$ is closed in $|\mathcal{T}|$, hence St x is open.

For (2), first take $\tau \in \mathcal{T}$ such that $x \in \tau$. Since for every $\sigma \in \mathcal{T}$ with $x \in \sigma$ we have $\sigma \cap \tau \preccurlyeq \tau$, and τ has only finitely many faces, car $x \in \mathcal{T}$.

Take an arbitrary $y \in |\mathcal{T}|$. If car y is a 0-dimensional simplex then y is a vertex of $\mathcal{T}$ and hence belongs to $\cup\{\mathrm{St}\ x: x \in \mathcal{T}^{(0)}\}$. Assume that car y is not 0-dimensional and let x be a vertex of car y. Observe that car y is the smallest simplex containing y, hence y does not belong to the boundary of car y. We claim that $y \in \mathrm{St}\ x$. If not, then there is a simplex $\tau \in \mathcal{T}$ containing y but not containing x. Then $\tau \cap$ car y is a proper face of car y containing y, contradiction.□

3.6.5. LEMMA: *Let* $\mathcal{T}$ *be a simplicial complex and Let* $K \subseteq |\mathcal{T}|$ *be compact. Then* K *is metrizable and is contained in the union of finitely many simplexes of* $\mathcal{T}$.

PROOF: That K is metrizable follows immediately from lemma 3.6.1(1) and exercise 1.1.4.

Suppose that K is not contained in the union of finitely many simplexes of $\mathcal{T}$. Then we can find an infinite collection $\mathcal{S} = \{\tau_n: n \in \mathbb{N}\}$ of simplexes in $\mathcal{T}$ and for every $n \in \mathbb{N}$ a point

$$(*) \qquad x_n \in (K \cap \tau_{n+1}) \setminus \cup_{i=1}^{n} \tau_n.$$

Since K is compact, and metrizable, we may assume without loss of generality that $x = \lim_{n \to \infty} x_n$ exists (and belongs to K of course). Let $\tau \in \mathcal{T}$ be an arbitrary simplex. Since $\tau_n \cap \tau$ is either empty or a face of τ, there exists $m \in \mathbb{N}$ such that

$$\bigcup_{n=1}^{\infty} \tau_n \cap \tau = \bigcup_{n=1}^{m} \tau_n \cap \tau,$$

from which it follows that

$$\{x_n: n \in \mathbb{N}\} \cap \tau \subseteq \{x_1, \cdots, x_m\}.$$

By the definition of the Whitehead topology, this implies that $\{x_n: n \in \mathbb{N}\}$ is closed in $|\mathcal{T}|$ (recall that $|\mathcal{T}|$ is T_1). Since the sequence $(x_n)_n$ converges, it therefore has to be eventually constant, which contradicts (*).□

One of the reasons that the Whitehead topology on a simplicial complex is relatively simple to deal with is that it is easy to check that certain functions are continuous.

3.6.6. LEMMA: *Let $\mathcal{T}$ be a simplicial complex and let* X *be a space. A function* f: $|\mathcal{T}| \to X$ *is continuous if and only if the restriction of* f *to every simplex* $\tau \in \mathcal{T}$ *is continuous on* τ.

PROOF: Suppose that f: $|\mathcal{T}| \to X$ is such that the restriction of f to every simplex $\tau \in \mathcal{T}$ is continuous. Let $U \subseteq X$ be open. For each $\tau \in \mathcal{T}$,

$$f^{-1}(U) \cap \tau = (f \mid \tau)^{-1}(U)$$

is open in τ. By the definition of the Whitehead topology, it therefore follows that $f^{-1}(U)$ is open in $|\mathcal{T}|$.□

3.6.7. COROLLARY: *Let $\mathcal{T}$ be a simplicial complex. Then $|\mathcal{T}|$ is a normal topological space, i.e. for every pair of disjoint closed subsets* A *and* B *of $|\mathcal{T}|$ there exists a continuous function* f: $|\mathcal{T}| \to I$ *such that* $f \mid A \equiv 0$ *and* $f \mid B \equiv 1$.

PROOF: Since $\mathcal{T}$ is countable, we can enumerate it as $\{\tau_n : n \in \mathbb{N}\}$. By corollary 1.4.15 there exists a continuous function $f_1: \tau_1 \to I$ such that $f_1 \mid (\tau_1 \cap A) \equiv 0$ and $f \mid (\tau_1 \cap B) \equiv 1$. By induction on n, we shall construct continuous functions $f_n: \bigcup_{i=1}^n \tau_i \to I$ such that

(1) $f_n \mid (A \cap \bigcup_{i=1}^n \tau_i) \equiv 0$,
(2) $f_n \mid (B \cap \bigcup_{i=1}^n \tau_i) \equiv 1$, and
(3) if $n > 1$ then f_n extends f_{n-1}.

Since we already defined f_1, assume f_i to be constructed for $1 \le i \le n$. We shall construct f_{n+1}. Let $A' = A \cap \bigcup_{i=1}^{n+1} \tau_i$ and $B' = B \cap \bigcup_{i=1}^{n+1} \tau_i$, respectively. Define $g: A' \cup B' \cup \bigcup_{i=1}^n \tau_i \to I$ by

$$g(x) = \begin{cases} 0 & (x \in A'), \\ 1 & (x \in B'), \\ f_n(x) & (x \in \bigcup_{i=1}^n \tau_i). \end{cases}$$

Then g is clearly continuous and extends f_n. Since $\bigcup_{i=1}^{n+1} \tau_i$ is compact, it is metrizable by lemma 3.6.5. Consequently, by corollary 1.4.14, there exists a continuous extension $f_{n+1}: \bigcup_{i=1}^{n+1} \tau_i \to I$ of g. It is clear that f_{n+1} is as required.

Now put $f = \bigcup_{n=1}^{\infty} f_n$. Then f is a well-defined function from $|\mathcal{T}|$ into I having the property that $f \mid A \equiv 0$ and $f \mid B \equiv 1$.

Moreover, f is continuous by lemma 3.6.6.□

These simple results enable us to prove the following

3.6.8. PROPOSITION: *Let $\mathcal{T}$ be a locally finite simplicial complex in* L. *Then* $|\mathcal{T}|$ *is a locally compact separable metrizable space.*

PROOF: We shall first prove that each point of $|\mathcal{T}|$ has an open neighborhood which is separable, metrizable, and has compact closure in $|\mathcal{T}|$. To this end, take an arbitrary $x \in |\mathcal{T}|$. By corollary 3.6.4(1), St x is open. Since $\mathcal{T}$ is locally finite, St x is contained in the union of finitely many simplexes of $\mathcal{T}$. Consequently, St x is contained in a subspace of $|\mathcal{T}|$ which is compact and therefore metrizable (lemma 3.6.5). We conclude that $|\mathcal{T}|$ is locally compact.

Since $\mathcal{T}$ is countable, $|\mathcal{T}|$ is the union of countably many compact subspaces. So $|\mathcal{T}|$ is Lindelöf and from the above it therefore follows that $|\mathcal{T}|$ can be covered by a countable family $\mathcal{B}$ consisting of open, separable metrizable subspaces of $|\mathcal{T}|$. For each $B \in \mathcal{B}$ let $\mathcal{F}(B)$ be a countable base for $\mathcal{B}$ consisting of open subsets of B. Then $\mathcal{F} = \bigcup_{B\in\mathcal{B}} \mathcal{F}(B)$ is countable, consists of open subsets of $|\mathcal{T}|$, and is easily seen to be a base for $|\mathcal{T}|$. So by corollary 3.6.7 $|\mathcal{T}|$ is a second countable regular T_1-space and is therefore a separable metrizable space. □

In exercise 3.6.5 we shall see that if $|\mathcal{T}|$ is metrizable then $\mathcal{T}$ is locally finite.

We shall now prove that the Whitehead topology on $|\mathcal{T}|$ is in fact "intrinsic", i.e. that it has nothing to do with the topology on L. To begin with, let us say that a subset F of $\mathcal{T}^{(0)}$, the 0-skeleton of $\mathcal{T}$, *spans a simplex in* $\mathcal{T}$ if there exists a simplex $\tau = |\{x_0,\cdots,x_k\}| \in \mathcal{T}$ such that $F = \{x_0,\cdots,x_k\}$.

3.6.9. LEMMA: *Let $\mathcal{T}$ be a simplicial complex in* L *and let* F *be a subset of* $\mathcal{T}^{(0)}$. *Then* F *spans a simplex in $\mathcal{T}$ if and only if* $\bigcap_{x\in F} \text{St } x \neq \emptyset$.

PROOF: Suppose that F spans a simplex τ in $\mathcal{T}$ and let b be the barycenter of τ. We claim that $b \in \bigcap_{x\in F} \text{St } x$. If not, then there exists $x \in F$ such that $b \notin \text{St } x$, or, equivalently, there is a simplex σ in $\mathcal{T}$ containing b but not containing x. Then $\sigma \cap \tau$ is a face of τ. However, b is not contained in a proper face of τ, which implies that $x \in \tau \subseteq \sigma$, contradiction. Conversely, assume that F is such that there exists $y \in \bigcap_{x\in F} \text{St } x$. We claim that F spans a face of car y. If for certain $x \in F$, $x \notin \text{car } y$ then by definition, $\text{St } x \cap \text{car } y = \emptyset$, which is a contradiction. Consequently, $F \subseteq \text{car } y$. Since $\mathcal{T}$ is a simplicial complex, and $F \subseteq \mathcal{T}^{(0)}$, each $x \in F$ is a vertex of car y. We conclude that F spans a simplex in $\mathcal{T}$. □

Now let L and E be normed linear spaces and let $\mathcal{T}$ and $\mathcal{S}$ be simplicial complexes in L and E, respectively. We say that $\mathcal{T}$ and $\mathcal{S}$ are *combinatorially equivalent* if there exists a bijection $f: \mathcal{T}^{(0)} \to \mathcal{S}^{(0)}$ such that $F \subseteq \mathcal{T}^{(0)}$ spans a simplex in $\mathcal{T}$ if and only if f(F) spans a simplex in $\mathcal{S}$. Let us also say that $|\mathcal{T}|$ and $|\mathcal{S}|$ are *simplicially homeomorphic* if there exists a homeomorphism $h: |\mathcal{T}| \to |\mathcal{S}|$ such that for all $\tau \in \mathcal{T}$ and $\sigma \in \mathcal{S}$, $h(\tau) \in \mathcal{S}$ and $h^{-1}(\sigma) \in \mathcal{T}$.

3.6.10. PROPOSITION: *Let* L *and* E *be normed linear spaces and let* $\mathcal{T}$ *and* $\mathcal{S}$ *be simplicial complexes in* L *and* E, *respectively. The following statements are equivalent:*

(a) $\mathcal{T}$ *and* $\mathcal{S}$ *are combinatorially equivalent,*

(b) $|\mathcal{T}|$ *and* $|\mathcal{S}|$ *are simplicially homeomorphic.*

PROOF: For (a) $\Rightarrow$ (b), let f: $\mathcal{T}^{(0)} \to \mathcal{S}^{(0)}$ be a bijection such that $F \subseteq \mathcal{T}^{(0)}$ spans a simplex in $\mathcal{T}$ if and only if f(F) spans a simplex in $\mathcal{S}$. For every $\tau = |\{x_0,\cdots,x_k\}| \in \mathcal{T}$, the set $\{f(x_0),\cdots,f(x_k)\}$ spans a simplex σ of $\mathcal{S}$. Define a homeomorphism $f_\tau: \tau \to \sigma$ by

$$f_\tau(\Sigma_{i=0}^k \alpha_i(x)x_i) = \Sigma_{i=0}^k \alpha_i(x)f(x_i);$$

here $\alpha_0(x),\cdots,\alpha_k(x)$ denote the affine coordinates with respect to τ of an arbitrarily chosen point $x \in \tau$ of course. It is easily seen that f_τ is indeed a homeomorphism, cf. theorem 3.2.4. Now define $\bar{f}: |\mathcal{T}| \to |\mathcal{S}|$ by $\bar{f} = \cup_{\tau\in\mathcal{T}} f_\tau$. Then $\bar{f}$ is continuous since the restriction of $\bar{f}$ to every simplex of $\mathcal{T}$ is continuous (lemma 3.6.6). It is easily seen that f is one-to-one and surjective and that $\bar{f}^{-1}$ is continuous for the same reason as $\bar{f}$ is. We conclude that f is a homeomorphism. Clearly both $\bar{f}$ and $\bar{f}^{-1}$ are "simplex preserving".

Conversely, let $\bar{f}: |\mathcal{T}| \to |\mathcal{S}|$ be a homeomorphism such as in (b). The restriction f of $\bar{f}$ to $\mathcal{T}^{(0)}$ is a bijection $\mathcal{T}^{(0)} \to \mathcal{S}^{(0)}$ such that $F \subseteq \mathcal{T}^{(0)}$ spans a simplex in $\mathcal{T}$ if and only if f(F) spans a simplex in $\mathcal{S}$. □

From the above proposition we conclude that the topology of a simplicial complex depends only on the combinatorial properties of its vertex set. For that reason when discussing a simplicial complex, it is no longer necessary to mention the normed linear space it is a subset of.

Let X be a space. We say that X is a *polytope* if there exists a locally finite simplicial complex $\mathcal{T}$ such that X and $|\mathcal{T}|$ are homeomorphic. In case $\mathcal{T}$ is *finite* we say that X is a *polyhedron*. Observe that each polyhedron is compact. Polytopes are very important since they are the "bridge" between abstract topological spaces and concrete ones.

If X is a polytope, say X is homeomorphic to $|\mathcal{T}|$, then we will find it sometimes convenient to not distinguish between X and $|\mathcal{T}|$. This will never cause confusion because the triangulation under consideration will always be explicitly defined but the reader should keep in mind that there usually exist many different triangulations of the same polytope. For example, let τ be a simplex in L; it is clear that each barycentric subdivision $sd^{(n)}\mathcal{F}(\tau)$ of τ is a triangulation.

If $|\mathcal{T}|$ is a polytope then a subset $Y \subseteq |\mathcal{T}|$ is called a *subpolytope* if there is a subcomplex $\mathcal{S} \subseteq \mathcal{T}$ such that $Y = |\mathcal{S}|$. Similarly for *subpolyhedron*.

3.6.11. THEOREM: *Each polyhedron is an* **ANR**.

PROOF: Let $\mathcal{T}$ be a finite simplicial complex. By induction on the cardinality of $\mathcal{T}$ we shall prove that $|\mathcal{T}|$ is an **ANR**. If $\mathcal{T}$ consists of one simplex only then $|\mathcal{T}|$ is an **AR** by theorem 3.2.4 and theorem 1.5.1. Suppose that for every simplicial complex $\mathcal{T}$ of cardinality at most $n-1 \geq 1$, $|\mathcal{T}|$ is an **ANR**, and let $\mathcal{T}$ be a simplicial complex of cardinality n. Since $\mathcal{T}$ is finite, there exists a simplex $\tau \in \mathcal{T}$ such that for every $\sigma \in \mathcal{T}$ with $\tau \preccurlyeq \sigma$, $\sigma = \tau$ (let τ be any simplex of maximal dimension). Put $\mathcal{S} = \mathcal{T}\setminus\{\tau\}$. Then $\mathcal{S}$ is a simplicial complex and $|\mathcal{T}| = \tau \cup |\mathcal{S}|$. In addition, clearly $\tau \cap |\mathcal{S}| = |\{\tau \cap \sigma\colon \sigma \in \mathcal{S}\}|$. Consequently, by our inductive assumptions and by theorem 1.5.9(1) we conclude that $|\mathcal{T}|$ is an **ANR**. □

In corollary 5.4.6 we shall prove that every polytope is an **ANR**, thereby generalizing theorem 3.6.11.

We shall now formulate a very important property of polyhedra.

3.6.12. THEOREM: *Let $\mathcal{T}$ be a finite simplicial complex. Then for every open cover $\mathcal{U}$ of $|\mathcal{T}|$ there exists $m \in \mathbb{N}$ such that $\mathrm{sd}^{(m)}\mathcal{T}$ refines $\mathcal{U}$, i.e. for every $\sigma \in \mathrm{sd}^{(m)}\mathcal{T}$ there exists $U \in \mathcal{U}$ with $\sigma \subseteq U$.*

PROOF: Fix $\tau \in \mathcal{T}$ for a moment. By corollary 3.3.6 and lemma 1.1.1 there exists $n(\tau) \in \mathbb{N}$ such that every $\sigma \in \mathrm{sd}^{(n(\tau))}\mathcal{F}(\tau)$ is contained in an element of $\mathcal{U}$. Let $m = \max\{n(\tau)\colon \tau \in \mathcal{T}\}$. Now take an arbitrary $\kappa \in \mathrm{sd}^{(m)}\mathcal{T}$. By the definition of $\mathrm{sd}^{(m)}\mathcal{T}$ it is clear that there exists $\tau \in \mathcal{T}$ with $\kappa \in \mathrm{sd}^{(m)}\mathcal{F}(\tau)$. Since every element of $\mathrm{sd}^{(m)}\mathcal{F}(\tau)$ is contained in an element of $\mathrm{sd}^{(n(\tau))}\mathcal{F}(\tau)$, κ is contained in an element of $\mathcal{U}$. □

The above theorem can be generalized as follows: for every polytope $|\mathcal{T}|$ and for every open cover $\mathcal{U}$ of $|\mathcal{T}|$ there exists a subdivision $\mathcal{S}$ of $\mathcal{T}$ such that each simplex $\sigma \in \mathcal{S}$ is contained in an element of $\mathcal{U}$. However, for noncompact polytopes, $\mathcal{S}$ generally cannot be chosen to be an iterated barycentric subdivision. For details, see Whitehead [149].

Now let X be a space and let $\mathcal{U}$ be a *countable* open cover of X. We say that a simplicial complex $\mathcal{T}$ is a *nerve* of $\mathcal{U}$ if $\mathcal{T}^{(0)}$ can be indexed as $\{x(U)\colon U \in \mathcal{U}\}$ such that for every $n \geq 0$,

(**) $$x(U_0),\cdots,x(U_n) \text{ spans a simplex in } \mathcal{T} \text{ iff } \bigcap_{i=0}^{n} U_i \neq \emptyset.$$

If $\mathcal{T}$ is a nerve of $\mathcal{U}$ then it is convenient to adopt the notation $\{x(U)\colon U \in \mathcal{U}\}$ for $\mathcal{T}^{(0)}$, where it is implicitly assumed that the "indexing" is such that (**) holds (see figure 3.6.1).

3.6.13. PROPOSITION: *Let X be a space. Each countable open cover of X has a nerve.*

PROOF: Let $\mathcal{U} = \{U_n: n \in \mathbb{N}\}$ be a countable open cover of X. For each $n \in \mathbb{N}$ let $x_n \in l^2$ be the vector all coordinates of which are 0, except the n-th coordinate which equals 1. The sequence $(x_n)_n$ is clearly linearly independent, hence geometrically independent. Define

$$\mathcal{F} = \{F \subseteq \mathbb{N}: F \text{ is finite and } \bigcap_{n \in F} U_n \neq \emptyset\}.$$

For $F \in \mathcal{F}$ let $\tau(F)$ be the simplex in l^2 spanned by $\{x_n: n \in F\}$. Put $\mathcal{T} = \{\tau(F): F \in \mathcal{F}\}$. It is easy to verify that $\mathcal{T}$ is a simplicial complex and that $\mathcal{T}$ is a nerve for $\mathcal{U}$. □

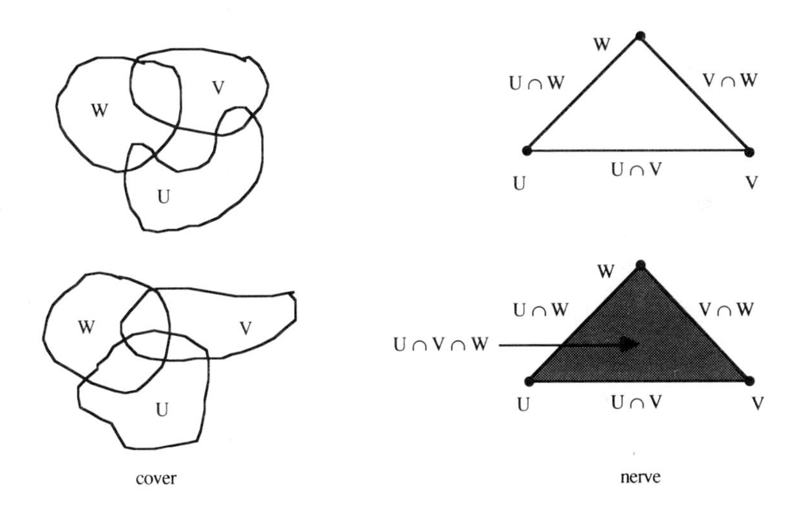

Figure 3.6.1.

Clearly, any two nerves of the same open cover are combinatorially equivalent (hence "isomorphic"), so by proposition 3.6.10 we can now speak of *the* nerve $N(\mathcal{U})$ of the cover $\mathcal{U}$. If L is a normed linear space containing a simplicial complex $\mathcal{T}$ which is combinatorially equivalent to the nerve $N(\mathcal{U})$ then we say that $N(\mathcal{U})$ can be *realized* in L.

Let $\mathcal{U}$ be a locally finite open cover of a space (X,d). As in §1.4, associate to $\mathcal{U}$ the following family of continuous functions:

$$\kappa_U(x) = \frac{d(x, X\backslash U)}{\Sigma_{V \in \mathcal{U}} d(x, X\backslash V)} \qquad (U \in \mathcal{U}).$$

Recall that these functions are called the κ-*functions with respect to the cover* $\mathcal{U}$ and that they are well-defined and continuous, cf. §1.4.

Let $N(\mathcal{U})$ be the nerve of $\mathcal{U}$ (recall that $\mathcal{U}$ is countable). We shall use the κ-functions with respect to $\mathcal{U}$ to define a canonical continuous function $\kappa: X \to |N(\mathcal{U})|$. This function is the "bridge" between the "abstract" space X and the "concrete" space $|N(\mathcal{U})|$ and is called the κ-*function* of the cover $\mathcal{U}$.

For $x \in X$ put $\mathcal{F}(x) = \{U \in \mathcal{U}: x \in U\}$. Observe that by the local finiteness condition, F(x) is finite, that $\kappa_U(x) = 0$ if $U \notin \mathcal{F}(x)$, and that $\Sigma_{U\in \mathcal{F}(x)} \kappa_U(x) = 1$. Let $\tau(x)$ be the simplex spanned by the vertices x(U), $U \in \mathcal{F}(x)$. Define $\kappa: X \to |N(\mathcal{U})|$ by

$$\kappa(x) = \Sigma_{U\in \mathcal{F}(x)} \kappa_U(x)\cdot x(U).$$

Observe that κ is well-defined and that $\kappa(x) \in \tau(x)$.

3.6.14. LEMMA: *Let* X *be a space and let* $\mathcal{U}$ *be a locally finite open cover of* X. *Then for each* $x \in X$, $\tau(x)$ *is the carrier of* $\kappa(x)$ *in* $N(\mathcal{U})$.

PROOF: As remarked above, $\tau(x)$ is a simplex of $N(\mathcal{U})$ that contains $\kappa(x)$. By unicity of affine coordinates (theorem 3.2.1) the $\kappa_U(x)$'s are the affine coordinates of $\kappa(x)$ with respect to $\{x(U): U \in \mathcal{F}(x)\}$. Now since for every $U \in \mathcal{F}(x)$, $\kappa_U(x) \neq 0$, $\tau(x)$ is the smallest simplex containing $\kappa(x)$, i.e. $\tau(x) = \text{car } \kappa(x)$.□

3.6.15. THEOREM: *Let* X *be a space and let* $\mathcal{U}$ *be a locally finite open cover of* X. *Then the* κ-*function* $\kappa: X \to |N(\mathcal{U})|$ *has the following properties:*

(1) κ *is continuous, and*

(2) *for every* $U \in \mathcal{U}$: $\kappa^{-1}(\text{St } x(U)) = U$.

PROOF: We shall first prove (2). Observe that for every $x \in X$ and $U \in \mathcal{U}$, $x \notin U$ iff $U \notin \mathcal{F}(x)$ iff $x(U) \notin \tau(x) = \text{car } \kappa(x)$ (lemma 3.6.14). Now if $x(U) \notin \tau(x)$ then since $\kappa(x) \in \tau(x)$, $\kappa(x) \notin \text{St } x(U)$. Also, if $\kappa(x) \notin \text{St } x(U)$ then there is a simplex $\tau \in N(\mathcal{U})$ which contains $\kappa(x)$ but not x(U). Since car $\kappa(x) \subseteq \tau$, this gives us that $x(U) \notin \text{car } \kappa(x) = \tau(x)$. We conclude that $x \notin U$ iff $\kappa(x) \notin \text{St } x(U)$.

For (1), let L be a normed linear space that realizes $N(\mathcal{U})$ (e.g. l^2, see the proof of proposition 3.6.13). Let $x \in X$. There is a neighborhood W of x meeting only finitely many elements of $\mathcal{U}$. Since the infinite sum in the definition of κ for points of W clearly reduces to a finite sum, and the functions κ_U are all continuous, by the continuity of the algebraic operations on L it follows that κ regarded as a mapping from X into L is continuous. Now let $\mathcal{W} = \{U \in \mathcal{U}: U \cap W \neq \emptyset\}$ and let Y be the union of all simplexes of $N(\mathcal{U})$ the vertices of which correspond to elements of the collection $\mathcal{W}$. If $x \in W$ then $\mathcal{F}(x) \subseteq \mathcal{W}$ which implies that $\kappa(x) \in \tau(x) \subseteq Y$. Since Y is compact, by lemma 3.6.1(1) it follows that the topology that Y inherits from L is the same as the topology that Y inherits from $|N(\mathcal{U})|$. We conclude that $\kappa: X \to |N(\mathcal{U})|$ is continuous at all points of W, and hence at x. □

Let X and Y be spaces and let $\mathcal{U}$ be an open cover of X. A continuous function $f: X \to Y$ is called a *$\mathcal{U}$-mapping* if there is an open cover $\mathcal{V}$ of Y such that $f^{-1}(\mathcal{V}) < \mathcal{U}$, cf. §1.3.

3.6.16. COROLLARY: *Let X be a space and let $\mathcal{U}$ be a locally finite open cover of X. Then $\kappa: X \to |N(\mathcal{U})|$ is a $\mathcal{U}$-mapping.*

PROOF: This follows directly from theorem 3.6.15(2) and corollary 3.6.4. □

It is unfortunately not the case that a cover $\mathcal{U}$ of a space X is locally finite if and only if the simplicial complex $N(\mathcal{U})$ is locally finite. Since, as we pointed out above, our main interest is in locally finite simplicial complexes, it is natural to wonder when $N(\mathcal{U})$ is locally finite.

First a definition. An open cover $\mathcal{U}$ of X is said to be *star-finite* if for every $U \in \mathcal{U}$ the set

$$\{V \in \mathcal{U}: V \cap U \neq \emptyset\}$$

is finite. Observe that a star-finite open cover is locally finite. Consequently, every star-finite cover is countable.

This definition leads us to the following

3.6.17. THEOREM: *Let X be a space and let $\mathcal{U}$ be an open cover of X. Then*

(1) *there exists an open refinement $\mathcal{V}$ of $\mathcal{U}$ such that $\mathcal{V}$ is star-finite, and*

(2) *$N(\mathcal{U})$ is locally finite if and only if $\mathcal{U}$ is star-finite.*

PROOF: For (1), first observe that without loss of generality, $\mathcal{U}$ is countable, say $\mathcal{U} = \{U_i: i \in \mathbb{N}\}$. By theorem 1.4.17, for every i there is a continuous function $f_i: X \to I$ such that $f_i^{-1}(0,1] = U_i$. Define $f: X \to I$ by

$$f(x) = \sum_{i=1}^{\infty} 2^{-i} f_i(x).$$

Since the series $\Sigma_{i=1}^{\infty} 2^{-i}$ converges, proposition 1.3.4 easily implies that f is continuous. As $\mathcal{U}$ covers X, $f(x) > 0$ for every $x \in X$.

For $k \in \mathbb{N}$ define

$$V_k = f^{-1}(\tfrac{1}{k},1] \text{ and } F_k = f^{-1}[\tfrac{1}{k},1],$$

respectively. The families $\{V_k: k \in \mathbb{N}\}$ and $\{F_k: k \in \mathbb{N}\}$ cover X. Put $F_0 = \emptyset$. For every $k \in \mathbb{N}$ and $1 \leq j \leq k$, put

$$U_{k,j} = U_j \cap (V_{k+1} \backslash F_{k-1}).$$

Observe that every $U_{k,j}$ is open; we shall prove that $\mathcal{U}^*$, the family of all $U_{k,j}$'s, is the required star-finite open refinement of $\mathcal{U}$.

We shall prove that $\mathcal{U}^*$ is a cover of X. To this end, take an arbitrary $x \in X$. Let k be the smallest integer k with the property that $x \in F_k$. Then $f(x) = \Sigma_{i=1}^{\infty} 2^{-i} f_i(x) \geq 1/k$ and since

$$\Sigma_{j=k+1}^{\infty} 2^{-j} f_j(x) \leq \Sigma_{j=k+1}^{\infty} 2^{-j} = 2^{-k} < \tfrac{1}{k},$$

it follows that $\Sigma_{j=1}^{k} 2^{-j} f_j(x) \neq 0$, i.e. there exists $j \leq k$ with $f_j(x) \neq 0$. So there exists $j \leq k$ such that $x \in U_j$. By the special choice of k we now get

$$x \in U_j \cap (F_k \backslash F_{k-1}) \subseteq U_{k,j}.$$

We shall now prove that $\mathcal{U}^*$ is star-finite. This is a triviality since for $j \leq k$ we have $U_{k,j} \subseteq V_{k+1} \subseteq F_{k+1}$, so that

$$U_{k,j} \cap U_{m,i} = \emptyset \text{ for } m \geq k+2 \text{ and } i \leq m,$$

from which the desired result follows immediately.

That $\mathcal{U}^* < \mathcal{U}$ is a triviality.

For (2), first assume that $N(\mathcal{U})$ is locally finite. If there exists $U \in \mathcal{U}$ such that the set $\{V \in \mathcal{U}: V \cap U \neq \emptyset\}$ is infinite then the vertex $x(U)$ of $N(\mathcal{U})$ is contained in infinitely many 1-dimensional simplexes of $N(\mathcal{U})$, which contradicts the local finiteness condition on $N(\mathcal{U})$. The verification of the converse implication is a triviality which we leave to the reader. □

We shall now prove that every space can be "approximated arbitrarily closely" by a polytope.

3.6.18. COROLLARY: *Let* X *be a space. For every open cover* $\mathcal{U}$ *of* X *there exists a polytope* P *and a* $\mathcal{U}$*-mapping* f: X → P. *If* X *is compact then* P *can be chosen to be a polyhedron.*

PROOF: Let $\mathcal{U}$ be an open cover of X. By theorem 3.6.17(1), there exists a star-finite open cover $\mathcal{V}$ of X which refines $\mathcal{U}$. Let $\kappa: X \to |N(\mathcal{V})|$ be the κ-mapping of $\mathcal{V}$. Since $\mathcal{V}$ is star-finite, $|N(\mathcal{V})|$ is locally finite by theorem 3.6.17(2). By corollary 3.6.16, κ is a $\mathcal{V}$-mapping. Since $\mathcal{V} < \mathcal{U}$, we are done.

If X is compact then $\mathcal{V}$ is a finite subcover of $\mathcal{U}$ and proceed as above. □

Exercises for §3.6.

1. Let P and Q be polyhedra. Prove that $P \times Q$ is a polyhedron. Prove that there is a triangulation of $P \times I$ such that $P \times \{0,1\}$ is a subpolyhedron of $P \times I$.

2. Let $\mathcal{T}$ be a simplicial complex. Prove that $|\mathcal{T}|$ is paracompact (we do not assume that $\mathcal{T}$ is locally finite).

3. Give an example of a space X and a locally finite open cover $\mathcal{U}$ of X such that $\mathcal{U}$ is not star-finite.

4. Prove that the torus $S^1 \times S^1$ is a polyhedron.

5. Let $\mathcal{T}$ be a simplicial complex that is not locally finite. Prove that $|\mathcal{T}|$ is not metrizable (this is the converse to proposition 3.6.8).

6. Let X be a space and let $\mathcal{U}$ and $\mathcal{V}$ be open covers of X such that $\mathcal{V} < \mathcal{U}$. For every $V \in \mathcal{V}$ pick $U(V) \in \mathcal{U}$ such that $V \subseteq U(V)$. This defines a function f_0: $|N(\mathcal{V})^{(0)}| \to |N(\mathcal{U})^{(0)}|$. Extend this function "linearly" over each simplex of $N(\mathcal{V})$ and prove that the resulting function from $|N(\mathcal{V})|$ to $|N(\mathcal{U})|$ is continuous.

7. Let $\mathcal{T}$ be a simplicial complex and $\mathcal{S}$ a subcomplex of $\mathcal{T}$. Prove that the topological interior and topological boundary of $|\mathcal{S}|$ in $|\mathcal{T}|$ are given by $\{x \in |\mathcal{S}| : x \in \sigma \in \mathcal{T} \Rightarrow \sigma \in \mathcal{S}\}$ and $\bigcup(\mathcal{T} \setminus \mathcal{S}) \cap |\mathcal{S}|$.

Notes and comments for chapter 3.

§4.

Theorem 3.4.3 is due to Sperner [131].

§5.

Lemma 3.5.1. can be found in Knaster, Kuratowski and Mazurkiewicz [83]. Theorem 3.5.2 is due to Brouwer [33]. The approach in this section is due to Knaster, Kuratowski and Mazurkiewicz [83]. Exercise 3.5.7 is due to Schauder [122].

§6.

Nerves of covers were introduced by Alexandrov [1], and κ-mappings by Kuratowski [87]. Theorem 3.6.17(1) is due to Morita [110]; the proof presented here was taken from Engelking [59, lemma 5.2.4].

4. Elementary Dimension Theory

Dimension theory enables us to assign to every topological space X a number

$$\dim X \in \{-1,0,1,\cdots\} \cup \{\infty\}$$

having, among others, the following properties

(1) if X and Y are homeomorphic spaces then $\dim X = \dim Y$, and
(2) $\dim \mathbb{R}^n = n$ for every $n \in \mathbb{N}$.

So dim X is a topological invariant of X, and by (2) it distinguishes between the euclidean spaces $\mathbb{R}^n$, $n \in \mathbb{N}$. A space X for which $\dim X < \infty$ is called *finite-dimensional,* and a space is *infinite-dimensional* if it is not finite-dimensional.

The aim of this chapter is to present some basic results from dimension theory. Some of the presented results are standard and are over seventy years old; they concern finite-dimensional spaces. However, during the last years significant contributions were made concerning the topology of *infinite-dimensional spaces*. Naturally, we shall also present some of these results in detail.

4.1. The Covering Dimension

Let X be a space and let Γ be an index set. The Theorem on Partitions 3.5.7 motivates the following definition: a family of pairs of disjoint closed sets $\tau = \{(A_i,B_i) : i \in \Gamma\}$ of X is called *es-*

sential if for every family $\{L_i : i \in \Gamma\}$, where L_i is a partition between A_i and B_i for every i, we have $\bigcap_{i\in\Gamma} L_i \neq \emptyset$; if τ is not essential then it is called *inessential*. So theorem 3.5.7 and corollary 3.5.8 show that I^n has an essential family of pairs of disjoint closed sets of cardinality n and Q has one of every (finite) cardinality and also one of infinite cardinality. It follows easily that every subfamily of an essential family is again essential.

4.1.1. THEOREM: *Let* $n \in \mathbb{N}$. *Then* I^n *has an essential family consisting of* n *pairs of disjoint closed sets, but every family consisting of at least* n+1 *pairs of disjoint closed sets is inessential.*

PROOF: By the above remarks we need only to consider the second part of the theorem.

Let A and B be disjoint closed subsets of I^n and let $E \subseteq I$ be a dense.

CLAIM: There is a partition D between A and B such that

$$D \subseteq \{x \in I^n : (\exists i \leq n)(x_i \in E)\}.$$

This is easy. Every point $x \in A$ has a neighborhood of the form $\prod_{i=1}^n (a_i,b_i)$ with $a_i,b_i \in E$ for every $i \leq n$ such that $\prod_{i=1}^n [a_i,b_i] \cap B = \emptyset$. There is a finite family $\mathcal{F}$ of these neighborhoods whose union covers A, and the boundary D of this union is contained in the union of the boundaries of the elements of $\mathcal{F}$. We conclude that D is the required partition between A and B.

Now let $\tau = \{(A_i,B_i) : i \leq n+1\}$ be a family consisting of n+1 pairs of disjoint closed subsets of I^n. There exist n+1 pairwise disjoint dense subsets of I. One can take for example

$$E_1 = (\sqrt{2} + \mathbb{Q}) \cap I,\ E_2 = (\sqrt{3} + \mathbb{Q}) \cap I,\ E_3 = (\sqrt{5} + \mathbb{Q}) \cap I, \cdots,$$

respectively (as usual, $\mathbb{Q}$ denotes the space of rational numbers in $\mathbb{R}$). By the above there exist partitions D_i between A_i and B_i such that

$$D_i \subseteq \{x \in I^n : (\exists i \leq n)(x_i \in E_i)\} \qquad (i \leq n+1).$$

Since the E_i's are pairwise disjoint, a straightforward verification yields $\bigcap_{i=1}^{n+1} D_i = \emptyset$. □

Observe that in theorem 4.1.1 we formulated a *topological* property of I^n shared by no I^m for $m \neq n$. In particular we obtain:

4.1.2. COROLLARY: *Let* $n,m \in \mathbb{N}$. *If* $n \neq m$ *then* I^n *is not homeomorphic to* I^m. □

These remarks suggest the following definition: for a space X define its *covering dimension*, $\dim X \in \{-1,0,1,\cdots\} \cup \{\infty\}$, by

$\dim X = -1$	$\Leftrightarrow$	$X = \emptyset$,
$\dim X \leq n,\ n \geq 0,$	$\Leftrightarrow$	X is nonempty and every family of n+1 pairs of disjoint closed subsets of X is inessential,
$\dim X = n$	$\Leftrightarrow$	$\dim X \leq n$ and $\dim X$ is not smaller than n,
$\dim X = \infty$	$\Leftrightarrow$	$\dim X \neq n$ for every $n \geq -1$.

Observe that $\dim X \geq n$ if and only if there is an essential family consisting of n pairs of disjoint closed subsets of X. So in a sense, there are n different "directions" in X. Also observe that it is not clear at all why we call dim X the *covering dimension* of X; the term *partition degree* seems to be more appropriate. We will explain our terminology later.

By theorem 4.1.1, $\dim I^n = n$ ($n \in \mathbb{N}$) and by corollary 3.5.8, $\dim Q = \infty$. A space X with $\dim X < \infty$ is called *finite-dimensional.* A space that is not finite-dimensional is called *infinite-dimensional.* It is easy to see that if X and Y are homeomorphic spaces then $\dim X = \dim Y$.

Without too much trouble, the proof of theorem 4.1.1 can be adapted to show that $\dim \mathbb{R}^n = n$ for every n. This equality however will turn out to follow trivially from theorem 4.1.1 and results to be derived in §4.3. For that reason we will not verify it here.

For later use we shall now study some elementary properties of essential families of pairs of disjoint closed sets.

Let X be a space. It will be convenient to let ρX denote the family of all closed subsets of X. In addition, for every $x \in X$, we put $d(x,\emptyset) = \infty$.

4.1.3. LEMMA: *Let* X *be a space and let* Y *be a subspace of* X. *The function* $\kappa: \rho Y \to \rho X$ *defined by* $\kappa(A) = \{x \in X: d(x,A) \leq d(x,Y\backslash A)\}$ *has the following properties:*

(1) $\kappa(\emptyset) = \emptyset$, $\kappa(Y) = X$,

(2) $\kappa(A) \cap Y = A$ *for every* $A \in \rho Y$,

(3) *if* $A,B \in \rho Y$ *and* $A \subseteq B$ *then* $\kappa(A) \subseteq \kappa(B)$, *and*

(4) *if* $A,B \in \rho Y$ *then* $\kappa(A \cup B) = \kappa(A) \cup \kappa(B)$.

PROOF: The straightforward verification that κ is well-defined is left to the reader. Clearly, (1), (2), and (3) hold. For (4), take $A,B \in \rho Y$. Observe that by (3), $\kappa(A) \cup \kappa(B) \subseteq \kappa(A \cup B)$. Now take an arbitrary point $x \in \kappa(A \cup B)$. Since $d(x,A \cup B) \in \{d(x,A),d(x,B)\}$, without loss of generality $d(x,A \cup B) = d(x,A)$. We shall prove that $x \in \kappa(A)$. Since $x \in \kappa(A \cup B)$,

$$d(x,A) = d(x,A \cup B) \leq d(x,Y\backslash(A \cup B)),$$

and trivially,

$$d(x,A) = d(x,A \cup B) \leq d(x,B).$$

We conclude that

$$d(x,A) \leq d(x,(Y\backslash(A \cup B)) \cup B) \leq d(x,Y\backslash A),$$

since $Y\backslash A \subseteq (Y\backslash(A \cup B)) \cup B$.□

Let X be a space. Subsets A and B of X are called *separated* if $\overline{A} \cap B = \emptyset = A \cap \overline{B}$. It is clear that if A and B are disjoint and are both closed, or both open, then A and B are separated. More interesting examples of separated sets are obtained in the following way. Let Y be a subspace of X and let U and V be disjoint subsets of Y that are open *in* Y. Then U and V are separated *in* X. For let U' be an open subset of X such that $U' \cap Y = U$. Then $U' \cap V = \emptyset$, i.e. $\overline{V} \cap U = \emptyset$. Similarly it follows that $\overline{U} \cap V = \emptyset$.

4.1.4. COROLLARY: *Let* A *and* B *be separated subsets of a space* X. *Then* A *and* B *can be separated by disjoint open subsets of* X.

PROOF: It is clear that A and B are closed in their union $A \cup B$. By lemma 4.1.3, there exist closed subsets A' and B' of X such that $A \subseteq A'$, $B \subseteq B'$ and $A' \cup B' = X$. Then $U = X\backslash B'$ and $V = X\backslash A'$ are disjoint open neighborhoods of A and B, respectively.□

The following simple result is fundamental for dimension theory.

4.1.5. LEMMA: *Let* Y *be a subspace of a space* X. *In addition, let* A *and* B *be disjoint closed subsets of* X. *If* U *and* V *are open neighborhoods of* A *and* B, *respectively, having disjoint closures, and if* S *is a partition in* Y *between* $Y \cap \overline{U}$ *and* $Y \cap \overline{V}$, *then there is a partition* T *in* X *between* A *and* B *such that* $T \cap Y \subseteq S$.

PROOF: Write $Y\backslash S$ as the disjoint union of two open (in Y) sets E and F such that

$$Y \cap \overline{U} \subseteq E \text{ and } Y \cap \overline{V} \subseteq F.$$

Since $E \cap V = \emptyset$ we obtain $\overline{E} \cap B = \emptyset$, and similarly $\overline{F} \cap A = \emptyset$. From this it follows easily that $A \cup E$ and $B \cup F$ are separated. By corollary 4.1.4 there exist disjoint open neighborhoods U' and V' of $A \cup E$ and $B \cup F$, respectively. Clearly, $T = X \backslash (U' \cup V')$ is as required.□

4.1.6. COROLLARY: *Let* Y *be a closed subspace of a space* X. *In addition, let* A *and* B *be disjoint closed subsets of* X. *If* S *is a partition in* Y *between* $Y \cap A$ *and* $Y \cap B$, *then there is a partition* T *in* X *between* A *and* B *such that* $T \cap Y \subseteq S$.

PROOF: Write $Y \backslash S$ as the disjoint union of two open (in Y) sets E and F such that

$$Y \cap A \subseteq E \text{ and } Y \cap B \subseteq F.$$

Observe that $S \cup F \cup B$ is closed in X and that $A \cap (S \cup F \cup B) = \emptyset$. By corollary 1.4.16 there is an open neighborhood U of A in X such that $\overline{U} \cap (S \cup F \cup B) = \emptyset$. By a similar argument it is possible to find an open neighborhood V of B in X such that $\overline{V} \cap (S \cup E \cup \overline{U}) = \emptyset$. By construction, S is a partition between $\overline{U} \cap Y$ and $\overline{V} \cap Y$. Now apply lemma 4.1.5.□

These simple results enable us to prove the following:

4.1.7. THEOREM: *Let* X *be a space, let* $\{(A_i,B_i): i \in \Gamma\}$ *be an essential family in* X, *and let* $\Gamma(0) \subseteq \Gamma$. *If for every* $i \in \Gamma(0)$, L_i *is a partition between* A_i *and* B_i *and* $L = \bigcap_{i\in \Gamma(0)} L_i$ *then* $\{(L \cap A_i, L \cap B_i): i \in \Gamma \backslash \Gamma(0)\}$ *is essential in* L.

PROOF: For $i \notin \Gamma(0)$ let E_i be a partition in L between $L \cap A_i$ and $L \cap B_i$. By corollary 4.1.6, for $i \notin \Gamma(0)$ we can find a partition F_i in X between A_i and B_i such that $F_i \cap L \subseteq E_i$. Then

$$\emptyset \neq \bigcap_{i\in \Gamma(0)} L_i \cap \bigcap_{i\in \Gamma\backslash\Gamma(0)} F_i \subseteq L \cap \bigcap_{i\in \Gamma\backslash\Gamma(0)} E_i = \bigcap_{i\in \Gamma\backslash\Gamma(0)} E_i,$$

which is as required.□

4.1.8. COROLLARY: *Let* X *be a space and let* $n \geq 0$. *If* $\dim X \geq n$ *then there exist disjoint closed subsets* A *and* B *of* X *such that if* L *is a partition between* A *and* B *then* $\dim L \geq n-1$,

PROOF: If $n = 0$ then the corollary is certainly true. So assume that $n \geq 1$. If $\dim X \geq n$ then $\dim X$ is not less than or equal to $n-1 \geq 0$, i.e. there exists an essential family $\tau = \{(A_i,B_i): i \leq n\}$ in X. Now let L be a partition between A_1 and B_1. Observe that $L \neq \emptyset$ since τ is essential. We consider two cases. If $n = 1$ then since $L \neq \emptyset$, $\dim L \geq 0$. If $n > 1$ then by theorem 4.1.7, L has an essential family of cardinality n-1 namely

$$\{(L\cap A_i, L\cap B_i): i = 2,\cdots,n\},$$

so that dim $L \geq n-1$.□

Exercises for §4.1.

1. Let $\tau = \{(A_i,B_i): i \leq n\}$ be a family of pairs of disjoint closed subsets in the space X. Prove that if there exist different indices $i,j \leq n$ such that $(A_i,B_i) = (A_j,B_j)$ then τ is inessential.

2. Let $\tau = \{(A_i,B_i): i \leq n\}$ be a family of pairs of disjoint closed subsets in the space X. Prove that if τ is essential then X can be mapped continuously *onto* I^n.

3. Let X be a compact space and let $\tau = \{(A_i,B_i): i\in \Gamma\}$ be an essential family in X. Prove that Γ is countable (Hint: Let $\mathcal{E}$ be a countable closed base for X which is closed under finite unions and finite intersections. Use the compactness of X to show that for every $i\in \Gamma$ there exist disjoint E and F in $\mathcal{E}$ with $A_i \subseteq E$ and $B_i \subseteq F$. Assume that Γ is uncountable. There have to be disjoint elements E and F in $\mathcal{E}$ and *distinct* indices i and j in Γ such that $A_i \cup A_j \subseteq E$ and $B_i \cup B_j \subseteq F$. Conclude that τ is inessential).

4. Prove that dim $\mathbb{Q} = 0$.

5. Prove that if X is connected then dim $X \geq 1$.

A space X is called *countable dimensional* if it can be written as the union of countably many zero-dimensional subspaces. In addition, a space X is called *strongly infinite-dimensional* if it has an infinite essential family. See §4.8 for more information on these notions.

6. Prove that if X is strongly infinite-dimensional then X is not countable dimensional. Give an example of a strongly infinite-dimensional space. Assuming that every finite-dimensional space is countable dimensional, give an example of an infinite-dimensional countable dimensional compact space.

7. Prove that dim $\mathbb{R}^n = n$ for every n. Use this to conclude that $\mathbb{R}^n$ and $\mathbb{R}^m$ are not homeomorphic for different n and m.

4.2. Zero-Dimensional Spaces and Applications

The spaces X with $\dim X = 0$ are usually called *zero-dimensional.* Recall that by definition,

$\dim X = 0 \quad \Leftrightarrow \quad X \neq \emptyset$ and if A and B are disjoint closed subsets of X then $\emptyset$ is a partition between A and B.

$\Leftrightarrow \quad X \neq \emptyset$ and if A and B are disjoint closed subsets of X then there is a clopen subset U of X containing A but missing B.

It turns out that zero-dimensional spaces are, in a sense, the "building blocks" of other spaces and for that reason we study them in a separate section in some detail.

The following characterization of zero-dimensional spaces is useful.

4.2.1. PROPOSITION: *Let* X *be a nonempty space. The following statements are equivalent:*

(a) X *is zero-dimensional,*

(b) *for every* $x \in X$ *and for every neighborhood* U *of* x *there exists a clopen subset* C *of* X *such that* $x \in C \subseteq U$,

(c) *the clopen subsets of* X *form a (open) basis for* X,

(d) X *has a countable basis consisting of clopen sets,*

(e) *each open cover* $\mathcal{U}$ *of* X *has an open refinement* $\mathcal{V}$ *such that the elements of* $\mathcal{V}$ *are pairwise disjoint (and therefore, clopen).*

PROOF: The implications (a) $\Rightarrow$ (b) $\Rightarrow$ (c) are trivialities.

For (c) $\Rightarrow$ (d), let $\mathcal{B}$ be a countable open basis for X. By (c) for each $B \in \mathcal{B}$ there is a family $\mathcal{U}(B)$ of clopen subsets of X such that $\bigcup\mathcal{U}(B) = B$. Since each $B \in \mathcal{B}$ is Lindelöf, being separable metric, without loss of generality we may assume that $\mathcal{U}(B)$ is countable. Now clearly $\mathcal{U} = \bigcup_{B\in\mathcal{B}} \mathcal{U}(B)$ is a countable basis for X consisting of clopen subsets of X.

For (d) $\Rightarrow$ (e), let $\mathcal{B}$ be any countable clopen basis for X and let $\mathcal{U}$ be an open cover of X. For each $x \in X$ there exists $B(x) \in \mathcal{B}$ such that $B(x)$ is contained in an element of $\mathcal{U}$. The cover $\{B(x): x \in X\}$ of X refines $\mathcal{U}$ and is countable; say it can be enumerated as $\{B_n: n \in \mathbb{N}\}$. For each $n \in \mathbb{N}$ define $V_n \subseteq X$ by

$$V_n = B_n \setminus \bigcup_{m<n} B_m.$$

It is easy to see that $\mathcal{V} = \{V_n: n \in \mathbb{N}\}$ consists of pairwise disjoint open sets and that each element of $\mathcal{V}$ is contained in an element of $\mathcal{U}$. The only "problem" is to prove that $\mathcal{V}$ covers X. For

$x \in X$ let $n(x)$ be the first natural number n with the property that $x \in B_n$. Then $x \in V_n$.

For (e) $\Rightarrow$ (a), let A and B be disjoint closed subsets of X. For each $x \in X$ there is a neighborhood $U(x)$ of x such that $U(x) \cap A = \emptyset$ or $U(x) \cap B = \emptyset$. By (e), the open cover $\mathcal{U} = \{U(x): x \in X\}$ has a disjoint open refinement $\mathcal{V}$. Now let $U = \bigcup\{V \in \mathcal{V}: V \cap A \neq \emptyset\}$. Observe that U is open and closed, since its complement is open, that U contains A and misses B. □

In the following result we summarize relevant information on zero-dimensional spaces.

4.2.2. THEOREM: (1) *If* X *is zero-dimensional and if* $Y \subseteq X$ *is nonempty then* Y *is zero-dimensional.*

(2) *If* X_n *is zero-dimensional for every* $n \in \mathbb{N}$ *then so is the product* $\prod_{n=1}^{\infty} X_n$.

(3) *If* X *is zero-dimensional then* X *can be imbedded in* $\{0,1\}^\infty$.

(4) *If* X *is zero-dimensional then* X *admits a zero-dimensional compactification.*

(5) *Let* X *be a zero-dimensional subspace of* Y. *Then for all disjoint closed subsets* A *and* B *of* Y *there exists a partition* S *between* A *and* B *in* Y *such that* $S \cap X = \emptyset$.

(6) *If a space* X *is the union of at most* n+1 *zero-dimensional subspaces then* $\dim X \leq n$.

PROOF: *We prove* (1).

This follows immediately from proposition 4.2.1(c).

We prove (2).

For each $n \in \mathbb{N}$ let $\mathcal{B}_n$ be a countable basis for X_n consisting of clopen sets. The family of sets of the form

$$E_1 \times E_2 \times \cdots \times E_m \times X_{m+1} \times X_{m+2} \times \cdots\cdots,$$

where $m \in \mathbb{N}$ and for $i \leq m$, $E_i \in \mathcal{B}_i$, is clearly a countable base for $\prod_{n=1}^{\infty} X_n$ consisting of clopen sets. An appeal to proposition 4.2.1(d) now yields the desired result.

We prove (3).

Let $\mathcal{B}$ be a countable basis for X consisting of clopen sets (proposition 4.2.1(d)). Enumerate $\mathcal{B}$ as $\{B_n: n \in \mathbb{N}\}$. For each $n \in \mathbb{N}$ define $f_n: X \to \{0,1\}$ by

$$f_n(x) = \begin{cases} 0 & (x \in B_n), \\ 1 & (x \notin B_n). \end{cases}$$

Then f_n is clearly continuous. It is easy to see that the evaluation map $e: X \to \{0,1\}^\infty$ defined by

$$e(x)_n = f_n(x) \qquad (n \in \mathbb{N}),$$

is an imbedding.

We prove (4).

By (3) we may assume that X is a subspace of $K = \{0,1\}^\infty$. Since K is compact and zero-dimensional by (2), the desired compactification of X is the closure of X in K.

We prove (5).

Let A and B be disjoint closed subsets of Y and let U and V be disjoint neighborhoods of A and B, respectively, such that $\overline{U} \cap \overline{V} = \emptyset$ (corollary 1.4.16). Since X is zero-dimensional, $\emptyset$ is a partition between $\overline{U} \cap X$ and $\overline{V} \cap X$ in X. Now apply lemma 4.1.5.

We prove (6).

Let $X = \bigcup_{i=1}^{n+1} X_i$ with $\dim X_i \leq 0$ for $i \leq n+1$. Let $\tau = \{(A_i,B_i): i \leq n+1\}$ be a family of n+1 pairs of disjoint closed subsets of X. By (5), for each $i \leq n+1$ there exists a partition L_i between A_i and B_i in X such that $L_i \cap X_i = \emptyset$. Then $\bigcap_{i=1}^{n+1} L_i = \emptyset$ so that τ is inessential. We conclude that $\dim X \leq n$, as required. □

We shall now characterize the zero-dimensional subspaces of the real line $\mathbb{R}$.

4.2.3. PROPOSITION: *A nonempty subspace* X *of* $\mathbb{R}$ *is zero-dimensional if and only if it does not contain any (nondegenerate) interval.*

PROOF: Assume that $X \subseteq \mathbb{R}$ is zero-dimensional. If X contains an interval E then X contains a subspace which is not zero-dimensional by connectivity (exercise 4.1.5). This contradicts theorem 4.2.2(1).

Now assume that $X \subseteq \mathbb{R}$ is such that it contains no intervals. Then the set $D = \mathbb{R} \setminus X$ is dense in $\mathbb{R}$. The collection

$$\{(d_1,d_2) \cap X: d_1 < d_2 \text{ and } d_1,d_2 \in D\}$$

is easily seen to be a basis for X consisting of clopen subsets of X. Consequently, X is zero-dimensional by proposition 4.2.1(c). □

This proposition gives us a rich supply of zero-dimensional spaces. For example, the rationals $\mathbb{Q}$, the irrationals $\mathbb{P}$, the product $\mathbb{Q} \times \mathbb{P}$, etc., are all zero-dimensional. The following example is of particular interest:

4.2.4. EXAMPLE: *The Cantor middle-third set* C.

From $I = [0,1]$ remove the interval $(\frac{1}{3},\frac{2}{3})$, i.e. the "middle-third" interval. From the remaining two intervals, again remove their "middle-thirds". Continue in this way. At stage i of the construction we have a disjoint family $\mathcal{F}_i$ of 2^i closed subintervals of I, each of length 3^{-i}. The union of $\mathcal{F}_i$ is denoted by H_i. Observe that H_i is a closed subset of I and hence is compact. The intersection of the H_i's is called *the Cantor middle-third set*, **C**.

Clearly, **C** is closed in I and hence compact. Also, **C** is zero-dimensional by proposition 4.2.3, for if (a,b) is a nondegenerate subinterval of $\mathbb{R}$ then there exists $i \in \mathbb{N}$ such that $3^{-i} < b-a$ which implies, by the fact that **C** can be covered by a disjoint family of intervals each of length 3^{-i}, that (a,b) cannot be contained in **C**.

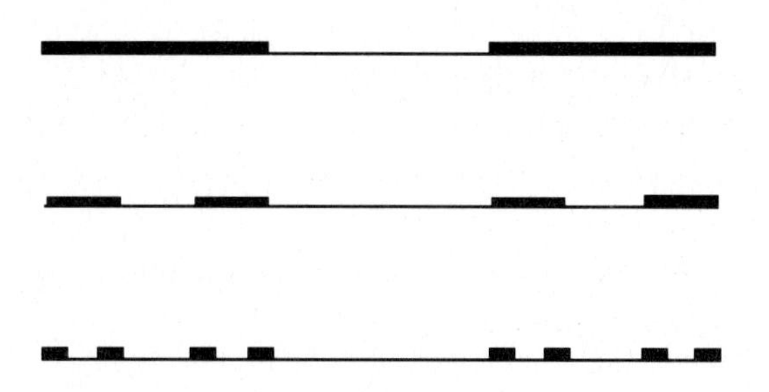

Figure 4.2.1.

Observe that if $F \in \mathcal{F}_i$ then $F \cap \mathbf{C} \neq \emptyset$. For at stage i+1 of the construction there exists an interval $A_1 \in \mathcal{F}_{i+1}$ such that $A_1 \subseteq F$. In addition, at stage i+2 of the construction, there exists an interval $A_2 \in \mathcal{F}_{i+2}$ such that $A_2 \subseteq A_1$. Proceeding in this way inductively, it is easy to construct a decreasing sequence $(A_n)_n$ of elements of $\mathcal{F}_{i+n}$ such that $A_n \subseteq F$ for every n. By compactness of I, $\emptyset \neq \bigcap_{n=1}^{\infty} A_n \subseteq \mathbf{C}$, as desired.

It is easy to see that **C** is the subspace of I consisting of all points that have a tryadic expansion in which the digit 1 does not occur, i.e., the set

$$\{x = \sum_{i=1}^{\infty} \frac{2x_i}{3^i} : x_i \in \{0,1\} \text{ for every } i\}.$$

We claim that the set **C** has no isolated points. If x belongs to **C** and $a < x < b$ then choose i so large that $3^{-i} < \min\{x-a, b-x\}$. Let $F \in \mathcal{F}_i$ be such that $x \in F$. Then $F \subseteq (a,b)$. There are precisely two disjoint elements of $\mathcal{F}_{i+1}$ that are contained in F. By the above, both these elements contain at least one point of **C**. We conclude that $(a,b) \cap \mathbf{C}$ contains at least two points.

The Cantor set **C** is a very interesting zero-dimensional space. There exists a simple topological characterization of **C** which we shall now present. If $\mathcal{U}$ is a collection of subsets of a space X then the *mesh* of $\mathcal{U}$ is the number

$$\text{mesh}(\mathcal{U}) = \sup\{\text{diam}(U)\colon U \in \mathcal{U}\},$$

cf. §3.3. We allow mesh($\mathcal{U}$) to be equal to ∞.

4.2.5. THEOREM: C *is (topologically) the unique nonempty zero-dimensional compact space without isolated points.*

PROOF: Since in example 4.2.4 we showed that zero-dimensional compact spaces without isolated points exist, it suffices to prove that if X and Y are such spaces then X and Y are homeomorphic. To this end, let X and Y be compact zero-dimensional spaces without isolated points. By induction on n, we shall construct elements $m(n) \in \mathbb{N}$ and finite disjoint clopen covers $\mathcal{U}_n = \{U_{n,i}\colon i \le m(n)\}$ and $\mathcal{V}_n = \{V_{n,i}\colon i \le m(n)\}$ of X and Y, respectively, consisting of nonempty sets, such that $m(1) = 1$, $\mathcal{U}_1 = \{X\}$, $\mathcal{V}_1 = \{Y\}$, and for $n \ge 2$,

(1) $\text{mesh}(\mathcal{U}_n) < 1/n$, $\text{mesh}(\mathcal{V}_n) < 1/n$,
(2) $\mathcal{U}_n < \mathcal{U}_{n-1}$, $\mathcal{V}_n < \mathcal{V}_{n-1}$, and
(3) for $i \le m(n-1)$ and $j \le m(n)$: $U_{n,j} \subseteq U_{n-1,i}$ iff $V_{n,j} \subseteq V_{n-1,i}$.

Suppose that we constructed the covers $\mathcal{U}_p$ and $\mathcal{V}_p$ for all $p \le n$, $n \ge 1$. Pick an arbitrary $i \le m(n)$. By proposition 4.2.1(e) there exist covers $\mathcal{E}$ and $\mathcal{F}$ of $U_{n,i}$ and $V_{n,i}$, respectively, consisting of pairwise disjoint open sets, such that

(4) $\mathcal{E} < \{B(x,\frac{1}{2(n+1)})\colon x \in U_{n,i}\}$, and
(5) $\mathcal{F} < \{B(y,\frac{1}{2(n+1)})\colon y \in V_{n,i}\}$.

By compactness of $U_{n,i}$ and $V_{n,i}$, the covers $\mathcal{E}$ and $\mathcal{F}$ are finite. We assume that each element of $\mathcal{E}$ as well as $\mathcal{F}$ is nonempty. We claim that, without loss of generality, $\mathcal{E}$ and $\mathcal{F}$ have the same cardinality. If not, suppose that e.g. $|\mathcal{E}| < |\mathcal{F}|$, pick an arbitrary $E \in \mathcal{E}$. Since X has no isolated points, there exist distinct $x,y \in E$. In addition, since X is zero-dimensional, its clopen sets form a base (proposition 4.2.1(c)) so there is a clopen subset $C \subseteq E$ such that $x \in C$ and $y \notin C$. So E can be split into two nonempty clopen sets, namely, C and $E\setminus C$. By repeating the same procedure with C and/or $E\setminus C$, etc., it follows that we can split E in as many nonempty clopen sets as we wish. So there is a disjoint clopen cover $\mathcal{G}$ of E consisting of $|\mathcal{F}| - |\mathcal{E}| + 1$ nonempty sets. Now replace $\mathcal{E}$ by $(\mathcal{E}\setminus\{E\}) \cup \mathcal{G}$.

We conclude that for $i \le m(n)$ there are covers $\mathcal{E}_i$ and $\mathcal{F}_i$ of $U_{n,i}$ and $V_{n,i}$, respectively, such that

(6) $|\mathcal{E}_i| = |\mathcal{F}_i|$, and

(7) $\mathcal{E}_i$ and $\mathcal{F}_i$ are pairwise disjoint and consist of nonempty open sets, each of diameter less than $\frac{1}{n+1}$.

Put $\mathcal{U}_{n+1} = \bigcup_{i=1}^{m(n)} \mathcal{E}_i$ and $\mathcal{V}_{n+1} = \bigcup_{i=1}^{m(n)} \mathcal{F}_i$. By construction, we can enumerate $\mathcal{U}_{n+1}$ and $\mathcal{V}_{n+1}$ in such a way that (3) is satisfied for n+1. This completes the inductive construction. Observe that $\bigcup_{n=1}^{\infty} \mathcal{U}_n$ is a base for X and, similarly, that $\bigcup_{n=1}^{\infty} \mathcal{V}_n$ is a base for Y.

Now if $x \in X$ then for each $n \in \mathbb{N}$ there is a unique $i(n,x) \leq m(n)$ such that $x \in U_{n,i(n,x)}$. By (3) it follows easily that the collection $\{V_{n,i(n,x)}\}_n$ is decreasing and since Y is compact, we get $\bigcap V_{n,i(n,x)} \neq \emptyset$. So by (1), $\bigcap V_{n,i(n,x)}$ consists of precisely one point. By interchanging the role of X and Y, these remarks imply that the function $f: X \to Y$ defined by

$$f(x) = y \Leftrightarrow \text{ for every } n \in \mathbb{N} \text{ and } i \leq m(n) \ (x \in U_{n,i} \text{ iff } y \in V_{n,i}),$$

is a bijection. Also f is continuous since for $n \in \mathbb{N}$ and $i \leq m(n)$, $f^{-1}(V_{n,i}) = U_{n,i}$. Similarly, f^{-1} is continuous. We conclude that f is a homeomorphism. □

A space homeomorphic to **C** is called *a Cantor set* from now on. The characterization of **C** allows us to identify "many" Cantor sets.

4.2.6. COROLLARY: *Let* X *be any compact nonempty zero-dimensional space. Then* $X \times \mathbf{C}$ *is a Cantor set.*

PROOF: $X \times \mathbf{C}$ is compact and zero-dimensional by theorem 4.2.2(2). Since **C** has no isolated points it follows that $X \times \mathbf{C}$ has no isolated points. An appeal to theorem 4.2.5 now yields the desired result. □

Since $\{0,1\}^\infty$ is compact and zero-dimensional by theorem 4.2.2(2), and has clearly no isolated points, we conclude that, in particular, $\mathbf{C} \approx \{0,1\}^\infty$. This consequence can also be verified directly: the function $f: \{0,1\}^\infty \to \mathbf{C}$ defined by

$$f(x) = \sum_{i=1}^{\infty} \frac{2x_i}{3^i}$$

is a homeomorphism.

By a simple diagonal argument it follows that $\{0,1\}^\infty$ is uncountable, so $\mathbf{C}$ is uncountable as well. Since by corollary 4.2.6, $\mathbf{C}$ is homeomorphic to $\mathbf{C} \times \mathbf{C}$, this remark implies that there is a family consisting of uncountably many pairwise disjoint Cantor sets in $\mathbb{R}$. Since there are only countably many rational numbers in $\mathbb{R}$ one of these Cantor sets misses $\mathbb{Q}$, i.e. there exists a Cantor set K in $\mathbb{R}$ consisting entirely of irrational numbers.

Let $\mathbb{P}$ denote the set of irrational numbers in $\mathbb{R}$. Since as was observed in theorem 4.2.2(3), every zero-dimensional space can be imbedded in $\{0,1\}^\infty$, we now obtain:

4.2.7. COROLLARY: *Every zero-dimensional space can be imbedded in* $\mathbf{C}$ *and also in* $\mathbb{P}$. □

We now aim at proving that for every compact space X there exists a continuous surjection from $\mathbf{C}$ onto X. This result is of interest in dimension theory, for it implies that there exists a continuous surjection from the zero-dimensional space $\mathbf{C}$ onto the 1-dimensional space I. So continuous functions can raise dimension.

4.2.8. LEMMA: *The function* $f: \mathbf{C} \times \mathbf{C} \to [-1,1]$ *defined by* $f(x,y) = x-y$ *is a continuous surjection.*

PROOF: Since $\mathbf{C} \subseteq I$, f is well-defined and clearly continuous. Now consider the set $F_1 = [0,\frac{1}{3}] \cup [\frac{2}{3},1]$, i.e. the first approximation to $\mathbf{C}$. It is a triviality to verify that $F_1 - F_1$, i.e. the set $\{x-y: x,y \in F_1\}$, is equal to $[-1,1]$. The second approximation F_2 to $\mathbf{C}$, namely the set $F_2 = [0,\frac{1}{9}] \cup [\frac{2}{9},\frac{1}{3}] \cup [\frac{2}{3},\frac{7}{9}] \cup [\frac{8}{9},1]$, has also the property that $F_2 - F_2 = [-1,1]$. By induction it follows that all approximations F_n, $n \in \mathbb{N}$, to $\mathbf{C}$ that were used to define $\mathbf{C}$, have the same property (the induction step can be performed by noting that $F_n = (\frac{1}{3}\cdot F_{n-1}) \cup (\frac{1}{3}\cdot F_{n-1} + \frac{2}{3})$, a union of two shrunken copies of F_{n-1}.). Since the F_n's decrease and $\mathbf{C} = \bigcap_{n=1}^\infty F_n$, an easy compactness-type argument shows that

$$\mathbf{C} - \mathbf{C} = [-1,1],$$

as required. □

We now come to the announced

4.2.9. THEOREM: *Every compact space is a continuous image of* $\mathbf{C}$.

PROOF: By lemma 4.2.8 it follows that there exists a continuous surjection $f: \mathbf{C}^\infty \to Q$. Now let X be any compact space. According to theorem 1.4.18, we may assume that $X \subseteq Q$. Put Y =

$f^{-1}(X)$. The restriction $g = f \mid Y: Y \to X$ is a continuous surjection. Observe that Y is zero-dimensional by theorem 4.2.2(2), (1). By corollary 4.2.6, $\mathbf{C} \times Y$ and $\mathbf{C}$ are homeomorphic. The required continuous surjection can therefore be obtained as the composition of the map g, the projection from $\mathbf{C} \times Y$ onto Y and a homeomorphism between $\mathbf{C}$ and $\mathbf{C} \times Y$.□

4.2.10. *Remark:* Theorem 4.2.9 can be used to construct *"space filling curves"*, i.e. continuous surjections from I onto I^2. For let $g: \mathbf{C} \to I^2$ be a continuous surjection (theorem 4.2.9). Then, by corollary 1.4.14, g can be extended to a continuous surjection $\bar{g}: I \to I^2$.

We say that a space X is *totally disconnected* if for all distinct $x,y \in X$ there exists a clopen set C in X such that $x \in C$ but $y \notin C$. It is clear that every zero-dimensional space is totally disconnected. In view of proposition 4.2.1(b), the question naturally arises whether every totally disconnected space is zero-dimensional. The answer to this question is in the negative, as the following example shows.

4.2.11. EXAMPLE: *Erdös' space.*

Put $E = \{x \in l^2: x_i \text{ is rational for every } i\}$, cf. §1.2. We claim that E is totally disconnected and that $\dim E \geq 1$ (we shall prove in example 4.5.10 that in fact, $\dim E = 1$). Lemma 1.2.6 shows that the topology on l^2 is finer than the topology that l^2 inherits from $\mathbb{R}^\infty$. Recall that $\mathbb{Q}$ denotes the space of rational numbers. We conclude that E admits a one-to-one continuous function into the product $\mathbb{Q}^\infty$, which is zero-dimensional by proposition 4.2.3 and theorem 4.2.2(2). This implies that E is totally disconnected.

We shall now prove that E is not zero-dimensional by showing that if $U \subseteq l^2$ is open and $\underline{0} \in U \subseteq \{x \in l^2 : \|x\| < 1\}$ then $U \cap E$ is not clopen in E. To this end, let U be such a neighborhood of $\underline{0}$. We shall inductively define a sequence $q_i \in \mathbb{Q}$, $i \in \mathbb{N}$, such that for every i,

(1) $$x_i = (q_1,\cdots,q_i,0,0,\cdots) \in U \text{ and } d(x_i, l^2 \setminus U) \leq \tfrac{1}{i}.$$

Let $q_1 = 0$ and assume the q_j to be defined for $j \leq i-1$, $i \geq 2$. For $0 \leq m \leq i$ put

$$x(i,m) = (q_1,\cdots,q_{i-1},m/i,0,0,\cdots).$$

Observe that $x(i,0) = x_{i-1} \in U$ and that $x(i,i) = (q_1,\cdots,q_{i-1},1,0,0,\cdots)$ has norm at least one and therefore does not belong to U. So there exists $0 \leq m_0 < i$ such that $x(i,m_0) \in U$ but $x(i,m_0+1) \notin U$. Now put $q_i = m_0/i$. Then q_i is as required since

$$x_i = (q_1,\cdots,q_i,0,0,\cdots) = x(i,m_0) \in U,$$

and

$$d(x_i, l^2\setminus U) \leq d(x(i,m_0),x(i,m_0+1)) = \|(0,0,\cdots,-1/i,0,0,\cdots)\| = 1/i.$$

This completes the inductive construction.

By (1) we have

$$\sum_{j=1}^{i} q_j^2 \leq 1 \text{ for every } i \in \mathbb{N},$$

and hence

$$\sum_{i=1}^{\infty} q_i^2 \leq 1.$$

We conclude that the point $q = (q_1,q_2,\cdots)$ belongs to l^2, and hence to E. Since clearly,

$$\lim_{i\to\infty} x_i = q \qquad (\text{in } l^2)$$

(lemma 1.2.6), it follows by (1) that on the one hand

q belongs to the closure of $U \cap E$ in E,

and on the other hand

$$d(q, l^2\setminus U) = \lim_{i\to\infty} d(q_i, l^2\setminus U) = 0, \text{ so } q \notin U.$$

We conclude that $U \cap E$ is not closed in E, hence not clopen. □

In §4.7 we shall see that totally disconnected spaces of every dimension exist.

Exercises for §4.2.

1. Let X be a space, let $A \subseteq X$ be closed, and assume that $X\setminus A$ is zero-dimensional. Prove that A is a retract of X (Hint: use lemma 1.4.12 and proposition 4.2.1(e)).

2. Let X be a space. Prove that X is zero-dimensional if and only if every closed subspace of X is a retract of X.

3. Let X be a space such that X can be written as $A \cup B$, with both A and B zero-dimensional and A closed. Prove that X is zero-dimensional.

4. Prove that every nonempty space with cardinality less than the cardinality of $\mathbb{R}$ is zero-dimensional (Hint: Use Urysohn's lemma).

5. Prove that every nonempty countable space is zero-dimensional without using exercise 4.

6. Let E be Erdös' space. Prove that E is homeomorphic to $E \times E$.

7. Let $\mathbb{P}$ be the space of irrational numbers in $\mathbb{R}$. Prove that $\mathbb{P}$ is topologically complete and nowhere locally compact (i.e. no point has a neighborhood with compact closure). Prove that $\mathbb{P}$ is topologically the unique topologically complete, nowhere locally compact, zero-dimensional space (Hint: Use proposition 4.2.1(e) and imitate the proof of theorem 4.2.5).

8. Prove that for every nonempty space X there exists a zero-dimensional space Y such that Y can be mapped onto X. (Hint: Let B be a compactification of X and apply theorem 4.2.9).

9. Prove that the space of rational numbers is topologically the unique countable space without isolated points (Hint: Observe that such a space is zero-dimensional and then to some extent imitate the proof of theorem 4.2.5; one has to be careful however, since there are no "general" results available now that ensure that certain decreasing sequences of closed sets have nonempty intersection).

Let X be a set. We call a metric d on X *non-Archimedean* if $d(x,y) \leq \max(d(x,z),d(y,z))$ for all $x,y,z \in X$.

10. Prove that for a space X the following statements are equivalent: (a) $\dim X = 0$, and (b) there is an admissible non-Archimedean metric on X.

11. Let X be a locally compact totally disconnected space. Prove that $\dim X \leq 0$.

4.3. Translation into Open Covers

We defined the covering dimension dim X in terms of properties of the family of all closed subsets of X. In this section we will see that dim X is also describable in terms of properties of the family of all open covers of X. As an application of the results derived here we present a simple proof that $\dim \mathbb{R}^n \leq n$. In combination with earlier results this yields $\dim \mathbb{R}^n = n$. We conclude that $\mathbb{R}^n$ and $\mathbb{R}^m$ are not homeomorphic if $n \neq m$.

Let X be a space and let $\mathcal{A}$ and $\mathcal{B} = \{B(A): A \in \mathcal{A}\}$ be families of subsets of X. We say that $\mathcal{B}$ is a *swelling* of $\mathcal{A}$ if

(1) for every $A \in \mathcal{A}$, $A \subseteq B(A)$, and
(2) for every finite $\mathcal{F} \subseteq \mathcal{A}$, $\bigcap\mathcal{F} = \emptyset$ iff $\bigcap_{A\in\mathcal{F}} B(A) = \emptyset$.

Observe that if $B(A_0), B(A_1) \in \mathcal{B}$ are distinct then so are A_0 and A_1, but not conversely.

4.3.1. PROPOSITION: *Let $\mathcal{F}$ be a locally finite collection of closed subsets of a space* X. *Then $\mathcal{F}$ has a swelling consisting of open subsets of* X.

PROOF: Since locally finite collections are countable (see the remark in §1.4 following the definition of locally finite collection), we can enumerate $\mathcal{F}$ as $\{F_i: i \in \mathbb{N}\}$. Without loss of generality, let $F_1 = \emptyset$. By induction on $i \in \mathbb{N}$ we shall construct an open neighborhood U_i of F_i such that for every i,

$$\mathcal{F}_i = \{\overline{U}_1,\cdots,\overline{U}_i,F_{i+1},F_{i+2},\cdots\}$$

is a swelling of $\mathcal{F}$. Put $U_1 = \emptyset$, and assume that for some i the sets $U_1,\cdots,U_i$ have been constructed. Put

$$\mathcal{B} = \{\bigcap\mathcal{E}: \mathcal{E} \text{ is a finite subcollection of } \mathcal{F}_i \text{ and } (\bigcap\mathcal{E}) \cap F_{i+1} = \emptyset\}.$$

Observe that $\mathcal{B}$ is locally finite and put $B = \bigcup\mathcal{B}$. Then B is closed by exercise 1.4.13 and clearly $B \cap F_{i+1} = \emptyset$. Consequently, by corollary 1.4.16 there exists an open neighborhood U_{i+1} of F_{i+1} the closure of which misses B. It is easily seen that U_{i+1} is as required.

We claim that $\{U_i: i \in \mathbb{N}\}$ is a swelling of $\mathcal{F}$. To this end, take $i(1),\cdots,i(n) \in \mathbb{N}$ and assume that $\bigcap_{j=1}^n F_{i(j)} = \emptyset$. Let $m = \max\{i(1),\cdots,i(n)\}$. By construction, $\{U_1,\cdots,U_m\}$ is a swelling of $\{F_1,\cdots,F_m\}$. We conclude that $\bigcap_{j=1}^n U_{i(j)} = \emptyset$. □

4.3.2. COROLLARY: *Let $\mathcal{F}$ be a locally finite family of closed subsets of a space* X. *Also, for every* $F \in \mathcal{F}$ *let* V(F) *be a neighborhood of* F. *Then there exists a swelling* $\{U(F): F \in \mathcal{F}\}$ *of $\mathcal{F}$ consisting of open subsets of* X *such that for every* $F \in \mathcal{F}$ we have $\overline{U(F)} \subseteq V(F)$.

PROOF: By proposition 4.3.1 there exists an "open" swelling $\{W(F): F \in \mathcal{F}\}$ of $\mathcal{F}$. By corollary 1.4.16 there exists for every $F \in \mathcal{F}$ an open neighborhood U(F) such that $\overline{U(F)} \subseteq V(F) \cap W(F)$. We claim that $\{U(F): F \in \mathcal{F}\}$ is as required. To this end, let $\mathcal{G} \subseteq \mathcal{F}$ be finite such that $\cap\mathcal{G} = \emptyset$. Then $\bigcap_{F\in \mathcal{G}} W(F) = \emptyset$ from which it follows that $\bigcap_{F\in \mathcal{G}} U(F) = \emptyset$.□

Let X be a space and let $\mathcal{A}$ and $\mathcal{B} = \{B(A): A \in \mathcal{A}\}$ be covers of X (not necessarily by open or closed sets). We say that $\mathcal{B}$ is a *shrinking* of $\mathcal{A}$ if for every $A \in \mathcal{A}$, $B(A) \subseteq A$. Again observe that if $B(A_0), B(A_1) \in \mathcal{B}$ are distinct then so are A_0 and A_1, but not conversely. We call $\mathcal{B}$ an *open* shrinking if $\mathcal{B}$ consists of open subsets of X. *Closed* shrinkings are defined similarly. Observe that each shrinking of a locally finite (star-finite) cover is again locally finite (star-finite).

4.3.3. PROPOSITION: *Let* X *be a space and let* $\mathcal{U}$ *be an open cover of* X. *Then* $\mathcal{U}$ *admits a closed shrinking.*

PROOF: First assume that $\mathcal{U}$ is countable. Enumerate $\mathcal{U}$ as $\{U_n: n \in \mathbb{N}\}$. By theorem 1.4.17, for each n there exists a continuous function $f_n: X \to I$ such that $f_n^{-1}((0,1]) = U_n$. Define $f: X \to I$ by

$$f(x) = \sum_{n=1}^{\infty} 2^{-n} f_n(x).$$

Since

$$\sum_{n=1}^{\infty} 2^{-n} = 1,$$

we get by proposition 1.3.4 that f is continuous, cf. the proof of theorem 3.6.17. As $\mathcal{U}$ covers X, $f(x) > 0$ for every $x \in X$. Now for $n \in \mathbb{N}$ put

$$A_n = \{x \in X: f_n(x) \geq \frac{f(x)}{2}\}.$$

By continuity of the functions f_n and f, it follows easily that each A_n is closed. Also, since $f(x) > 0$ for all x, $A_n \subseteq U_n$.

We claim that the A_n's cover X. To this end, assume that there exists $x \in X$ such that for every n, $x \notin A_n$. Then for every n,

$$f_n(x) < \frac{f(x)}{2},$$

so that

$$f(x) = \sum_{n=1}^{\infty} 2^{-n} f_n(x) \leq \tfrac{1}{2} \cdot \sum_{n=1}^{\infty} 2^{-n} f(x) = \frac{f(x)}{2},$$

which is a contradiction since $f(x) > 0$.

Now let $\mathcal{U}$ be an arbitrary open cover of X. Let $\mathcal{V}$ be a countable subcover of $\mathcal{U}$. By the above, there exists a closed shrinking $\mathcal{W} = \{W(V)\colon V \in \mathcal{V}\}$ of $\mathcal{V}$. For each $U \in \mathcal{U}$ define a subset $E(U) \subseteq X$ by

$$E(U) = \begin{cases} W(U) & (U \in \mathcal{V}), \\ \emptyset & (U \notin \mathcal{V}). \end{cases}$$

Since $\mathcal{W}$ covers X, $\mathcal{E} = \{E(U)\colon U \in \mathcal{U}\}$ is clearly a closed shrinking of $\mathcal{U}$.□

4.3.4. COROLLARY: *Let* X *be a space such that* $\dim X \leq n < \infty$. *Then for every countable open cover* $\mathcal{U}$ *of* X *and for every subcollection* $\mathcal{F} \subseteq \mathcal{U}$ *of cardinality* $n+2$ *there exists an open shrinking* $\mathcal{V} = \{V(U)\colon U \in \mathcal{U}\}$ *of* $\mathcal{U}$ *having the following properties:*

(1) *for every* $U \in \mathcal{U}$, $V(U) \subseteq \overline{V(U)} \subseteq U$,

(2) $\bigcap_{U \in \mathcal{F}} V(U) = \emptyset$.

PROOF: Enumerate $\mathcal{U}$ as $\{U_i\colon i \in \mathbb{N}\}$. Without loss of generality, $\mathcal{F} = \{U_1,\cdots,U_{n+2}\}$. By proposition 4.3.3 there exists a closed shrinking $\{B_i\colon i \in \mathbb{N}\}$ of $\mathcal{U}$. By corollary 1.4.16 for every i there exists an open set E_i such that $B_i \subseteq E_i \subseteq \overline{E_i} \subseteq U_i$. For $i \leq n+1$ define $A_i = X \backslash E_i$. Then $\{(A_i,B_i)\colon i \leq n+1\}$ is a family of $n+1$ pairs of disjoint closed subsets of X. Since $\dim X \leq n$, there exist open sets $V_i, W_i \subseteq X$ such that

(*) $\quad A_i \subseteq V_i,\ B_i \subseteq W_i,\ V_i \cap W_i = \emptyset$, and

(**) $\quad \bigcap_{i=1}^{n+1} X \backslash (V_i \cup W_i) = \emptyset.$

Observe that by by (**),

$$\bigcup_{i=1}^{n+1} (V_i \cup W_i) = X.$$

For $m \geq n+2$, put $W_m = E_m \cap \bigcup_{i=1}^{n+1} V_i$. Observe that the closure of W_m is contained in U_m for every $m \geq n+2$. For convenience, put $E = \bigcup_{m=n+2}^{\infty} E_m$. Then

$$\bigcup_{m=1}^{\infty} W_m = \bigcup_{m=1}^{n+1} W_m \cup \bigcup_{m=n+2}^{\infty} (E_m \cap \bigcup_{i=1}^{n+1} V_i)$$

$$= \bigcup_{m=1}^{n+1} W_m \cup (E \cap \bigcup_{i=1}^{n+1} V_i)$$
$$= (\bigcup_{m=1}^{n+1} W_m \cup E) \cap (\bigcup_{m=1}^{n+1} W_m \cup \bigcup_{i=1}^{n+1} V_i)$$
$$= \bigcup_{m=1}^{n+1} W_m \cup E$$
$$\supseteq \bigcup_{m=1}^{n+1} B_m \cup \bigcup_{m=n+2}^{\infty} B_m$$
$$= X.$$

We conclude that $\mathcal{W} = \{W_i: i \in \mathbb{N}\}$ is an open shrinking of the cover $\mathcal{U}$, such that for every i,

$$\overline{W}_i \subseteq U_i.$$

Moreover, from (*) we get

$$\bigcap_{i=1}^{n+2} W_i = \bigcap_{i=1}^{n+1} W_i \cap (E_{n+2} \cap \bigcup_{i=1}^{n+1} V_i)$$
$$\subseteq \bigcap_{i=1}^{n+1} W_i \cap \bigcup_{i=1}^{n+1} V_i$$
$$= \emptyset,$$

as required. □

Let $\mathcal{U}$ be a cover of a space X and let $n \geq 0$ (we do not assume $\mathcal{U}$ to be open). We say that the *order* of $\mathcal{U}$ is at most n, $\mathrm{ord}(\mathcal{U}) \leq n$, iff for every $x \in X$,

$$|\{U \in \mathcal{U}: x \in U\}| \leq n+1.$$

We now come to the following result.

4.3.5. THEOREM: *Let* X *be a nonempty space and let* $n \geq 0$. *The following statements are equivalent:*

(a) $\dim X \leq n$,

(b) *for every open cover* $\mathcal{U}$ *of* X *there exists a locally finite closed refinement* $\mathcal{V}$ *of* $\mathcal{U}$ *such that* $\mathrm{ord}(\mathcal{V}) \leq n$,

(c) *for every open cover* $\mathcal{U}$ *of* X *there exists an open refinement* $\mathcal{V}$ *of* $\mathcal{U}$ *with* $\mathrm{ord}(\mathcal{V}) \leq n$,

(d) *for every open cover* $\mathcal{U}$ *of* X *there exists a closed shrinking* $\mathcal{V}$ *of* $\mathcal{U}$ *with* $\mathrm{ord}(\mathcal{V}) \leq n$,

(e) *for every open cover* $\mathcal{U}$ *of* X *there exists an open shrinking* $\mathcal{V}$ *of* $\mathcal{U}$ *with* $\mathrm{ord}(\mathcal{V}) \leq n$,

(f) *for every finite open cover* $\mathcal{U}$ *of* X *there exists a closed shrinking* $\mathcal{V}$ *of* $\mathcal{U}$ *such that* $\mathrm{ord}(\mathcal{V}) \leq n$,

(g) *for every finite open cover* $\mathcal{U}$ *of* X *there exists an open shrinking* $\mathcal{V}$ *of* $\mathcal{U}$ *such that* $\mathrm{ord}(\mathcal{V}) \leq n$.

PROOF: *We prove* (a) $\Rightarrow$ (b).

Since a refinement of a refinement is a refinement, by lemma 1.4.1 we may assume that $\mathcal{U}$ is locally finite. Enumerate $\mathcal{U}$ as $\{U_{0,i}: i \in \mathbb{N}\}$. In addition, let $\{F(j): j \in \mathbb{N}\}$ enumerate the collection of all subsets of $\mathbb{N}$ of cardinality precisely n+2. By corollary 4.3.4, for $j \in \mathbb{N}$ there exists an open cover $\mathcal{V}_j = \{U_{j,i}: i \in \mathbb{N}\}$ of X having the following properties

(1) for each i, $U_{j,i} \subseteq \overline{U}_{j,i} \subseteq U_{j-1,i}$, and

(2) $\bigcap_{i \in F(j)} U_{j,i} = \emptyset$.

Now for each $i \in \mathbb{N}$ define

$$S_i = \bigcap_{j=1}^{\infty} \overline{U}_{j,i} .$$

CLAIM: The collection $\mathcal{V} = \{S_i: i \in \mathbb{N}\}$ covers X.

Take an arbitrary $x \in X$. Since $\mathcal{U}$ is locally finite, x is contained in only finitely many elements of $\mathcal{U}$. So there exists $m \in \mathbb{N}$ such that $x \notin \bigcup_{i>m} U_{0,i}$. We conclude that for every $j \in \mathbb{N}$ there exists an index $k(j) \leq m$ such that $x \in U_{j,k(j)}$. Consequently, there exists $k \leq m$ such that x belongs to infinitely many of the $U_{j,k}$. By (1) this implies that $x \in S_k$, as required.

We conclude that $\mathcal{V}$ is a closed shrinking of $\mathcal{U}$ and hence is locally finite. Moreover, $\mathrm{ord}(\mathcal{V}) \leq n$ by (1) and (2).

We prove (b) $\Rightarrow$ (c).

By (b) there exists a locally finite closed refinement $\mathcal{S}$ of $\mathcal{U}$ such that $\mathrm{ord}(\mathcal{S}) \leq n$. For each $S \in \mathcal{S}$ pick $U(S) \in \mathcal{U}$ containing S. By corollary 4.3.2 there exists a swelling $\mathcal{V} = \{V(S): S \in \mathcal{S}\}$ of $\mathcal{S}$ such that for every $S \in \mathcal{S}$, $V(S) \subseteq U(S)$. Since $\mathrm{ord}(\mathcal{S}) \leq n$, $\mathrm{ord}(\mathcal{V}) \leq n$, so $\mathcal{V}$ is as required.

We prove (c) $\Rightarrow$ (e).

By (c) there exists an open refinement $\mathcal{W}$ of $\mathcal{U}$ such that $\mathrm{ord}(\mathcal{W}) \leq n$. For each $W \in \mathcal{W}$ let $U(W) \in \mathcal{U}$ be such that $W \subseteq U(W)$. Now for each $U \in \mathcal{U}$ define

$$V(U) = \bigcup\{W \in \mathcal{W}: U(W) = U\}.$$

Clearly, $\mathcal{V} = \{V(U): U \in \mathcal{U}\}$ is an open shrinking of $\mathcal{U}$ and it therefore suffices to prove that $\mathrm{ord}(\mathcal{V}) \leq n$. Take pairwise distinct $V(U_1),\cdots,V(U_{n+2}) \in \mathcal{V}$, and assume that $x \in \bigcap_{i=1}^{n+2} V(U_i)$. For each $i \leq n+2$ there exists $W_i \in \mathcal{W}$ such that $U(W_i) = U_i$ and $x \in W_i$. Since the U_i's are pairwise distinct, this implies that the W_i's are pairwise distinct. Since $\mathrm{ord}(\mathcal{W}) \leq n$, we now get

$x \in \bigcap_{i=1}^{n+2} W_i = \emptyset$, which is a contradiction.

We prove (e) $\Rightarrow$ (d).

Observe that if $\mathcal{A}$ is a cover of X such that $\mathrm{ord}(\mathcal{A}) \leq n$ and $\mathcal{B}$ is a shrinking of $\mathcal{A}$ then $\mathrm{ord}(\mathcal{B}) \leq n$. So (d) follows directly from (e) and proposition 4.3.4.

We prove (d) $\Rightarrow$ (f).

This is a triviality.

We prove (f) $\Rightarrow$ (g).

This follows immediately from corollary 4.3.2.

We prove (g) $\Rightarrow$ (a).

Let $\{(A_i, B_i): i \leq n+1\}$ be a family of n+1 pairs of disjoint closed subsets of X. Observe that the collection

$$\{X \backslash A_1, X \backslash A_2, \cdots, X \backslash A_{n+1}, X \backslash \bigcup_{i=1}^{n+1} B_i\}$$

covers X. By (g) there consequently exists an open cover $\mathcal{V} = \{V_i: i \leq n+2\}$ of X such that

(1) $V_i \subseteq X \backslash A_i \qquad (i \leq n+1)$,

(2) $V_{n+2} \subseteq X \backslash \bigcup_{i=1}^{n+1} B_i$,

(3) $\mathrm{ord}(\mathcal{V}) \leq n$.

Let $\{F_i: i \leq n+2\}$ be a closed shrinking of $\mathcal{V} = \{V_i: i \leq n+2\}$ (proposition 4.3.3). Now for $i \leq n+2$, define $\vec{A}_i$ and $\vec{B}_i$ by

$$\vec{A}_i = (F_{n+2} \backslash V_i) \cup A_i,$$
$$\vec{B}_i = F_i \cup B_i.$$

CLAIM 1: For every $i \leq n+2$, $\vec{A}_i \cap \vec{B}_i = \emptyset$.

This is easy. Simply observe that $F_i \subseteq V_i$, that $F_{n+2} \cap B_i = \emptyset$, and that

$$A_i \cap \vec{B}_i = A_i \cap (F_i \cup B_i) = \emptyset.$$

CLAIM 2: $\bigcup_{i=1}^{n+1} F_i \subseteq \bigcup_{i=1}^{n+1} \vec{B}_i$ and $F_{n+2} \subseteq \bigcup_{i=1}^{n+1} \vec{A}_i$.

The first part of the claim is a triviality and for the second part observe that $F_{n+2} \subseteq V_{n+2}$ and that by (3), $V_{n+2} \cap \bigcap_{i=1}^{n+1} V_i = \emptyset$, from which it follows that

$$F_{n+2} = F_{n+2}\setminus(\bigcap_{i=1}^{n+1} V_i) \subseteq \bigcup_{i=1}^{n+1} \vec{A}_i.$$

By corollary 1.4.16 there exists for every $i \leq n+1$ a partition L_i between $\vec{A}_i$ and $\vec{B}_i$. Since $A_i \subseteq \vec{A}_i$ and $B_i \subseteq \vec{B}_i$ for every i, L_i is also a partition between A_i and B_i. By claim 2 we obtain

$$\bigcap_{i=1}^{n+1} L_i \subseteq X\setminus(\bigcup_{i=1}^{n+1}(\vec{A}_i \cup \vec{B}_i)) \subseteq X\setminus(F_{n+2} \cup \bigcup_{i=1}^{n+1} F_i) = X\setminus X = \emptyset,$$

as required. □

4.3.6. *Remarks:* The above theorem explains the earlier terminology *covering dimension.* The vague idea behind dimension theory is that one wishes to cover a given space with finitely many "small" open sets in such a way that as few of the open sets as possible have points in common. If $\dim X \leq n$ then it is possible to do that in such a way that at most n+1 elements intersect. If $\dim X \geq n$ then no matter how the "small" cover is chosen, there are always at least n+1 elements having a point in common. It is interesting and intriguing that these ideas can be used to distinguish between various spaces topologically.

Naturally, the reader wonders about the relation between the dimension of a space X and the various dimensions of its subspaces. We finish this section by deriving two results about this relation.

Recall that if $\mathcal{U}$ is a collection of subsets of X then $\overline{\mathcal{U}}$ denotes the collection $\{\overline{U}: U \in \mathcal{U}\}$.

4.3.7. THEOREM ("The Countable Closed Sum Theorem"): *Let* X *be a space and let* $\mathcal{F}$ *be a countable closed cover of* X *such that for every* $F \in \mathcal{F}$, $\dim F \leq n$. *Then* $\dim X \leq n$.

PROOF: It is clear that without loss of generality we may assume that $n < \infty$. Enumerate $\mathcal{F}$ as $\{F_i: i \in \mathbb{N}\}$ and put $F_0 = \emptyset$. Let $\mathcal{U}$ be a finite open cover of X. By induction on $i \geq 0$ we shall construct an open cover $\mathcal{U}(i)$ of X such that the following conditions are satisfied:

(1) $\mathcal{U}(0) = \mathcal{U}$,
(2) if $0 \leq j < i$ then $\overline{\mathcal{U}(i)}$ is a shrinking of $\mathcal{U}(j)$,
(3) $\text{ord}(\overline{\mathcal{U}(i)} \cap F_i) \leq n$.

Observe that conditions (2) and (3) are satisfied for i = 0. Now assume that we completed the construction for all j with $0 \leq j < i$. Since $\dim F_i \leq n$, by assumption, the open cover $\mathcal{U}(i-1) \cap F_i$ of F_i has an open shrinking $\mathcal{V} = \{V(U): U \in \mathcal{U}(i-1)\}$ of order at most n (theorem 4.3.5(g)). For each $U \in \mathcal{U}(i-1)$ put

$$W(U) = (U\backslash F_i) \cup V(U).$$

It is clear that $\mathcal{W} = \{W(U): U \in \mathcal{U}(i-1)\}$ is an open shrinking of $\mathcal{U}(i-1)$ such that $\operatorname{ord}(\mathcal{W} \cap F_i) \leq n$. An easy application of proposition 4.3.3 and corollary 4.3.2 now show that there exists an open cover $\mathcal{U}(i)$ of X satisfying (2) and (3). This completes the inductive construction.

Let the cardinality of $\mathcal{U}$ be k. It is clear that for $i \in \mathbb{N}$ we can enumerate $\mathcal{U}(i)$ as $\{U_{m,i}: m \leq k\}$ such that for $0 \leq i < j$ and $m \leq k$ we have $\overline{U}_{m,j} \subseteq U_{m,i}$. Now for each $x \in X$ there exists $m(x) \leq k$ such that x belongs to infinitely many of the $U_{m(x),i}$. So by construction we have $x \in \bigcap_{i=1}^{\infty} U_{m(x),i}$. By (2) and (3) we now conclude that

$$\{\bigcap_{i=1}^{\infty} U_{m,i}: m \leq k\} = \{\bigcap_{i=1}^{\infty} \overline{U}_{m,i}: m \leq k\}$$

is a closed shrinking of $\mathcal{U}$ and has order at most n. Consequently, $\dim X \leq n$ by theorem 4.3.5(f).□

Since $\mathbb{R}^n$ is a countable union of homeomorphs of I^n, theorems 4.1.1 and 4.3.7 now immediately yield that $\dim \mathbb{R}^n \leq n$. In the remaining part of this section we shall present, among other things, two additional proofs of this inequality.

4.3.8. THEOREM ("The Subspace Theorem"): *Let* A *be a subspace of a space* X. *Then* $\dim A \leq \dim X$.

PROOF: If $\dim X = \infty$ then there is nothing to prove. So without loss of generality assume that $n = \dim X < \infty$.

First assume that A is closed. Then observe that if τ is an essential family in A then τ is also an essential family in X. From this it follows immediately that $\dim A \leq n$.

Next, assume that A is open. For each n, put $A_i = \{x \in A: d(x, X\backslash A) \leq 1/i\}$. Then each A_i is closed and the union of the A_i's is equal to A. That $\dim A \leq n$ now follows from the above and the Countable Closed Sum Theorem 4.3.7.

Finally, assume that A is an arbitrary subspace of X. In addition, let $\mathcal{U}$ be a cover of A by sets that are open in A. For each $U \in \mathcal{U}$ pick an open subset V(U) of X with $V(U) \cap A = U$. Put $V = \bigcup_{U \in \mathcal{U}} V(U)$. Then V is an open subset of X and hence, by the above, $\dim V \leq n$. Since $\mathcal{V} = \{V(U): U \in \mathcal{U}\}$ covers V, theorem 4.3.5 yields the existence of an open refinement $\mathcal{W}$ of $\mathcal{V}$ with $\operatorname{ord}(\mathcal{W}) \leq n$. Then $\mathcal{X} = \{W \cap A: W \in \mathcal{W}\}$ is an open refinement of $\mathcal{U}$ with $\operatorname{ord}(\mathcal{X}) \leq n$. Another application of theorem 4.3.5 now gives us that $\dim A \leq n$, as required.□

A different proof of theorem 4.3.8 shall be given in corollary 4.5.12(b).

Observe that since $\mathbb{R}^n$ is homeomorphic to $(0,1)^n$ (exercise 1.1.8), by theorems 4.1.1 and 4.3.8, $\dim \mathbb{R}^n \le n$.

For all $n \in \mathbb{N}$ and $0 \le m \le n$ let

$$\mathfrak{R}_{n,m} = \{x \in \mathbb{R}^n\colon \text{exactly } m \text{ coordinates of } x \text{ are rational}\}.$$

The results in this section can be used to prove that various spaces are zero-dimensional. We shall demonstrate this by proving that for all $n \in \mathbb{N}$ and $0 \le m \le n$, $\dim \mathfrak{R}_{n,m} = 0$. This result will be used to present yet another proof that $\dim \mathbb{R}^n \le n$.

4.3.9. PROPOSITION: *For all* $n \in \mathbb{N}$ *and* $0 \le m \le n$, $\mathfrak{R}_{n,m}$ *is zero-dimensional.*

PROOF: Let $A = \{i(1),\cdots,i(m)\}$ be a set of m indices in $\{1,2,\cdots,n\}$ and let $q_1,\cdots,q_m \in \mathbb{Q}$. Put

$$X = \{x \in \mathbb{R}^n\colon x_{i(j)} = q_j \text{ for every } j \le m\}.$$

Observe that X is a closed subspace of $\mathbb{R}^n$ and that

$$X \cap \mathfrak{R}_{n,m} = \{x \in \mathbb{R}^n\colon x_{i(j)} = q_j \text{ for every } j \le m \text{ and } x_i \in \mathbb{P} \text{ for } i \notin A\}.$$

Consequently, $X \cap \mathfrak{R}_{n,m}$ is a closed subspace of $\mathfrak{R}_{n,m}$ which is homeomorphic to the product of $n-m$ copies of $\mathbb{P}$. We conclude that $X \cap \mathfrak{R}_{n,m}$ is zero-dimensional by proposition 4.2.3 and theorem 4.2.2(2). Since $\mathfrak{R}_{n,m}$ is the union of countably many sets of the form $X \cap \mathfrak{R}_{n,m}$, we conclude that $\dim \mathfrak{R}_{n,m} = 0$ by theorem 4.3.7.□

Observe that $\mathbb{R}^n = \mathfrak{R}_{n,0} \cup \mathfrak{R}_{n,1} \cup \cdots \cup \mathfrak{R}_{n,n}$. By proposition 4.3.9 and theorem 4.2.2(6) this again yields that $\dim \mathbb{R}^n \le n$.

We now come to what is sometimes called the "Fundamental Theorem of Dimension Theory".

4.3.10. THEOREM: *For all* $n \in \mathbb{N}$, $\dim \mathbb{R}^n = \dim I^n = \dim S^n = n$.

PROOF: By theorem 4.1.1, $\dim I^n = n$. By theorem 4.3.9 this implies that $\dim \mathbb{R}^n \ge n$. Since $\dim \mathbb{R}^n \le n$, we obtain $\dim \mathbb{R}^n = n$. Also, S^n is the union of two homeomorphs of I^n, so an appeal to theorem 4.3.7 gives us that $\dim S^n \le n$. It now follows easily that $\dim S^n = n$ as well.□

Theorem 4.3.10 is of fundamental importance since it confirms our geometric intuition that $\mathbb{R}^n$,

I^n and S^n have to be n-dimensional. Since covering dimension is a *topological* notion, we now obtain the following

4.3.11. COROLLARY ("The Classification Theorem (part 2)"): *If* $n \neq m$ *then* $\mathbb{R}^n$ *and* $\mathbb{R}^m$ *are not homeomorphic.* □

Exercises for §4.3.

1. Prove that there exists a finite open cover $\mathcal{U}$ of I such that $|N(\mathcal{U})|$ cannot be imbedded in the plane (Hint: Apply theorem 2.4.8.).

4.4. The Imbedding Theorem

The aim of this section is to prove that every space X with $\dim X \leq n$ can be imbedded in $\mathbb{R}^{2n+1}$. This is a result of fundamental importance.

For $n \geq 0$, define

$$\mathfrak{N}_n = \{x \in \mathbb{R}^{2n+1}: \text{at most } n \text{ coordinates of } x \text{ are rational}\}.$$

Observe that $\mathfrak{N}_0$ is the space of irrational numbers $\mathbb{P}$. The space $\mathfrak{N}_n$ is called *Nöbeling's universal* n*-dimensional space.*

4.4.1. LEMMA: *For* $n \geq 0$, $\dim \mathfrak{N}_n = n$ *and* $\mathfrak{N}_n$ *is the union of* n+1 *zero-dimensional subspaces.*

PROOF: Since clearly $\mathfrak{N}_n = \mathfrak{R}_{2n+1,0} \cup \mathfrak{R}_{2n+1,1} \cup \cdots \cup \mathfrak{R}_{2n+1,n}$, an application of proposition 4.3.9 gives us that $\mathfrak{N}_n$ is the union of n+1 zero-dimensional subspaces. Consequently, by theorem 4.2.2(6), $\dim \mathfrak{N}_n \leq n$. We claim that $\mathfrak{N}_n$ contains a homeomorph of I^n. This is easy. Define an imbedding $i: I^n \to \mathfrak{N}_n$ by

$$i(x_1,\cdots,x_n) = (x_1,\cdots,x_n,\sqrt{2},\sqrt{2},\cdots,\sqrt{2}).$$

By theorems 4.3.8 and 4.3.10 we now obtain $\dim \mathfrak{N}_n \geq n$. □

For reasons that cannot be explained now, we do not aim at imbeddings in $\mathbb{R}^{2n+1}$ but at imbeddings in $\mathfrak{N}_n$. This will not complicate our argumentation.

Recall that a *hyperplane* in a linear space is a translated linear subspace. A hyperplane is called m-*dimensional* if it is a translated linear subspace which is spanned by m linearly independent vectors.

Naturally, we endow $\mathbb{R}^m$, $m \in \mathbb{N}$, with the metric d derived from its standard "euclidean" norm, cf. §1.2.

A finite subset $F = \{x_1,\cdots,x_n\}$ of $\mathbb{R}^m$ is said to be *in general position* if for every collection $i(0) < i(1) < \cdots < i(k)$ of at most m+1 indices in $\{1,\cdots,n\}$, the set $x_{i(0)},\cdots,x_{i(k)}$ is geometrically independent, cf. chapter 3.

4.4.2. LEMMA: *Let* $G = \{x_1,\cdots,x_n\}$ *and* $F = \{y_1,\cdots,y_k\}$ *be subsets of* $\mathbb{R}^m$ *such that* G *is in general position, and let* $\varepsilon > 0$. *Then for each* $i \leq k$ *there exists a point* $x_{n+i} \in \mathbb{R}^m$ *such that*

(1) $\|y_i - x_{n+i}\| < \varepsilon$,

(2) $\{x_1,\cdots,x_n,x_{n+1},\cdots,x_{n+k}\}$ *is in general position.*

PROOF: To begin with, let us establish the following:

CLAIM: Let $Z = \{z_1,\cdots,z_n\} \subseteq \mathbb{R}^m$ be in general position, and let $\varepsilon > 0$. Then for each $z \in \mathbb{R}^m$ there exists a point $z_{n+1} \in \mathbb{R}^m$ such that

(1) $\|z - z_{n+1}\| < \varepsilon$,

(2) $\{z_1,\cdots,z_{n+1}\}$ is in general position.

For every nonempty subset $F \subseteq \{1,\cdots,n\}$ of cardinality at most m, let v(F) be the unique $(|F|-1)$-dimensional hyperplane spanned by $\{z_i: i \in F\}$, cf. §3.1. Each v(F) is certainly a nowhere dense closed subset of $\mathbb{R}^m$. Since Z is finite, there exists a point $z_{n+1} \in \mathbb{R}^m \setminus \bigcup\{v(F): F \subseteq \{1,\cdots,n\}, |F| \leq m\}$ such that $d(z,z_{n+1}) < \varepsilon$. We claim that $\{z_1,\cdots,z_{n+1}\}$ is in general position. Since Z is, we need only prove that if $F \subseteq \{1,\cdots,n\}$ has cardinality at most m then $\{z_i: i \in F\} \cup \{z_{n+1}\}$ is geometrically independent. This however is clear since by construction, $z_{n+1} \notin v(F)$ (theorem 3.1.5).

Since the lemma clearly follows by a repeated application of the claim, we are done. □

Now for the remaining part of this section, let X be a fixed space with $\dim X \leq n$, $n \geq 0$.

4.4.3. PROPOSITION: *Let* $f \in C(X,\mathbb{R}^{2n+1};d)$, *let* $\varepsilon > 0$, *and let* H *be an at most* n-*dimensional hyperplane in* $\mathbb{R}^{2n+1}$. *For every finite open cover* $\mathcal{U}$ *of* X *there exists* $g \in C(X,\mathbb{R}^{2n+1};d)$ *such that*

(1) $d(f,g) < \varepsilon$,
(2) $\overline{g(X)} \cap H = \emptyset$,
(3) g *is a* $\mathcal{U}$*-mapping*.

PROOF: Since f is bounded, $\overline{f(X)}$ is compact. Consequently, there exists a finite open cover $\mathcal{V}$ of $\overline{f(X)}$ such that $\text{mesh}(\mathcal{V}) < \varepsilon/4$. Let $\mathcal{W}$ be the common refinement of $\mathcal{U}$ and $f^{-1}(\mathcal{V})$. Observe that $\mathcal{W}$ is finite. Since $\dim X \leq n$, there exists by theorem 4.3.5(g) a finite open shrinking $\mathcal{E}$ of $\mathcal{W}$ of order at most n. The essential properties of the cover $\mathcal{E}$ are the following ones:

(4) $\mathcal{E}$ has order at most n,
(5) $\mathcal{E} < \mathcal{U}$, and
(6) for each $E \in \mathcal{E}$, $\text{diam}(f(E)) < \varepsilon/4$.

Let $\mathcal{E} = \{E_1,\cdots,E_m\}$. Now for each $i \leq m$, pick an arbitrary point $z_i \in f(E_i)$. There is a geometrically independent set $F = \{e_1,\cdots,e_k\} \subseteq \mathbb{R}^{2n+1}$ of cardinality at most n+1 such that F spans H. By lemma 4.4.2, for each $i \leq m$ there exists $x_i \in \mathbb{R}^{2n+1}$ such that

(7) $d(x_i,z_i) < \varepsilon/4$, and
(8) $\{e_1,\cdots,e_k,x_1,\cdots,x_m\}$ is in general position.

Let $\kappa_1,\cdots,\kappa_m\colon X \to I$ denote the κ-functions with respect to the cover $\mathcal{E} = \{E_1,\cdots,E_m\}$, cf. §1.4. Define $g\colon X \to \mathbb{R}^{2n+1}$ by

$$g(x) = \sum_{i=1}^{m} \kappa_i(x)\cdot x_i.$$

Then g is clearly continuous and we claim that it is as required. First observe that g is bounded because its range is contained in the simplex spanned by the vectors $\{x_1,\cdots,x_k\}$ which is compact by lemma 1.2.2(2).

For the remaining part of the proof, fix an arbitrary $x \in X$ and let $F = \{i \leq m\colon x \in E_i\}$. Observe that by (4), $|F| \leq n+1$, which implies by (8) that the set $\vec{F} = \{x_i\colon i \in F\}$ is geometrically independent. Consequently, the $\kappa_i(x)$, $i \leq m$, are the affine coordinates of g(x) with respect to $\vec{F}$. Observe that $\kappa_i(x) > 0$ for every $i \leq m$.

CLAIM 1: $d(f(x),g(x)) < \varepsilon/2$.

Observe that $\kappa_i(x) = 0$ if $i \notin F$. It follows by (6) and (7) that $\|f(x)-x_i\| < \varepsilon/2$ for every $i \in F$.

Consequently,

$$\|f(x)-g(x)\| = \|\Sigma_{i\in F}\kappa_i(x)\cdot f(x) - \Sigma_{i\in F}\kappa_i(x)\cdot x_i\| \le \Sigma_{i\in F}\kappa_i(x)\|f(x)-x_i\| < \Sigma_{i\in F}\kappa_i(x)\cdot\tfrac{\varepsilon}{2} = \tfrac{\varepsilon}{2}.$$

By claim 1 we now obtain $d(f,g) \le \varepsilon/2 < \varepsilon$.

CLAIM 2: Let $i_0 \in F$. If $A \subseteq \{e_1,\cdots,e_k,x_1,\cdots,x_m\}\setminus\{x_{i_0}\}$ and $|A| \le n+1$ then $g(x) \notin \mathrm{aff}(A)$.

Assume the contrary. As observed above, $|\vec{F}| \le n+1$. Consequently, $|\vec{F} \cup A| \le (n+1)+(n+1) = 2n+2$. By (8) this implies that $F \cup A$ is geometrically independent. By proposition 3.1.4, $g(x) \in \mathrm{aff}(A \cap \vec{F})$. Since the $\kappa_i(x)$, $i \le m$, are the affine coordinates of $g(x)$ with respect to $\{x_i: i \in F\}$, $i_0 \notin A$ and $\kappa_{i_0}(x) > 0$, by unicity of affine coordinates (theorem 3.2.1), this is a contradiction.

Since $k \le n+1$, by claim 2 we obtain $g(x) \notin \mathrm{aff}(\{e_1,\cdots,e_k\}) = H$. Since x was arbitrary this implies $g(X) \cap H = \emptyset$. So it remains to prove that g is a $\mathcal{U}$-mapping. To this end, put

$$B = \mathbb{R}^{2n+1}\setminus\bigcup\{\mathrm{conv}(A): A \subseteq \{x_1,\cdots,x_m\}, |A| \le n+1 \text{ and for some } i \in F,\ x_i \notin A\}.$$

By claim 2 and lemma 1.2.2, B is an open neighborhood of x in $\mathbb{R}^{2n+1}$.

CLAIM 3: $g^{-1}(B) \subseteq \bigcap_{i\in F} E_i$.

Pick arbitrary $z \in g^{-1}(B)$ and $i_0 \in F$. Let $F_0 = \{i \le m: z \in E_i\}$. Again by (4), $|F_0| \le n+1$ and by the definition of g, $g(z) \in \mathrm{conv}(A)$, where $A = \{x_i: i \in F_0\}$. Since $g(z) \in B$, $i_0 \in F_0$.

Since by (5) $\mathcal{E} < \mathcal{U}$, this completes the proof. □

We now come to the main result in this section.

4.4.4. THEOREM ("The Imbedding Theorem"): *Let* X *be a space with* dim X ≤ n, n ≥ 0. *Then there exists an imbedding* $i: X \to \mathfrak{N}_n$ *such that* i(X) *has compact closure in* $\mathfrak{N}_n$.

PROOF: Let $A = \{i(1),\cdots,i(n+1)\}$ be a set of n+1 indices in $\{1,2,\cdots,2n+1\}$ and let $q_1,\cdots,q_{n+1} \in \mathbb{Q}$. Put

$$H = \{x \in \mathbb{R}^{2n+1}: x_{i(j)} = q_j \text{ for every } j \le n+1\}.$$

Observe that H is an n-dimensional hyperplane of $\mathbb{R}^{2n+1}$ and that $H \cap \mathfrak{N}_n = \emptyset$. In addition, each point of $\mathbb{R}^{2n+1}\backslash\mathfrak{N}_n$ has at least n+1 rational coordinates and is therefore contained in a hyperplane of the form H. We conclude that $\mathbb{R}^{2n+1}\backslash\mathfrak{N}_n$ is the union of countably many n-dimensional hyperplanes, say $\{L_i: i \in \mathbb{N}\}$.

We may assume that X is a subspace of the Hilbert cube Q (theorem 1.4.18). By compactness, Q has a sequence of *finite* open covers $\{\mathcal{V}_i: i \in \mathbb{N}\}$ such that for every i, $\text{mesh}(\mathcal{V}_i) < 1/i$. Put $\mathcal{U}_i = \mathcal{V}_i \cap X$ $(i \in \mathbb{N})$.

Now consider the function space $C(X,\mathbb{R}^{2n+1};d)$. For every $i \in \mathbb{N}$, put

$$\mathcal{C}_i = \{f \in C(X,\mathbb{R}^{2n+1};d): f \text{ is a } \mathcal{U}_i\text{-map and } \overline{f(X)} \cap L_i = \emptyset\}.$$

CLAIM 1: $\mathcal{C}_i$ is dense in $C(X,\mathbb{R}^{2n+1};d)$ for every i.

This follows directly from proposition 4.4.3.

CLAIM 2: $\mathcal{C}_i$ is open in $C(X,\mathbb{R}^{2n+1};d)$ for every i.

Take an arbitrary $f \in \mathcal{C}_i$. Since f is bounded, $\overline{f(X)}$ is compact. Therefore, since $\overline{f(X)} \cap L_i = \emptyset$, there exists $\varepsilon > 0$ such that

$$(1) \qquad \{y \in \mathbb{R}^{2n+1}: d(y,\overline{f(X)}) \leq \varepsilon\} \cap L_i = \emptyset.$$

Now since f is a $\mathcal{U}_i$-map, every y in the compact set $\overline{f(X)}$ has a neighborhood V_y in $\mathbb{R}^{2n+1}$ such that $f^{-1}(V_y)$ is contained in an element of $\mathcal{U}_i$. Put $\mathcal{V} = \{V_y: y \in \overline{f(X)}\}$. By lemma 1.1.1 there exists $\delta > 0$ such that every $A \subseteq \mathbb{R}^{2n+1}$ with $\text{diam}(A) < 3\delta$ and which moreover intersects $\overline{f(X)}$ is contained in an element of $\mathcal{V}$. Let $\gamma = \min\{\delta/3,\varepsilon\}$. We claim that the open ball about f with radius γ is contained in $\mathcal{C}_i$. To this end, take $g \in C(X,\mathbb{R}^{2n+1}; d)$ such that $d(f,g) < \gamma$. We shall first prove that $\overline{g(X)} \cap L_i = \emptyset$. This is easy. Take an arbitrary $x \in X$. Then $d(f(x),g(x)) < \varepsilon$, which implies that g(x) is contained in the compact set $\{y \in \mathbb{R}^{2n+1}: d(y,\overline{f(X)}) \leq \varepsilon\}$. Consequently by (1), $\overline{g(X)} \cap L_i = \emptyset$. We next prove that g is a $\mathcal{U}_i$-map. Take an arbitrary $x \in X$ and consider $B = g^{-1}(B(g(x),\delta/3))$. If $z \in B$ then $d(g(x),g(z)) < \delta/3$ and since $d(f,g) < \delta/3$, also $d(g(x),f(x)) < \delta/3$ and $d(g(z),f(z)) < \delta/3$. Consequently, $d(f(z),f(x)) < d(f(z),g(z)) + d(g(z),g(x)) + d(g(x),f(x)) < \delta/3 + \delta/3 + \delta/3 = \delta$. We conclude that

$$g^{-1}(B(g(x),\delta/3)) \subseteq f^{-1}(B(f(x),\delta)).$$

Since $B(f(x),\delta)$ is contained in an element of $\mathcal{V}$ it follows that $f^{-1}(B(f(x),\delta))$ is contained in an element of $\mathcal{U}_i$, so we are done.

Now by corollary 1.3.5, $C(X,\mathbb{R}^{2n+1};d)$ is a topologically complete space, and is therefore a Baire space. Consequently, claims 1 and 2 imply that

$$\mathcal{C} = \bigcap_{i=1}^{\infty} \mathcal{C}_i$$

is dense in $C(X,\mathbb{R}^{2n+1};d)$.

Now take any function $f \in \mathcal{C}$. Then $\overline{f(X)} \cap \bigcup_{i=1}^{\infty} L_i = \emptyset$, i.e. $\overline{f(X)} \subseteq \mathfrak{N}_n$. We claim that f is an imbedding.

We shall prove that for every $x \in X$ and every neighborhood W of x there exists a neighborhood V of f(x) such that $f^{-1}(V) \subseteq W$. From this it follows that f is one-to-one and that $f: X \to f(X)$ is closed, i.e., f is an imbedding.

Take an arbitrary $x \in X$ and a neighborhood W of x. Let $\varepsilon > 0$ be such that $B(x, \varepsilon) \subseteq W$. There exists $i \in \mathbb{N}$ with $1/i < \varepsilon$. Since f is a $\mathcal{U}_i$-map, there exists a neighborhood V of f(x) such that $f^{-1}(V)$ is contained in an element U of $\mathcal{U}_i$. So $x \in U$ and since $\text{diam}(U) < 1/i < \varepsilon$, we conclude that $f^{-1}(V) \subseteq U \subseteq B(x, \varepsilon) \subseteq W$, as required. □

4.4.5. *Remark:* Notice that we in fact proved the stronger statement that if $\dim X \le n$, $n \ge 0$, then any bounded map $f: X \to \mathbb{R}^{2n+1}$ can be approximated arbitrarily closely by an imbedding.

4.4.6. *Remark:* Theorem 4.4.4 is "best possible". By theorem 4.1.1, $\dim I = 1$. By the Countable Closed Sum Theorem 4.3.7, it therefore follows that the covering dimension of the space X in theorem 2.4.8 is equal to 1. In that theorem we proved that X cannot be imbedded in the plane. It is possible to prove that the covering dimension of the union of all n-faces of a (2n+2)-dimensional simplex is equal to n and cannot be imbedded in $\mathbb{R}^{2n}$ for all n, see Flores [63].

4.4.7. COROLLARY ("The Compactification Theorem"): *If* $\dim X \le n$, $n \ge 0$, *then* X *has a compactification* γX *such that* $\dim \gamma X \le n$.

PROOF: Apply theorem 4.4.4 and lemma 4.4.1. □

4.4.8. COROLLARY: *Let* X *be a nonempty space and let* $n \ge 0$. *The following statements are equivalent:*

(a) $\dim X \le n$,

(b) X *is the union of at most* n+1 *zero-dimensional subspaces.*

PROOF: Suppose that dim $X \leq n$. By theorem 4.4.4 we may assume that $X \subseteq \mathfrak{N}_n$. By lemma 4.4.1, $\mathfrak{N}_n$ is the union of n+1 zero-dimensional subspaces. Now apply theorem 4.2.2(1).

Conversely, assume that X is the union of at most n+1 zero-dimensional subspaces. Then apply theorem 4.2.2(6) to conclude that dim $X \leq n$. □

Exercises for §4.4.

1. Give an example of a space X, an open cover $\mathcal{U}$ of X, and an imbedding $f: X \to X$ such that f is not a $\mathcal{U}$-mapping.

2. Let X be a space with dim $X \leq n$, $n \geq 0$, let Y be compact, and let $f: X \to Y$ be continuous. Prove that there is a compactification γX of X such that dim $\gamma X \leq n$, while moreover the function f can be extended to a continuous function $g: \gamma X \to Y$.

3. Let X be a space. Prove that every continuous function $f: X \to Q$ can be approximated arbitrarily closely by an imbedding.

4. Prove that for every $n \geq 0$ there exists a compact space X_n such that (1) dim $X_n = n$, and (2) every space Y with dim $Y \leq n$ can be imbedded in X_n.

5. Let X be a compact space without isolated points such that $0 \leq \dim X \leq n < \infty$. Prove that there exists a continuous surjection $f: \mathbf{C} \to X$ such that each fiber of f has cardinality at most n+1 (Hint: By proposition 1.3.7, $\mathcal{S}(\mathbf{C},X)$ is closed in $C(\mathbf{C},X)$ and hence is a Baire space. Prove that for every $k \in \mathbb{N}$ the set $\mathcal{C}_k = \{f \in \mathcal{S}(\mathbf{C},X): (\exists x \in X)(\exists x_1,\cdots,x_{n+2} \in f^{-1}(x)$: if $i \neq j$ then $d(x_i,x_j) \geq 1/k)\}$ is closed and nowhere dense in $\mathcal{S}(\mathbf{C},X)$).

6. Let X be a compact space and let $f: \mathbf{C} \to X$ be a continuous surjection such that each fiber of f has cardinality at most n+1. Prove that dim $X \leq n$.

7. Let X be a space which is not compact. Prove that X has a compactification γX such that dim $\gamma X = \infty$.

4.5. The Inductive Dimension Functions ind and Ind; Equality of all Dimension Functions

There are two other dimension functions that are important in dimension theory, namely, the *small* and the *large* inductive dimension functions ind and Ind, respectively. It turns out that for a given space X these functions and the dimension function dim take the same value. The functions ind and Ind are important because in certain situations it is easier to deal with ind or Ind than with dim. For example, it is a triviality to verify that for all X and Y, $\operatorname{ind}(X \times Y) \leq \operatorname{ind} X + \operatorname{ind} Y$; by equality of ind and dim it therefore follows that $\dim(X \times Y) \leq \dim X + \dim Y$. However, to verify this straight from the definition of dim is unpleasant. The aim of this section is to study the dimension functions ind and Ind and to conclude equality of all three dimension functions.

We shall now give the definition of ind, which differs from the definition of dim in the sense that it is an *inductive* definition.

Let X be a space. Define $\operatorname{ind} X \in \{-1,0,1,\cdots\} \cup \{\infty\}$ by

$\operatorname{ind} X = -1$	$\Leftrightarrow$	$X = \emptyset$,
$\operatorname{ind} X \leq n,\ n \geq 0,$	$\Leftrightarrow$	if for every $x \in X$ and every closed subset A of X not containing x there exists a partition L between $\{x\}$ and A such that $\operatorname{ind} L \leq n-1$,
$\operatorname{ind} X = n$	$\Leftrightarrow$	$\operatorname{ind} X \leq n$ and ind X is not smaller than n,
$\operatorname{ind} X = \infty$	$\Leftrightarrow$	$\operatorname{ind} X \neq n$ for every $n \geq -1$,

The number ind X is called the *small* inductive dimension of X. It is clear that if X and Y are homeomorphic spaces then ind X = ind Y.

4.5.1. PROPOSITION: *Let* X *be a space. Then* dim X = 0 *iff* ind X = 0.

PROOF: This is a triviality since ind X = 0 iff the clopen subsets of X form a base. Now apply proposition 4.2.1.□

We shall now derive a few properties of the dimension function ind. The following triviality shall be used several times in the forthcoming: if X is a space, $Y \subseteq X$, $y \in Y$ and A is a (relatively) closed subset of Y not containing y, then y does not belong to the closure of A in X.

4.5.2. LEMMA: *Let* X *be a space, and let* A *be a subspace of* X. *Then* $\operatorname{ind} A \leq \operatorname{ind} X$.

PROOF: There is nothing to prove if $\operatorname{ind} X = \infty$, so we assume that $\operatorname{ind} X < \infty$. Again, there is nothing to prove if $\operatorname{ind} X = -1$. So assume the lemma to be true for all spaces Y with $\operatorname{ind} Y \leq n-1$, $n \geq 0$, and assume that $\operatorname{ind} X \leq n$. Take $x \in A$ and let C be a closed subset of A not containing x. Since $x \notin \overline{C}$ there is a partition L in X between x and $\overline{C}$ such that $\operatorname{ind} L \leq n-1$. Then $L \cap A$ is a partition between x and C in A so that, by our inductive assumption, $\operatorname{ind}(L \cap A) \leq n-1$. We conclude that $\operatorname{ind} A \leq n$.□

This result enables us to prove the following:

4.5.3. THEOREM ("The Addition Theorem"): *If* A *and* B *are subspaces of a space* X *then*

$$\operatorname{ind}(A \cup B) \leq \operatorname{ind} A + \operatorname{ind} B + 1.$$

PROOF: If ind A = ∞ or ind B = ∞ then there is nothing to prove. So we assume that ind A and ind B are both finite. We induct on ind A + ind B. If ind A = ind B = -1 then again there is nothing to prove. So assume that the theorem is true for all subspaces E and F of X having ind E + ind F $\leq$ n-1, n $\geq$ -1. Let α = ind A and β = ind B and let $\alpha + \beta = n$. We shall prove that $\operatorname{ind}(A \cup B) \leq n+1$. Take an arbitrary $x \in A \cup B$, say $x \in A$, and let C be a closed subset of $A \cup B$ not containing x. Since $x \notin \overline{C}$, there is a closed neighborhood E of $\overline{C}$ such that $x \notin E$. Let L be a partition in A between $\{x\}$ and $E \cap A$ such that ind L $\leq \alpha - 1$. Write A\L as the union of two relatively open disjoint sets A_0 and A_1 such that $x \in A_0$ and $E \cap A \subseteq A_1$. There exists an open neighborhood W of x in X such that $\overline{W} \cap A \subseteq A_0$. Now observe that L is a partition between $\overline{W} \cap A$ and $E \cap A$. By lemma 4.1.5, there exists a partition S in X between $\{x\}$ and $\overline{C}$ such that $S \cap A \subseteq L$. Now observe that by lemma 4.5.2, $\operatorname{ind}(S \cap A) \leq \alpha - 1$. Similarly, $\operatorname{ind}(S \cap B) \leq \beta$ since ind B = β. Consequently, we obtain by our inductive assumption,

$$\operatorname{ind}(S \cap (A \cup B)) \leq \alpha - 1 + \beta + 1 = \alpha + \beta = n.$$

Since $S \cap (A \cup B)$ is a partition in $A \cup B$ between $\{x\}$ and C, this shows that $\operatorname{ind}(A \cup B) \leq n + 1$.□

This theorem has the following consequence.

4.5.4. COROLLARY: *If* X *can be written as the union of* n+1 *zero-dimensional subspaces then* ind X $\leq$ n.

PROOF: Since for every space Y, dim Y = 0 iff ind Y = 0 (proposition 4.5.1) the result follows immediately from theorem 4.5.3.□

We want to prove the equality of dim and ind. Before being able to do that, we need to derive a preliminary lemma.

4.5.5. LEMMA: *Let* X *be space and let* n $\geq$ 0. *The following two statements are equivalent:*

(a) *For every* $x \in X$ *and for every open neighborhood* U *of* x *there exists a partition* L *between* $\{x\}$ *and* $X\setminus U$ *such that* $\dim L \le n\text{-}1$;

(b) *For every pair* A *and* B *of disjoint closed subsets of* X *there exists a partition* L *between* A *and* B *such that* $\dim L \le n\text{-}1$.

PROOF: First observe that (b) $\Rightarrow$ (a) is a triviality. For (a) $\Rightarrow$ (b), let A and B be disjoint closed subsets of X. By assumption, for each $x \in X$ there exist open sets $U(x)$ and $V(x)$ such that

(3) $x \in U(x)$, $U(x) \cap V(x) = \emptyset$,

(4) if $L(x) = X\setminus(U(x) \cup V(x))$ then $\dim L(x) \le n\text{-}1$, and

(5) $\overline{U(x)} \cap A = \emptyset$ or $\overline{U(x)} \cap B = \emptyset$.

The cover $\{U(x)\colon x \in X\}$ has a countable subcover, say $\mathcal{U}$. Put $L = \bigcup\{L(x)\colon U(x) \in \mathcal{U}\}$. Observe that by theorem 4.3.7 and (4) we have $\dim L \le n\text{-}1$. Let $\mathcal{U}_0 = \{U \in \mathcal{U}\colon \overline{U} \cap A \ne \emptyset\}$ and $\mathcal{U}_1 = \mathcal{U}\setminus\mathcal{U}_0$. Enumerate $\mathcal{U}_0$ as $\{U_{0,i}\colon i \in \mathbb{N}\}$ and $\mathcal{U}_1$ as $\{U_{1,i}\colon i \in \mathbb{N}\}$, respectively. For each $i \in \mathbb{N}$ put

(6) $E_i = U_{0,i}\setminus\bigcup_{j<i}\overline{U}_{1,j}$ and $F_i = U_{1,i}\setminus\bigcup_{j\le i}\overline{U}_{0,j}$.

By (5) and (6) it follows easily that $A \subseteq E = \bigcup_{i=1}^{\infty}E_i$, $B \subseteq F = \bigcup_{i=1}^{\infty}F_i$, and $E \cap F = \emptyset$. Also observe that both E and F are open. Consequently, $G = X\setminus(E \cup F)$ is a partition between A and B. Now take an arbitrary point $y \in G$. Let i be the first natural number with the property that $y \in \overline{U}_{0,i} \cup \overline{U}_{1,i}$. Suppose first that

$$y \in \overline{U}_{0,i}.$$

If $y \in U_{0,i}$ then $y \in E_i$, which is not the case. Consequently, $y \notin U_{0,i}$, which implies that $y \in L$. Now suppose that $y \notin \overline{U}_{0,i}$. If $y \in U_{1,i}$ then $y \in F_i$, contradiction. Consequently, $y \notin U_{1,i}$, which implies that in this case also $y \in L$.

We conclude that $G \subseteq L$. Since $\dim L \le n\text{-}1$, by theorem 4.3.8, $\dim G \le n\text{-}1$. Therefore, G is the required partition between A and B. □

4.5.6. THEOREM: *For every space* X, $\dim X = \operatorname{ind} X$.

PROOF: We shall first prove that $\dim X \le \operatorname{ind} X$. First, notice that if $\operatorname{ind} X = \infty$ then there is nothing to prove. So assume that $\operatorname{ind} X < \infty$. We induct on $\operatorname{ind} X$. If $\operatorname{ind} X = 0$ then apply proposition 4.5.1. Now assume that the inequality holds for all spaces Y with $\operatorname{ind} Y \le n\text{-}1$, $n \ge 1$, and let $\operatorname{ind} X = n$. Then for each point $x \in X$ and for every open neighborhood U of x there exists a partition L between $\{x\}$ and $X\setminus U$ such that ind $L \leq n-1$. By our inductive hypothesis, all these partitions have covering dimension at most n-1. We conclude that dim $X \leq n$ by lemma 4.5.5 and corollary 4.1.8.

We shall now prove that ind $X \leq$ dim X. If dim $X \in \{-1,\infty\}$ then this is a triviality. So assume that dim X = n, $n \geq 0$. By corollary 4.4.8, X is the union of at most n+1 zero-dimensional subspaces. Consequently, ind $X \leq n =$ dim X by corollary 4.5.4. □

We shall now present the definition of the *large inductive dimension function* Ind. For every space X let

Ind X = -1	$\Leftrightarrow$	$X = \emptyset$,
Ind $X \leq n$, $n \geq 0$,	$\Leftrightarrow$	If for every pair of disjoint closed subsets A and B of X there exists a partition L between A and B such that Ind L $\leq n-1$,
Ind X = n	$\Leftrightarrow$	Ind $X \leq n$ and Ind X is not smaller than n,
Ind $X = \infty$	$\Leftrightarrow$	Ind $X \neq n$ for every $n \geq -1$.

Again it is easy to see that if X and Y are homeomorphic spaces then Ind X = Ind Y. We shall prove that Ind = dim, thereby establishing the announced equality.

4.5.7. LEMMA: *Let* X *be a space. Then* ind $X \leq$ Ind X.

PROOF: If Ind $X = \infty$ or Ind X = -1 then there is nothing to prove, so assume that the lemma is true for all spaces Y with $-1 \leq$ Ind $Y \leq n-1$, $0 \leq n < \infty$, and assume that Ind X = n. Let $x \in X$ and let U be an open neighborhood of x. Since $\{x\}$ and $X\setminus U$ are disjoint closed sets in X, there is a partition L between them such that Ind $L \leq n-1$. By our inductive assumption we have ind $L \leq$ Ind $L \leq n-1$. We conclude that ind $X \leq n =$ Ind X. □

We now come to the announced

4.5.8. THEOREM ("The Coincidence Theorem"): *For every space* X *we have* dim X = ind X = Ind X.

PROOF: We shall first prove that Ind $X \leq$ dim X. If dim $X = \infty$ or dim X = -1 then there is nothing to prove, so assume that the lemma is true for all spaces Y with $-1 \leq$ dim $Y \leq n-1$, $0 \leq n < \infty$, and assume that dim X = n. Since ind X = dim X (theorem 4.5.6) for every $x \in X$ and every closed set A in X with $x \notin A$, there exists a partition L between $\{x\}$ and A such that ind L

$\leq$ n-1. By another application of theorem 4.5.6 it follows that all these partitions have covering dimension at most n-1. Consequently, by lemma 4.5.5, for every pair of disjoint closed subsets A and B of X there exists a partition L between A and B such that $\dim L \leq n-1$. By our inductive hypothesis it follows that these partitions all have large inductive dimension at most n-1. We conclude that $\operatorname{Ind} X \leq n = \dim X$, as required.

Now apply theorem 4.5.6 and lemma 4.5.7. □

Let X and Y be spaces. It is natural to wonder about the relation between dim X, dim Y and $\dim(X \times Y)$. Since $\dim \mathbb{R}^n = n$ for every n, one would expect the relation $\dim(X \times Y) = \dim X + \dim Y$ to hold for all X and Y. Unfortunately, this is not true, see example 4.5.10. We shall now prove "half" of the expected equality.

4.5.9. THEOREM ("The Product Theorem"): *Let* X *and* Y *be nonempty spaces. Then*

$$\dim(X \times Y) \leq \dim X + \dim Y.$$

PROOF: By the Coincidence Theorem 4.5.8, it suffices to prove that $\operatorname{ind}(X \times Y) \leq \operatorname{ind} X + \operatorname{ind} Y$. First observe that without loss of generality, $0 \leq \operatorname{ind} X, \operatorname{ind} Y < \infty$. We shall prove the theorem by induction on ind X + ind Y. If ind X + ind Y = 0 then apply theorem 4.2.2(2). So assume that the theorem is true for all spaces X and Y with $\operatorname{ind} X + \operatorname{ind} Y \leq n-1$, $n \geq 1$. Assume that $\operatorname{ind} X = \alpha$, $\operatorname{ind} Y = \beta$ and that $\alpha + \beta = n$. Now take a point $(x,y) \in X \times Y$ and a closed subset $A \subseteq X \times Y$ such that $(x,y) \notin A$. There are neighborhoods U and V of x and y, respectively, such that $\overline{U \times V} \cap A = \emptyset$. Let L_x be a partition between $\{x\}$ and $X \setminus U$ such that $\operatorname{ind} L_x \leq \alpha - 1$. Similarly, let L_y be a partition between $\{y\}$ and $Y \setminus V$ such that $\operatorname{ind} L_y \leq \beta - 1$. Now consider $Z = (L_x \times Y) \cup (X \times L_y)$.

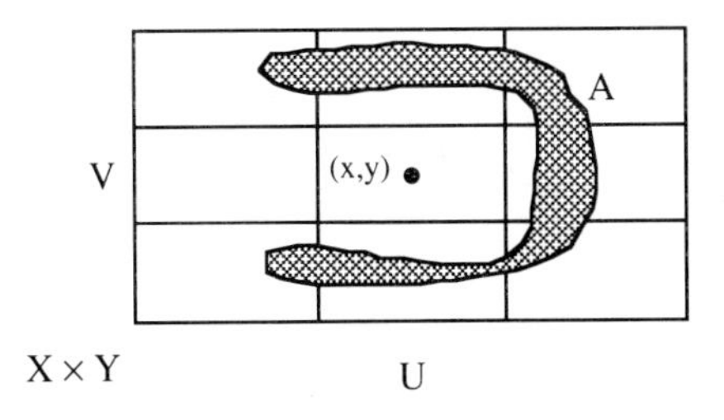

Figure 4.5.1.

By our inductive assumption, $\operatorname{ind}(L_x \times Y) \leq \operatorname{ind} L_x + \operatorname{ind} Y \leq \alpha - 1 + \beta = \alpha + \beta - 1$. Similarly, it follows that $\operatorname{ind}(X \times L_y) \leq \alpha + \beta - 1$. By the Coincidence Theorem 4.5.8 and the Countable

Closed Sum Theorem 4.3.7, it follows that ind $Z \leq \alpha + \beta - 1$. Now take open sets E_1 and E_2 of X such that $x \in E_1$, $X \backslash U \subseteq E_2$, $E_1 \cap E_2 = \emptyset$, and $E_1 \cup E_2 = X \backslash L_x$. Similarly, take open sets F_1 and F_2 of X such that $y \in F_1$, $Y \backslash V \subseteq F_2$, $F_1 \cap F_2 = \emptyset$, and $F_1 \cup F_2 = X \backslash L_y$. It is easy to see that the boundary B of $E_1 \times F_1$ is contained in Z, see figure 4.5.1. So by lemma 4.5.2, B has small inductive dimension at most $\alpha + \beta - 1$. We conclude that

$$\operatorname{ind}(X \times Y) \leq \alpha + \beta = \operatorname{ind} X + \operatorname{ind} Y,$$

as required.□

4.5.10. EXAMPLE: *There exists a space* E *with* $\dim E = 1$ *such that* $\dim(E \times E) = 1$.

E is the space of example 4.2.11, i.e. Erdös' space. We first claim that E and $E \times E$ are homeomorphic, cf. exercise 4.2.6. The function $f: E \times E \to E$ defined by

$$f(q,r) = (q_1,r_1,q_2,r_2,\cdots,q_n,r_n,q_{n+1},r_{n+1},\cdots),$$

is easily seen to be a homeomorphism. So it suffices to prove that $\dim E = 1$. In example 4.2.11 we showed that $\dim E \geq 1$. We shall prove here that ind $E \leq 1$. By theorem 4.5.8 this gives us the required result.

First observe that E is a subgroup of l^2, and hence is a *topological group*. By homogeneity, it therefore suffices to prove that ind $E \leq 1$ *at* the zero of E. We claim that for every $\varepsilon > 0$, if $S_\varepsilon = \{x \in l^2 : \|x\| = \varepsilon\} \cap E = \{x \in E : \|x\| = \varepsilon\}$ then $\dim S_\varepsilon = 0$. By lemma 1.2.6, S_ε is homeomorphic to a subspace of the countable infinite product of rationals, which is zero-dimensional by proposition 4.2.3 and theorem 4.2.2(2). By theorem 4.2.2(1) we therefore conclude that $\dim S_\varepsilon = 0$, as required.□

4.5.11. *Remarks:* If X and Y are compact spaces and $\dim X \leq 1$ then $\dim(X \times Y) = \dim X + \dim Y$, [78]; see also [60, problem 1.9(E)]. This result is "best possible". There exist compact spaces X and Y such that $\dim X = \dim Y = 2$, while $\dim(X \times Y) = 3$, [118]; see also [84]. Finally, examples such as in example 4.5.10 exist for every dimension, [14].

We now summarize the results obtained so far in the following

4.5.12. COROLLARY: (1) *Let* X *be a nonempty space and let* $n \geq 0$. *The following statements are equivalent:*

(a) $\dim X \leq n$,

(b) *for every subspace* A *of* X, $\dim A \le n$,

(c) X *has a compactification* γX *such that* $\dim \gamma X \le n$,

(d) *for every pair of disjoint closed subsets* A *and* B *of* X *there exists a partition* L *between* A *and* B *such that* $\dim L \le n-1$,

(e) X *is the union of at most* n+1 *zero-dimensional subspaces.*

(2) *If* X *is a space and if* A *and* B *are subspaces of* X *then* $\dim(A \cup B) \le \dim A + \dim B + 1$ *and* $\dim(A \times B) \le \dim A + \dim B$.

PROOF: For (1), apply theorem 4.5.8 in combination with lemma 4.5.2, corollary 4.4.7 and corollary 4.4.8. For (2), apply theorems 4.5.8, 4.5.9 and 4.5.3.□

We finish this section with three characterizations of dimension, two of which are in the spirit of theorem 4.3.5 and one of which is "combinatorial".

Let τ be an n-dimensional simplex in a linear space L. By exercise 3.5.8(d), τ is homeomorphic to I^n and hence by theorem 4.3.10, $\dim \tau = n$. Now let $X = |\mathcal{T}|$ be a polytope. Since $\mathcal{T}$ is countable, by the Countable Closed Sum Theorem 4.3.7 and the above remark we get $\dim X = \sup\{\dim \tau : \tau \in \mathcal{T}\} = \sup\{k : (\exists \tau \in \mathcal{T})(\tau \text{ is a k-simplex})\}$. We conclude that the *topological* dimension of X equals its *"combinatorial"* dimension.

4.5.13. THEOREM: *Let* X *be a nonempty space and let* $n \ge 0$. *The following statements are equivalent:*

(a) $\dim X \le n$,

(b) *for every open cover* $\mathcal{U}$ *of* X *there exists a locally finite open refinement* $\mathcal{V}$ *of* $\mathcal{U}$ *such that such that* $\text{ord}(\mathcal{V}) \le n$,

(c) *for every open cover* $\mathcal{U}$ *of* X *there exists a star-finite open refinement* $\mathcal{V}$ *of* $\mathcal{U}$ *such that such that* $\text{ord}(\mathcal{V}) \le n$,

(d) *for every open cover* $\mathcal{U}$ *of* X *there exist a polytope* P *such that* $\dim P \le n$ *and a* $\mathcal{U}$-*mapping* $f: X \to P$.

PROOF: *We prove* (a) ⇒ (c).

Without loss of generality, $\mathcal{U}$ is star-finite (theorem 3.6.17(1)). Now by theorem 4.3.5 there exists an open shrinking $\mathcal{V}$ of $\mathcal{U}$ such that $\text{ord}(\mathcal{V}) \le n$. Since each shrinking of a star-finite cover is clearly star- finite, we are done.

We prove (c) ⇒ (b) ⇒ (a).

That (c) ⇒ (b) is a triviality and that (b) ⇒ (a) follows from theorem 4.3.5.

We prove (c) ⇒ (d).

Let $\mathcal{U}$ be an open cover of X. By (c) there exists a (countable) star-finite open refinement $\mathcal{V}$ of

$\mathcal{U}$ such that ord($\mathcal{V}$) $\leq$ n. Let P = $|N(\mathcal{V})|$. Then κ: X $\rightarrow$ P is a $\mathcal{V}$-mapping by corollary 3.6.16, and hence a $\mathcal{U}$-mapping. In addition, P is a polytope by theorem 3.6.17. Finally, since ord($\mathcal{V}$) $\leq$ n, N($\mathcal{V}$) has no simplexes of dimension greater than n. By the Countable Closed Sum Theorem 4.3.7 it follows that dim P $\leq$ n.

We prove (d) $\Rightarrow$ (a).

Let $\mathcal{U} = \{U_i: i \in I\}$ be an open cover of X. By (d) there exists a polytope P such that dim P $\leq$ n and a $\mathcal{U}$-mapping f: X $\rightarrow$ P. For each i let E_i be the set of all y $\in$ P for which there exists an open neighborhood V_y such that $f^{-1}(V_y)$ is contained in U_i. Then $\mathcal{E} = \{E_i: i \in I\}$ is an open cover of P and $f^{-1}(E_i) \subseteq U_i$ for every i. Since dim P $\leq$ n, theorem 4.3.5 implies the existence of an open shrinking $\mathcal{V} = \{V_i: i \in I\}$ of $\mathcal{E}$ such that ord($\mathcal{V}$) $\leq$ n. It is clear that the collection $\mathcal{W} = \{f^{-1}(V_i): i \in I\}$ is an open shrinking of $\mathcal{U}$ such that ord($\mathcal{W}$) $\leq$ n. By theorem 4.3.5 we therefore conclude that dim X $\leq$ n.□

Exercises for §4.5.

1. Prove that Erdös' space E is not topologically complete. Give an example of a 1-dimensional totally disconnected topologically complete space F such that F $\times$ F is 1-dimensional (Hint: Consider the subspace of Hilbert space consisting of all points all of whose coordinates are irrational).

2. Let X be a totally disconnected space which is not compact and let γX be a compactification of X. Prove that dim $\gamma X \backslash X \geq$ dim X, i.e. X cannot be compactified by adding a set of dimension smaller than dim X.

4.6. Mappings into Spheres

In the previous sections we presented several characterizations of dimension, some "internal" and some "external", cf. theorem 4.5.13. In this section we shall prove that a space X is at most n-dimensional if and only if every continuous function f : A $\rightarrow$ S^n defined on a closed subspace A of X can be extended over X, thereby deducing another important "external" characterization of dimension. As an application of this result we shall present a proof of the Brouwer Invariance of Domain Theorem, cf. exercise 2.3.4.

4.6.1. LEMMA: *Let* X *be a space and let* A_1 *and* A_2 *be disjoint closed subsets of* X *such that* $0 \leq \dim (X \backslash (A_1 \cup A_2)) \leq n$. *Then there exists a partition* L *in* X *between* A_1 *and* A_2 *such that* dim L $\leq$ n-1.

PROOF: Put $A = A_1 \cup A_2$. By corollary 1.4.16 there exist open subsets U and V of X such that $A_1 \subseteq U$, $A_2 \subseteq V$ and $\overline{U} \cap \overline{V} = \emptyset$. Since $\dim(X\backslash A) \leq n$ and $\overline{U} \cap (X\backslash A)$ and $\overline{V} \cap (X\backslash A)$ are disjoint closed subsets of $X\backslash A$, by corollary 4.5.12 there exist disjoint open sets E and F in $X\backslash A$ such that

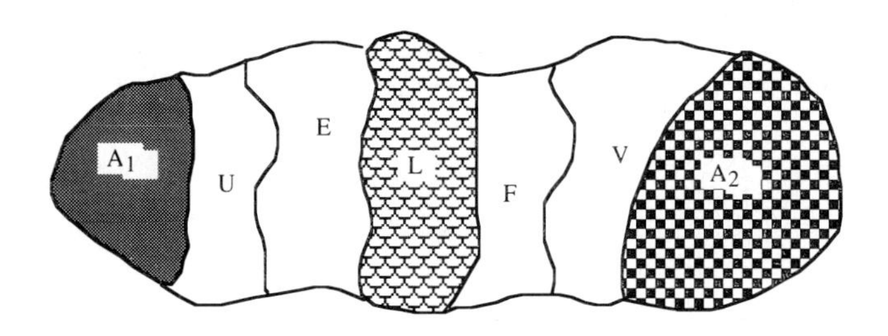

Figure 4.6.1.

(3) $\overline{U} \cap (X\backslash A) \subseteq E$,
(4) $\overline{V} \cap (X\backslash A) \subseteq F$,
(5) if $L = (X\backslash A)\backslash(E \cup F)$ then $\dim L \leq n-1$.

Since $A_1 \cup E = U \cup E$, and E is open in X being open in the subspace $X\backslash A$, we conclude that $A_1 \cup E$ is open in X. Similarly, $A_2 \cup F$ is open in X. Consequently, L is closed in X and moreover is a partition between A_1 and A_2 with $\dim L \leq n-1$ by (5). So L is as required. □

4.6.2. COROLLARY: *Let* X *be a space and let* A_1 *and* A_2 *be closed subspaces of* X *such that* $0 \leq \dim(X\backslash(A_1 \cup A_2)) \leq n$. *Then there exist closed subspaces* X_1 *and* X_2 *of* X *such that*

(1) $A_1 = X_1 \cap (A_1 \cup A_2)$,
(2) $A_2 = X_2 \cap (A_1 \cup A_2)$,
(3) $\dim((X_1 \cap X_2)\backslash(A_1 \cup A_2)) \leq n-1$,
(4) $X_1 \cup X_2 = X$.

PROOF: Put $Y = X\backslash(A_1 \cap A_2)$. By lemma 4.6.1 there exists a closed set L in Y with $\dim L \leq n-1$, such that $Y\backslash L$ can be written as the disjoint union of two open sets E and F such that $A_1 \cap Y \subseteq E$ and $A_2 \cap Y \subseteq F$. It is easy to see that $X_1 = A_1 \cup E \cup L$ and $X_2 = A_2 \cup F \cup L$ are as required. □

These simple results enable us to derive the following:

4.6.3. THEOREM: *Let* X *be a space and let* A *be a closed subspace of* X *such that* $0 \leq \dim(X\setminus A) \leq n$. *Then every continuous function* $g: A \to S^n$ *can be extended to a continuous function* $\bar{g}: X \to S^n$.

PROOF: We shall prove the theorem by induction on $n \geq 0$. Suppose first that $n = 0$ and recall that $S^0 = \{-1,1\}$. By lemma 4.6.1 there exists a clopen set $C \subseteq X$ such that $g^{-1}(-1) \subseteq C$ and $g^{-1}(1) \subseteq X\setminus C$. Define $\bar{g}: X \to S^0$ by

$$\bar{g}(x) = \begin{cases} -1 & (x \in C), \\ 1 & (x \notin C). \end{cases}$$

An easy check shows that $\bar{g}$ is as required.

Now assume that the theorem is true for n-1, $n \geq 1$, and let $g: A \to S^n$. Define $S_1^n = \{x \in \mathbb{R}^{n+1}: x_1 \geq 0\}$ and $S_2^n = \{x \in \mathbb{R}^{n+1}: x_1 \leq 0\}$, respectively. Observe that S_1^n and S_2^n are both homeomorphic to I^n and that $S_1^n \cap S_2^n$ is homeomorphic to S^{n-1}. Now put $A_1 = f^{-1}(S_1^n)$ and $A_2 = f^{-1}(S_2^n)$, respectively. Let X_1 and X_2 be such as in corollary 4.6.2. By our inductive assumption, we can extend $g \mid A_1 \cap A_2$ to a continuous function $h: X_1 \cap X_2 \to S_1^n \cap S_2^n$. Now for i = 1,2 define $g_i: A_i \cup (X_1 \cap X_2) \to S_i^n$ by $g_i = (g \mid A_i) \cup h$. Since S_i^n is homeomorphic to I^n, we can extend g_i, i = 1,2, to a continuous function $\bar{g}_i: X_i \to S_i^n$ (corollary 1.5.5). Clearly, $\bar{g} = \bar{g}_1 \cup \bar{g}_2: X \to S^n$ is a continuous extension of g.□

4.6.4. THEOREM: *Let* X *be a space and let* $n \geq 0$. *The following statements are equivalent:*

(a) $\dim X \leq n$,

(b) *for every closed subset* A *of* X, *every continuous function* $f: A \to S^n$ *can be continuously extended over* X.

PROOF: For (a) ⇒ (b) apply the previous theorem and corollary 4.5.12(b). For (b) ⇒ (a), let $\tau = \{(A_i,B_i): i \leq n+1\}$ be a family of pairs of disjoint closed subsets of X. We shall prove that τ is inessential. To this end, for every $i \leq n+1$ let $\alpha_i: X \to I$ be a Urysohn function such that $\alpha_i \mid A_i \equiv 0$ and $\alpha_i \mid B_i \equiv 1$ (corollary 1.4.15). Define $\alpha: X \to I^{n+1}$ by $\alpha(x) = (\alpha_1(x),\cdots,\alpha_{n+1}(x))$. Put $A = \bigcup_{i=1}^{n+1}(A_i \cup B_i)$ and $\beta = \alpha \mid A$. Then $\beta(A)$ is contained in the boundary B of I^{n+1} which is homeomorphic to S^n by exercise 3.5.8(c). By assumption, there exists a continuous extension $\gamma: X \to B$ of β. Now for every $i \leq n+1$ put $E_i = (\pi_i \circ \gamma)^{-1}(\frac{1}{2})$, where $\pi_i: I^{n+1} \to I$ denotes the projection onto the i-th factor of I^{n+1}. Then clearly E_i is a partition between A_i and B_i for every i such that $\bigcap_{i=1}^{n+1} E_i = \emptyset$. Consequently, τ is inessential, as required.□

Application: The Brouwer Invariance of Domain Theorem

To begin with, we shall first present an "internal" characterization of the boundary points of an arbitrary closed subset of a fixed $\mathbb{R}^n$.

4.6.5. PROPOSITION: *Let* $n \in \mathbb{N}$ *and let* X *be a closed subspace of* $\mathbb{R}^n$. *Then* $x \in X$ *belongs to the boundary* Bd(X) *of* X *in* $\mathbb{R}^n$ *if and only if* x *has arbitrarily small neighborhoods* U *in* X *such that every continuous mapping* $g: X\backslash U \to S^{n-1}$ *can be extended continuously over* X.

PROOF: First assume that $x \in \text{Bd}(X)$. Without loss of generality, $x = (0,0,\cdots,0)$. If V is a neighborhood of x in X then there is $\varepsilon > 0$ such that $B(x,\varepsilon) \cap X \subseteq V$. Consequently, it suffices to consider a "spherical" neighborhood $U = B(x,\varepsilon) = \{y \in \mathbb{R}^n: \|y\| < \varepsilon\}$ of x and a continuous function $g: X\backslash U \to S^{n-1}$. Without loss of generality assume that $\varepsilon = 1$. Since x belongs to the boundary of X, there is a point $y \in U\backslash X$. For each $p \in D(x,1)\backslash\{y\}$ let $\tau(p)$ denote the "projection" of p on S^{n-1} from y.

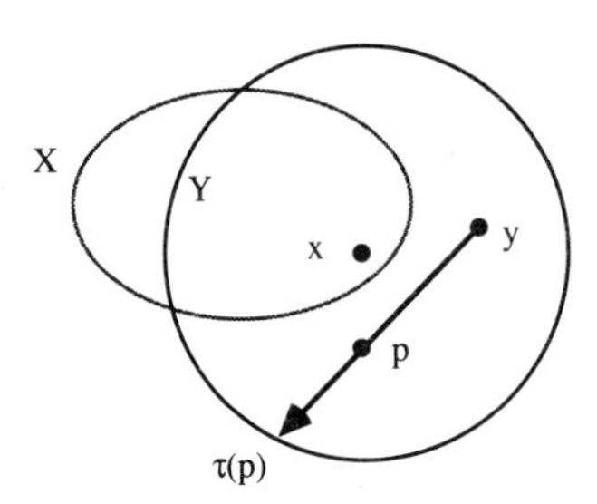

Figure 4.6.2.

CLAIM: g can be extended to a continuous function $f: (X\backslash U) \cup (D(x,1)\backslash\{y\}) \to S^{n-1}$.

First observe that $Y = (X\backslash U) \cap S^{n-1}$ is a closed subspace of S^{n-1}. Put $h = g \mid Y$. Since $\dim S^{n-1} \le n-1$ (theorem 4.3.10), the function h can be extended to a continuous function $\bar{h}: S^{n-1} \to S^{n-1}$ by theorem 4.6.4. Now define $f: (X\backslash U) \cup (D(x,1)\backslash\{y\}) \to S^{n-1}$ by

$$f = g \cup (\bar{h} \circ \tau).$$

Then f is clearly as required.

Now $\bar{g} = f \mid X$ is the desired extension of g.

Conversely, let $x \in X$ and assume that x is an interior point of X. We shall derive a contradiction. There is $\varepsilon > 0$ such that $B = D(x,\varepsilon) \subseteq X$. Let U be an open neighborhood of x in X.

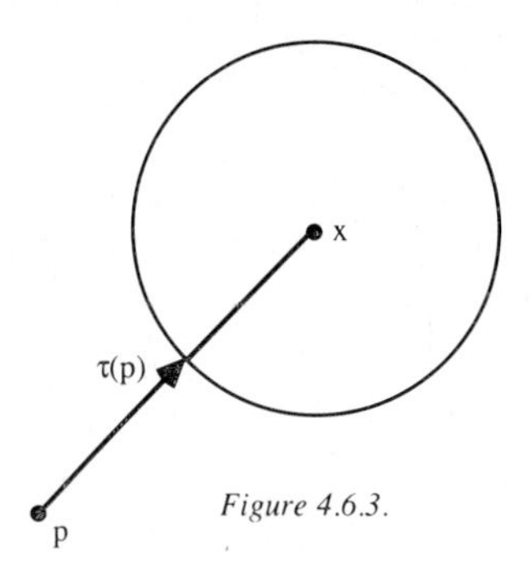

Figure 4.6.3.

We identify S^{n-1} and the boundary of B. For each $p \in X\backslash U$ let $\tau(p)$ denote the "projection" of p on S^{n-1} from x. Then τ cannot be extended over X since this would yield a retraction from B onto its boundary, which is impossible by theorem 3.5.5. □

4.6.6. COROLLARY: *Let* $n \in \mathbb{N}$ *and let* X *and* Y *be closed subspaces of* $\mathbb{R}^n$. *If* $f: X \to Y$ *is a homeomorphism then* $f(Bd(X)) = Bd(Y)$. □

We are now in a position to present a proof of the following interesting:

4.6.7. THEOREM ("Brouwer Invariance of Domain Theorem"): *Let* $n \in \mathbb{N}$ *and let* U *be an open subset of* $\mathbb{R}^n$. *Then for each continuous injective function* $f: U \to \mathbb{R}^n$ *the following hold:*

(1) f(U) *is open in* $\mathbb{R}^n$, *and*

(2) $f: U \to f(U)$ *is a homeomorphism.*

PROOF: We shall first prove (1). Take an arbitrary $x \in U$. We shall prove that f(x) belongs to the interior of f(U). There exists $\varepsilon > 0$ such that $B = \{y \in \mathbb{R}^n: \|x - y\| \leq \varepsilon\}$ is contained in U. Since B is compact, by exercise 1.1.4, $f \mid B$ is a homeomorphism. By corollary 4.6.6 we conclude that f(x) belongs to the interior of f(B) and hence to the interior of f(U).

We shall now prove (2). This is easy. If V is an open subset of U then (1) applied to $f \mid V$ shows that f(V) is open in $\mathbb{R}^n$ and hence in f(U). We conclude that $f: U \to f(U)$ is an open mapping and therefore by injectiveness is a homeomorphism. □

The question naturally arises whether something like theorem 4.6.7 also can be derived for a larger class of spaces, e.g. for *infinite-dimensional* normed linear spaces. As to be expected, this is not possible. For the first part of the theorem, this can be demonstrated quite easily by considering Hilbert space l^2. The function f: $l^2 \to l^2$ defined by

$$f(x_1,x_2,\cdots) = (0,x_1,x_2,\cdots),$$

is easily seen to be an imbedding of l^2 such that its range is nowhere dense. It can be shown that theorem 4.6.7(1) does not hold for *any* infinite-dimensional normed linear space, but the proof of this is much more complicated than the above triviality.

We now turn to the second part of theorem 4.6.7.

4.6.8. THEOREM: *Let* L *be an infinite-dimensional normed linear space. There exists a bijective continuous function* f: L $\to$ L *such that* f *is not a homeomorphism.*

PROOF: Let S be the unit sphere in L. Since L is infinite-dimensional, S is not compact by exercise 1.2.16. Consequently, by exercise 1.4.10 there exists a continuous function λ: S $\to$ (0,1] such that inf $\lambda(S) = 0$. Now define f: L $\to$ L by the formulas

$$\begin{cases} f(y) = \lambda(\frac{y}{\|y\|})\cdot y & (y \neq \underline{0}), \\ f(\underline{0}) = \underline{0}. \end{cases}$$

CLAIM: If $x \in L$ and $\alpha \in [0,\infty)$ then $f(\alpha x) = \alpha f(x)$.

This is a triviality. If $x = \underline{0}$ or $\alpha = 0$ then there is nothing to prove. In addition, if $x \neq \underline{0}$ and $\alpha \neq 0$ then

$$f(\alpha x) = \lambda(\frac{\alpha x}{\|\alpha x\|})\cdot \alpha x = \lambda(\frac{\alpha x}{\alpha\|x\|})\cdot\alpha x = \alpha f(x).$$

We shall now prove that f is continuous. It is clear that this need only be checked at the origin. To this end, let $(x_n)_n$ be a sequence in $L\backslash\{\underline{0}\}$ such that $\lim_{n\to\infty} x_n = \underline{0}$. Then

$$\|f(x_n)\| = \lambda(\frac{x_n}{\|x_n\|})\cdot\|x_n\| \leq 1\cdot\|x_n\| \to 0 \qquad (n \to \infty).$$

We conclude that $\lim_{n\to\infty} f(x_n) = \underline{0}$, as required.

We shall next prove that f is injective. To this end, take $x,y \in L$ and assume that $f(x) = f(y)$. Without loss of generality, $x \neq \underline{0}$. Then $f(x) \neq \underline{0}$ so that $f(y) \neq \underline{0}$ from which it follows that $y \neq \underline{0}$. Consequently, since $f(x) = f(y)$, we have

$$\lambda(\frac{x}{\|x\|})\cdot x = \lambda(\frac{y}{\|y\|})\cdot y,$$

and since $\lambda(S) \subseteq (0,1]$ this implies that there exists $\alpha > 0$ such that $x = \alpha y$. By the claim, this yields $f(x) = \alpha f(y)$, i.e. $\alpha = 1$. We conclude that $x = y$.

Now take $y \in L \setminus \{\underline{0}\}$. Put

$$x = \frac{y}{\lambda(\frac{y}{\|y\|})}.$$

An easy application of the claim gives us that $f(x) = y$.

We shall now prove that f is not a homeomorphism by showing that f^{-1} is not continuous. Since $\inf \lambda(S) = 0$, we can choose a sequence $(x_n)_n$ in S such that

$$\lim_{n\to\infty} \lambda(x_n) = 0.$$

It follows that

$$\|f(x_n)\| = \lambda(x_n)\cdot\|x_n\| \to 0\cdot 1 = 0,$$

so $f(x_n) \to \underline{0}$. However, $\|x_n\| = 1$ for every n, so $x_n \not\to \underline{0}$. □

Exercises for §4.6.

1. Prove that no continuum can be partitioned into countably many nonempty pairwise disjoint closed sets (This is called the Sierpiński Theorem) (Hint: First prove that if X is a continuum and B is a nonempty closed subset of X then every component of B meets the boundary of B (by the boundary of B we mean the boundary of B in X). Then observe that if X is a continuum which is the union of countably many pairwise disjoint closed sets $X_1, X_2, \cdots$, such that at least two of the X_i are nonempty, then there is a continuum C in X such that $C \cap X_1 = \emptyset$, while moreover at least two of the $C \cap X_1, C \cap X_2, \cdots$, are nonempty).

2. Let $n \geq -1$ and let X be a compact space which can be written as the union of closed sets X_i, $i \in \mathbb{N}$, such that for all distinct $i,j \in \mathbb{N}$, $\dim(X_i \cap X_j) \leq n$. Prove that every continuous function $f: X_1 \to S^n$ can be extended over X (Hint: Prove a stronger statement than this exercise by induction on n, use the same technique as in the proof of theorem 4.6.3, and apply exercise 4.6.1).

3. Let A be a closed subspace of a space X such that $\dim(X\setminus A) \leq n$, let $0 \leq k \leq n$, and let $f: A \to S^k$ be continuous. Prove that there exists a closed subspace B of X such that $A \cap B = \emptyset$ and $\dim B \leq n-k-1$, while moreover the function f can be extended over $X\setminus B$ (Hint: Prove this exercise by induction on k+n and use the same tech-

nique as in the proof of theorem 4.6.3).

4.7. Totally Disconnected Spaces

In this section we shall prove that totally disconnected spaces of every dimension exist, cf. examples 4.2.11 and 4.5.10 where we constructed an example of a 1-dimensional totally disconnected space. We shall introduce hyperspaces, prove the existence of certain "G_δ-selections" and characterize the class of all topologically complete spaces. After this preparatory work, the construction of the examples is rather simple.

Let X be a compact space. The collection of all *non-empty* closed subsets of X shall be denoted by 2^X. It will be convenient to endow 2^X with a suitable topology. For $E,F \in 2^X$ define the *Hausdorff distance* $d_H(E,F)$ between E and F by

$$d_H(E,F) = \inf\{\varepsilon > 0: E \subseteq B(F,\varepsilon) \text{ and } F \subseteq B(E,\varepsilon)\}.$$

Since X is compact, it is clear that $d_H(E,F) \in [0,\infty)$.

4.7.1. LEMMA: d_H: $2^X \times 2^X \to [0,\infty)$ *is a metric.*

PROOF: We shall only present a proof of the triangle inequality; the other elementary verifications are left to the reader.

Now take $E,F,G \in 2^X$. Let $\varepsilon > 0$ and put $\delta = \varepsilon/2$. By definition we have

$$(1) \qquad E \subseteq B(F,\delta+d_H(E,F)) \text{ and } F \subseteq B(G,\delta+d_H(F,G)).$$

Take an arbitrary point $x \in E$. By (1) there exist $y \in F$ such that $d(x,y) < \delta+d_H(E,F)$. Again by (1) there exists $z \in G$ such that $d(y,z) < \delta+d_H(F,G)$. We conclude that $d(x,z) < \varepsilon + d_H(E,F) + d_H(F,G)$. Therefore, since x was an arbitrary point of E,

$$E \subseteq B(G,\varepsilon+d_H(E,F)+d_H(F,G)).$$

Similarly,

$$G \subseteq B(E,\varepsilon+d_H(E,F)+d_H(F,G)).$$

Since ε was an arbitrary positive number, we obtain

$$d_H(E,G) \leq d_H(E,F)+d_H(F,G),$$

as required. □

The metric d_H is called the *Hausdorff metric* on 2^X. We endow 2^X with the topology derived from this metric and call it the *hyperspace* of X.

4.7.2. PROPOSITION: *If* X *is compact then so is* 2^X.

PROOF: Let $(A_i)_i$ be a sequence in 2^X. We shall prove that it has a convergent subsequence.

Let $\mathcal{B} = \{B_n: n \in \mathbb{N}\}$ be an open basis for X with $B_1 = X$. By induction on n we shall construct a decreasing sequence X_n of closed subsets of X with the following properties:

(1) $X_1 = X$,
$(2)_n$ every neighborhood of X_n contains infinitely many A_i's, and
$(3)_n$ if $X_n \cap B_n \neq \emptyset$ then $X_n \backslash B_n$ has a neighborhood V containing at most finitely many A_i's.

X_1 clearly satisfies $(2)_1$ and $(3)_1$. Suppose that X_n has been constructed satisfying $(2)_n$ and $(3)_n$. There are two cases to consider. Suppose first that every neighborhood of $X_n \backslash B_{n+1}$ contains infinitely many A_i's. Then define $X_{n+1} = X_n \backslash B_{n+1}$. If this is not the case then put $X_{n+1} = X_n$.

Now define

$$D = \bigcap_{n=1}^{\infty} X_n.$$

Observe that $D \neq \emptyset$ since X is compact and by (2) every X_n is nonempty. Since the X_n's decrease, by compactness of X it follows that every neighborhood V of D contains some X_n from which it follows that V contains infinitely many A_i's. For every $m \in \mathbb{N}$ let $i(m) \in \mathbb{N}$ be such that $i(m) > i(m-1)$ and $A_{i(m)} \subseteq B(D,1/m)$.

CLAIM : For each $\varepsilon > 0$ there exists $N \in \mathbb{N}$ such that $D \subseteq B(A_{i(m)},\varepsilon)$ for every $m \geq N$.

We argue by contradiction. Suppose that such an N does not exist. Then there is an infinite subset E of $\mathbb{N}$ such that for every $m \in E$, $D\backslash B(A_{i(m)},\varepsilon) \neq \emptyset$, say $x_m \in D\backslash B(A_{i(m)},\varepsilon)$. By compactness of D, replacing E if necessary by an infinite subset, we may assume without loss of generality that $x_m \to x$, $m \to \infty$. Let $B_n \in \mathcal{B}$ be such that $x \in B_n$ and $\text{diam}(B_n) < \varepsilon$.

Let V be a neighborhood of $X_n \setminus B_n$ such as in $(3)_n$. Then $V \cup B_n$ is a neighborhood of D and therefore contains all but finitely many of the $A_{i(m)}$. We conclude that all but finitely many of the $A_{i(m)}$ intersect B_n. Now pick $m \in \mathbb{N}$ so large that $x_m \in B_n$ while moreover $A_{i(m)}$ intersects B_n. Since $\mathrm{diam}(B_n) < \varepsilon$ and $x_m \notin B(A_{i(m)},\varepsilon)$, we arrived at a contradiction.

We conclude that the sequence $(A_{i(m)})_m$ converges to D in 2^X. □

We now will prove the existence of certain "G_δ-selections".

Suppose that X is compact and that $f: X \to Y$ is a continuous surjection. The set-valued function $F: Y \Rightarrow X$ defined by $F(y) = f^{-1}(y)$ in general unfortunately does not admit a *continuous* selection, cf. §1.4. We shall show however that there always exists a selection for F the range of which is a G_δ-subset of X.

As usual, for each $n \in \mathbb{N}$ let $\pi_n: Q \to [-1,1]_n$ denote the projection. Now for any nonempty compact subset X of Q inductively define points $p(X,n) \in [-1,1]$ as follows:

$$(1)\ p(X,1) = \min \pi_1(X),$$

$$(2)\ p(X,n+1) = \min \pi_{n+1}\Big(X \cap \bigcap_{i=1}^{n} \pi_i^{-1}(p(X,i))\Big) \qquad (n \geq 1).$$

We claim that these points are well-defined. Clearly, p(X,1) is well-defined. Assume that p(X,n) is well-defined for certain n. There consequently exists a point

$$x \in X \cap \bigcap_{i=1}^{n-1} \pi_i^{-1}(p(X,i))$$

such that

$$\pi_n(x) = p(X,n).$$

Consequently,

$$x \in X \cap \bigcap_{i=1}^{n-1} \pi_i^{-1}(p(X,i))$$

from which it follows by compactness that

$$p(X,n+1) = \min \pi_{n+1}\Big(X \cap \bigcap_{i=1}^{n-1} \pi_i^{-1}(p(X,i))\Big)$$

is well-defined.

We constructed a point $p(X) = (p(X,1),p(X,2),\cdots,p(X,n),\cdots)$ in Q. Since the collection

$$\{X \cap \bigcap_{i=1}^{n} \pi_i^{-1}(p(X,i)) : n \in \mathbb{N}\}$$

has the finite intersection property and since

$$\bigcap_{i=1}^{\infty} \pi_i^{-1}(p(X,i)) = \{p(X)\},$$

we conclude that $p(X) \in X$.

4.7.3. THEOREM: *Let* X *be compact and let* $f: X \to Y$ *be a continuous surjection. There exists a* G_δ*-subset* S *of* X *which intersects each fiber of* f *in precisely one point.*

PROOF: By theorem 1.4.18 we may assume that X is a subspace of Q. Now define

$$S = \{p(f^{-1}(y)) : y \in Y\}.$$

We shall prove that S is a G_δ-subset of X. For $n,m \in \mathbb{N}$ define the following sets:

$$U_{n,m} = \{x \in X : \forall z \in X \text{ such that } x_i = z_i \text{ for } i \leq n-1 \text{ and } z_n \leq x_n - 1/m \text{ we have } f(x) \neq f(z)\}.$$

We first prove that the $U_{n,m}$'s are open by establishing that their complements are closed. To this end, for each $j \in \mathbb{N}$ take $x(j) \in X \setminus U_{n,m}$ and assume that the sequence $(x(j))_j$ converges to a point $x \in X$. For each j there exists z(j) in X such that $x(j)_i = z(j)_i$ for every $i \leq n-1$, $z(j)_n \leq x(j)_n - 1/m$ and $f(x(j)) = f(z(j))$. By compactness, without loss of generality we assume that $\{z(j)\}_j$ converges to a point z in X. By continuity of the functions π_j, $j \leq n$, and f we get

$$x_i = z_i \text{ for every } i \leq n-1,\ z_n \leq x_n - 1/m \text{ and } f(x) = f(z), \text{ i.e.}$$

$x \in X \setminus U_{n,m}$. We conclude that $X \setminus U_{n,m}$ is closed.

We next claim that for all n and m, $S \subseteq U_{n,m}$. Take an arbitrary point $y \in Y$. Fix arbitrary $n,m \in \mathbb{N}$. Take a point $z \in X$ such that $p(f^{-1}(y))_i = z_i$ for all $i \leq n-1$ and $z_n \leq p(f^{-1}(y))_n - 1/m$. Assume that $z \in f^{-1}(y)$. Then by construction, $p(f^{-1}(y))_n \leq z_n$. This is a contradiction. Therefore $z \notin f^{-1}(y)$, as required.

We finally claim that the intersection U of all the $U_{n,m}$'s is equal to S. Since S intersects every fiber of f in precisely one point, and as we just observed, $S \subseteq U$, it suffices to prove that U intersects every fiber of f in at most one point. To this end, take an arbitrary $y \in Y$ and assume that there are distinct points $x(1),x(2) \in U \cap f^{-1}(y)$. Let $n \in \mathbb{N}$ be the first index such that $x(1)_n \neq x(2)_n$. Without loss of generality assume that $x(1)_n < x(2)_n$. There exists $i \in \mathbb{N}$ such that $x(1)_n \leq x(2)_n - 1/i$. Since $x(2) \in U_{n,i}$ and $x(1)_m = x(2)_m$ for all $m \leq n-1$, we get $f(x(1)) \neq f(x(2))$, which is a contradiction. □

Our next aim is to characterize the class of all topologically complete spaces.

4.7.4. THEOREM: *Let* X *be a space. The following statements are equivalent:*

(a) *if* Y *is any space containing* X *then* X *is a* G_δ*-subset of* Y,

(b) *there is a topologically complete space* Y *such that* X *is homeomorphic to a* G_δ*-subset of* Y,

(c) X *can be imbedded in* $\mathbb{R}^\infty$ *as a closed subspace,*

(d) X *is topologically complete.*

PROOF: To begin with, let us establish the following

CLAIM: Let X be a space and let S be a G_δ-subset of X. Then S can be imbedded in $X \times \mathbb{R}^\infty$ as a closed subset.

There exist closed sets G_i, $i \in \mathbb{N}$, in X such that

$$X \setminus \bigcup_{i=1}^\infty G_i = S.$$

For $i \in \mathbb{N}$, let $f_i: S \to \mathbb{R}$ be defined by

$$f_i(x) = \frac{1}{d(x,G_i)}.$$

In addition, define $f: S \to X \times \mathbb{R}^\infty$ by

$$f(x) = (x,f_1(x),f_2(x),\cdots\cdots).$$

Since the inclusion $S \to X$ is an imbedding, so is f. We shall prove that f(S) is closed in $X \times \mathbb{R}^\infty$. To this end, let $(x_n)_n$ be a sequence in S such that $(f(x_n))_n$ converges to a point $(a,b) \in X \times \mathbb{R}^\infty$. Then $(x_n)_n$ converges to a. Suppose that $a \notin S$, say $a \in G_i$. Since $(x_n)_n$ converges to a we have $\lim_{n\to\infty} d(x_n,G_i) = 0$ from which it follows that $\lim_{n\to\infty} f_i(x_n) = \infty$.

So by continuity of f, the i-th coordinate of b must be equal to ∞. This is a contradiction and we conclude that $a \in S$. Then by continuity of f we have

$$f(a) = \lim_{n\to\infty} f(x_n) = (a,b),$$

so $(a,b) \in f(S)$.

We prove (a) $\Rightarrow$ (b).

By exercise 1.4.9 we may assume that X is a subspace of $\mathbb{R}^\infty$. Since $\mathbb{R}^\infty$ is topologically complete, cf. the remarks following lemma 1.2.1, this establishes (b).

We prove (b) $\Rightarrow$ (d).

By the claim, X is homeomorphic to a closed subspace of the product $Y \times \mathbb{R}^\infty$. Since the product of two topologically complete spaces is topologically complete, we are done.

We prove (d) $\Rightarrow$ (a).

Let Y be a space containing X. Without loss of generality we may assume that X is dense in Y. Since the closure of X is clearly a G_δ-subset of Y, it suffices to prove that X is a G_δ-subset of its closure. So assume, for convenience, that X is dense in Y. Let d be an admissible complete metric on X. For each $n \in \mathbb{N}$ let $\mathcal{B}_n$ be a cover of X by open subsets of X each of d-diameter less than 1/n. There exists for each $n \in \mathbb{N}$ a collection $\mathcal{E}_n$ of open subsets of Y such that $\mathcal{E}_n \cap X = \mathcal{B}_n$. Let E_n be the union of the family $\mathcal{E}_n$. We claim that X is equal to the intersection of the E_n's. To this end, take an arbitrary point $y \in \bigcap_{n=1}^\infty E_n$. Since X is dense in Y, there exists a sequence $(x_i)_i$ in X whose limit is y. This sequence is d-Cauchy. To this end, take $\varepsilon > 0$ and let $n \in \mathbb{N}$ be so large that $1/n < \varepsilon$. There exists $E \in \mathcal{E}_n$ such that $y \in E$. Since $(x_i)_i$ converges to y, all but finitely many of the x_i belong to $E \cap X$, which has d-diameter less than $1/n < \varepsilon$ by construction. So the sequence $(x_i)_i$ is indeed d-Cauchy which implies that $y \in X$ because d is complete.

We prove (a) $\Rightarrow$ (c).

Since, as observed above, we can think of X as being a subspace of $\mathbb{R}^\infty$, it follows from (a) and the claim that X can be imbedded as a closed subspace of $\mathbb{R}^\infty \times \mathbb{R}^\infty \approx \mathbb{R}^\infty$.

We prove (c) $\Rightarrow$ (b).

This is a triviality. □

Before we are able to construct the examples, we need to derive some elementary results about essential families and quasi-components (see below) first.

4.7.5. THEOREM: *Let* X *be a space and let* $\{(A_i,B_i): i \in \Gamma\}$ *be an essential family of pairs of disjoint closed subsets of* X. *Suppose that* $Y \subseteq X$ *is such that* $Y \cap \bigcap_{i\in\Gamma} L_i \neq \emptyset$ *for any choice of partitions* L_i *of* A_i *and* B_i *in* X. *For each* $i \in \Gamma$ *let* U_i *and* V_i *be disjoint closed neighborhoods of*

A_i *and* B_i, *respectively. Then* $\{(U_i \cap Y, V_i \cap Y): i \in \Gamma\}$ *is essential in* Y.

PROOF: For $i \in \Gamma$ let E_i be a partition between $U_i \cap Y$ and $V_i \cap Y$ in Y. By lemma 4.1.5 for $i \in \Gamma$ there exist partitions F_i between A_i and B_i in X such that $F_i \cap Y \subseteq E_i$. By assumption,

$$\emptyset \neq Y \cap \bigcap_{i \in \Gamma} F_i \subseteq \bigcap_{i \in \Gamma} E_i,$$

which is as required. □

Let X be a space. A subset of X is called *clopen* if it is both closed and open. The *quasi-component* of a point x in X is the intersection of all clopen subsets of X which contain the point x. It is easy to see that the quasi-components of X form a decomposition of X into pairwise disjoint closed sets.

4.7.6. LEMMA: *Let* Q *be a quasi-component of the compact space* X. *Then for every neighborhood* U *of* Q *there exists a clopen subset* E *of* X *such that* $Q \subseteq E \subseteq U$.

PROOF: Let $\mathcal{E}$ be the family of all clopen neighborhoods of Q. By assumption, $\bigcap \mathcal{E} = Q$. Observe that the intersection of finitely many elements of $\mathcal{E}$ again belongs to $\mathcal{E}$. Without loss of generality we assume that U is open. Then $(X \setminus U) \cap \bigcap \mathcal{E} = \emptyset$, from which it follows by compactness that U must contain an element of $\mathcal{E}$. □

Let $\mathcal{P}$ be a decomposition of X into pairwise disjoint closed subsets. We say that $\mathcal{P}$ is *upper semicontinuous* provided that for every closed subset A of X, $\bigcup\{P \in \mathcal{P}: P \cap A \neq \emptyset\}$ is closed in X.

4.7.7. PROPOSITION: *Let* X *be compact. Then*

(1) *the component of* $x \in X$ *coincides with the quasi-component of* x,

(2) *the collection of components is an upper semicontinuous decomposition of* X.

PROOF: For (1), let C(x) and Q(x) denote the component and the quasi-component of x, respectively. It is clear that C(x) is a subset of Q(x), since if A is any clopen subset of X then either $C(x) \cap A = \emptyset$ or $C(x) \subseteq A$. So it suffices to prove that Q(x) is connected. To this end, assume that Q(x) is not connected. We shall derive a contradiction. We can write Q(x) as $E \cup F$, where both E and F are nonempty closed subsets of X, and $E \cap F = \emptyset$. Without loss of generality, we may assume that $x \in E$. There exist disjoint open neighborhoods U and V of E and F, respectively (corollary 1.4.16). Since $U \cup V$ is open and contains Q(x), by lemma 4.7.6 there exists a clopen subset C of X such that $Q(x) \subseteq C \subseteq U \cup V$. Observe that

$$C \cap U = C \setminus V$$

is both open and closed. But this implies that $Q(x) \subseteq C \cap U$ and consequently, $F = \emptyset$, which is a contradiction.

For (2), let $\mathcal{C}$ denote the collection of all components of X and let A be closed. We shall prove that if $\mathcal{P} = \{C \in \mathcal{C}: C \cap A \neq \emptyset\}$ then $\bigcup\mathcal{P}$ is closed in X. Take $x \notin \bigcup\mathcal{P}$. Then by the above, $Q(x) \cap A = \emptyset$, hence by lemma 4.7.6, for some clopen C in X, $Q(x) \subseteq C \subseteq X\setminus A$. But then $C \cap \bigcup\mathcal{P} = \emptyset$, so that $x \notin \overline{\bigcup\mathcal{P}}$. □

A *continuum* is a compact connected space. Let X be a space and let A and B be disjoint closed subsets of X. By a *continuum from* A *to* B we mean a continuum C in X meeting A as well as B.

4.7.8. PROPOSITION: *Let* X *be a compact space, let* $\{(A_i,B_i): i \in \Gamma\}$ *be an essential family of pairs of disjoint closed subsets of* X *and let* $\Gamma(0) \subseteq \Gamma$. *If for each* $i \in \Gamma(0)$, L_i *is a partition between* A_i *and* B_i *in* X *and* $n \in \Gamma\setminus\Gamma(0)$ *then* $L = \bigcap_{i \in \Gamma(0)} L_i$ *contains a continuum from* A_n to B_n.

PROOF: Suppose that L does not contain a continuum from A_n to B_n. Let $H = A_n \cap L$ and $G = B_n \cap L$, respectively. Put

$$\mathcal{E} = \{C: C \text{ is a component of } L \text{ and } C \cap H \neq \emptyset\},\ E = \bigcup\mathcal{E},$$

and

$$\mathcal{F} = \{C: C \text{ is a component of } L \text{ and } C \cap G \neq \emptyset\},\ F = \bigcup\mathcal{F},$$

respectively. From proposition 4.7.7 it follows that both E and F are closed. Also, by assumption, $E \cap F = \emptyset$. Take an arbitrary $C \in \mathcal{E}$. Since $L\setminus F$ is an open neighborhood of C in L, by lemma 4.7.6 there exists a clopen (in L) set U_C which contains C but misses F. By compactness, finitely many U_C's cover E. We conclude that $\emptyset$ is a partition in L between H and G. By corollary 4.1.6 there is a partition T in X between A_n and B_n such that $T \cap L = \emptyset$. But this contradicts the fact that $\{(A_i,B_i): i \in \Gamma(0)\} \cup \{(A_n,B_n)\}$ is essential. □

4.7.9. COROLLARY: *Let* X *be a compact space, let* $\{(A_i,B_i): i \in \Gamma\}$ *be an essential family of pairs of disjoint closed subsets of* X *and let* $n \in \Gamma$. *Suppose that* $Y \subseteq X$ *is such that* Y *meets every continuum from* A_n *to* B_n. *For each* $i \in \Gamma\setminus\{n\}$ *let* U_i *and* V_i *be disjoint closed neighborhoods of* A_i *and* B_i, *respectively. Then* $\{(U_i \cap Y, V_i \cap Y): i \in \Gamma\setminus\{n\}\}$ *is essential in* Y.

PROOF: By proposition 4.7.8, if L_i is a partition between A_i and B_i for $i \in \Gamma\setminus\{n\}$ then $Y \cap \bigcap_{i\in\Gamma\setminus\{n\}} L_i \neq \emptyset$. Now apply theorem 4.7.5. □

All these unrelated results allow us to prove quite easily the following:

4.7.10. THEOREM: *For each* $n \geq 0$ *there exists a topologically complete totally disconnected space of dimension* n. *There also exists a strongly infinite-dimensional topologically complete totally disconnected space.*

PROOF: We shall construct the examples simultaneously. Consider $X = J \times J^n$, where $n \in \mathbb{N} \cup \{\infty\}$. Let $\pi: X \to J$ be the projection and put $A = \pi^{-1}(-1)$ and $B = \pi^{-1}(1)$, respectively. Define

$$\mathcal{C} = \{C \in 2^X: C \text{ is a continuum from A to B}\}.$$

CLAIM 1: $\mathcal{C}$ is a closed subspace of 2^X.

Take an arbitrary $E \in 2^X\setminus\mathcal{C}$. First assume that E is not a continuum. Then E can be written as the union of two nonempty pairwise disjoint closed sets, say E_1 and E_2. Let $\varepsilon > 0$ be such that $D(E_1,\varepsilon) \cap D(E_2,\varepsilon) = \emptyset$. It is easy to see that every $F \in 2^X$ with $d_H(E,F) < \varepsilon$ is contained in $D(E_1,\varepsilon) \cup D(E_2,\varepsilon)$ and intersects $D(E_1,\varepsilon)$ as well as $D(E_2,\varepsilon)$. Therefore, the ball (in 2^X) about E of radius ε contains no continua and therefore misses $\mathcal{C}$. Next assume that E misses A. Then let $\delta > 0$ be such that $D(E,\delta) \cap A = \emptyset$. It is clear that every $F \in 2^X$ with $d_H(E,F) < \delta$ is contained in $D(E,\delta)$ and therefore also misses A. We conclude that the ball (in 2^X) about E of radius δ consists entirely of sets that do not intersect A and consequently this ball misses $\mathcal{C}$. If E misses B then argue similarly.

Since 2^X is compact by proposition 4.7.2, the claim implies that $\mathcal{C}$ is compact. Let $\alpha: \mathbf{C} \to \mathcal{C}$ be a continuous surjection (theorem 4.2.9). Now put

$$Y = \bigcup\{\pi^{-1}(t) \cap \alpha(t): t \in \mathbf{C}\}.$$

CLAIM 2: Y is compact and $\pi(Y) = \mathbf{C}$.

We shall first prove that $\pi(Y) = \mathbf{C}$. Clearly, $\pi(Y) \subseteq \mathbf{C}$. Take $t \in \mathbf{C}$. Then $\alpha(t)$ is a continuum from A to B. Consequently, $\pi(\alpha(t))$ is a continuum in J from -1 to 1 and therefore is equal to J. So there exists $y \in \alpha(t)$ with $\pi(y) = t$. Since this y belongs to Y, this proves that $\mathbf{C} \subseteq \pi(Y)$.

We shall next show that Y is compact. To this end, let $(y_i)_i$ be a sequence in Y converging to a point $y \in X$. Then $t_i = \pi(y_i)$ belongs to **C** for every i and $(t_i)_i$ converges to $t = \pi(y)$. Consequently, t belongs to **C**. Since α is continuous, the sequence of continua $(\alpha(t_i))_i$ converges to $\alpha(t)$ in $\mathcal{C}$. We claim that $y \in \alpha(t)$. If this is not the case then we can find $\varepsilon > 0$ such that $y \notin D(\alpha(t),\varepsilon)$. Since $(\alpha(t_i))_i$ converges to $\alpha(t)$, for all but finitely many i, $y_i \in \alpha(t_i) \subseteq D(\alpha(t),\varepsilon)$ which implies that $y \in D(\alpha(t),\varepsilon)$; contradiction. We conclude that $y \in Y$.

Now by theorem 4.7.3 there exists a G_δ-subset S of Y which intersects each fiber of $\pi \mid Y$ in precisely one point. Since Y is compact (claim 2), Y is topologically complete. So S is topologically complete as well by theorem 4.7.4.

CLAIM 3: S intersects each continuum from A to B.

Take an arbitrary $C \in \mathcal{C}$. Since α is surjective, there exists $t \in \mathbf{C}$ such that $\alpha(t) = C$. Observe that $(\pi \mid Y)^{-1}(t) = \pi^{-1}(t) \cap Y = \pi^{-1}(t) \cap \alpha(t) = \pi^{-1}(t) \cap C \subseteq C$. Since S intersects each fiber of $\pi \mid Y$ we now get $S \cap C \neq \emptyset$.

Observe that S can be mapped by a one-to-one mapping onto the zero-dimensional space **C**. This implies that S is totally disconnected, cf. example 4.2.11.

Now assume that n is finite. Then $\dim S \geq n$ by theorem 3.5.7 and corollary 4.7.9. Also $\dim S \leq n$ since S is contained in $\mathbf{C} \times J^n$, which is n-dimensional by theorems 4.3.10 and 4.5.9. We conclude that $\dim S = n$.

Now if $n = \infty$ then S is strongly infinite-dimensional by corollaries 3.5.8 and 4.7.9. □

Exercises for §4.7.

1. Prove that 2^I, the hyperspace of I, is infinite-dimensional and connected (in §8.4 we shall prove that in fact 2^I is homeomorphic to the Hilbert cube).

2. Let X be a compact space such that $\dim X \geq n+1$. Prove that X contains a totally disconnected G_δ-subset S such that $\dim S \geq n$.

3. Let X and Y be totally disconnected. Prove that $X \times Y$ is totally disconnected.

4. Prove that there exists a countable dimensional infinite-dimensional topologically complete totally disconnect-

ed space.

5. Prove that each compact subspace of a totally disconnected space is zero-dimensional.

6. Let X be a compact space and define i: $X \to 2^X$ by $i(x) = \{x\}$. Prove that i is an imbedding.

7. Let X be a compact space. Prove that the function diam: $2^X \to [0,\infty)$ sending each element of 2^X onto its diameter, is continuous.

8. Let X be compact and let $E,F \in 2^X$. Prove that $d_H(E,F) = \max\{\sup_{x\in E} d(x,F), \sup_{x\in F} d(x,E)\}$.

Let $\mathcal{P}$ be a decomposition of a space X into pairwise disjoint closed sets. In addition, let f: $X \to \mathcal{P}$ be the function sending $x \in X$ to the unique element of $\mathcal{P}$ containing x. Call a subset $\mathcal{U}$ of $\mathcal{P}$ open if and only if $f^{-1}(\mathcal{U})$ is open in X, i.e. we endow $\mathcal{P}$ with the *quotient* topology derived from X and f. With this topology we denote $\mathcal{P}$ by $X/\mathcal{P}$ and call it X *modulo* $\mathcal{P}$.

9. Let X be a space and let $\mathcal{P}$ be a decomposition of X into pairwise disjoint closed sets. Prove that the above topology is indeed a topology on $\mathcal{P}$. Prove that $\mathcal{P}$ is upper semicontinuous if and only if the natural map f: X $\to X/\mathcal{P}$ is a closed map. Give an example that $X/\mathcal{P}$ need not be metrizable. Prove that if $\mathcal{P}$ is upper semicontinuous and consists of compact sets then $X/\mathcal{P}$ is separable and metrizable.

4.8. Various kinds of Infinite-Dimensionality

Recall that a space X is *strongly infinite-dimensional* if it has an infinite essential family of pairs of disjoint closed sets. By corollary 3.5.8, the Hilbert cube Q is an example of a strongly infinite-dimensional space. A space X is called *weakly infinite-dimensional* if it is infinite-dimensional but not strongly infinite-dimensional, i.e. for every family $\{(A_i,B_i): i \in \mathbb{N}\}$ of pairs of disjoint closed subsets of X there exist partitions D_i between A_i and B_i such that $\bigcap_{i=1}^{\infty} D_i = \emptyset$.

4.8.1. PROPOSITION: *Let* $X \subseteq Q$ *be weakly infinite-dimensional. Then for every sequence* $\{(A_i,B_i): i \in \mathbb{N}\}$ *of pairs of disjoint closed subsets of* Q *there exist partitions* D_i *between* A_i *and* B_i *in* Q *such that* $\bigcap_{i=1}^{\infty} D_i \cap X = \emptyset$.

PROOF: For every $i \in \mathbb{N}$, let U_i and V_i be disjoint closed neighborhoods of A_i and B_i, respectively (corollary 1.4.16). Since X is weakly infinite-dimensional, the sequence $\{(U_i \cap X, V_i \cap X), i \in \mathbb{N}\}$ is inessential. Consequently, there exist partitions S_i between $U_i \cap X$ and

$V_i \cap X$ in X such that the intersection of the S_i's is empty. By lemma 4.1.5 there exist partitions D_i, $i \in \mathbb{N}$, between A_i and B_i in Q such that $D_i \cap X \subseteq S_i$. Then clearly,

$$\bigcap_{i=1}^{\infty} D_i \cap X \subseteq \bigcap_{i=1}^{\infty} S_i = \emptyset,$$

as required. □

Finally, recall that a space X is called *countable dimensional* if it can be written as the union of countably many zero-dimensional subspaces. So every finite-dimensional space is countable dimensional by corollary 4.4.8, but not conversely, cf. example 4.8.2. Observe that every subspace of a countable dimensional space is again countable dimensional.

4.8.2. EXAMPLE: *There exists an infinite-dimensional countable dimensional compact space.*

This is a triviality. Let X be the one-point compactification of the topological sum of the spaces I, I^2, I^3, $\cdots$. Then X is infinite-dimensional by theorem 4.3.13 and corollary 4.5.12(b). In addition, X is countable dimensional by corollary 4.5.12(e). □

4.8.3. PROPOSITION: *Every countable dimensional space is weakly infinite-dimensional.* □

PROOF: Let X be countable dimensional. Then X is the union of countably many zero-dimensional subspaces, say X_i, $i \in \mathbb{N}$. Let $\{(A_i,B_i)\colon i \in \mathbb{N}\}$ be a family of pairs of disjoint closed subsets of X. By theorem 4.2.2(5) for every i there exists a partition D_i between A_i and B_i such that $D_i \cap X_i = \emptyset$. Since the union of the X_i's equals X, the intersection of the D_i's is empty, which is as required. □

4.8.4. COROLLARY: *No strongly infinite-dimensional space is countable dimensional.* □

4.8.5. COROLLARY: *The Hilbert cube cannot be written as the union of countably many zero-dimensional subspaces.* □

In view of the above results, the question naturally arises whether the converse to proposition 4.8.3 holds. This question was known as *Alexandrov's problem* for decades, [1, §4, Hypothesis]. We shall answer this problem in the negative. However, we need to derive the following result first.

4.8.6. THEOREM: *Every topologically complete space X has a compactification γX such that the remainder $\gamma X \setminus X$ is countable dimensional.*

PROOF: We may assume that X is a subspace of the Hilbert cube Q (theorem 1.4.18). Since X is topologically complete, there is a decreasing sequence $(U_n)_n$ of open sets in Q such that

$$(1) \qquad X = \bigcap_{n=1}^{\infty} U_n$$

(theorem 4.7.4). Since Q is compact, for each $\varepsilon > 0$ there is a finite open cover of Q with mesh less than ε. Consequently, for each n there exist finitely many open sets $U_{n,1},\cdots,U_{n,m_n}$ such that

$$(2) \ U_{n,1} \cup \cdots \cup U_{n,m_n} = U_n,$$
$$(3) \ \text{mesh}(\{U_{n,1},\cdots,U_{n,m_n}\}) < \tfrac{1}{n}.$$

For each $n \in \mathbb{N}$ and $m \leq m_n$, define $f_{n,m}: Q \to I$ by

$$f_{n,m}(x) = d(x, X \setminus U_{n,m})$$

and define $f: Q \to Q$ coordinatewise as follows:

$$f(x) = (f_{1,1}(x),\cdots,f_{1,m_1}(x),f_{2,1}(x),\cdots,f_{2,m_2}(x),\cdots\cdots).$$

By (3) it follows easily that $f \mid X$ is an imbedding. We claim that

$$f(Q)\setminus f(X) \subseteq K_\omega = \{x \in Q: (\exists\, n \in \mathbb{N})(\forall\, m \geq n)(x_m = 0)\}.$$

Take an arbitrary point $f(q) \in f(Q)\setminus f(X)$. Since $q \notin X$, by (1) there exists an $N \in \mathbb{N}$ such that $q \notin U_N$. Consequently, since the U_n's decrease, for every $n \geq N$ and $m \leq m_n$ we have

$$f_{n,m}(q) = 0,$$

as required.

Now let γX be the closure of f(X) in f(Q). Then γX is a compactification of X the remainder $\gamma X \setminus X$ of which is contained in K_ω. For each $n \in \mathbb{N}$ put

$$K_n = \{x \in Q: (\forall\, m \geq n)(x_m = 0)\}.$$

It is clear that K_n is homeomorphic to J^n, $n \in \mathbb{N}$, and hence is the union of n+1 zero-dimensional subspaces (corollary 4.4.8). Since K_ω is equal to the union of the K_n's, it follows that K_ω, as well as its subspace $\gamma X \backslash X$, is countable dimensional. □

The topological completeness of the spaces under discussion in the above theorem is essential. It can be shown that $\Omega = \{x \in \mathbb{R}^\infty : x_i = 0 \text{ for all but finitely many } i\}$ is an example of a countable dimensional space the remainder of each compactification of which is strongly infinite-dimensional, see exercise 4.8.7.

We shall now present the solution to Alexandrov's problem.

4.8.7. EXAMPLE: *There exists a weakly infinite-dimensional compact space that is not countable dimensional.*

By theorem 4.7.10 there exists a topologically complete strongly infinite-dimensional totally disconnected space, say S. Let $X = \gamma S$ be a compactification of S with countable dimensional remainder (theorem 4.8.6). We claim that X is the required example. To this end, first observe that X is not countable dimensional since by corollary 4.8.4 its subspace S is not countable dimensional. Now let $\tau = \{(A_i,B_i): i \geq 0\}$ be a sequence of pairs of disjoint closed subsets of X. Write $X\backslash S$ as the union of countably many zero-dimensional subspaces, say X_i, $i \in \mathbb{N}$. By theorem 4.2.2(5), for each $i \in \mathbb{N}$ there exists a partition D_i between A_i and B_i such that $D_i \cap X_i = \emptyset$. We conclude that the compact set $D = \bigcap_{i=1}^\infty D_i$ is contained in S, and S being totally disconnected, therefore has to be zero-dimensional (exercise 4.7.5). By another appeal to theorem 4.2.2(5), there exists a partition D_0 between A_0 and B_0 such that $D_0 \cap D = \emptyset$. We conclude that $\bigcap_{i=0}^\infty D_i = \emptyset$. □

An infinite-dimensional space X is called *hereditarily infinite-dimensional* if every nonempty subspace of X is either zero-dimensional or infinite-dimensional. The Hilbert cube is certainly not hereditarily infinite dimensional since it contains cells of every possible dimension. It seems very unlikely that hereditarily infinite-dimensional spaces can exist. However, such spaces exist that moreover are compact. We finish this section by presenting an example of a hereditarily infinite-dimensional space.

As usual, for every $n \in \mathbb{N}$, $\pi_n: Q \to [-1,1]_n$ denotes the projection.

4.8.8. LEMMA: *Let* $K \subseteq [-1,1]$ *be a Cantor set, let* $\xi \in \mathbb{N}$ *and let* $\sigma \subseteq \mathbb{N}\backslash\{\xi\}$ *be an infinite set the complement of which is also infinite. For each* $j \in \sigma$ *there exists a partition* S_j *between the opposite faces* $W_j(-1)$ *and* $W_j(1)$ *(cf. §3.4) of* Q *such that each subset* $X \subseteq \bigcap_{j\in\sigma} S_j$ *with* $K \subseteq \pi_\xi(X)$ *is infinite-dimensional.*

PROOF: Without loss of generality we may assume that $\xi = 1$ and σ is the set of even natural numbers. For each $i \in \mathbb{N}$ let

$$C_i = \pi_i^{-1}[-1,-\tfrac{1}{2}] \text{ and } D_i = \pi_i^{-1}[\tfrac{1}{2},1],$$

respectively. Now put

$$\Lambda = \{f \in C(Q,Q): (\forall i)(f^{-1}(W_i(-1)) = C_{2i} \text{ and } f^{-1}(W_i(1)) = D_{2i})\}.$$

Then Λ is a separable metric space since $C(Q,Q)$ is one by proposition 1.3.3. By exercise 4.2.8 there exists a subspace T of K such that T can be mapped onto Λ, say by the map τ. Let

$$E = T \times [-1,1]_2 \times [-1,1]_3 \times \cdots \subseteq Q$$

and define $\Phi: E \to Q$ by

$$\Phi(x) = \tau(x_1)(x).$$

Then Φ is clearly continuous.

CLAIM 1: For each i, $(\pi_i \circ \Phi)^{-1}(0)$ is a partition in E between $C_{2i} \cap E$ and $D_{2i} \cap E$.

Fix $i \in \mathbb{N}$. Then, $\Phi^{-1}(W_i(-1)) = \{x \in E: \Phi(x) \in W_i(-1)\} = \{x \in E: \tau(x_1)(x) \in W_i(-1)\} = \{x \in E: x \in \tau(x_1)^{-1}(W_i(-1))\} = E \cap C_{2i}$, and similarly, $\Phi^{-1}(W_i(1)) = E \cap D_{2i}$. Now since $\pi_i^{-1}(0)$ is a partition between $W_i(-1)$ and $W_i(1)$, the claim follows.

Since $W_{2i}(-1)$ and $W_{2i}(1)$ are in the interior of C_{2i} and D_{2i}, respectively, there exists by lemma 4.1.5 a partition S_{2i} between $W_{2i}(-1)$ and $W_{2i}(1)$ in Q such that

$$(*) \qquad S_{2i} \cap E \subseteq (\pi_i \circ \Phi)^{-1}(0).$$

We claim that

$$S = \bigcap_{i=1}^{\infty} S_{2i}$$

is as required.

CLAIM 2: $S \cap E \subseteq \Phi^{-1}(0,0,\cdots)$.

This immediately follows from (*).

Now let M be a subspace of S such that $K \subseteq \pi_1(M)$. We shall prove that M is strongly infinite-dimensional. Striving for a contradiction, suppose that this is not the case, i.e. that M is weakly infinite-dimensional. By proposition 4.8.1 for $i \in \mathbb{N}$ there exist partitions M_i in Q between C_{2i} and D_{2i} such that $\bigcap_{i=1}^{\infty} M_i \cap M = \emptyset$. For $i \in \mathbb{N}$ let f_i: $Q \to [-1,1]_i$ be a continuous function such that

$$f_i^{-1}(-1) = C_{2i},\ f_i^{-1}(1) = D_{2i},\ \text{and}\ f_i^{-1}(0) = M_i,$$

cf. the proof of theorem 2.1.12. Define f: $Q \to Q$ by

$$f(x) = (f_1(x), f_2(x), \cdots).$$

CLAIM 3: $f \in \Lambda$ and $M \cap f^{-1}(0,0,\cdots) = \emptyset$.

That f is continuous is trivial. Now take $i \in \mathbb{N}$. Then $f^{-1}(W_i(-1)) = \{x \in Q\colon f(x)_i = -1\} = C_{2i}$ by the definition of f_i. Similarly one proves that $f^{-1}(W_i(1)) = D_{2i}$. We conclude that $f \in \Lambda$. Finally, it is clear that

$$f^{-1}(0,0,\cdots) \cap M = \bigcap_{i=1}^{\infty} M_i \cap M = \emptyset.$$

Now since τ is onto, there exists $p \in T$ such that $\tau(p) = f$. Since $T \subseteq \pi_1(M)$, there exists $x \in M$ such that $x_1 = p$. Then $x \in S \cap E$ from which it follows by claim 2 that

$$f(x) = \tau(p)(x) = \tau(x_1)(x) = \Phi(x) = (0,0,\cdots).$$

But this contradicts claim 3.□

This result enables us to construct our final example.

4.8.9. EXAMPLE: *There exists a hereditarily infinite-dimensional compact space.*

To begin with, we shall prove the following:

CLAIM: There exists a collection $\{K_i: i \in \mathbb{N}\}$ of pairwise disjoint Cantor sets in [-1,1] such that each nondegenerate subinterval of [-1,1] contains one of the K_i.

Let $\{(r_i,t_i): i \in \mathbb{N}\}$ enumerate all nondegenerate open subintervals of [-1,1] with rational endpoints. Since each non-degenerate subinterval of [-1,1] contains one of the (r_i,t_i), it suffices to construct a pairwise disjoint family of Cantor sets $\{K_i: i \in \mathbb{N}\}$ such that for every i, $K_i \subseteq (r_i,t_i)$. Suppose that for certain $i \in \mathbb{N}$ the Cantor sets K_j have been constructed for $j \leq i$. By proposition 4.2.3, $K = \cup_{j=1}^{i} K_j$ is nowhere dense in [-1,1]. Consequently, there exists a nondegenerate interval

$$[p,q] \subseteq (r_{i+1},s_{i+1})\backslash K.$$

Since [p,q] is homeomorphic to [0,1], example 4.2.4 implies that there exists a Cantor set K_{i+1} in [p,q]. Then K_{i+1} is clearly as required.

Now for each i choose pairwise disjoint Cantor sets $K_{i1}, K_{i2}, \cdots$ in $[-1,1]_i$ such as in the claim. In addition, let σ_{ik} $(i,k \in \mathbb{N})$ be a collection of pairwise disjoint infinite subsets of $\mathbb{N}\backslash\{1\}$ such that

(1) $\quad i \notin \sigma_{ik} = \emptyset$ for all i and k.

Fix $i,k \in \mathbb{N}$. By lemma 4.8.8, for every $j \in \sigma_{ik}$ we can choose a partition S_j in Q between $W_j(-1)$ and $W_j(1)$ such that each subset in the intersection

$$S_{ik} = \cap\{S_j: j \in \sigma_{ik}\}$$

whose projection onto the i-th axis contains K_{ik}, is infinite-dimensional. Now put

$$S = \cap_{i,k} S_{ik}.$$

We claim that S is hereditarily infinite-dimensional. Clearly S is a compact subspace of Q. Since $1 \notin \cup_{i,k} \sigma_{ik}$, by corollary 3.5.8 S intersects every partition between $W_1(-1)$ and $W_1(1)$. By theorem 4.7.5 it therefore follows that S is not zero-dimensional. Now let M be an arbitrary nonempty subspace of S. There are two cases to consider.

CASE 1: For each $i \in \mathbb{N}$, $\pi_i(M)$ is zero-dimensional.

Then M is contained in the set

$$\prod_{i=1}^{\infty} \pi_i(M),$$

which is zero-dimensional by theorem 4.2.2(2). We conclude that dim M = 0 by theorem 4.2.2(1).

CASE 2: There exists $i \in \mathbb{N}$ such that $\pi_i(M)$ is not zero-dimensional.

Then $\pi_i(M)$ contains a nondegenerate subinterval of $[-1,1]_i$ by proposition 4.2.3. Consequently, by construction, $\pi_i(M)$ contains some K_{ij} which implies that M is (strongly) infinite-dimensional because M is contained in S_{ij}.□

Exercises for §4.8.

1. Prove that if a space X is the union of countably many weakly infinite-dimensional subspaces then X is weakly infinite-dimensional. Conclude that the Hilbert cube is not the union of countably many weakly infinite-dimensional subspaces.

2. Let X be a strongly infinite-dimensional compact space. Prove that Y contains a hereditarily infinite-dimensional compact subspace.

3. Let X and Y be spaces, let $A \subseteq X$ and let $f: A \to Y$ be continuous. Prove that $G = \{x \in X: \forall \varepsilon > 0\ \exists \delta > 0$ such that $\operatorname{diam}(f(B(x,\delta)) \cap A) < \varepsilon\}$ is a G_δ-subset of X containing A. In addition, prove that if (Y,d) is complete then f can be extended to a continuous function $g: G \to Y$.

4. Let X and Y be topologically complete spaces, let $A \subseteq X$ and $B \subseteq Y$ be subspaces, and let $f: A \to B$ be a homeomorphism. Prove that there exist G_δ-subsets S and T of X and Y, respectively, such that $A \subseteq S$ and $B \subseteq T$ such that moreover f can be extended to a homeomorphism $g: S \to T$ (Hint: Apply exercise 4.8.3) (This is called the Lavrentiev Theorem).

Let $\Omega = \{x \in \mathbb{R}^\infty: x_i = 0$ for all but finitely many $i\}$.

5. Prove that Ω is countable dimensional.

6. Prove that if G is a G_δ-subset of $\mathbb{R}^\infty$ containing Ω then $G\backslash\Omega$ contains a copy of the Hilbert cube.

7. Prove that if $\gamma\Omega$ is a compactification of Ω then the remainder $\gamma\Omega\backslash\Omega$ contains a copy of the Hilbert cube and is therefore strongly infinite-dimensional (Hint: Apply exercises 4.8.4 and 4.8.6).

Notes and comments for chapter 4.

Our historical comments concerning the sections 1 through 6 are taken from Engelking [60], where the reader can find much more information than in these notes.

The first definition of a dimension function is due to Brouwer [35]. His definition agrees with the now standard definitions in the class of all locally connected compact spaces.

§1.

Corollary 4.1.2 is due to Brouwer [35]. The covering dimension dim in terms of finite open covers was formally introduced in Čech [43]. Our definition of dim is usually called "The Theorem on Partitions" and was proved in Eilenberg and Otto [58]. Lemma 4.1.3 is due to Kuratowski [90, §15, XIII, p. 122]. The other results in §1 are well-known.

§2.

The Cantor middle-third set **C** was introduced by Cantor [41]. Theorem 4.2.5 is due to Brouwer [32] and theorem 4.2.9 is due to Alexandrov and Urysohn [4]. Example 4.2.11 was established in Erdös [62]. Exercise 4.2.7 is due to Alexandrov and Urysohn [3]. Exercise 4.2.9 is due to Sierpiński [128]. Exercise 4.2.10 is due to Hausdorff [71] and de Groot [68].

§3.

The proof of theorem 4.3.5 consists mostly of modifications of arguments that can be found in Engelking [60, Chapter 1]. Theorem 4.3.7 for the case of the small inductive dimension function ind (which takes the same value as dim, cf. theorem 4.5.8) was proved for compact spaces by Menger [102] and by Urysohn [140]; finally, it was established in full generality by Tumarkin [138] and Hurewicz [76]. The Subspace Theorem 4.3.8 is a triviality for the small inductive dimension function ind. Theorem 4.3.10 and corollary 4.3.11 are due to Brouwer [35].

§4.

That every n-dimensional space can be imbedded in $\mathbb{R}^{2n+1}$ is due independently to Nöbeling [113], Pontrjagin and Tolstowa [119] and Lefschetz [93]; that $\mathfrak{N}_n$ is universal for n-dimensional spaces is due to Nöbeling [113]. Corollary 4.4.7 is due to Hurewicz [77]. Exercises 4.4.5 and 4.4.6 are due to Kuratowski [86] and Hurewicz [75], respectively.

§5.

The small inductive dimension function ind was first defined by Urysohn [139] and Menger [101]. The Addition Theorem 4.5.3 is due, for compact spaces, to Urysohn [140] and for general spaces to Tumarkin [138] and Hurewicz [76]. The large inductive dimension function Ind is related to Brouwer [35] but formally first defined by Čech [42]. The Coincidence Theorem 4.5.8 is partly due to Hurewicz [77], [75], Brouwer [36] , Menger [102], Urysohn [140] and Tumarkin [138]. Theorem 4.5.9 is due to Menger [103]. Theorem 4.5.13 is due to Dowker [52] and Kuratowski [87].

§6.

Theorem 4.6.4 is basically due to Alexandrov [2]; contributions were also made by Hurewicz [79] and Hurewicz and Wallman [80]. Theorem 4.6.7 is due to Brouwer [34]. The proof presented here was taken from Hurewicz and Wallman [80]. Theorem 4.6.8 is due to D.W. Curtis and was published in van Mill [106]. Exercise 4.6.1 is usually called "The Sierpiński Theorem" and was proved by Sierpiński in [127]. Its generalization exercise 4.6.2 is due to Dijkstra [51]. Exercise 4.6.3 is due to Borsuk [28].

§7.

Hyperspaces were first considered in the early 1900's in the work of Hausdorff and Vietoris. For general information on hyperspaces see Nadler [112]. It is not clear where Proposition 4.7.2 was stated for the first time; the proof presented here seems to be new (it is considerably simpler than the proof of [112, theorem 0.8]). Theorem 4.7.3 is due to Bourbaki [30, p. 144, Exercise 9a]. The proof presented here is new and is inspired by Parthasarathy [114, theorem 4.1]. Proposition 4.7.8 can be found in Rubin, Schori and Walsh [120]. The first topologically complete 1-dimensional totally disconnected space was constructed by Sierpiński [129]. Theorem 4.7.10 is due to Mazurkiewicz [100]. The proof presented here is due to Rubin, Schori and Walsh [120] with a little help from Pol [116].

§8.

Theorem 4.8.6 is due to Lelek [95]. The simple proof presented here was discovered recently by Engelking and Pol [61]. Example 4.8.7 is due to Pol [116]. In addition, Example 4.8.9 is due to Walsh [143] (the first example of a compact space all subcompacta of which are either zero-dimensional or infinite-dimensional, is due to Henderson [73]). The construction of Example 4.8.9 however is due to Pol [117]. Exercise 4.8.4 is due to Lavrentiev [91].

5. Elementary ANR Theory

In this chapter we shall present several elementary results from **ANR** theory. Among other things, we shall derive a characterization of **ANR**'s which enables us to prove that the hyperspace of the unit interval is an **AR**.

5.1. Some Properties of ANR's

In this section we shall derive a few useful properties of **ANR**'s. We have to introduce some terminology first.

Let Y be a space and let $\mathcal{U}$ be an open cover of Y. Two continuous functions f,g: $X \to Y$ are called *$\mathcal{U}$-close* if for every $x \in X$ there exists $U \in \mathcal{U}$ such that $\{f(x),g(x)\} \subseteq U$.

Again, let Y be a space and let $\mathcal{U}$ be an open cover of Y. A homotopy H: $X \times I \to Y$ is said to be *limited by* $\mathcal{U}$ provided that for any $x \in X$ there exists $U \in \mathcal{U}$ such that

$$H(\{x\} \times I) \subseteq U.$$

Two continuous functions f,g: $X \to Y$ are called *$\mathcal{U}$-homotopic*, abbreviated $f \underset{\mathcal{U}}{\simeq} g$, if there is a homotopy H: $X \times I \to Y$ which is limited by $\mathcal{U}$ and connects f and g, i.e. $H_0 = f$ and $H_1 = g$.

5.1.1. THEOREM: *Let* X *be an* **ANR**. *Then for every open cover* $\mathcal{U}$ *of* X *there exists an open refinement* $\mathcal{V}$ *of* $\mathcal{U}$ *such that for every space* Y, *any two* $\mathcal{V}$*-close maps* f,g: $Y \to X$ *are* $\mathcal{U}$*-homotopic.*

PROOF: By (the remark following) lemma 1.2.3 we may assume that X is a closed subspace of a convex subset C of a normed linear space L. Observe that L is locally convex. Since X is an **ANR**, there are a neighborhood V of X in C and a retraction r: V → X. Put $\mathcal{E} = \{r^{-1}(U): U \in \mathcal{U}\}$. Then $\mathcal{E}$ is an open cover of V, and since V is open in C, there is an open cover $\mathcal{F}$ of V having the following properties:

(1) $\mathcal{F} < \mathcal{E}$, and

(2) $\mathcal{F}$ consists of convex sets.

Let $\mathcal{V} = \mathcal{F} \cap X$. We claim that $\mathcal{V}$ is the required open cover of X. To this end, let Y be a space, and assume that f,g: Y → X are continuous and $\mathcal{V}$-close. Define a homotopy G: Y × I → C by

$$G(y,t) = tg(y) + (1-t)f(y).$$

Then G is continuous and clearly connects f and g. Since by (2) the elements of $\mathcal{F}$ are convex, by the special definition of G it follows that for every $y \in Y$ there exists $F \in \mathcal{F}$ such that

$$G(\{y\} \times I) \subseteq F. \tag{3}$$

We conclude that G can be considered to be a mapping from Y × I into V and that moreover G is an $\mathcal{F}$-homotopy. Now define H: Y × I → X by

$$H = r \circ G.$$

Then H connects f and g since r is a retraction. Moreover, by (3) and (1) it follows easily that H is limited by $\mathcal{U}$. □

5.1.2. *Remark:* It is an open problem whether the property of **ANR**'s stated in theorem 5.1.1 in fact *characterizes* the class of all **ANR**'s. This is a difficult problem which has been open for decades. A slightly stronger property than the one above is in fact characteristic, [74, IV theorem 1.2].

In many applications one needs homotopies "with control". We will therefore present a "controlled" version of the Borsuk Homotopy Extension Theorem, cf. theorem 1.6.3.

5.1.3. THEOREM: *Let* X *be an* **ANR** *and let* $\mathcal{U}$ *be an open cover of* X. *Then for every closed subset* A *of a space* Y *and for every homotopy* H: A × I → X *which is limited by* $\mathcal{U}$, *if* H_0

can be extended to a continuous function h_0: $Y \to X$ *then there exists a homotopy* $\overline{H}$: $Y \times I \to X$ *such that*

(1) $\overline{H}$ *is limited by* $\mathcal{U}$,

(2) $\overline{H}_0 = h_0$ *and* $\overline{H} \mid (A \times I) = H$.

PROOF: Let H, A and Y be as in the formulation of the theorem. By theorem 1.6.3 there exists a homotopy F: $Y \times I \to X$ such that $F_0 = h_0$ and $F \mid (A \times I) = H$. For each $a \in A$ there exists $U_a \in \mathcal{U}$ containing $H(\{a\} \times I)$. There consequently exists a neighborhood E_a of a in Y such that

$$F(E_a \times I) \subseteq U_a,$$

cf. the proof of lemma 1.6.2. Put $E = \bigcup_{a \in A} E_a$. Then E is a neighborhood of A in Y. By corollary 1.4.15 there is a Urysohn function λ: $Y \to I$ such that $\lambda \mid A \equiv 1$ and $\lambda \mid Y \setminus E \equiv 0$. Now define $\overline{H}$: $Y \times I \to X$ as follows:

$$\overline{H}(y,t) = F(y,\lambda(y)\cdot t).$$

Then $\overline{H}$ is clearly continuous and $\overline{H} \mid (A \times I) = H$. In addition,

$$\overline{H}(y,0) = F(y,0) = h_0(y)$$

for every $y \in Y$ so $\overline{H}_0 = h_0$. It remains to prove that $\overline{H}$ is limited by $\mathcal{U}$. This is a triviality. Take an arbitrary $y \in Y$. If $y \in E$ then there exists $a \in A$ such that $y \in E_a$. Consequently,

$$\overline{H}(\{y\} \times I) \subseteq F(\{y\} \times I) \subseteq U_a.$$

Now if $y \notin E$ then $\lambda(y) = 0$ from which it follows that

$$\overline{H}(\{y\} \times I) = \{F(y,0)\} = \{h_0(y)\}$$

consists of precisely one point and is therefore also contained in an element of $\mathcal{U}$. □

We shall now derive another important property of **ANR**'s. First we need to introduce some terminology. Let X be a space and let $\mathcal{U}$ be an open cover of X. In addition, let $\mathcal{T}$ be a locally finite simplicial complex and let $\mathcal{S}$ be a subcomplex of $\mathcal{T}$ containing all the vertices of $\mathcal{T}$. A *partial realization* of $\mathcal{T}$ in X relative to $(\mathcal{S},\mathcal{U})$ is a continuous function

$$f: |\mathcal{S}| \to X$$

such that for every $\sigma \in \mathcal{T}$ there exists $U \in \mathcal{U}$ such that

$$f(\sigma \cap |\mathcal{S}|) \subseteq U.$$

In case $\mathcal{S} = \mathcal{T}$ we say that f is a *full realization* of $\mathcal{T}$ in X relative to $\mathcal{U}$.

At first glance the following result does not seem to be very appealing. However, it turns out to be of fundamental importance in the process of finding a *usable* topological characterization of **ANR**'s.

5.1.4. THEOREM: *Let* X *be an* **ANR**. *Then for every open cover* $\mathcal{U}$ *of* X *there exists an open refinement* $\mathcal{V}$ *of* $\mathcal{U}$ *such that for every locally finite simplicial complex* $\mathcal{T}$ *and every subcomplex* $\mathcal{S}$ *of* $\mathcal{T}$, *containing all the vertices of* $\mathcal{T}$, *and every partial realization* $f: |\mathcal{S}| \to X$ *of* $\mathcal{T}$ *in* X *relative to* $(\mathcal{S},\mathcal{V})$ *there exists a full realization* $g: |\mathcal{T}| \to X$ *of* $\mathcal{T}$ *in* X *relative to* $\mathcal{U}$ *having the following properties:*

(1) g *extends* f, *i.e. if* $p \in |\mathcal{S}|$ *then* $g(p) = f(p)$, *and*

(2) *for every* $\sigma \in \mathcal{S}$ *and for every* $V \in \mathcal{V}$ *with* $f(\sigma \cap |\mathcal{S}|) \subseteq V$ *there exists* $U \in \mathcal{U}$ *such that* $g(\sigma) \cup V \subseteq U$.

PROOF: We can take the same refinement of $\mathcal{U}$ as in the proof of theorem 5.1.1. Recall that by (the remark following) lemma 1.2.3 we may regard X to be a closed subspace of a convex subset C of a locally convex linear space L. Since X is an **ANR**, there are a neighborhood V of X in C and a retraction $r: V \to X$. Put $\mathcal{E} = \{r^{-1}(U): U \in \mathcal{U}\}$. Then $\mathcal{E}$ is an open cover of V, and since V is open in C, there is an open cover $\mathcal{F}$ of V having the following properties:

(1) $\mathcal{F} < \mathcal{E}$, and

(2) $\mathcal{F}$ consists of convex sets.

As in the proof of theorem 5.1.1, put $\mathcal{V} = \mathcal{F} \cap X$. We claim that $\mathcal{V}$ is the required open cover of X.

To this end, let $\mathcal{T}$ be a locally finite simplicial complex, let $\mathcal{S}$ be a subcomplex of $\mathcal{T}$ containing all the vertices of $\mathcal{T}$, and let $f: |\mathcal{S}| \to X$ be a partial realization of $\mathcal{T}$ in X relative to $(\mathcal{S},\mathcal{V})$. For each $\sigma \in \mathcal{T}$ let σ^* denote the convex hull of $f(\sigma \cap |\mathcal{S}|)$ in L; observe that by (2) σ^* is contained in an element of $\mathcal{F}$. For each $n \geq 0$, let

$$\mathcal{K}_n = \mathcal{T}^{(n)} \cup \mathcal{S}$$

(here $\mathcal{T}^{(n)}$ denotes the n-skeleton of $\mathcal{T}$ of course, see §3.4). Then $\mathcal{K}_n$ is a subcomplex of $\mathcal{T}$. By induction on $n \geq 0$ we shall construct a continuous function $f_n: |\mathcal{K}_n| \to V$ such that the following conditions are satisfied:

(3) $f_0 = f$,
(4) if $n \geq 1$ then f_n extends f_{n-1},
(5) for every $\sigma \in \mathcal{T}$, $f_n(\sigma \cap |\mathcal{K}_n|) \subseteq \sigma^*$.

Observe that f_0 is well-defined since $\mathcal{T}^{(0)} \subseteq \mathcal{S}$. Now assume that for certain $n \geq 0$ we constructed the function f_{n-1}. We will define f_n on every simplex $\sigma \in \mathcal{K}_n$ separately and will conclude by applying lemma 3.6.6 that the union of all the constructed functions is continuous. We have to be careful of course since we want the union of all the functions to be well-defined. So take an arbitrary simplex $\sigma \in \mathcal{K}_n = \mathcal{T}^{(n)} \cup \mathcal{S}$. We want to define f_n on σ. There are two cases to consider. First assume that $\sigma \in \mathcal{K}_{n-1}$. Then (4) and (3) tell us that f_n is already defined on σ. Suppose therefore that $\sigma \notin \mathcal{K}_{n-1}$, i.e. $\sigma \in \mathcal{T}^{(n)} \setminus \mathcal{K}_{n-1}$. Then $\sigma \setminus \partial\sigma$, the (geometric) interior of σ, does not intersect $|\mathcal{K}_{n-1}|$. Since $\partial\sigma$ is a subset of $|\mathcal{K}_{n-1}|$, and f_n is not defined yet at any point of $\sigma \setminus \partial\sigma$, all we have to do is to extend the restriction $f_{n-1} | \partial\sigma: \partial\sigma \to \sigma^*$ over σ. Since σ^* is convex, this can be done by theorem 1.5.1. This defines f_n on σ. Observe that the union of all constructed functions is a function since the geometric interiors of all simplexes in $\mathcal{K}_n$ form a pairwise disjoint collection (exercise 3.3.1) and that it is continuous by lemma 3.6.6. Since (5) is easily seen to be satisfied, this completes the inductive construction.

Now define $\phi: |\mathcal{T}| \to V$ by

$$\phi(x) = f_n(x) \qquad (x \in \mathcal{T}^{(n)}).$$

Observe that ϕ is clearly well-defined by (4), that the range of ϕ is contained in V by (5) and that ϕ extends f by (3). In addition, by another appeal to lemma 3.6.6 it follows that g is continuous.

Finally define $g: |\mathcal{T}| \to X$ by

$$g = r \circ \phi.$$

Since r is a retraction and ϕ extends f, g extends f as well.

Now take an arbitrary $\sigma \in \mathcal{T}$ and an element $W \in \mathcal{V}$ such that $f(\sigma \cap |\mathcal{S}|) \subseteq W$. There exists $n \geq 0$ such that $\sigma \in \mathcal{T}^{(n)}$. By (5) and the definition of ϕ we obtain $\phi(\sigma) \subseteq \sigma^*$. There exists $F \in \mathcal{F}$ such that $F \cap X = W$; observe that $\sigma^* \subseteq F$. By (1) there exists $U \in \mathcal{U}$ such that $F \subseteq r^{-1}(U)$. We conclude that

$$g(\sigma) \cup W = r(\phi(\sigma)) \cup W \subseteq r(\sigma^*) \cup W \subseteq r(F) \cup W \subseteq U.$$

Consequently, g is the desired extension of f. □

5.1.5. COROLLARY: *Let* X *be an* **ANR**. *Then for every open cover* $\mathcal{U}$ *of* X *there exists an open refinement* $\mathcal{V}$ *of* $\mathcal{U}$ *such that for every locally finite simplicial complex* $\mathcal{T}$ *and every subcomplex* $\mathcal{S}$ *of* $\mathcal{T}$ *containing all the vertices of* $\mathcal{T}$, *every partial realization of* $\mathcal{T}$ *in* X *relative to* $(\mathcal{S},\mathcal{V})$ *can be extended to a full realization of* $\mathcal{T}$ *in* X *relative to* $\mathcal{U}$. □

Let X and Y be spaces. We say that X *dominates* Y if there are two maps

$$g: Y \to X \text{ and } f: X \to Y,$$

such that the composition $f \circ g: Y \to Y$ is homotopic to the identity on Y. The space X is called a *dominating space* for Y.

5.1.6. LEMMA: *A space* X *is contractible if and only if a one-point space dominates* X.

PROOF: Suppose first that X is contractible and let $H: X \times I \to X$ be a homotopy contracting X to a point, say p. Now let $g: X \to \{p\}$ be the obvious function and let $f: \{p\} \to X$ be the inclusion. Then H connects the identity on X with $f \circ g$, i.e. $\{p\}$ dominates X.

Conversely, assume that there exists a point p and functions $g: X \to \{p\}$ and $f: \{p\} \to X$ such that $f \circ g$ is homotopic to the identity on X. Since $f \circ g$ is clearly a constant function we conclude that X is contractible. □

As remarked above, one sometimes needs homotopies "with control". For that reason we define a "controlled version" of the concept of domination. Again let X and Y be spaces and let $\mathcal{U}$ be an open cover of Y. We say that X *$\mathcal{U}$-dominates* Y if there are two maps

$$g: Y \to X \text{ and } f: X \to Y,$$

such that the composition $f \circ g: Y \to Y$ is $\mathcal{U}$-homotopic to the identity on Y. The space X is called a *$\mathcal{U}$-dominating space* for Y.

Before we can state our last main result in this section, we need to derive the following lemma.

5.1.7. LEMMA: *Let* X *be a space and let* $\mathcal{U}$ *be an open cover of* X. *Then there exists an open refinement* $\mathcal{V}$ *of* $\mathcal{U}$ *such that*

(1) $\mathcal{V}$ *is star-finite, and*

(2) $\mathcal{V}$ *is a star-refinement of* $\mathcal{U}$.

PROOF: By lemma 1.4.1 we may assume that $\mathcal{U}$ is locally finite. So in particular, $\mathcal{U}$ is countable. Enumerate $\mathcal{U}$ as $\{U_n: n \in \mathbb{N}\}$. By proposition 4.3.3 there exists a closed shrinking $\mathcal{F} = \{F_n: n \in \mathbb{N}\}$ of $\mathcal{U}$. Fix $x \in X$ for a moment. Since $\mathcal{U}$ is locally finite and $\mathcal{F}$ is a shrinking of $\mathcal{U}$, the set

$$A_x = \{n \in \mathbb{N}: x \in F_n\}$$

is finite. By exercise 1.4.13 it therefore follows that

$$W_x = \bigcap_{n \in A_x} U_n \setminus (\bigcup_{n \notin A_x} F_n)$$

is open.

Put $\mathcal{W} = \{W_x: x \in X\}$. Then $\mathcal{W}$ is an open cover of X. We claim that $\{St(\{x\},\mathcal{W}): x \in X\} < \mathcal{U}$. To this end, take $x,y \in X$ and $n \in A_x$. If $x \in W_y$ then clearly $A_x \subseteq A_y$ from which it follows that $W_y \subseteq U_n$. We conclude that $St(\{x\},\mathcal{W}) \subseteq U_n$, as required.

Repeating this argument for $\mathcal{W}$ we get an open cover $\mathcal{W}_1$ of X such that for every $x \in X$ there is a $W \in \mathcal{W}$ with $St(\{x\},\mathcal{W}_1) \subseteq W$. Now let $W_1 \in \mathcal{W}_1$. The reader can readily verify that $St(W_1,\mathcal{W}_1) \subseteq St(\{x\},\mathcal{W})$ for every $x \in W_1$ so that $St(W_1,\mathcal{W}_1) \subseteq U$ for some $U \in \mathcal{U}$.

Now by theorem 3.6.17(1) there is a star-finite refinement $\mathcal{V}$ of $\mathcal{W}_1$. Then $\mathcal{V}$ is clearly as required. □

We now come to the last main result in this section.

5.1.8. THEOREM: *Let* X *be a (compact)* **ANR**. *Then for every open cover* $\mathcal{U}$ *of* X *there exists a (polyhedron) polytope* P *such that* P $\mathcal{U}$*-dominates* X.

PROOF: Let $\mathcal{V}$ be an open refinement of $\mathcal{U}$ with the properties stated in theorem 5.1.1. By theorem 5.1.4 there exists an open refinement $\mathcal{W}$ of $\mathcal{V}$ such that for every locally finite simplicial complex $\mathcal{T}$, for every subcomplex $\mathcal{S}$ containing all the vertices of $\mathcal{T}$, and for every partial realization $f: |\mathcal{S}| \to X$ of $\mathcal{T}$ in X relative to $(\mathcal{S},\mathcal{W})$ there exists a full realization $g: |\mathcal{T}| \to X$ of $\mathcal{T}$ in X relative to $\mathcal{V}$ having the following properties:

(1) g extends f, and

(2) for every $\sigma \in \mathcal{S}$ and for every $W \in \mathcal{W}$ with $f(\sigma \cap |\mathcal{S}|) \subseteq W$ there exists $V \in \mathcal{V}$

such that $g(\sigma) \cup W \subseteq V$.

By lemma 5.1.7 there exists a countable open cover $\mathcal{A}$ of X such that $\mathcal{A}$ is star-finite and a star-refinement of $\mathcal{W}$ (in case that X is compact, $\mathcal{A}$ is finite). Put $P = |N(\mathcal{A})|$ and let $\kappa: X \to P$ be the κ-mapping of $\mathcal{A}$. Now for every vertex $x(A) \in N(\mathcal{A})^{(0)}$ choose an arbitrary point $f(x(A)) \in A$. This defines a function

$$f: |N(\mathcal{A})^{(0)}| \to X.$$

By corollary 3.6.3, $|N(\mathcal{A})^{(0)}|$ is a discrete topological space, hence f is continuous.

CLAIM 1: f is a partial realization of $N(\mathcal{A})$ in X relative to $(N(\mathcal{A})^{(0)}, \mathcal{W})$.

This is easy. Let $\sigma = |\{x(A_0), x(A_1), \cdots, x(A_n)\}|$ be a simplex in $N(\mathcal{A})$. Then

$$A_0 \cap A_1 \cap \cdots \cap A_n \neq \emptyset.$$

Since $\mathcal{A}$ is a star-refinement of $\mathcal{W}$ there exists $W \in \mathcal{W}$ such that

$$A_0 \cup A_1 \cup \cdots \cup A_n \subseteq W.$$

Consequently,

$$f(\sigma \cap N(\mathcal{A})^{(0)}) = f(\{x(A_0), x(A_1), \cdots, x(A_n)\}) \subseteq A_0 \cup A_1 \cup \cdots \cup A_n \subseteq W.$$

Now by the special choice of $\mathcal{W}$, the function f can be extended to a continuous function $g: |N(\mathcal{A})| \to X$ satisfying (1) and (2).

CLAIM 2: The identity 1_X on X and the function $g \circ \kappa$ are $\mathcal{V}$-close.

Take an arbitrary $x \in X$. Since $\mathcal{A}$ is star-finite, there are only finitely many elements of $\mathcal{A}$ that contain x, say $A_0, A_1, \cdots, A_n$. Then $\sigma = |\{x(A_0), \cdots, x(A_n)\}|$ is a simplex in $N(\mathcal{A})$, and by lemma 3.6.14, $\kappa(x) \in \sigma$. There exists $W \in \mathcal{W}$ such that

$$A_0 \cup A_1 \cup \cdots \cup A_n \subseteq W.$$

Then $x \in W$ and since $f(x(A_i)) \in A_i$ for every $0 \le i \le n$, we have

$$f(\sigma \cap |N(\mathcal{A})^{(0)}|) \subseteq W.$$

By (2) there exists $V \in \mathcal{V}$ such that $g(\sigma) \cup W \subseteq V$. Since $\kappa(x) \in \sigma$ we have $g(\kappa(x)) \in V$ and since $W \subseteq V$ we have $x \in V$. We conclude that $\{x, g(\kappa(x))\} \subseteq V$.

By the special choice of $\mathcal{V}$ we now conclude that 1_X and $g \circ \kappa$ are $\mathcal{U}$-homotopic, i.e. P $\mathcal{U}$-dominates X. □

Exercises for §5.1.

1. Let r: $X \to Y$ be a retraction. Prove that X dominates Y.

2. Let X be an **ANR** with $\dim X \leq n$. Prove that for every open cover $\mathcal{U}$ of X there exists a polytope P such that P $\mathcal{U}$-dominates X and $\dim P \leq n$.

3. Let X be a compact **ANR** with $\dim X = n < \infty$. Prove that there is a finite open cover $\mathcal{U}$ of X such that for every polyhedron P that $\mathcal{U}$-dominates X we have $\dim P \geq n$.

4. Let X be a compact **ANR**. Prove that for every $\varepsilon > 0$ there exists $\delta > 0$ such that for every space Y and for every closed subspace A of Y and for all maps $f,g: A \to X$ with $d(f,g) < \delta$, if f has a continuous extension $\bar{f}: Y \to X$ then g has a continuous extension $\bar{g}: Y \to X$ such that $d(\bar{f},\bar{g}) < \varepsilon$.

5.2. A Characterization of ANR's and AR's

The aim of this section is to present a purely topological characterization of the class of all **ANR**'s and of the class of all **AR**'s. We will demonstrate the usefulness of our characterization in §5.3 by showing that the hyperspace of each *Peano continuum,* i.e. a locally connected continuum, is an **AR**.

Here is the announced characterization.

5.2.1. THEOREM: *Let* X *be a space. The following statements are equivalent:*

(a) X *is an* **ANR**, *and*

(b) *for every open cover* $\mathcal{U}$ *of* X *there exists an open refinement* $\mathcal{V}$ *of* $\mathcal{U}$ *such that for every locally finite simplicial complex* $\mathcal{T}$ *and every subcomplex* $\mathcal{S}$ *of* $\mathcal{T}$ *containing all the vertices of* $\mathcal{T}$, *every partial realization of* $\mathcal{T}$ *in* X *relative to* $(\mathcal{S},\mathcal{V})$ *can be extended to a full* *realization of $\mathcal{T}$ in X relative to $\mathcal{U}$.*

Observe that by corollary 5.1.5 we only need to verify the implication (b) ⇒ (a). This will be done by proving a series of lemmas.

Let X be a space. It will be convenient to let τX denote the family of all open subsets of X.

5.2.2. LEMMA: *Let* X *be a space and let* Y *be a subspace of* X. *The function* $\sigma\colon \tau Y \to \tau X$ *defined by* $\sigma(A) = \{x \in X\colon d(x,A) < d(x,Y\setminus A)\}$ *(where, by convention,* $d(x,\emptyset) = \infty$*) has the following properties:*

(1) $\sigma(\emptyset) = \emptyset$, $\sigma(Y) = X$,

(2) $\sigma(A) \cap Y = A$ *for every* $A \in \tau Y$,

(3) *if* $A,B \in \tau Y$ *then* $A \subseteq B$ *iff* $\sigma(A) \subseteq \sigma(B)$, *and*

(4) *if* $A,B \in \tau Y$ *then* $\sigma(A \cap B) = \sigma(A) \cap \sigma(B)$.

PROOF: Let $\kappa\colon \rho Y \to \rho X$ be as in lemma 4.1.3. Then for every $U \in \tau Y$, $\sigma(U) = X\setminus\kappa(Y\setminus U)$. An easy check shows that σ is as required. □

Now let X be a space having the realization property stated in theorem 5.2.1(b). By lemma 1.2.3. we may assume that X is a closed subspace of a convex subset C of a normed linear space L. We shall prove that X is a neighborhood retract of C. An application of exercise 1.5.1 and theorem 1.5.1 then yields the desired result.

It will be convenient to fix some notation. Let $\sigma\colon \tau X \to \tau C$ be the function of lemma 5.2.2. In addition, by applying our assumptions on X and lemma 5.1.7, by induction on $n \geq 0$ it is a triviality to construct two sequences of open (= open in X) covers $\mathcal{A}_n$ and $\mathcal{B}_n$ of X as follows:

(a) $\mathcal{A}_0 = \{X\}$,

(b) $\mathcal{B}_n < \mathcal{A}_n$ and for every locally finite simplicial complex $\mathcal{T}$ and every subcomplex $\mathcal{S}$ of $\mathcal{T}$ containing all the vertices of $\mathcal{T}$, every partial realization of $\mathcal{T}$ in X relative to $(\mathcal{S},\mathcal{B}_n)$ can be extended to a full realization of $\mathcal{T}$ in X relative to $\mathcal{A}_n$,

(c) for $n \geq 1$, $\mathcal{A}_n$ is a star-refinement of $\mathcal{B}_{n-1}$,

(d) $\text{mesh}(\mathcal{A}_n) < \frac{1}{n}$ $(n \in \mathbb{N})$.

Finally, by lemma 1.4.12 there exists a countable open cover $\mathcal{V}$ of $C\setminus X$ and a sequence of points $(b_V)_{V\in\mathcal{V}}$ in X with the following properties:

(e) if $y \in V \in \mathcal{V}$ then $d(y,b_V) \leq 2d(y,X)$, and

(f) if $V_n \in \mathcal{V}$ for every n and $\lim_{n\to\infty} d(V_n,X) = 0$ then $\lim_{n\to\infty} \text{diam}(V_n) = 0$.

By an appeal to theorem 3.6.17(1) we find that there exists a countable open cover $\mathcal{U}$ of $C\backslash X$ such that

(g) $\mathcal{U}$ is star-finite and refines $\mathcal{V}$.

Without loss of generality we assume that every $U \in \mathcal{U}$ is nonempty. For every $U \in \mathcal{U}$ pick $V(U) \in \mathcal{V}$ such that $U \subseteq V(U)$ and let $a_U = b_{V(U)}$. Then clearly

(h) for every $y \in U \in \mathcal{U}$, $d(y,a_U) \leq 2d(y,X)$, and
(i) if $U_n \in \mathcal{U}$ for every n and $\lim_{n\to\infty} d(U_n,X) = 0$ then $\lim_{n\to\infty} \operatorname{diam}(U_n) = 0$.

5.2.3. LEMMA: *Let* $x \in X$ *and let* E *be a neighborhood of* x *in* C. *There exists a neighborhood* F *of* x *in* C *such that if* $U \in \mathcal{U}$ *meets* F *then* $U \subseteq E$.

PROOF: Let $\varepsilon > 0$ be such that $B(x,\varepsilon) \subseteq E$. If the lemma is not true then for every n there exists $U_n \in \mathcal{U}$ such that

(1) $U_n \cap B(x,\varepsilon/n) \neq \emptyset$, and
(2) $U_n \backslash E \neq \emptyset$.

By (1), $\lim_{n\to\infty} d(U_n,X) = 0$ so by (i), $\lim_{n\to\infty} \operatorname{diam}(U_n) = 0$. This easily contradicts (1) and (2).□

This simple result is used in the proof of the following lemma.

5.2.4. LEMMA: *For each* $n \geq 0$ *there exists an open neighborhood* H_n *of* X *in* C *such that* $H_0 = C$ *and for* $n \geq 1$,
$(1)_n$ $\overline{H}_n \subseteq B(X,1/n)$,
$(2)_n$ $\overline{H}_n \subseteq H_{n-1}$,
$(3)_n$ *if* $U \in \mathcal{U}$ *meets* $\overline{H}_n$ *then there exists* $A \in \mathcal{A}_n$ *such that* $U \subseteq \sigma(A) \cap H_{n-1}$.

PROOF: We shall construct the H_n's by induction on $n \geq 0$. Suppose that $n \geq 1$ and that H_{n-1} has been constructed. By corollary 1.4.16 there exists a neighborhood V of X in C such that

$$X \subseteq V \subseteq \overline{V} \subseteq H_{n-1} \cap B(X,1/n).$$

For each $x \in X$ there exists $A_x \in \mathcal{A}_n$ such that $x \in A_x$. Then $\sigma(A_x)$ is a neighborhood of x in C and consequently by lemma 5.2.3 there exists a neighborhood B_x of x in C such that if $U \in \mathcal{U}$ intersects B_x then U is contained in $\sigma(A_x) \cap V$. Put

$$H_n = \cup_{x \in X} B_x.$$

We claim that H_n is as required.

Observe that $(1)_n$ and $(2)_n$ are trivially satisfied. For the verification of $(3)_n$ assume that $U \in \mathcal{U}$ meets $\overline{H}_n$. Since U is open, $U \cap H_n \neq \emptyset$, so there exists x such that $U \cap B_x \neq \emptyset$ from which it follows that $U \subseteq \sigma(A_x) \cap V$. Since $V \subseteq H_{n-1}$, we are done.□

Now fix $U \in \mathcal{U}$ for a moment and put

$$E = \{n \geq 0: U \cap \overline{H}_n \neq \emptyset\}.$$

Since $H_0 \neq \emptyset$, $E \neq \emptyset$. We claim that E is finite. Striving for a contradiction, assume that E is infinite. Then by lemma 5.2.4(1), $d(U,X) = 0$ from which it follows by (i) that $\text{diam}(U) = 0$. Since U is nonempty, we find that U consists of a single point which belongs to X since $d(U,X) = 0$; contradiction.

Now for every $U \in \mathcal{U}$ define

$$n(U) = \max\{n \geq 0: U \cap \overline{H}_n \neq \emptyset\}.$$

Let $\kappa: C \setminus X \to |N(\mathcal{U})|$ be the κ-function of the cover $\mathcal{U}$, cf. §3.5. For every $U \in \mathcal{U}$ select a point $z_U \in U$ and by lemma $5.2.4(3)_{n(U)}$, an $A_U \in \mathcal{A}_{n(U)}$ such that $U \subseteq \sigma(A_U)$.

5.2.5. LEMMA: *For each* $U \in \mathcal{U}$ *there exists a point* $y_U \in A_U$ *such that* $d(z_U,y_U) \leq 2d(z_U,X)$.

PROOF: Let $U \in \mathcal{U}$. If $a_U \in A_U$ then we are done by (h). So assume that $a_U \notin A_U$. Since $z_U \in U \subseteq \sigma(A_U)$ it follows by the definition of σ that $d(z_U,A_U) < d(z_U,X \setminus A_U)$. Since $a_U \notin A_U$ by (h) this implies that $d(z_U,A_U) < d(z_U,a_U) \leq 2d(z_U,X)$. From this the lemma follows immediately.□

Recall that by convention the vertex of $N(\mathcal{U})$ corresponding to $U \in \mathcal{U}$ is denoted by $x(U)$, cf. §3.6. Now we define $\Phi: |N(\mathcal{U})^{(0)}| \to X$ as follows:

$$\Phi(x(U)) = y_U.$$

Since $|N(\mathcal{U})^{(0)}|$ is a discrete topological space (corollary 3.6.3) it follows that Φ is continuous.
We now aim at extending Φ over a large part of $|N(\mathcal{U})|$. For each $m \geq 0$ put

$$B_m = \overline{H}_m \setminus H_{m+1},$$

and let

$$\mathcal{P}_m = \{|\{x(U_{i(0)}),\cdots,x(U_{i(k)})\}| \in N(\mathcal{U}): (\forall 0 \leq j \leq k)(U_{i(j)} \cap B_m \neq \emptyset)\}.$$

Clearly $\mathcal{P}_m$ is a subcomplex of $N(\mathcal{U})$.

5.2.6. LEMMA: *If* $|m - n| \geq 2$ *then* $\mathcal{P}_n \cap \mathcal{P}_m = \emptyset$.

PROOF: This follows directly from lemma 5.2.4(3).□

5.2.7. LEMMA: *For each* $m > 0$, $\Phi \mid |(\mathcal{P}_m)^{(0)}|$ *is a partial realization of* $\mathcal{P}_m$ *in* X *relative to* $((\mathcal{P}_m)^{(0)},\mathcal{B}_{m-1})$.

PROOF: Let $|\{x(U_{i(0)}),\cdots,x(U_{i(k)})\}|$ be a simplex in $\mathcal{P}_m$. Then by the definition of $\mathcal{P}_m$, for each $0 \leq j \leq k$, $U_{i(j)} \cap \overline{H}_m \neq \emptyset$, from which it follows that $m \leq n(U_{i(j)})$. Consequently, (b) and (c) imply that for each $0 \leq j \leq k$,

$$\mathcal{A}_{n(U_{i(j)})} < \mathcal{A}_m,$$

so that by lemma 5.2.4(3) there exists $A_j \in \mathcal{A}_m$ such that

$$U_{i(j)} \subseteq \sigma(A_{U_{i(j)}}) \subseteq \sigma(A_j).$$

Observe that by the definition of Φ this implies that for every $0 \leq j \leq k$,

$$\Phi(x(U_{i(j)})) \in A_{U_{i(j)}} \subseteq A_j.$$

Since $|\{x(U_{i(0)}),\cdots,x(U_{i(k)})\}|$ is a simplex in $N(\mathcal{U})$,

$$U_{i(0)} \cap \cdots \cap U_{i(k)} \neq \emptyset,$$

so that

$$\sigma(A_0) \cap \cdots \cap \sigma(A_k) \neq \emptyset,$$

from which it follows by lemma 5.2.2(1),(4) that

$$A_0 \cap \cdots \cap A_k \neq \emptyset.$$

Since by (c) $\mathcal{A}_m$ is a star-refinement of $\mathcal{B}_{m-1}$, there exists an element of $\mathcal{B}_{m-1}$ which contains all A_j and hence all $\Phi(x(U_{i(j)}))$. □

By (b) we may extend for every $m \geq 1$ the partial realization Φ of $\mathcal{P}_{2m}$ in X relative to $((\mathcal{P}_{2m})^{(0)}, \mathcal{B}_{2m-1})$ to a full realization

$$\Psi_{2m}: |\mathcal{P}_{2m}| \to X,$$

relative to $\mathcal{A}_{2m-1}$.

Now for $m \geq 0$ consider the subcomplexes $\mathcal{P}_{2m+1}$ of $N(\mathcal{U})$. By lemma 5.2.6, $\mathcal{P}_{2m+1}$ meets only $\mathcal{P}_{2m}$ and $\mathcal{P}_{2m+2}$. Define

$$\mathcal{T}_{2m+1} = (\mathcal{P}_{2m+1})^{(0)} \cup (\mathcal{P}_{2m+1} \cap \mathcal{P}_{2m}) \cup (\mathcal{P}_{2m+1} \cap \mathcal{P}_{2m+2}).$$

Clearly, $\mathcal{T}_{2m+1}$ is a subcomplex of $\mathcal{P}_{2m+1}$, containing every vertex of $\mathcal{P}_{2m+1}$. Define a function $\Phi_{2m+1}: |\mathcal{T}_{2m+1}| \to X$ by

$$\Phi_{2m+1}(x) = \begin{cases} \Psi_{2m}(x) & (x \in |\mathcal{P}_{2m}|), \\ \Phi(x) & (x \in |(\mathcal{P}_{2m+1})^{(0)}|), \\ \Psi_{2m+2}(x) & (x \in |\mathcal{P}_{2m+2}|). \end{cases}$$

Since the functions Ψ_{2m} and Ψ_{2m+2} extend $\Phi \mid |(\mathcal{P}_{2m})^{(0)}|$ and $\Phi \mid |(\mathcal{P}_{2m+1})^{(0)}|$, respectively, and since by lemma 3.6.2(1) the sets $|(\mathcal{P}_{2m+1})^{(0)}|$, $|\mathcal{P}_{2m+1} \cap \mathcal{P}_{2m}|$, $|\mathcal{P}_{2m+1} \cap \mathcal{P}_{2m+2}|$ are closed in $|N(\mathcal{U})|$, the function Φ_{2m+1} is well-defined and continuous.

5.2.8. LEMMA: *For each* $m \geq 1$, Φ_{2m+1} *is a partial realization of* $\mathcal{P}_{2m+1}$ *in* X *relative to* $(\mathcal{T}_{2m+1}, \mathcal{B}_{2m-2})$.

PROOF: Let $m \geq 1$ and let $\tau \in \mathcal{P}_{2m+1}$. By lemma 5.2.7 there exists an element $B \in \mathcal{B}_{2m}$ such that

$$\Phi_{2m+1}(\tau \cap |N(\mathcal{U})^{(0)}|) \subseteq B.$$

Since by (b) $\mathcal{B}_{2m} < \mathcal{A}_{2m}$ there also exists $A \in \mathcal{A}_{2m}$ such that

$$\Phi_{2m+1}(\tau \cap |N(\mathcal{U})^{(0)}|) \subseteq A.$$

Let τ' be a face of τ such that $\tau' \in \mathcal{P}_{2m+1} \cap \mathcal{P}_{2m}$. By construction there exists $A_{\tau'} \in \mathcal{A}_{2m-1}$ such that

$$\Phi_{2m+1}(\tau') \subseteq A_{\tau'}.$$

Observe that $A \cap A_{\tau'} \neq \emptyset$. Now let τ'' be a face of τ such that $\tau'' \in \mathcal{P}_{2m+1} \cap \mathcal{P}_{2m+2}$. By construction there also exists $A_{\tau''} \in \mathcal{A}_{2m+1}$ such that

$$\Phi_{2m+1}(\tau'') \subseteq A_{\tau''}.$$

As above, $A \cap A_{\tau''} \neq \emptyset$. Now since $\mathcal{A}_{2m+1} < \mathcal{A}_{2m}$, $\mathcal{A}_{2m} < \mathcal{A}_{2m-1}$, and by (c), $\mathcal{A}_{2m-1}$ is a star-refinement of $\mathcal{B}_{2m-2}$, we are done. □

So by lemma 5.2.8 and (b), we may extend for each $m \geq 1$ the function Φ_{2m+1} to a full realization

$$\Psi_{2m+1}\colon |\mathcal{P}_{2m+1}| \to X,$$

relative to the covering $\mathcal{A}_{2m-2}$.

5.2.9. LEMMA: *For each* $n \geq 0$, $\kappa(B_n) \subseteq |\mathcal{P}_n|$.

PROOF: Take an arbitrary $x \in B_n$. By (g) there exist only finitely many elements of $\mathcal{U}$ that contain x, say $U_{i(0)},\cdots,U_{i(k)}$. The simplex

$$\tau = |\{x(U_{i(0)}),\cdots,x(U_{i(k)})\}|$$

is an element of $\mathcal{P}_n$. By lemma 3.6.14, $\kappa(x)$ belongs to τ, and since τ is a subset of $|\mathcal{P}_n|$ we are done. □

For each $n \geq 0$ let

$$P_n = \bigcup_{m=n}^{\infty} |\mathcal{P}_m|.$$

Define $\Psi: P_2 \to X$ by

$$\Psi(x) = \Psi_m(x) \qquad (x \in |\mathcal{P}_m|;\ m \geq 2).$$

Then Ψ is continuous by lemma 3.6.6.

Notice that for every $m \geq 3$, $\Psi \mid |\mathcal{P}_m|$ is a full realization of $|\mathcal{P}_m|$ with respect to $\mathcal{A}_{m-3}$.

Finally, define $r: \overline{H}_2 \to X$ by

$$r(x) = \begin{cases} x & (x \in X), \\ (\Psi \circ \kappa)(x) & (x \in \overline{H}_2 \setminus X). \end{cases}$$

Then r is well-defined by lemma 5.2.9. We claim that r is a retraction. Since by theorem 3.6.15, κ is continuous and X is closed in C, it remains to check the continuity of r at the points of the boundary Bd X of X.

5.2.10. LEMMA: r *is continuous at every point of* Bd X.

PROOF: Let $x \in \text{Bd } X$ and let $(t_n)_n$ be a sequence in $\overline{H}_2 \setminus X$ such that $\lim_{n\to\infty} t_n = x$. We shall prove that $\lim_{n\to\infty} r(t_n) = x$. To this end, for every n pick $U_n \in \mathcal{U}$ containing t_n. In two steps we shall prove that

$$\lim_{n\to\infty} r(t_n) = \lim_{n\to\infty} \Psi(\kappa(t_n)) = x.$$

CLAIM 1: $\lim_{n\to\infty} \Psi(x(U_n)) = x$.

Since $\lim_{n\to\infty} d(U_n,X) = 0$, by (i) it follows that $\lim_{n\to\infty} \text{diam}(U_n) = 0$. From this we conclude that

$$\lim_{n\to\infty} z_{U_n} = x.$$

By lemma 5.2.5 we therefore obtain

$$\lim_{n\to\infty} d(z_{U_n}, y_{U_n}) \leq 2 \lim_{n\to\infty} d(z_{U_n}, X) = 0,$$

so that by the definition of Ψ,

$$\lim_{n\to\infty} \Psi(x(U_n)) = \lim_{n\to\infty} y_{U_n} = \lim_{n\to\infty} z_{U_n} = x,$$

as desired.

As in §3.6 for every $y \in \overline{H}_2\setminus X$ let $F(y) = \{U \in \mathcal{U}: y \in U\}$ and let $\tau(y)$ be the simplex of $N(\mathcal{U})$ spanned by the (finitely many) vertices $x(U)$, $U \in F(y)$. By lemma 3.6.14, $\tau(y)$ is the carrier of $\kappa(y)$ in $N(\mathcal{U})$, so in particular, $\kappa(y) \in \tau(y)$.

CLAIM 2: $\lim_{n\to\infty} d(\Psi(\kappa(t_n)),\Psi(x(U_n))) = 0.$

For every $n \in \mathbb{N}$ let $k(n) \in \mathbb{N}$ be such that $t_n \in B_{k(n)}$ (without loss of generality, for every n, $k(n) \geq 3$). By lemma 5.2.4(2) and the fact that the sequence $(t_n)_n$ converges to $x \in X$, it follows easily that $\lim_{n\to\infty} k(n) = \infty$. Observe that for every n, $\tau(t_n) \in \mathcal{P}_{k(n)}$. By construction, for every n there exists $A_n \in \mathcal{A}_{k(n)-3}$ such that

$$\Psi(\tau(t_n)) \subseteq A_n,$$

and since $\{x(U_n),\kappa(t_n)\} \subseteq \tau(t_n)$ we therefore obtain

$$\{\Psi(x(U_n)),\Psi(\kappa(t_n))\} \subseteq \Psi(\tau(t_n)) \subseteq A_n.$$

From $\lim_{n\to\infty} k(n) = \infty$ we find by (d) that

$$\lim_{n\to\infty} \operatorname{diam}(A_n) = 0,$$

so we are done. □

We shall now derive some applications. Call a space X *homotopically trivial* if for every $n \geq 0$, every continuous function $f: S^n \to X$ can be continuously extended over B^{n+1}.

5.2.11. LEMMA: *Every contractible space is homotopically trivial.*

PROOF: Let X be a contractible space and let $H: X \times I \to X$ be a homotopy contracting X to a single point, say p. In addition, let $n \geq 0$ and let $g: S^n \to X$ be continuous. Define $\bar{g}: B^{n+1} \to X$ by

$$\begin{cases} \bar{g}(x) = H(g(\frac{x}{\|x\|}), 1-\|x\|) & (x \neq 0), \\ \bar{g}(\underline{0}) = p. \end{cases}$$

An easy check shows that $\bar{g}$ is the required continuous extension of g. □

There exist elementary examples of homotopically trivial spaces that are not contractible, see exercise 5.2.1.

We now present the following consequence of theorem 5.2.1.

5.2.12. THEOREM: *Let* X *be a locally connected space which has an open base* $\mathcal{B}$ *such that for every finite subfamily* $\mathcal{F}$ *of* $\mathcal{B}$:

if $\bigcap\mathcal{F}$ *is nonempty then every component of* $\bigcap\mathcal{F}$ *is homotopically trivial.*

Then X *is an* **ANR**.

PROOF: First observe that if C is a component of an element $B \in \mathcal{B}$ then C is path-connected. Since $\mathcal{B}$ is a base, this implies that every open subset of X is locally path-connected.

Let $\mathcal{U}$ be an open cover of X. Since $\mathcal{B}$ is an open base for the topology of X, there exists a refinement $\mathcal{V}$ of $\mathcal{U}$ consisting entirely of elements of $\mathcal{B}$. By lemma 5.1.7 there exists an open star-refinement $\mathcal{W}$ of $\mathcal{V}$. Since X is locally connected, the family of all components of elements of $\mathcal{W}$ is an open refinement of $\mathcal{W}$ and is clearly also a star-refinement of $\mathcal{V}$. We therefore assume without loss of generality that the elements of $\mathcal{W}$ are connected. By the above remark and lemma 2.4.2 this implies that every element of $\mathcal{W}$ is path-connected.

We aim at applying theorem 5.2.1. To this end, let $\mathcal{T}$ be a locally finite simplicial complex, let $\mathcal{S}$ be a subcomplex of $\mathcal{T}$ containing all the vertices of $\mathcal{T}$, and let $f\colon |\mathcal{S}| \to X$ be a partial realization of $\mathcal{T}$ in X relative to $(\mathcal{S},\mathcal{W})$. For each $\sigma \in \mathcal{T}$ pick $W_\sigma \in \mathcal{W}$ such that $f(\sigma \cap |\mathcal{S}|) \subseteq W_\sigma$. We first want to extend f to a continuous function $g\colon |\mathcal{T}^{(1)} \cup \mathcal{S}| \to X$ such that g is a partial realization of $\mathcal{T}$ in X relative to $(\mathcal{T}^{(1)} \cup \mathcal{S},\mathcal{V})$. As to be expected, we shall construct g "simplex-wise". To this end, take an arbitrary $\sigma \in \mathcal{T}^{(1)} \cup \mathcal{S}$. If $\sigma \in \mathcal{S}$ then f is already defined on all of σ: put $g_\sigma = f \mid \sigma$. If $\sigma \notin \mathcal{S}$ then $\sigma \in \mathcal{T}^{(1)} \setminus \mathcal{T}^{(0)}$. Since $\mathcal{S}$ contains all the vertices of $\mathcal{T}$, $\sigma \cap |\mathcal{S}| = \partial\sigma$. Consequently, by the fact that W_σ is path-connected, there exists an extension $g_\sigma\colon \sigma \to W_\sigma$ of $f \mid \partial\sigma$. Now define $g\colon |\mathcal{T}^{(1)} \cup \mathcal{S}| \to X$ as follows:

$$g(x) = g_\sigma(x) \qquad (x \in \sigma \in \mathcal{T}^{(1)} \cup \mathcal{S}).$$

Since $\mathcal{T}^{(1)} \cup \mathcal{S}$ is a simplicial complex, g is well-defined. Also, g is continuous by lemma 3.6.6.

We claim that g is a partial realization of $\mathcal{T}$ in X relative to $(\mathcal{T}^{(1)} \cup \mathcal{S}, \mathcal{V})$. To this end, take an arbitrary simplex $\sigma \in \mathcal{T}$. First observe that

$$g(\sigma \cap |\mathcal{T}^{(1)} \cup \mathcal{S}|) = g(\sigma \cap |\mathcal{T}^{(1)}|) \cup f(\sigma \cap |\mathcal{S}|).$$

Consequently, by construction it follows that

$$g(\sigma \cap |\mathcal{T}^{(1)} \cup \mathcal{S}|) \subseteq \bigcup\{W_\tau : \tau \text{ is a 1-dimensional face of } \sigma\} \cup W_\sigma.$$

Again since $\mathcal{S}$ contains all the vertices of $\mathcal{T}$, each of the W_τ's intersects W_σ. Since $\mathcal{W}$ is a star-refinement of $\mathcal{V}$, there consequently exists $V_\sigma \in \mathcal{V}$ such that

$$g(\sigma \cap |\mathcal{T}^{(1)} \cup \mathcal{S}|) \subseteq V_\sigma,$$

which is as required.

Now let $\sigma \in \mathcal{T}$ be an arbitrary simplex and put

$$\mathcal{E}(\sigma) = \{\tau \in \mathcal{T} : \sigma \preccurlyeq \tau\}.$$

Since by the local finiteness of $\mathcal{T}$ each vertex of σ is contained in at most finitely many simplexes in $\mathcal{T}$, it is clear that $\mathcal{E}(\sigma)$ is finite. Define

$$F(\sigma) = \bigcap_{\tau \in \mathcal{E}(\sigma)} V_\tau.$$

CLAIM: For each $\sigma \in \mathcal{T}$, each component of $F(\sigma)$ is homotopically trivial and if $\sigma \preccurlyeq \tau$ then $F(\sigma) \subseteq F(\tau)$. Moreover, if $\sigma \in \mathcal{T}^{(1)} \cup \mathcal{S}$ then $g(\sigma) \subseteq F(\sigma)$.

That each component of $F(\sigma)$ is homotopically trivial, follows by the assumptions on $\mathcal{B}$ and the fact that the $\mathcal{E}(\sigma)$'s are finite. Also, if $\sigma \preccurlyeq \tau$ then $\mathcal{E}(\tau) \subseteq \mathcal{E}(\sigma)$ from which it follows that $F(\sigma) \subseteq F(\tau)$. Now take an arbitrary $\sigma \in \mathcal{S} \cup \mathcal{T}^{(1)}$ and $\tau \in \mathcal{E}(\sigma)$. Then

$$g(\sigma) \subseteq g(\tau \cap |\mathcal{S} \cup \mathcal{T}^{(1)}|) \subseteq V_\tau.$$

We conclude that $g(\sigma) \subseteq F(\sigma)$.

Now by induction on $n \geq 1$ we shall construct continuous functions $f_n : |\mathcal{T}^{(n)} \cup \mathcal{S}| \to X$ such that the following conditions are satisfied:

(1) $f_1 = g$ and f_{n+1} extends f_n,
(2) for every $\sigma \in \mathcal{T}^{(n)} \cup \mathcal{S}$, $f_n(\sigma) \subseteq F(\sigma)$.

Assume that for certain $n \geq 2$, f_{n-1} has been defined. We shall construct f_n "simplex-wise". To this end, take an arbitrary $\sigma \in \mathcal{T}^{(n)} \cup \mathcal{S}$. If $\sigma \in \mathcal{T}^{(n-1)} \cup \mathcal{S}$ then f_{n-1} is already defined on all of σ: put $f_\sigma = f_{n-1} \mid \sigma$. If $\sigma \notin \mathcal{T}^{(n-1)} \cup \mathcal{S}$ then $\sigma \in \mathcal{T}^{(n)} \setminus \mathcal{T}^{(n-1)}$. Observe that $\sigma \cap |\mathcal{T}^{(n-1)} \cup \mathcal{S}| = \partial\sigma$. Let τ be a proper face of σ. Then $\tau \in \mathcal{T}^{(n-1)}$, which implies, by our inductive assumption and the claim, that

$$f_{n-1}(\tau) \subseteq F(\tau) \subseteq F(\sigma).$$

Consequently,

$$f_{n-1}(\partial\sigma) \subseteq F(\sigma).$$

Since $n \geq 2$ and $\sigma \notin \mathcal{T}^{(n-1)}$, σ is at least two-dimensional. Consequently, $\partial\sigma$ is connected from which it follows that $f_{n-1}(\partial\sigma)$ is connected. Let C be the component of $F(\sigma)$ which contains $f_{n-1}(\partial\sigma)$. By the claim it follows that C is homotopically trivial. The function $f_{n-1} \mid \partial\sigma$ can therefore be extended to a continuous function

$$f_\sigma: \sigma \to C \subseteq F(\sigma).$$

Now define $f_n: |\mathcal{T}^{(n)} \cup \mathcal{S}| \to X$ as follows:

$$f_n(x) = f_\sigma(x) \qquad (x \in \sigma \in \mathcal{T}^{(n)} \cup \mathcal{S}).$$

Since $\mathcal{T}^{(n)} \cup \mathcal{S}$ is a simplicial complex, f_n is well-defined. Also, f_n is continuous by lemma 3.6.6. We conclude that f_n is as required.

This completes the inductive construction of the functions f_n.

As to be expected, define $h: |\mathcal{T}| \to X$ by

$$h(x) = f_n(x) \qquad (x \in |\mathcal{T}^{(n)} \cup \mathcal{S}|).$$

It is clear that by (1), h is well-defined. By another appeal to lemma 3.6.6 it follows that h is continuous. By construction, h extends f and by (2), h is a full realization of $\mathcal{T}$ in X relative to $\mathcal{V}$ and therefore to $\mathcal{U}$. □

5.2.13. *Remark:* Since the intersection of an arbitrary family of convex sets in a normed linear space is again convex, it seems that theorem 5.2.12 gives a new proof of theorem theorem 1.5.1. This is not true however, since theorem 1.5.1 was used in the proof of theorem 5.2.1.

Naturally, theorem 5.2.1 suggests the question whether there exists a characterization of the class of **AR**'s in the same spirit. In corollary 1.6.7 we showed that a space X is an **AR** if and only if X is a contractible **ANR**. Consequently, it seems that our question has already been answered. However, deciding whether a given space is contractible is sometimes quite a complicated task, cf. corollary 3.5.6. Within the class of **ANR**'s it turns out that contractibility and homotopy triviality are equivalent notions. Since for a given space X, verifying that X is homotopically trivial is usually much easier than verifying that X is contractible, we shall present a proof of our assertion in full detail. We first derive the following

5.2.14. LEMMA: *Let* X *be a space. The following statements are equivalent:*

(a) X *is homotopically trivial, and*

(b) *for every locally finite simplicial complex* $\mathcal{T}$ *and every subcomplex* $\mathcal{S}$ *of* $\mathcal{T}$*, every continuous function* $g: |\mathcal{S}| \to X$ *can be extended to a continuous function* $\bar{g}: |\mathcal{T}| \to X$.

PROOF: The implication (b) $\Rightarrow$ (a) is a triviality. Simply observe that there is a homeomorphism from B^n onto an n-dimensional simplex σ taking S^{n-1} onto the geometric boundary $\partial\sigma$ of σ (exercise 3.5.8(c)).

The implication (a) $\Rightarrow$ (b) is also very simple to prove. Let us first show that without loss of generality we may assume that $\mathcal{S}$ contains all the vertices of $\mathcal{T}$. Since $|\mathcal{T}^{(0)}|$ is a discrete topological space by corollary 3.6.3, and since both $|\mathcal{S}|$ and $|\mathcal{T}^{(0)}|$ are closed in $|\mathcal{T}|$ by lemma 3.6.2(1), we can extend g over $|\mathcal{S} \cup \mathcal{T}^{(0)}|$ in an arbitrary way.

Now, by induction on $n \geq 0$ by a standard procedure we shall construct a continuous function $g_n: |\mathcal{S} \cup \mathcal{T}^{(n)}| \to X$ such that the following condition is satisfied:

$$g_0 = g \text{ and } g_{n+1} \text{ extends } g_n.$$

Since $\mathcal{S}$ contains all the vertices of $\mathcal{T}$, we may indeed put $g_0 = g$. Assume that g_{n-1} has been constructed for certain $n \geq 1$. As previously in this section, we shall construct g_n "simplex-wise". To this end, take an arbitrary $\sigma \in \mathcal{S} \cup \mathcal{T}^{(n)}$. If $\sigma \in \mathcal{S} \cup \mathcal{T}^{(n-1)}$ put $g_\sigma = g_{n-1} | \sigma = g | \sigma$. So assume that $\sigma \in \mathcal{T}^{(n)} \setminus (\mathcal{S} \cup \mathcal{T}^{(n-1)})$. Now we use our assumption on X to extend the function $g_{n-1} | \partial\sigma: \partial\sigma \to X$ to a continuous function $g_\sigma: \sigma \to X$. Define $g_n: |\mathcal{S} \cup \mathcal{T}^{(n)}| \to X$ by

$$g_n(x) = g_\sigma(x) \qquad (x \in \sigma \in \mathcal{S} \cup \mathcal{T}^{(n)}).$$

As in the previous proofs in this section it follows that g_n is well-defined and continuous. This completes the inductive construction of the functions g_n, $n \in \mathbb{N}$.

Now define $\bar{g}: |\mathcal{T}| \to X$ as follows:

$$\bar{g}(x) = g_n(x) \qquad (x \in \sigma \in \mathcal{T}^{(n)}).$$

Then $\bar{g}$ is clearly as required. □

We shall now present a proof of the following

5.2.15. THEOREM: *Let* X *be a space. The following statements are equivalent:*

(a) X *is an* **AR**,

(b) X *is a contractible* **ANR**, *and*

(c) X *is a homotopically trivial* **ANR**.

PROOF: Observe that the implications (a) ⇒ (b) ⇒ (c) are trivialities, see corollary 1.6.7 and lemma 5.2.11. So let X be a homotopically trivial **ANR**. We shall prove that X is an **AR**. Observe that X has the realization property stated in theorem 5.2.1(b).

It will be convenient to think of X as the space X in the proof of theorem 5.2.1. We adopt all notation and terminology introduced in that proof. We shall prove that the retraction $r: \overline{H}_2 \to X$ can be extended over C so that the desired result follows from theorem 1.5.1.

Recall that

$$P_2 = \bigcup_{m=2}^{\infty} |\mathcal{P}_m|.$$

Define

$$Q = |\mathcal{P}_0| \cup |\mathcal{P}_1| \cup |\mathcal{P}_2|.$$

Observe that by lemma 5.2.6,

$$P_2 \cap Q = |\mathcal{P}_2|,$$

and also that by definition, $P_2 \cup Q = P_0$. Now since $\mathcal{T} = \mathcal{P}_0 \cup \mathcal{P}_1 \cup \mathcal{P}_2$ is a simplicial complex, and $\mathcal{P}_2$ is a subcomplex of $\mathcal{T}$, by lemma 5.2.14 we can extend $\Psi \mid |\mathcal{P}_2|$ to a continuous function

$$\vartheta: Q \to X.$$

By lemma 5.2.9 it follows that $\kappa(C\backslash X) \subseteq Q$. Consequently, the function $\bar{r}: C \to X$ defined by

$$\bar{r}(x) = \begin{cases} r(x) & (x \in \overline{H}_2), \\ (\vartheta \circ \kappa)(x) & (x \in C\backslash H_2), \end{cases}$$

is well-defined if for every $x \in D = \overline{H}_2\backslash H_2$, $r(x) = (\vartheta \circ \kappa)(x)$. To verify this, take an arbitrary $x \in D$. Then $x \in B_2$. Consequently, lemma 5.2.9 gives us that $\kappa(x) \in |\mathcal{P}_2|$. By the definition of ϑ we therefore have $r(x) = (\Psi \circ \kappa)(x) = (\vartheta \circ \kappa)(x)$, which is as required. Finally, $\bar{r}$ is continuous since the sets $\overline{H}_2$ and $C\backslash H_2$ are closed in C, and the restrictions $\bar{r} \mid \overline{H}_2$ and $\bar{r} \mid C\backslash H_2$ are both continuous. □

Exercises for §5.2.

1. Give an example of a homotopically trivial compact space which is not contractible (Hint: consider the well-known sin(1/x)-curve, cf. exercise 1.5.4).

2. Let $\mathcal{T}$ be a locally finite simplicial complex such that $|\mathcal{T}|$ is homotopically trivial. Prove that $|\mathcal{T}|$ is contractible.

5.3. Hyperspaces and the AR-Property

As in §4.7, for every compact space X we let 2^X denote its hyperspace. We shall apply the results in §5.2 to prove that if X is a *Peano continuum*, i.e. a connected and locally connected compact space, then 2^X is an **AR**. This demonstrates the power of the techniques derived in the previous two sections. In §8.4 we shall in fact prove that the hyperspace of each nondegenerate Peano continuum is homeomorphic to the Hilbert cube; this is the Curtis-Schori-West Hyperspace Theorem. Since the Hilbert cube is an **AR**, at first glance there seems no reason for proving the **AR**-property for hyperspaces here. However, that the hyperspace of each Peano continuum is an **AR** is a step in our proof of the Curtis-Schori-West Hyperspace Theorem.

We need to derive a few easy results on hyperspaces first.

Let X and Y be compact spaces and let $f: X \to Y$ be continuous. By exercise 4.7.6 we may regard X and Y to be (canonical) subspaces of 2^X and 2^Y, respectively. There exists a canonical continuous extension 2^f of f over 2^X, which is called the *hyperspace map* of f. Define $2^f: 2^X \to 2^Y$ by

$$2^f(A) = f(A).$$

We shall prove below that hyperspace maps are continuous. They are of crucial importance in the process of analyzing the structure of hyperspaces.

The following simple result shall be used a few times in the remaining part of this section.

5.3.1. LEMMA: *Let* X *be a compact space and let* $(A_n)_n$ *be a sequence in* 2^X *converging to an element* $A \in 2^X$. *Then* $A = \{x \in X: (\forall n \in \mathbb{N})(\exists x_n \in A_n)(\lim_{n\to\infty} x_n = x)\}$.

PROOF: Let x be a point in X for which there exists a sequence $(x_n)_n$ with $x_n \in A_n$ for every n and $x = \lim_{n\to\infty} x_n$. We shall prove that $x \in A$. Let $\varepsilon > 0$. Since $(A_n)_n$ converges to A, there exists $N \in \mathbb{N}$ such that $A_m \subseteq B(A,\varepsilon)$ for every $m \geq N$. Since $(x_n)_n$ converges to x, this implies that $d(x,A) \leq \varepsilon$. Consequently, $d(x,A) = 0$ and the fact that A is closed now gives us that that $x \in A$, as required.

Pick an arbitrary $x \in A$. For each $n \in \mathbb{N}$ there exists a point $x_n \in A_n$ such that $d(x,x_n) \leq 2d(x,A_n)$. We claim that the sequence $(x_n)_n$ converges to x. To this end, let $\varepsilon > 0$. Since $(A_n)_n$ converges to A, there exists $N \in \mathbb{N}$ such that for each $m \geq N$, $A \subseteq B(A_m,\varepsilon)$. From this it follows that for $m \geq N$, $d(x,A_m) < \varepsilon$, i.e. $d(x,x_n) < 2\varepsilon$. We are done.□

5.3.2. LEMMA: *Let* X *and* Y *be compact spaces and let* $f: X \to Y$ *be continuous. Then the hyperspace map* $2^f: 2^X \to 2^Y$ *is continuous.*

PROOF: This is easy. Given $\varepsilon > 0$ find $\delta > 0$ such that if $d(x,y) < \delta$ then $d(f(x),f(y)) < \varepsilon$. Then also, if $d_H(D,E) < \delta$ then $d_H(f(D),f(E)) < \varepsilon$.□

5.3.3. COROLLARY: *Let* X *be a compact space and let* $A \subseteq X$ *be closed. Then the natural inclusion* $2^A \to 2^X$ *is a topological imbedding.*

PROOF: Since 2^A is compact (proposition 4.7.2), this follows immediately from lemma 5.3.2 and exercise 1.1.4.□

Let X be compact and let $A \subseteq X$ be closed. By corollary 5.3.3 we may identify 2^A and $\{B \in 2^X: B \subseteq A\}$. We find it sometimes convenient to do that.

The following lemma is a triviality, the proof of which is left as an exercise to the reader.

5.3.4. LEMMA: *Let* X *be a compact space and let* $A_1,\cdots,A_n$ *be nonempty subsets of* X. *Then the function* $f: \prod_{i=1}^n A_i \to 2^X$ *defined by* $f(a_1,\cdots,a_n) = \{a_1,\cdots,a_n\}$ *is continuous.*□

We shall now consider a more complicated "hyperspace map". Let X be a compact space. Recall that 2^X is compact (proposition 4.7.2). Consider the space 2^{2^X}, i.e. the hyperspace of the hyperspace of X. Since 2^X is compact, so is 2^{2^X}. Each point $\mathcal{A} \in 2^{2^X}$ is a closed subset of 2^X and therefore can be regarded as a "closed" family of closed subsets of X. We shall prove that the union of such a family is closed in X. From this it follows that the "union-operator"

$$\cup: 2^{2^X} \to 2^X,$$

defined by

$$\cup(\mathcal{A}) = \cup\mathcal{A},$$

is well-defined. We shall also prove that this operator is continuous.

5.3.5. LEMMA: *Let* X *be a compact space and let* $\mathcal{A}$ *be an element of* 2^{2^X}. *Then* $\cup\mathcal{A}$ *is closed in* X.

PROOF: Take a sequence $(x_n)_n$ in $\cup\mathcal{A}$ converging to a point $x \in X$. For each n there exists $A_n \in \mathcal{A}$ such that $x_n \in A_n$. By compactness of $\mathcal{A}$, regarded as subspace of 2^X, we may assume without loss of generality that the sequence $(A_n)_n$ converges to a point $A \in \mathcal{A}$. From lemma 5.3.1 it follows that $x \in A \subseteq \cup\mathcal{A}$. □

So from lemma 5.3.5 we conclude that for compact X, the union-operator $\cup: 2^{2^X} \to 2^X$ is well-defined. We shall now prove that it is continuous.

5.3.6. PROPOSITION: *Let* X *be a compact space. Then the union-operator* $\cup: 2^{2^X} \to 2^X$ *is continuous.*

PROOF: Assume that $(\mathcal{A}_n)_n$ is a sequence in 2^{2^X}, converging to a point $\mathcal{A} \in 2^{2^X}$. Let $\varepsilon > 0$.

We shall first prove that the set $E = \{n \in \mathbb{N}: \cup\mathcal{A}_n$ is not contained in $B(\cup\mathcal{A},\varepsilon)\}$ is finite. To the contrary, assume that E is infinite. For every $n \in E$ pick $x_n \in A_n \in \mathcal{A}_n$ such that $x_n \notin B(\cup\mathcal{A},\varepsilon)$. By compactness of 2^X we may assume without loss of generality that the sequence $(A_n)_{n\in E}$ converges to a point $A \in 2^X$. By lemma 5.3.1, A belongs to $\mathcal{A}$. By compactness of X, we may also assume without loss of generality that the sequence $(x_n)_{n\in E}$ converges to $x \in X$. Another appeal to lemma 5.3.1 yields that $x \in A$. Since $(x_n)_{n\in E}$ converges to x, $x \notin B(\cup\mathcal{A},\varepsilon)$. This is a contradiction however since $x \in A \subseteq \cup\mathcal{A} \subseteq B(\cup\mathcal{A},\varepsilon)$.

We shall next prove that the set $F = \{n \in \mathbb{N}: \cup\mathcal{A}$ is not contained in $B(\cup\mathcal{A}_n,\varepsilon)\}$ is finite. To the contrary, assume that F is infinite. For every $n \in F$ pick $x_n \in \cup\mathcal{A}$ such that $x_n \notin B(\cup\mathcal{A}_n,\varepsilon)$. By lemma 5.3.5, $\cup\mathcal{A}$ is compact, so that without loss of generality we may assume that the sequence $(x_n)_{n\in F}$ converges to a point $x \in \cup\mathcal{A}$. Pick $A \in \mathcal{A}$ such that $x \in A$. Since $(\mathcal{A}_n)_n$ converges to $\mathcal{A}$, lemma 5.3.1 implies that for $n \in F$ there exist $A_n \in \mathcal{A}_n$ such that $(A_n)_{n\in F}$ converges to A. By another appeal to lemma 5.3.1 we find that for every $n \in F$ there exists a point $y_n \in A_n$ such that $(y_n)_{n\in F}$ converges to x. Since for every $n \in F$, $A_n \in \mathcal{A}_n$ we obtain $y_n \in \cup\mathcal{A}_n$ from which it follows that $d(x_n,y_n) \geq \varepsilon > 0$. This is a contradiction however, since the sequences $(x_n)_{n\in F}$ and $(y_n)_{n\in F}$ both converge to x.

We conclude that there exists $N \in \mathbb{N}$ such that for all $m \geq N$, $\cup\mathcal{A}_m$ is contained in $B(\cup\mathcal{A},\varepsilon)$ and $\cup\mathcal{A}$ is contained in $B(\cup\mathcal{A}_m,\varepsilon)$. Consequently, $(\cup\mathcal{A}_n)_n$ converges to $\cup\mathcal{A}$. □

5.3.7. COROLLARY: *Let* X *and* Y *be compact spaces and let* $g: X \to 2^Y$ *be continuous. Then the function* $\bar{g}: 2^X \to 2^Y$ *defined by*

$$\bar{g}(A) = \cup_{x\in A}\, g(x)$$

is continuous.

PROOF: The hyperspace map $2^g: 2^X \to 2^{2^Y}$ is continuous by lemma 5.3.2. In addition, by proposition 5.3.6, the union-operator $\cup: 2^{2^Y} \to 2^Y$ is also continuous. Since $\bar{g} = \cup \circ 2^g$, we are done. □

By a similar reasoning we obtain

5.3.8. COROLLARY: *Let* X *be a compact space and let* $\mathcal{A}_1,\cdots,\mathcal{A}_n$ *be nonempty subsets of* 2^X. *Then the function* $f: \prod_{i=1}^n \mathcal{A}_i \to 2^X$ *defined by* $f(A_1,\cdots,A_n) = A_1 \cup \cdots \cup A_n$ *is continuous.*

PROOF: Again, let $\cup: 2^{2^X} \to 2^X$ denote the union-operator. By lemma 5.3.4 we conclude that the function $g: \prod_{i=1}^n \mathcal{A}_i \to 2^{2^X}$ defined by

$$g(A_1,\cdots,A_n) = \{A_1,\cdots,A_n\}$$

is continuous. Since $f = \cup \circ g$, we are done. □

We now turn to a different aspect of hyperspace theory. Again, let X be a compact space. For every finite collection $\mathcal{A}$ of subsets of X put

$$<\mathcal{A}> = \{B \in 2^X: B \subseteq \cup\mathcal{A} \text{ and for every } A \in \mathcal{A}, A \cap B \neq \emptyset\}.$$

5.3.9. LEMMA: *Let* X *be a compact space. Then*

(1) *if* $\mathcal{A}$ *is a finite family of open subsets of* X *then* $<\mathcal{A}>$ *is open in* 2^X,

(2) *the collection* $\mathcal{B}(X) = \{<\mathcal{F}>: \mathcal{F}$ *is a finite family of open subsets of* X$\}$ *is an open base for* 2^X, *and*

(3) $\mathcal{B}(X)$ *is closed under finite intersections.*

PROOF: For (1), let $\mathcal{A}$ be a finite family of open subsets of X and let $D \in <\mathcal{A}>$. For each $A \in \mathcal{A}$ pick an arbitrary point $x_A \in A \cap D$. Since $\mathcal{A}$ is finite and consists of open sets, there exists $\varepsilon > 0$ such that

(4) for every $A \in \mathcal{A}$, $B(x_A,\varepsilon) \subseteq A$, and

(5) $B(D,\varepsilon) \subseteq \cup\mathcal{A}$.

Now take $E \in 2^X$ such that $d_H(D,E) < \varepsilon$. By (5), $E \subseteq \cup\mathcal{A}$. Fix an arbitrary $A \in \mathcal{A}$. Since $D \subseteq B(E,\varepsilon)$, there exists $x \in E$ such that $d(x_A,x) < \varepsilon$. By (4), x belongs to A, i.e. E intersects A. This proves that the ball (in 2^X) about D of radius ε is contained in $<\mathcal{A}>$.

For (2), take arbitrary $D \in 2^X$ and $\varepsilon > 0$. By compactness, there is a finite family $\mathcal{A}$ of open subsets of X such that

(6) $D \subseteq \cup\mathcal{A} \subseteq B(D,\varepsilon)$,

(7) for every $A \in \mathcal{A}$, $A \cap D \neq \emptyset$, and

(8) for every $A \in \mathcal{A}$, $\mathrm{diam}(A) < \varepsilon$.

So $D \in <\mathcal{A}>$. Observe that $<\mathcal{A}>$ is open by (1). We claim that $<\mathcal{A}> \subseteq B(D,\varepsilon)$. To this end, take an arbitrary $E \in <\mathcal{A}>$. By (6), we need only show that $D \subseteq B(E,\varepsilon)$. But this is trivial. For take an arbitrary $x \in D$. By (6) there exists $A \in \mathcal{A}$ containing x. Since $E \in <\mathcal{A}>$, $E \cap A \neq \emptyset$, which implies by (8) that there exists $x' \in E$ with $d(x,x') < \varepsilon$. We conclude that $x \in B(E,\varepsilon)$.

For (3), take arbitrary finite collections $\mathcal{F}_0 = \{U_1,\cdots,U_n\}$ and $\mathcal{F}_1 = \{V_1,\cdots,V_m\}$ of open subsets of X. Put $U = \cup\mathcal{F}_0$ and $V = \cup\mathcal{F}_1$, respectively. An easy verification shows that

$$<\mathcal{F}_0> \cap <\mathcal{F}_1> = <\{U \cap V_1,\cdots,U \cap V_m, U_1 \cap V,\cdots,U_n \cap V\}>,$$

as required. □

We now are in a position to derive the following:

5.3.10. PROPOSITION: *Let* X *be a compact space. Then*

(1) X *is connected iff* 2^X *is connected, and*

(2) X *is locally connected iff* 2^X *is locally connected.*

PROOF: *We prove (1).*

First suppose that X is connected. For each $n \in \mathbb{N}$ define

$$\mathcal{F}_n(X) = \{A \in 2^X: |A| \le n\}.$$

Observe that $\mathcal{F}_1(X) \subseteq \mathcal{F}_2(X) \subseteq \cdots \subseteq \mathcal{F}_n(X) \subseteq \cdots$, and that by lemma 5.3.9(2),

$$\mathcal{F}_\infty(X) = \bigcup_{n=1}^{\infty} \mathcal{F}_n(X)$$

is dense in 2^X. Consequently, it suffices to prove that $\mathcal{F}_\infty(X)$ is connected, and for this it suffices to prove that every $\mathcal{F}_n(X)$ is connected. However, this follows directly by lemma 5.3.4 since for each n, $\mathcal{F}_n(X)$ is the image of X^n under the map $f_n: X^n \to 2^X$ defined by

$$f_n(x_1,\cdots,x_n) = \{x_1,\cdots,x_n\}.$$

Conversely, assume that 2^X is connected. We shall prove that X is connected. Striving for a contradiction, assume that X is not connected. Write X as $U \cup V$, where U and V are disjoint nonempty open sets. Clearly,

$$2^X = <\{U\}> \cup <\{V\}> \cup <\{U,V\}>.$$

In addition, $<\{U\}>$, $<\{V\}>$, and $<\{U,V\}>$ are disjoint, nonempty and open by lemma 5.3.9(1). This contradicts the connectivity of 2^X.

We prove (2).

Now assume that X is locally connected. Take $A \in 2^X$ and let $\mathcal{U}$ be a neighborhood of A in 2^X. By lemma 5.3.9(2) there exists a finite family $\mathcal{V}$ of open subsets of X such that $A \in <\mathcal{V}> \subseteq \mathcal{U}$. Since X is compact and locally connected, there exists a finite family $\mathcal{W}$ of *connected* open subsets of X with the following properties:

(1) $A \subseteq \bigcup\mathcal{W}$ and A meets every $W \in \mathcal{W}$, and

(2) for every $W \in \mathcal{W}$ there exists $V \in \mathcal{V}$ such that $\overline{W} \subseteq V$.

Observe that $A \in \langle\mathcal{W}\rangle \subseteq \langle\overline{\mathcal{W}}\rangle \subseteq \langle\mathcal{V}\rangle \subseteq \mathcal{U}$. We claim that $\langle\overline{\mathcal{W}}\rangle$ is connected, which suffices because it is a neighborhood of A by lemma 5.3.9(1). Let $\overline{\mathcal{W}} = \{E(1),\cdots,E(n)\}$. By corollary 5.3.3 we may regard each $2^{E(i)}$ to be a closed subset of 2^X. By part (1) of the lemma, $2^{E(i)}$ is connected for every i. Define f: $\prod_{i=1}^n 2^{E(i)} \to 2^X$ by $f(A_1,\cdots,A_n) = A_1 \cup \cdots \cup A_n$. Then f is continuous by corollary 5.3.8. So by connectivity of each $2^{E(i)}$, the range $\mathcal{R}$ of f is connected. Since $\mathcal{R}$ and $\langle\overline{\mathcal{W}}\rangle$ are clearly equal, this proves our claim and hence completes the proof of the local connectivity of 2^X.

Conversely, assume that 2^X is locally connected. We identify X and the subspace $\{\{x\}: x \in X\}$ of 2^X, cf. exercise 4.7.6. Let $x \in X$ and let U be an open neighborhood of x in X. Since $\langle\{U\}\rangle$ is an open neighborhood of x in 2^X (lemma 5.3.9(1)), by local connectivity of 2^X there is a connected neighborhood $\mathcal{C}$ of x in 2^X such that $x \in \mathcal{C} \subseteq \langle\{U\}\rangle$. Then $V = \mathcal{C} \cap X$ is a neighborhood of x in X and clearly $V \subseteq C = \cup\mathcal{C} \subseteq U$. We claim that C is connected. Suppose that this is not true. Let E and F be disjoint relatively open nonempty subsets of C such that $C = E \cup F$. By lemma 5.2.2 there exist disjoint open subsets E' and F' in X such that $E' \cap C = E$ and $F' \cap C = F$, respectively. Now observe that $\langle\{E'\}\rangle$, $\langle\{F'\}\rangle$ and $\langle\{E',F'\}\rangle$ are disjoint nonempty open subsets of 2^X (lemma 5.3.9(1)), that $\mathcal{C} \cap \langle\{E'\}\rangle \neq \emptyset$, $\mathcal{C} \cap \langle\{F'\}\rangle \neq \emptyset$, and that $\mathcal{C} \subseteq \langle\{E'\}\rangle \cup \langle\{F'\}\rangle \cup \langle\{E',F'\}\rangle$. This contradicts the connectivity of $\mathcal{C}$. □

Before we come to the announced **AR**-property result for hyperspaces, we need to derive two more technical tools.

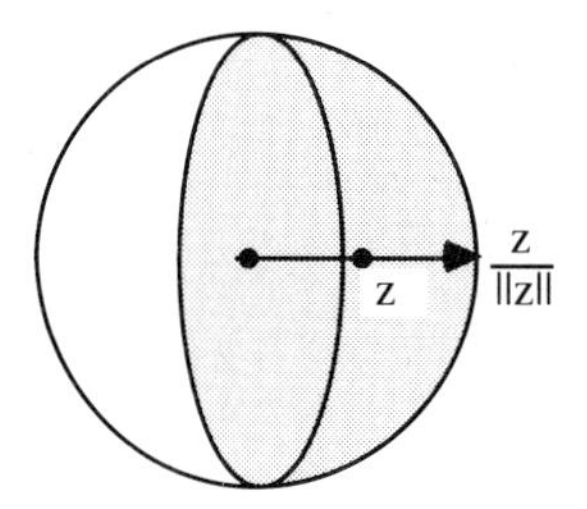

Figure 5.3.1.

5.3.11. PROPOSITION: *For each* $n \geq 1$ *there exists a continuous function*

$$f_n: B^{n+1} \to 2^{S^n}$$

such that for every $x \in S^n$, $f_n(x) = \{x\}$.

PROOF: Define $f_n: B^{n+1} \to 2^{S^n}$ by

$$\begin{cases} f_n(0) = \{S^n\}, \\ f_n(z) = \{x \in S^n: d(x, \frac{z}{\|z\|}) \leq 2 - 2\cdot\|z\|\} \quad (x \neq 0). \end{cases}$$

It is easy to visualize this function e.g. for $n = 2$ (see figure 5.3.1). From the pictures it is clear that f_n is continuous. Since it is trivial that $f_n(x) = \{x\}$ for every $x \in S^n$, we are done.□

Our final technical tool has nothing to do with hyperspaces and is interesting in its own right. Let X be a space and let $a,b \in X$. A *simple chain connecting* a and b is a collection $U_1,\cdots,U_n$ of open subsets of X such that

(1) $a \in U_1 \setminus \cup_{i=2}^{n} U_i$,
(2) $b \in U_n \setminus \cup_{i=1}^{n-1} U_i$,
(3) $U_i \cap U_j \neq \emptyset$ iff $|i - j| \leq 1$.

5.3.12. LEMMA: *Let* X *be a connected space and let* $\mathcal{U}$ *be an open cover of* X. *For any two points* x *and* y *in* X *there exists a simple chain connecting* x *and* y *consisting of elements of* $\mathcal{U}$.

PROOF: Let V be the set of all points in X which are connected to x by a simple chain of elements of $\mathcal{U}$. Then V is clearly open and $x \in V$. We shall prove that V is closed. The connectivity of X then implies that $V = X$.

Take an arbitrary $v \in \overline{V}$. There exists $U \in \mathcal{U}$ containing v; pick a point $t \in U \cap V$. Then x and t can be connected by a simple chain $U_1,\cdots,U_n$ of elements of $\mathcal{U}$. Observe that $U \cap U_n \neq \emptyset$ and define

$$k = \min\{i \leq n: U \cap U_i \neq \emptyset\}.$$

If v belongs to U_k then $U_1,\cdots,U_k$ is a simple chain from x to v. If v does not belong to U_k then $U_1,\cdots,U_k,U$ does the job for us.□

We now come to our last technical tool.

5.3.13. THEOREM: *Let* X *be a Peano continuum. If* U *is a connected open subset of* X *and if* x *and* y *are distinct elements of* U *then there exists an imbedding* $\Phi: I \to U$ *such that* $\Phi(0) = x$ *and* $\Phi(1) = y$.

PROOF: For each $n \in \mathbb{N}$ we shall construct a simple chain $U_{n,1},\cdots,U_{n,m(n)}$ from x to y having the following properties

(1) for each $i \leq m(1)$, $\overline{U}_{1,i} \subseteq U$,

(2) for each $i \leq m(n)$, $U_{n,i}$ is connected and $\mathrm{diam}(U_{n,i}) < \frac{1}{n}$,

(3) for all $i_1 \leq i_2 \leq m(n+1)$ there exist $j_1 \leq j_2 \leq m(n)$ such that $\overline{U}_{n+1,i_1} \subseteq U_{n,j_1}$ and $\overline{U}_{n+1,i_2} \subseteq U_{n,j_2}$.

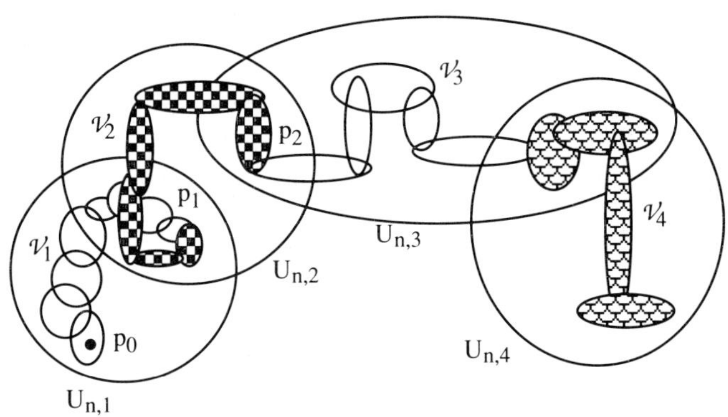

Figure 5.3.2.

The existence of $U_{1,1},\cdots,U_{1,m(1)}$ follows easily from lemma 5.3.12. Assume that the sets $U_{n,1},\cdots,U_{n,m(n)}$ have been constructed for certain n. For $1 \leq j < m(n)$ pick a point $p_j \in U_{n,j} \cap U_{n,j+1}$ and put $p_0 = x$ and $p_{m(n)} = y$. By another application of lemma 5.3.12 for each $1 \leq j < m(n)$ there exists a simple chain $\mathcal{V}_j$ from p_{j-1} to p_j in $U_{n,j}$ consisting of connected sets such that

(4) if $V \in \mathcal{V}_j$ then $\overline{V} \subseteq U_{n,j}$,

(5) for every $V \in \mathcal{V}_j$, $\mathrm{diam}(V) < \frac{1}{n+1}$.

Unfortunately, we cannot simply join these chains together, because of doubling back (see figure 5.3.2). We can obtain the desired simple chain by the following procedure. For every $1 \leq j \leq m(n)$ let $\mathcal{V}_j = \{V_{j,1},\cdots,V_{j,n(j)}\}$. Put

$$\pi(1) = \min\{\kappa \leq n(1): (\exists\lambda \leq n(2))(V_{1,\kappa} \cap V_{2,\lambda} \neq \emptyset)\}.$$

Let $\lambda = \max\{\kappa \leq n(2): V_{1,\pi(1)} \cap V_{2,\kappa} \neq \emptyset\}$. Replace $\mathcal{V}_1$ by $\{V_{1,1},\cdots,V_{1,\pi(1)}\}$ and $\mathcal{V}_2$ by $\{V_{2,\lambda},\cdots,V_{2,n(2)}\}$, respectively. Now repeat this with the "new" $\mathcal{V}_2$ and the "old" $\mathcal{V}_3$, etc. At the end of the process, the union of the "new" $\mathcal{V}_j$'s is clearly the required simple chain. This completes the inductive construction.

For every $n \in \mathbb{N}$ put

$$C_n = \bigcup_{i=1}^{m(n)} \overline{U}_{n,i},$$

and let

$$Y = \bigcap_{n=1}^{\infty} C_n.$$

We claim that Y is homeomorphic to the closed unit interval I. Observe that Y is a compact subset of X and that $x, y \in Y$.

CLAIM 1: Y is a continuum.

This is clear since by construction for each n, C_n is a continuum and by (3), $C_{n+1} \subseteq C_n$.

CLAIM 2: For every $z \in Y \setminus \{x,y\}$ there exist closed sets L_z and U_z of Y such that

(6) $L_z \cup U_z = Y$,
(7) $x \in L_z$, $y \in U_z$, and
(8) $L_z \cap U_z = \{z\}$.

Let $z \in Y \setminus \{x,y\}$. For each n, at least one and at most two of the $U_{n,i}$'s contain z. Let A_n be the union of all the $U_{n,i}$'s preceding these and B_n be the union of all the $U_{n,i}$'s following these. Put

$$A = Y \cap \bigcup_{n=1}^{\infty} A_n \qquad \text{and} \qquad B = Y \cap \bigcup_{n=1}^{\infty} B_n.$$

Observe that both A and B are open subsets of Y. Now if $d \in Y \setminus \{z\}$ then (2) easily implies that for certain n, $d \in A_n \cup B_n$. So $A \cup B = Y \setminus \{z\}$. We claim that $A \cap B = \emptyset$. To the contrary, assume that for $n, p \in \mathbb{N}$ there exists $d \in A_n \cap B_p$. Without loss of generality, $n \leq p$. Let $i_1 \leq m(p)$ be such that $z \in U_{p,i_1}$. Since $d \in B_p$ there exists i such that $i_1 < i \leq m(p)$ and $d \in U_{p,i}$. By (3) there exist $j_1 \leq j \leq m(n)$ such that

$$U_{p,i_1} \subseteq U_{n,j_1} \text{ and } U_{p,i} \subseteq U_{n,j}.$$

Suppose that $z \in U_{n,j}$. Then by construction, $U_{n,j} \cap (A_n \cup B_n) = \emptyset$, which is impossible since $d \in A_n \cap U_{n,j}$. Therefore $z \notin U_{n,j}$. Now since

$$z \in U_{p,i_1} \subseteq U_{n,j_1} \text{ and } j_1 \leq j,$$

we obtain $U_{n,j} \subseteq B_n$ which is a contradiction since $d \in U_{n,j} \cap A_n$ and $A_n \cap B_n = \emptyset$.

Now put $L_z = A \cup \{z\}$ and $U_z = B \cup \{z\}$, respectively.

CLAIM 3: For every $z \in Y \setminus \{x,y\}$, L_z and U_z are continua.

Suppose that for certain $z \in Y \setminus \{x,y\}$ e.g. L_z is not a continuum. Then L_z can be written as the union of two disjoint nonempty relatively open sets, say E and F. Without loss of generality, $z \in E$. Then F is contained in A and since A is open in Y, this easily implies that F is open in Y. However, since F is closed in L_z and L_z is closed in Y, it also follows that F is closed in Y. Consequently, F is a nonempty clopen subset of Y, which contradicts the fact that Y is connected (claim 1).

Define an order $\preccurlyeq$ on $Y \setminus \{x,y\}$ by putting: $d \preccurlyeq e$ iff $L_d \subseteq L_e$. Observe that by a simple connectivity argument one obtains $d \preccurlyeq e$ iff $U_e \subseteq U_d$. Finally, for every $d \in Y$ put $x \preccurlyeq d \preccurlyeq y$.

CLAIM 4: $\preccurlyeq$ is a linear order on Y.

This follows easily by (6), (7) and (8) and claim 3.

Let D be the set of all rational numbers between 0 and 1, i.e. $D = \mathbb{Q} \cap (0,1)$.

CLAIM 5: There is a subset $E \subseteq D$ and a one-to-one function $f: E \to Y \setminus \{x,y\}$ such that

(9) for all $d,e \in E$, $f(d) \preccurlyeq f(e)$ iff $d \leq e$,

(10) E is dense in I, and

(11) f(E) is dense in Y.

Let $\mathcal{B} = \{B_n : n \in \mathbb{N}, n \text{ even}\}$ be a basis for Y consisting of nonempty open sets. In addition, let $\mathcal{F} = \{F_n : n \in \mathbb{N}, n \text{ odd}\}$ be a basis for D, also consisting of nonempty open sets. By induction on n we shall pick a point $e_n \in D$ and a point $f(e_n) \in Y$ such that the following conditions are satisfied:

(12) if n is odd then $e_n \in F_n$,

(13) if n is even then $f(e_n) \in B_n$, and

(14) $f \mid \{e_1, \cdots, e_n\}$ is one-to-one and order preserving (in the sense of (9)).

For $n = 1$ pick an arbitrary point $e_1 \in F_1$ and an arbitrary point $f(e_1) \in Y \setminus \{x,y\}$. Observe that by connectivity of Y such a choice is possible. Now suppose that we completed the construction of the points $e_1,\cdots,e_n$ and $f(e_1),\cdots,f(e_n)$. There are two cases to consider of course. Suppose first that n+1 is odd. Pick an arbitrary point $e_{n+1} \in F_{n+1}\setminus\{e_1,\cdots,e_n\}$ and define $G = \{m \le n\colon e_m \le e_{n+1}\}$ and $H = \{m \le n\colon e_m \ge e_{n+1}\}$, respectively. Let $a = \max\{f(e_m)\colon m \in G\}$ and $b = \min\{f(e_m)\colon m \in H\}$. Observe that $L_a \cap U_b = \emptyset$ and that L_a and U_b are both closed (claim 2) so that by connectivity of Y there exists a point $f(e_{n+1})$ in Y strictly between a and b (with respect to the order $\preccurlyeq$). It is easy to see that e_{n+1} and $f(e_{n+1})$ are as required. If n+1 is even then proceed by a similar argument.

Now fix $t \in I$ for a moment and put

$$\mathcal{A}_t = \{L_{f(d)}\colon d \in E \text{ and } t < d\} \cup \{U_{f(d)}\colon d \in E \text{ and } d < t\}.$$

CLAIM 6: $\mathcal{A}_t$ has the finite intersection property and $\cap\mathcal{A}_t$ consists of precisely one point.

That $\mathcal{A}_t$ has the finite intersection property follows easily from the definition of $\preccurlyeq$. Consequently, the compactness of Y implies that $\cap\mathcal{A}_t \neq \emptyset$ (observe that every L_d and U_d is closed in Y). Now suppose that there exist two distinct points a and b in $\cap\mathcal{A}_t$. Without loss of generality, $a \preccurlyeq b$. Since both L_a and U_b are closed in Y and clearly disjoint, by connectivity and by the fact that f(E) is dense, there exists $d \in E\setminus\{t\}$ such that f(d) lies strictly between a and b. Now if $t < d$ then $L_{f(d)} \in \mathcal{A}_t$ which implies that $b \in \cap\mathcal{A}_t \subseteq L_{f(d)}$, which is a contradiction. However, if $d < t$ then $U_{f(d)} \in \mathcal{A}_t$ which implies that $a \in \cap\mathcal{A}_t \subseteq U_{f(d)}$, which is also a contradiction.

Now define a function $g\colon I \to Y$ by

$$\{g(t)\} = \cap\mathcal{A}_t.$$

CLAIM 7: g is a homeomorphism.

First observe that g is order-preserving. Therefore, since $g \mid E = f$ and f is one-to-one, it follows from the denseness of E in I that g is injective. Also observe that the range of g is dense in Y since it contains f(E). By compactness of I it therefore suffices to prove that g is continuous (exercise 1.1.4). To this end, let $U \subseteq Y$ be open, and take an arbitrary point $t \in g^{-1}(U)$. Then $g(t) \in U$ so that by compactness of Y, claim 6 implies that there exist $d_0,d_1 \in E$ with $d_0 < t < d_1$ and

$L_{f(d_1)} \cap U_{f(d_0)} \subseteq U$.

A straightforward verification shows that $(d_0,d_1) \subseteq g^{-1}(U)$, i.e. $g^{-1}(U)$ is a neighborhood of t.□

We now finally come to the main result in this section.

5.3.14. THEOREM: *Let* X *be a compact space. The following statements are equivalent:*

(a) X *is a Peano continuum,*

(b) 2^X *is a Peano continuum,*

(c) 2^X *is an* **AR**.

PROOF: The equivalence (a) ⇔ (b) was proved in propositions 4.7.2 and 5.3.10. Since (c) ⇒ (b) is a triviality (exercise 1.5.3), it remains to verify the implication (b) ⇒ (c). We aim at applying theorem 5.2.12 to prove that 2^X is an **ANR**. Consider the base $\mathcal{B}(X)$ for 2^X identified in lemma 5.3.9(2). Since $\mathcal{B}(X)$ is closed under finite intersections (lemma 5.3.9(3)), it suffices to prove that every component of an element of $\mathcal{B}(X)$ is homotopically trivial. To this end, let $U_1,\cdots,U_n \subseteq X$ be nonempty and open, and consider a component C of $B = <\{U_1,\cdots,U_n\}>$. By theorem 5.3.13, C is path-connected. So for the verification that C is homotopically trivial, it suffices to consider a continuous function $g\colon S^n \to C$, where $n \geq 1$. By corollary 5.3.7, the function $\bar{g}\colon 2^{S^n} \to 2^X$ defined by

$$\bar{g}(A) = \cup_{x\in A}\, g(x)$$

is continuous. We claim that the range of $\bar{g}$ is contained in C. We shall first prove that the range of $\bar{g}$ is contained in B. To this end, take an arbitrary A in the hyperspace of S^n and let $x \in A$. Since by assumption g(x) meets every U_i and $\bar{g}(A)$ contains g(x), we conclude that $\bar{g}(A)$ meets every U_i. Now pick an arbitrary $y \in A$. Then by assumption, g(y) is contained in the union of the U_i's. From this and from the definition of $\bar{g}$ it follows that $\bar{g}(A)$ is contained in the union of the U_i's. Consequently, $g(A) \in B$. Next observe that $\bar{g}$ extends g. Consequently, by the fact that the hyperspace of S^n is connected (proposition 5.3.10), we conclude that the range of $\bar{g}$ is contained in C. By proposition 5.3.11, there exists a continuous function $f_n\colon B^{n+1} \to 2^{S^n}$ such that $f_n(x) = \{x\}$ for every $x \in S^n$. Let $g\colon B^{n+1} \to 2^X$ be the composition $\bar{g} \circ f_n$. An easy check shows that g is the required extension of f. By theorem 5.2.12 it therefore follows that 2^X is an **ANR**. However, implicitly we also proved that 2^X is homotopically trivial itself, since clearly $2^X = <\{X\}>$ and 2^X is connected. Consequently, theorem 5.2.15 implies that 2^X is an **AR**.□

Exercises for §5.3.

1. Let X be a locally connected compactum and let f: $X \to Y$ be a continuous surjection. Prove that Y is locally connected.

2. Let X be a path-connected space. Prove that for all $x,y \in X$ there exists an imbedding $\Phi: I \to X$ such that $\Phi(0) = x$ and $\Phi(1) = y$.

Let X be a compact space and put $C(X) = \{A \in 2^X: A \text{ is connected}\}$.

3. Let X be a compact space. Prove that C(X) is a closed subspace of 2^X.

4. Prove that C(I) is homeomorphic to I^2.

5. Prove that C(X) is connected iff X is connected.

6. Prove that C(X) is locally connected iff X is locally connected.

7. Prove that C(X) is an **AR** iff X is a Peano continuum.

8. Prove that $C(I^2)$ is infinite-dimensional.

5.4. Open Subspaces of ANR's

In this section we shall prove that every open subspace of an **ANR** is again an **ANR** and that every space that admits an open cover by **ANR**'s is itself an **ANR**. As an application we conclude that every polytope is an **ANR**. Along the way, we also derive the result that the cone over any **ANR** is an **AR**.

5.4.1. THEOREM: *Let* X *be an* **ANR** *and let* U *be an open subspace of* X. *Then* U *is an* **ANR**.

PROOF: Let Y be a space, $A \subseteq Y$ be closed, and let $g: A \to U$ be continuous. Since X is an **ANR**, there exists an open neighborhood V of A such that g can be extended to a continuous function $f: V \to X$. Since f is continuous and extends g, and since U is open in X, we conclude that $W = f^{-1}(U)$ is an open neighborhood of A. Clearly $\bar{g} = f \restriction W: W \to U$ is the required extension of g. □

Observe that $[-1,0) \cup (0,1]$ is not an **AR** but is an open subspace of the **AR** $[-1,1]$. We now aim at proving that a "local" **ANR** is an **ANR**. First we need to derive a few elementary results.

Let X be a space. Recall that the cone $\Delta(X)$ over X is $(X \times [0,1)) \cup \{\infty\}$ topologized as follows: points of the form (x,t) have their usual product neighborhoods and a basic neighborhood of ∞ has the form

$$(X \times (s,1)) \cup \{\infty\},$$

where $0 < s < 1$, cf. the exercises of §1.6. The following trivialities shall be used a few times in the sequel. If A is a subspace of X, then the subspace $(A \times [0,1)) \cup \{\infty\}$ of $\Delta(X)$ is the cone over A. Moreover, if $A \subseteq X$ is closed then $\Delta(A)$ is a closed subspace of $\Delta(X)$.

5.4.2. THEOREM: *Let X be a space. Then X is an* **ANR** *if and only if* $\Delta(X)$ *is an* **AR**.

PROOF: Assume that X is an **ANR**. $\Delta(X)$ is contractible, so by corollary 1.6.7 or theorem 5.2.15 it is enough to prove that $\Delta(X)$ is an **ANR**. Naturally, we aim at applying the characterization theorem 5.2.1. To that end, let $\mathcal{U}$ be an open cover of $\Delta(X)$. Fix an element $U \in \mathcal{U}$ such that $\infty \in U$. There clearly exists a contractible open neighborhood A of ∞ such that $A \subseteq U$. Now let B be a neighborhood of ∞ with $\overline{B} \subseteq A$. There exists an open refinement $\mathcal{U}_0$ of $\mathcal{U}$ such that for every $U_0 \in \mathcal{U}_0$ with $U_0 \cap B \neq \emptyset$, we have $U_0 \subseteq A$. Since X and [0,1) are **ANR**'s, by theorem 1.5.8 we conclude that $X \times [0,1)$ is an **ANR**. So there exists an open cover $\mathcal{V}$ of $X \times [0,1)$ such that for every locally finite simplicial complex $\mathcal{T}$ and every subcomplex $\mathcal{S}$ of $\mathcal{T}$ containing all the vertices of $\mathcal{T}$, every partial realization of $\mathcal{T}$ in $X \times [0,1)$ relative to $(\mathcal{S},\mathcal{V})$ can be extended to a full realization of $\mathcal{T}$ in $X \times [0,1)$ relative to $\mathcal{U}_0 \cap (X \times [0,1))$. Put $\mathcal{W} = \mathcal{V} \cup \{B\}$. Then $\mathcal{W}$ is an open cover of $\Delta(X)$. Now let $\mathcal{T}$ be a locally finite simplicial complex, let $\mathcal{S}$ be a subcomplex of $\mathcal{T}$ containing all the vertices of $\mathcal{T}$, and let $f: |\mathcal{S}| \to \Delta(X)$ be a partial realization of $\mathcal{T}$ in $\Delta(X)$ relative to $(\mathcal{S},\mathcal{W})$. Define

$$\mathcal{T}_0 = \{\sigma \in \mathcal{T}: (\exists V \in \mathcal{V})(f(\sigma \cap |\mathcal{S}|) \subseteq V)\} \text{ and } \mathcal{T}_1 = \mathcal{T} \setminus \mathcal{T}_0.$$

In addition, put $\mathcal{S}_0 = \mathcal{S} \cap \mathcal{T}_0$. It is clear that $\mathcal{T}_0$ is a subcomplex of $\mathcal{T}$ and that $\mathcal{S}_0$ contains all the vertices of $\mathcal{T}_0$. Also, $g = f \restriction |\mathcal{S}_0|$ is a partial realization of $\mathcal{T}_0$ relative to $(\mathcal{S}_0,\mathcal{V})$. Consequently, g can be extended to a full realization $\bar{g}$ of $\mathcal{T}_0$ in $X \times [0,1)$ relative to $\mathcal{U}_0 \cap (X \times [0,1))$.

Now put $\mathcal{S}_1 = \mathcal{S} \cup \mathcal{T}_0$ and define $f_1: |\mathcal{S}_1| \to \Delta(X)$ by

$$f_1(x) = \begin{cases} f(x) & (x \in |\mathcal{S}|), \\ \bar{g}(x) & (x \in |\mathcal{T}_0|). \end{cases}$$

It is clear that f_1 is well-defined and continuous. By a standard argument, by induction on $n \geq 0$, we shall construct a continuous function $\bar{f}_{n+1}: |\mathcal{S}_1 \cup \mathcal{T}^{(n)}| \to \Delta(X)$ having the following properties:

(1) $\bar{f}_1 = f_1$ and for $n \geq 1$, $\bar{f}_{n+1}$ extends $\bar{f}_n$,
(2) for every $\sigma \in \mathcal{T}^{(n)} \cap \mathcal{T}_1$, $\bar{f}_{n+1}(\sigma) \subseteq A$.

Observe that since $\mathcal{S}$ contains all the vertices of $\mathcal{T}$, our choice $\bar{f}_1 = f_1$ is possible. Now assume that for certain $n \geq 0$, the function $\bar{f}_{n+1}$ has been constructed. Naturally, we shall construct $\bar{f}_{n+2}$ "simplex-wise". To this end, take an arbitrary simplex $\sigma \in \mathcal{S}_1 \cup \mathcal{T}^{(n+1)}$. If $\sigma \in \mathcal{S}_1 \cup \mathcal{T}^{(n)}$ then put $f_\sigma = \bar{f}_{n+1} \mid \sigma$. If $\sigma \in \mathcal{T}^{(n+1)} \setminus (\mathcal{S}_1 \cup \mathcal{T}^{(n)})$, then σ belongs to $\mathcal{T}_1$, from which it follows that $f(\sigma \cap |\mathcal{S}|) \subseteq B$.

CLAIM: $\bar{f}_{n+1}(\partial\sigma) \subseteq A$.

Let τ be a proper face of σ. There are two cases to consider.

CASE 1: $\tau \in \mathcal{T}_0$.

Then there exists an element $U \in \mathcal{U}_0 \cap (X \times [0,1))$ such that $\bar{f}_{n+1}(\tau) = f_1(\tau) \subseteq U$. Since, as was just observed, f_1 maps every vertex of τ in B, we conclude that $U \cap B \neq \emptyset$. Consequently, by the special choice of $\mathcal{U}_0$ we obtain, $\bar{f}_{n+1}(\tau) \subseteq A$.

CASE 2: $\tau \notin \mathcal{T}_0$.

Then by our inductive hypothesis (2) we get $\bar{f}_{n+1}(\tau) \subseteq A$.

Now since A is contractible, there exists a continuous extension $f_\sigma: \sigma \to A$ of $\bar{f}_{n+1} \mid \partial\sigma$ (lemma 5.2.11). Define $\bar{f}_{n+2}: |\mathcal{S}_1 \cup \mathcal{T}^{(n+1)}| \to \Delta(X)$ by

$$\bar{f}_{n+2}(x) = f_\sigma(x) \qquad (x \in \sigma \in \mathcal{S}_1 \cup \mathcal{T}^{(n+1)}).$$

An application of lemma 3.6.6 establishes the required continuity of $\bar{f}_{n+2}$. This completes the inductive construction.

Now define $\bar{f}: |\mathcal{T}| \to \Delta(X)$ by

$$\bar{f}(x) = \bar{f}_{n-1}(x) \qquad (x \in |\mathcal{T}^{(n)}|).$$

Applying lemma 3.6.6 again gives us that $\bar{f}$ is a full realization of $\mathcal{T}$ in $\Delta(X)$ relative to $\mathcal{U}$. Since $\bar{f}$ clearly extends f, we are done.

Conversely, assume that $\Delta(X)$ is an **AR**. Since $X \times [0,1)$ is an open subset of $\Delta(X)$, theorem 5.4.1 implies that $X \times [0,1)$ is an **ANR**. Theorem 1.5.8 now yields that X is an **ANR**.□

We now come to the following

5.4.3. PROPOSITION: *Let* X *be a space and assume that* X *can be written as* $U \cup V$, *where both* U *and* V *are open in* X. *If* U *and* V *are* **ANR**'s *then so is* X.

PROOF: We shall prove that the cone over X is an **AR** so that X is an **ANR** by theorem 5.4.2. By proposition 4.3.3 there exist closed sets E and F in X such that $E \subseteq U$, $F \subseteq V$ and $E \cup F = X$. Observe that $\Delta(E) \subseteq \Delta(U)$, $\Delta(F) \subseteq \Delta(V)$, and $\Delta(E) \cup \Delta(F) = \Delta(X)$.

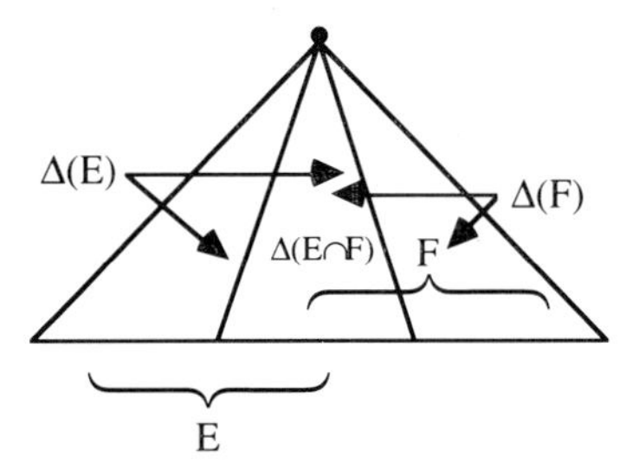

Figure 5.4.1.

Now let Y be a space, let $A \subseteq Y$ be closed, and let $f: A \to \Delta(X)$ be continuous. Put $E' = f^{-1}(\Delta(E))$ and $F' = f^{-1}(\Delta(F))$, respectively. By lemma 4.1.3 there exist closed sets E'' and F'' in Y such that $E'' \cap A = E'$, $F'' \cap A = F'$ and $E'' \cup F'' = Y$. Since $U \cap V$ is an open subspace of U, it is an **ANR** by theorem 5.4.1. Consequently, theorem 5.4.2 implies that the function $f \mid E' \cap F'$: $E' \cap F' \to \Delta(E \cap F)$ can be extended to a continuous function $g: E'' \cap F'' \to \Delta(U \cap V)$. Now define $f_E: E' \cup (E'' \cap F'') \to \Delta(U)$ by

$$f_E(x) = \begin{cases} f(x) & (x \in E'), \\ g(x) & (x \in E'' \cap F''). \end{cases}$$

Then f_E is obviously continuous. By another application of theorem 5.4.2, there exists a continuous extension f_1: E" $\to \Delta(U)$ of f_E. Similarly define f_F and a continuous extension f_2: F" $\to \Delta(V)$ of f_F. Now define $\bar{f}$: $Y \to \Delta(X)$ in the obvious way, namely,

$$\bar{f}(x) = \begin{cases} f_1(x) & (x \in E''), \\ f_2(x) & (x \in F''). \end{cases}$$

Then $\bar{f}$ is the required continuous extension of f.□

5.4.4. LEMMA: *Let* X *be a space having an open cover* $\mathcal{U}$ *by pairwise disjoint* **ANR**'s. *Then* X *is an* **ANR**.

PROOF: This is a triviality. Assume that X is a closed subspace of a space Y. By lemma 5.2.2, for every $U \in \mathcal{U}$ there exists an open subset V(U) of Y such that

(1) $V(U) \cap X = U$, and
(2) the collection $\{V(U): U \in \mathcal{U}\}$ is pairwise disjoint.

Note that U is closed in V(U) for every $U \in \mathcal{U}$, so by our assumptions on $\mathcal{U}$, for every $U \in \mathcal{U}$ there exist an open subset W(U) of V(U) which contains U and a retraction r_U: $W(U) \to U$. Observe that the W(U)'s are open in Y since the V(U)'s are. Now put $W = \bigcup_{U \in \mathcal{U}} W(U)$. Then W is a neighborhood of X in Y and the function r: $W \to X$ defined by

$$r(x) = r_U(x) \qquad (x \in W(U),\ U \in \mathcal{U})$$

is a retraction.□

We now come to the following

5.4.5. THEOREM: *Let* X *be a space and suppose that* X *admits an open cover* $\mathcal{U}$ *consisting of* **ANR**'s. *Then* X *is an* **ANR**.

PROOF: Without loss of generality assume that $\mathcal{U}$ is countable. Enumerate $\mathcal{U}$ as $\{U_n: n \in \mathbb{N}\}$. Since by proposition 5.4.3 the union of finitely many elements of $\mathcal{U}$ is an **ANR**, we may assume without loss of generality that for every n, $U_n \subseteq U_{n+1}$. Now for every n, put

$$V_n = \{x \in X: d(x, X \backslash U_n) > 1/n\}.$$

It is clear that each V_n is open, that $\overline{V}_n \subseteq V_{n+1}$, and that $\bigcup_{n=1}^{\infty} V_n = X$.
Finally define

$$R_n = \begin{cases} V_n & (n = 1,2), \\ V_n \setminus \overline{V}_{n-2} & (n \geq 3). \end{cases}$$

Then clearly $X = \bigcup_{n=1}^{\infty} R_n$ and for $|m-n| \geq 2$, $R_n \cap R_m = \emptyset$. Also, R_n is an open subset of the **ANR** U_n. So by theorem 5.4.1, we conclude that R_n is an **ANR**. Consequently, lemma 5.4.4 implies that

$$E = \bigcup_{n=1}^{\infty} R_{2n-1} \text{ and } F = \bigcup_{n=1}^{\infty} R_{2n}$$

are open **ANR** subspaces of X. Since their union equals X, we infer by proposition 5.4.3 that X is an **ANR**.□

As announced in §3.6, we now get

5.4.6. COROLLARY: *Every polytope is an* **ANR**.

PROOF: Let $\mathcal{T}$ be a locally finite simplicial complex and consider $P = |\mathcal{T}|$. Take an arbitrary $x \in |\mathcal{T}^{(0)}|$. Since $\mathcal{T}$ is locally finite, the collection $\mathcal{S} = \{\tau \in \mathcal{T}: x \in \tau\}$ is finite. Consequently, $\mathcal{P} = \{\sigma \in \mathcal{T}: (\exists \tau \in \mathcal{S})(\sigma \preccurlyeq \tau)\}$ is a finite subcomplex of $\mathcal{T}$. From theorem 3.6.11 we conclude that $|\mathcal{P}|$ is an **ANR**. Since St x, the star of x, is contained in $|\mathcal{P}|$ and is open in P by corollary 3.6.4, we conclude from theorem 5.4.1 that St x is an **ANR**. Since by corollary 3.6.4, $\{\text{St } x: x \in |\mathcal{T}^{(0)}|\}$ is an open cover of P, theorem 5.4.5 implies that P is an **ANR**.□

Exercises for §5.4.

1. Give an example of a space X which can be written as $E \cup F$, where both E and F are closed **ANR** subspaces of X, such that X is not an **ANR**.

2. Let X be the union of two open subspaces U and V such that U, V and $U \cap V$ are **AR**'s. Prove that X is an **AR**.

3. Let X be a space having a point x such that (1) x has arbitrarily small homotopically trivial neighborhoods, and (2) $X \setminus \{x\}$ is an **ANR**. Prove that X is an **ANR**.

5.5. Characterization of Finite-Dimensional ANR's and AR's

In this section we shall characterize the class of all *finite-dimensional* **ANR**'s and **AR**'s. The characterizations presented here are considerably simpler to state than the ones in §5.2.

Let X be a space and let $0 \le n < \infty$. We say that X is *connected in dimension n*, abbreviated $\mathbf{C^n}$, provided that for every $0 \le m \le n$, every continuous function f: $S^m \to X$ extends to a continuous function $\bar{f}$: $B^{m+1} \to X$. So the homotopically trivial spaces are precisely those spaces that are connected in every dimension. In addition, we say that X is *locally connected in dimension n*, abbreviated $\mathbf{LC^n}$, provided that for every $x \in X$ and for every neighborhood U of x and for every $0 \le m \le n$ there exists a neighborhood V of x such that every continuous function f: $S^m \to V$ extends to a continuous function $\bar{f}$: $B^{m+1} \to U$. We shall prove that if dim X = $n < \infty$ then X is an **ANR** if and only if X is $\mathbf{LC^n}$ and also that X is an **AR** if and only if X is both $\mathbf{LC^n}$ and $\mathbf{C^n}$.

The proof of the following lemma is precisely the same as the proof of lemma 5.2.11 and is therefore left as an exercise to the reader.

5.5.1. LEMMA: *Let* X *be a space. If* X *is locally contractible then* X *is* $\mathbf{LC^n}$ *for every* n.□

5.5.2. PROPOSITION: *Let* X *be a space and let* $0 \le n < \infty$. *The following statements are equivalent:*

(a) X *is* $\mathbf{LC^n}$,

(b) *for every open cover* $\mathcal{U}$ *of* X *there exists an open refinement* $\mathcal{V}$ *of* $\mathcal{U}$ *such that for every locally finite simplicial complex* $\mathcal{T}$ *with* dim $|\mathcal{T}| \le n+1$ *and every subcomplex* $\mathcal{S}$ *of* $\mathcal{T}$ *containing all the vertices of* $\mathcal{T}$, *every partial realization of* $\mathcal{T}$ *in* X *relative to* $(\mathcal{S},\mathcal{V})$ *can be extended to a full realization of* $\mathcal{T}$ *in* X *relative to* $\mathcal{U}$.

PROOF: *We prove (b)* $\Rightarrow$ *(a).*

Take an arbitrary $x \in X$ and a neighborhood U of X. There exists an open neighborhood W of x such that $\overline{W} \subseteq U$. Consider the open cover $\mathcal{U} = \{U, X\backslash\overline{W}\}$ of X. By assumption, for the open cover $\mathcal{U}$ there exists an open refinement $\mathcal{V}$ such as in (b). Pick $V \in \mathcal{V}$ such that $x \in V$. Fix an integer $m \le n$, and let σ be an arbitrary (m+1)-dimensional simplex in $\mathbb{R}^{m+1}$. It will be convenient to identify S^m and $\partial\sigma$ (exercise 3.5.8(c)). Now let $\mathcal{F}(\sigma)$ be the simplicial complex consisting of all the faces of σ and let $\mathcal{S}$ be the subcomplex consisting of all proper faces. Consider a continuous function f: $\partial\sigma \to W \cap V$ and observe that f is a partial realization of $\mathcal{F}(\sigma)$ in X relative to $(\mathcal{S},\mathcal{V})$. By assumption, f can be extended to a full realization $\bar{f}$: $\sigma = |\mathcal{F}(\sigma)| \to X$ relative to $\mathcal{U}$. Consequently, there exists $U' \in \mathcal{U}$ with $\bar{f}(\sigma) \subseteq U'$. Since $\bar{f}$ extends f, $U' \cap W \ne \emptyset$, i.e. U' = U.

We prove (a) $\Rightarrow$ *(b).*

Let $\mathcal{U}$ be an open cover of X. By downward induction, we shall construct open covers $\mathcal{U}_n, \mathcal{V}_n, \mathcal{U}_{n-1}, \mathcal{V}_{n-1}, \cdots, \mathcal{U}_0, \mathcal{V}_0$ of X having the following properties:

(1) $\mathcal{U}_n = \mathcal{U}$,

(2) if $0 \le i \le n$ then for every $V \in \mathcal{V}_i$ and for every continuous function $f: S^i \to V$ there exist $U \in \mathcal{U}_i$ with $V \subseteq U$ and a continuous extension $\bar{f}: B^{i+1} \to U$ of f,

(3) if $0 \le i \le n-1$ then $\mathcal{U}_i$ is a star-refinement of $\mathcal{V}_{i+1}$

(Observe that (2) implies $\mathcal{V}_i < \mathcal{U}_i$). The construction of these covers is a triviality. By lemma 5.1.7 there is never trouble with the construction of the $\mathcal{U}_n$'s. Let us construct the cover $\mathcal{V}_n$. Take an arbitrary $x \in X$ and pick $U_x \in \mathcal{U}_n$ containing x. Since X is $\mathbf{LC^n}$ there is an open neighborhood V_x of x such that $V_x \subseteq U_x$ while moreover every continuous function $f: S^n \to V_x$ can be extended to a continuous function $\bar{f}: B^{n+1} \to U_x$. The cover $\mathcal{V}_n = \{V_x : x \in X\}$ is clearly as required. The construction of the other $\mathcal{V}_i$'s is similar.

Now put $\mathcal{V} = \mathcal{V}_0$, let $\mathcal{T}$ be a locally finite simplicial complex with dim $|\mathcal{T}| \le n+1$, let $\mathcal{S}$ be a subcomplex of $\mathcal{T}$ containing all the vertices of $\mathcal{T}$, and let $f: |\mathcal{S}| \to X$ be a partial realization of $\mathcal{T}$ in X relative to $(\mathcal{S},\mathcal{V})$. By induction, we shall construct for every $0 \le i \le n$ a continuous function $f_i: |\mathcal{S} \cup \mathcal{T}^{(i)}| \to X$ having the following properties:

(4) $f_0 = f$ and for $1 \le i \le n$, f_i extends f_{i-1},

(5) for $0 \le i \le n$, f_i is a partial realization of $\mathcal{T}$ in X relative to $(\mathcal{S} \cup \mathcal{T}^{(i)}, \mathcal{V}_i)$.

Since $\mathcal{S}$ contains all the vertices of $\mathcal{T}$, our choice $f_0 = f$ is as required. Now assume that for certain $1 \le i \le n$ the function f_{i-1} has been constructed. As in the previous sections, we shall construct f_i "simplex-wise". Let σ be an element of $\mathcal{T}^{(i)} \setminus (\mathcal{S} \cup \mathcal{T}^{(i-1)})$. By (5) there exists an element $V \in \mathcal{V}_{i-1}$ such that $f_{i-1}(\partial\sigma) \subseteq V$. So by (2) there exist $U \in \mathcal{U}_{i-1}$ with $V \subseteq U$ and a continuous extension $f_\sigma: \sigma \to U$ of the function $f_{i-1} \mid \partial\sigma$. Now define $f_i : |\mathcal{S} \cup \mathcal{T}^{(i)}| \to X$ as follows:

$$f_i(x) = \begin{cases} f_{i-1}(x) & (x \in |\mathcal{S} \cup \mathcal{T}^{(i-1)}|), \\ f_\sigma(x) & (x \in \sigma \in \mathcal{T}^{(i)} \setminus (\mathcal{S} \cup \mathcal{T}^{(i-1)})). \end{cases}$$

By lemma 3.6.6 f_i is continuous. We now claim that f_i is a partial realization of $\mathcal{T}$ in X relative to $(\mathcal{S} \cup \mathcal{T}^{(i)}, \mathcal{V}_i)$. To this end, take an arbitrary simplex $\sigma \in \mathcal{T}$. By (5) there exists $V \in \mathcal{V}_{i-1}$ such that $f_{i-1}(\sigma \cap |\mathcal{S} \cup \mathcal{T}^{(i-1)}|) \subseteq V$. Now let τ be a face of σ such that $\tau \in \mathcal{T}^{(i)}$. By construction there exists $U \in \mathcal{U}_{i-1}$ such that $f_i(\tau) \subseteq U$. Observe that V intersects U. Consequently,

$$f_i(\sigma \cap |\mathcal{S} \cup \mathcal{T}^{(i)}|) \subseteq \mathrm{St}(V, \mathcal{U}_{i-1}).$$

Now since $\mathcal{V}_{i-1} < \mathcal{U}_{i-1}$ and by (3) $\mathcal{U}_{i-1} \overset{*}{<} \mathcal{V}_i$, we are done. This completes the inductive construction.

With the same technique it is clear that we can extend f_n to a continuous function

$$f_{n+1}: |\mathcal{S} \cup \mathcal{T}^{(n+1)}| \to X$$

such that for every $\sigma \in \mathcal{T}^{(n+1)}$ there exists $U \in \mathcal{U}_n = \mathcal{U}$ such that $f_{n+1}(\sigma) \subseteq U$. However, by assumption $\mathcal{T}^{(n+1)} = \mathcal{T}$ so that f_{n+1} is the required full realization of $\mathcal{T}$ in X relative to $\mathcal{U}$. □

We now come to the main result in this section.

5.5.3. THEOREM: *Let* X *be a space and let* $0 \le n < \infty$. *The following statements are equivalent:*

(a) X *is* **LC^n**,

(b) *for every space* Y *and for every closed subspace* A *of* Y *with* $\dim(Y\setminus A) \le n+1$, *every continuous function* $f: A \to X$ *can be continuously extended over a neighborhood of* A,

(c) *for every* $x \in X$ *and for every neighborhood* U *of* x *there exists a neighborhood* V *of* x *with* $V \subseteq U$ *such that for every space* Y *with* $\dim Y \le n$, *for every continuous function* $f: Y \to V$ *there exists a homotopy* $H: Y \times I \to U$ *such that* $H_0 = f$ *and* H_1 *is a constant function.*

PROOF: *We prove (a) ⇒ (b).*

Let Y be a space, let $A \subseteq Y$ be closed such that $\dim(Y\setminus A) \le n+1$ and let $f: A \to X$ be continuous. Without loss of generality we assume that $X \cap Y = \emptyset$.

CLAIM: The set $C = (Y\setminus A) \cup X$ can be topologized in such a way that

(1) X is a closed subspace of C,

(2) $Y\setminus A$ is a subspace of C, and

(3) the function $\bar{f}: Y \to C$ defined by

$$\bar{f}(x) = \begin{cases} z & (z \in Y\setminus A), \\ f(z) & (z \in A). \end{cases}$$

is continuous.

Let $\sigma: \tau A \to \tau Y$ be a function such as in lemma 5.2.2 and let $\mathcal{B}$ be a countable open basis for X. In addition, let $\mathcal{F}$ be a countable open basis for $Y\setminus A$. For every $B \in \mathcal{B}$ and $n \in \mathbb{N}$

define $B(n) \subseteq C$ by

$$B(n) = \{y \in (Y \backslash A) \cap \sigma(f^{-1}(B)) : d(y,A) < 1/n\} \cup B.$$

An easy check shows that the collection $\mathcal{F} \cup \{B(n) : B \in \mathcal{B}, n \in \mathbb{N}\}$ serves as a basis for the required topology on C.

We shall now prove that X is a neighborhood retract of C. Once this has been established, by part (3) of the claim we are done of course.

The proof of our assertion is precisely the same as the proof of theorem 5.2.1. Only two minor changes should be made. We shall adopt the notation and the terminology introduced in the proof of theorem 5.2.1, we shall indicate the required changes and we shall leave the precise verification to the reader. By proposition 5.5.2 we have a realization property available up to dimension n+1. So in the construction of the covers $\mathcal{A}_n$ and $\mathcal{B}_n$, one should replace "locally finite simplicial complex $\mathcal{T}$" by "locally finite simplicial complex $\mathcal{T}$ with $\dim |\mathcal{T}| \leq n+1$" everywhere. The special open cover $\mathcal{U}$ of $C \backslash X$ used in the proof of theorem 5.2.1 should have the additional property that its order does not exceed n+1. However, by the fact that $\dim(C \backslash X) \leq n+1$, this can be achieved by theorem 4.3.5(e). This implies that the nerve $N(\mathcal{U})$ has no simplexes of dimension greater than n+1. With these two small changes, the proof of theorem 5.2.1 can now be copied literally.

We prove (b) ⇒ (c).

Suppose that (c) is not true at the point p. Then there exists $\varepsilon > 0$ such that for every $i \in \mathbb{N}$ there exist a space Y_i with $\dim Y_i \leq n$ and a continuous function $f_i: Y_i \to B(p,\varepsilon/i)$ that is not homotopic within $B(p,\varepsilon)$ to a constant function. By theorem 1.4.18 we may assume that Y_i is a subspace of a space Q_i which is homeomorphic to the Hilbert cube. Define

$$Y = \left(\sum_{i=1}^{\infty}(Q_i \times I)\right) \cup \{\infty\},$$

where "Σ" means topological sum and "$\cdots \cup \{\infty\}$" means the one-point compactification of the locally compact space "$\cdots$". In addition, define A' and Y' by

$$A' = \left(\sum_{i=1}^{\infty}(Y_i \times \{0\}) \cup (Y_i \times \{1\})\right) \cup \{\infty\},$$

and

$$Y' = \left(\sum_{i=1}^{\infty}(Y_i \times I)\right) \cup \{\infty\},$$

and consider them to be subspaces of Y. Observe that every neighborhood of ∞ in Y' contains all but finitely many of the $Y_i \times I$. Now define f: $A' \to X$ as follows

$$\begin{cases} f(x,0) = f_i(x) & (x \in Y_i), \\ f(x,1) = p & (x \in Y_i), \\ f(\infty) = p. \end{cases}$$

Then f is clearly continuous. Observe that by theorems 4.5.9 and 4.3.8 we have $\dim(Y'\backslash A') \leq n+1$ so that by assumption there exists a neighborhood W of A' in Y' such that f can be extended to a continuous function $\bar{f}$: $W \to X$. Then $V = \bar{f}^{-1}(B(p,\varepsilon))$ is a neighborhood of ∞ in Y and therefore contains all but finitely many of the $Y_i \times I$, $i \in \mathbb{N}$. But this implies that for all but finitely many $i \in \mathbb{N}$ the function f_i is homotopic within $B(p,\varepsilon)$ to a constant function. This is a contradiction.

We prove (c) ⇒ (a).

Since $\dim S^m = m$ for all m (theorem 4.3.10), this implication is a triviality. □

5.5.4. THEOREM: *Let* X *be a space and let* $0 \leq n < \infty$. *The following statements are equivalent:*

(a) X *is* **LC^n^** *and* **C^n^**,

(b) *for every space* Y *and for every closed subspace* A *of* Y *with* $\dim(Y\backslash A) \leq n+1$, *every continuous function* f: $A \to X$ *can be continuously extended over* Y,

(c) X *is* **LC^n^** *and for every space* Y *with* $\dim Y \leq n$, *every continuous function* f: $Y \to X$ *is nullhomotopic.*

PROOF: The proof of (a) ⇒ (b) is precisely the same as the proof of the implication (c) ⇒ (a) in the proof of theorem 5.2.15 and is therefore left as an exercise to the reader.

We shall now prove (b) ⇒ (c). That X is **LC^n^** follows from theorem 5.5.3. Let Y be a space with $\dim Y \leq n$ and assume that f: $Y \to X$ is continuous. Since by theorems 4.5.9 and 4.3.7, clearly $\dim \Delta(Y) \leq n+1$ (here $\Delta(Y)$ denotes the cone over Y of course), by (b) we conclude that f can be extended to a continuous function $\bar{f}$: $\Delta(Y) \to X$. The contractibility of $\Delta(Y)$ (exercise 1.6.4) implies that $\bar{f}$ is nullhomotopic. From this it follows easily that f is nullhomotopic as well.

The proof of (c) ⇒ (a) is a triviality of course since $\dim S^m = m$ for every m (theorem 4.3.10). □

These results have the following corollaries.

5.5.5. COROLLARY: *Let* X *be a space. The following statements are equivalent:*

(a) X *is a Peano continuum,*

(b) *there is a continuous surjection* f: $I \to X$.

PROOF: Assume first that X is a Peano continuum. By theorem 4.2.9, there exists a continuous surjection g: $C \to X$. In addition, by theorem 5.3.13, X is $\mathbf{LC^0}$ and $\mathbf{C^0}$. Consequently, an easy application of theorem 5.5.4 yields that g can be extended to a continuous function f: $I \to X$.

The simple proof of (b) $\Rightarrow$ (a) is left as an exercise to the reader. □

Before we can state our characterization theorem for finite-dimensional **ANR**'s and **AR**'s, we need to derive one more simple lemma, cf. exercise 1.6.6.

5.5.6. LEMMA: *Let* $0 \le n < \infty$ *and let* A *be a subset of* S^n. *Then* $B^{n+1}\backslash A$ *is an* **AR**.

PROOF: Put $Z = B^{n+1}\backslash A$. Let Y be a space, let B be a closed subset of Y, and let f: $B \to Z$ be continuous. Since B^{n+1} is an **AR** by theorem 1.5.1, there exists a continuous extension g: $Y \to B^{n+1}$ of f. Let d be an admissible metric for Y which is bounded by 1, i.e. $\mathrm{diam}(Y) \le 1$ (cf. exercise 1.1.7). Now define $\bar{f}$: $Y \to Z$ by

$$\bar{f}(y) = (1 - d(y,B))\cdot g(y).$$

An easy check shows that $\bar{f}$ is the required continuous extension of f. □

We now come to the following characterization theorem:

5.5.7. THEOREM: *Let* X *be a space, let* $0 \le n < \infty$ *and let* $\dim X \le n$. *Then*

(1) X *is an* **ANR** *iff* X *is locally contractible iff* X *is* $\mathbf{LC^n}$, *and*

(2) X *is an* **AR** *iff* X *is* $\mathbf{LC^n}$ *and* $\mathbf{C^n}$.

PROOF: We shall first present a proof of (1). To this end, first observe that if X is an **ANR** then X is locally contractible (exercise 1.6.2) and consequently $\mathbf{LC^n}$ by lemma 5.5.1. Conversely, assume that X is $\mathbf{LC^n}$. Then X is locally contractible by theorem 5.5.3(c). Consequently, X is $\mathbf{LC^m}$ for every m (lemma 5.5.1). Now let $m = 2n+1$. By theorem 4.4.4, we may assume that X is a subspace of S^m. Put $Z = \{x \in B^{m+1}: \|x\| < 1\} \cup X$. Then Z is an **AR** by lemma 5.5.6 and X is clearly closed in Z. Since $\dim\{x \in B^{m+1}: \|x\| < 1\} \le m+1$, by theorem 5.5.3 it follows that X is a neighborhood retract of Z. Since Z is an **AR**, exercise 1.5.1 implies that X is an **ANR**.

We shall now present a proof of (2). To this end, assume that X is $\mathbf{LC^n}$ and $\mathbf{C^n}$. By theorem

5.5.4 we conclude that X is contractible and hence $\mathbf{C^m}$ for every m (lemma 5.2.11). Now define Z as in the proof for (1). By theorem 5.5.4(b) there is a retraction r: $Z \to X$ from which we conclude that X is an **AR**, as required. □

5.5.8. *Remark:* Unfortunately, a characterization of infinite-dimensional **ANR**'s and **AR**'s as simple as theorem 5.5.7 does not seem to be possible. Borsuk constructed an example of a contractible and locally contractible compactum that is not an **ANR**. For details see Borsuk [29] and Hu [74].

Exercises for §5.5.

1. Let X be a space and $0 \le n < \infty$. Prove that X is $\mathbf{LC^n}$ if and only if for every open cover $\mathcal{U}$ of X there exists an open refinement $\mathcal{V}$ of $\mathcal{U}$ such that for every space Y with dim $Y \le n$, every two $\mathcal{V}$-close maps f,g: $Y \to X$ are $\mathcal{U}$-homotopic.

5.6. Adjunction Spaces of Compact A(N)R's

Let X and Y be compact spaces, let $A \subseteq X$ be closed and let f: $A \to Y$ be continuous. Consider the topological sum $Z = X \oplus Y$ and for every $x \in f(A)$ identify the set $f^{-1}(x) \cup \{x\}$ to a single point (see figure 5.6.1). The resulting quotient space shall be denoted by $X \cup_f Y$ and is called the *adjunction space* obtained by *adjoining* X to Y by means of the map f: $A \to Y$. Observe that the space $X \cup_f Y$ is separable and metrizable, cf. exercise 4.7.9.

If p: $Z \to X \cup_f Y$ is the natural quotient map then the restriction $p \mid Y$ is a closed imbedding. Consequently, we may think of Y as being a closed subspace of $X \cup_f Y$.

The aim of this section is to prove the following important:

5.6.1. THEOREM: *Let* X *and* Y *be compact* **A(N)R**'s. *In addition, let* $A \subseteq X$ *be a compact* **A(N)R** *and let* f: $A \to Y$ *be continuous. Then* $X \cup_f Y$ *is an* **A(N)R**.

Before we present the proof of our theorem, we derive the following lemmas first.

5.6.2. LEMMA: *Let* X *be an* **ANR** *and let* A *be a closed* **ANR** *subspace of* X. *Then for every open cover* $\mathcal{U}$ *of* X *there exists a* $\mathcal{U}$-*homotopy* H: $X \times I \to X$ *having the following properties:*

(1) H_0 *is the identity,*

(2) *for every* $t \in I$ *and* $a \in A$, $H_t(a) = a$,

(3) *there is an open neighborhood* U *of* A *in* X *such that* $H_1(U) = A$.

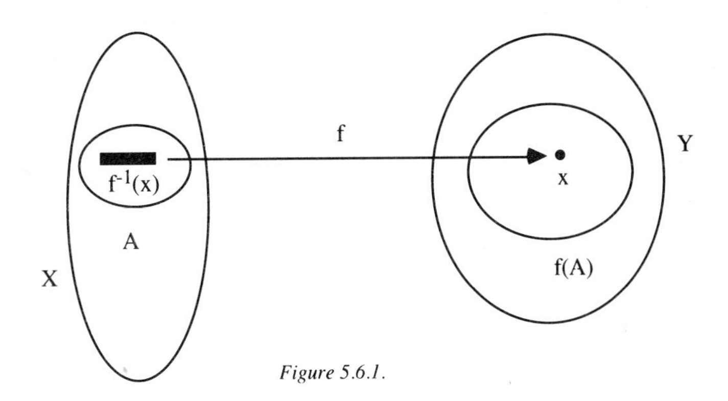

Figure 5.6.1.

PROOF: Since A is a closed **ANR** subspace of X, there exist a closed neighborhood V of A in X and a retraction r: V → A. Consider the closed subspace

$$P = (X \times \{0\}) \cup (A \times I) \cup (V \times \{1\})$$

of the product $X \times I$.

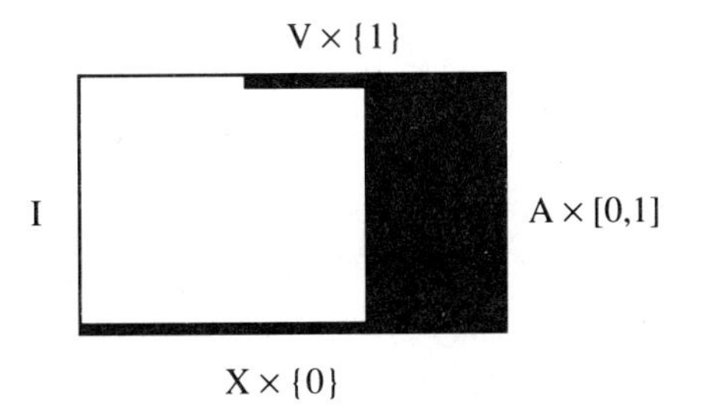

Figure 5.6.2.

Define a function f: P → X by

$$\begin{cases} f(x,0) = x & (x \in X), \\ f(x,t) = x & ((x,t) \in A \times I), \\ f(x,1) = r(x) & (x \in V). \end{cases}$$

It is clear that f is well-defined and continuous. Since X is an **ANR** there are a neighborhood N of P in $X \times I$ and a continuous extension $g: N \to X$ of f. By compactness of I, there is an open neighborhood W of A in X such that $W \times I \subseteq N$ (cf. the proof of lemma 1.6.2). We may assume without loss of generality that $W \subseteq V$ and that for every $x \in W$ the image $g(\{x\} \times I)$ is contained in an element of the open cover $\mathcal{U}$ (observe that for every $a \in A$, $g(\{a\} \times I)$ is a single point). There are an open neighborhood U of A in X such that

$$A \subseteq U \subseteq \overline{U} \subseteq W.$$

and a Urysohn function $\lambda: X \to I$ such that $\lambda \mid (X \backslash W) \equiv 0$ and $\lambda \mid \overline{U} \equiv 1$ (corollary 1.4.15). Now define the desired homotopy $H: X \times I \to X$ by

$$H(x,t) = g(x, t \cdot \lambda(x)).$$

An easy check shows that H is as required. □

Let X be a space and let $\mathcal{U}$ be an open cover of X. A *$\mathcal{U}$-deformation* is a $\mathcal{U}$-homotopy $H: X \times I \to X$ such that H_0 is the identity on X. If $\mathcal{U}$ is the cover of all open subsets of diameter at most $\varepsilon > 0$ then, for convenience, a $\mathcal{U}$-deformation is called an *ε-deformation*.

5.6.3. LEMMA: *Let* C *be an* **A(N)R**. *If* A *is a closed subset of* C *then the following statements are equivalent:*

(a) A *is an* **A(N)R**,

(b) *for each open cover $\mathcal{U}$ of* A *there exists a $\mathcal{U}$-deformation* $H: A \times I \to A$ *such that* H_1 *can be extended over (a neighborhood of* A *in)* C.

PROOF: The proof of (a) ⇒ (b) is a triviality since for any open cover $\mathcal{U}$ the homotopy $H(x,t) = x$ $(t \in I)$ is as required.

For the proof of (b) ⇒ (a), choose a sequence $(H(n))_n$ of 2^{-n}-deformations of A such that each $H(n)_1$ can be extended over (a neighborhood of A in) C. Let U be a neighborhood of A in C such that $H(1)_1$ can be extended to a continuous function $h: U \to A$ (in the **AR**-case, take U = C). Since C is an **A(N)R**, it suffices to prove that A is a retract of U (exercise 1.5.1).

To this end, for $n \in \mathbb{N}$ define $s_n = 1 - 2^{-(n-1)}$. We shall inductively construct a sequence of maps $\Phi_n: U \times I \to A$, $n \in \mathbb{N}$, having the following properties:

(1) $\Phi_1(u,t) = h(u)$ $\qquad (u \in U, t \in I)$,

(2) $\Phi_{n+1} \mid U \times [0,s_n] = \Phi_n \mid U \times [0,s_n]$ $\qquad (n \in \mathbb{N})$,

(3) $d(\Phi_{n+1} \mid A \times I, \Phi_n \mid A \times I) \leq 2^{-(n-1)}$ $\qquad (n \in \mathbb{N})$,

$$(4)\ \Phi_n(a,s) = H(n)_1(a) \qquad (n \in \mathbb{N},\ a \in A,\ s \geq s_n).$$

Since Φ_1 satisfies condition $(4)_1$, we can assume that the functions $\Phi_1,\cdots,\Phi_n$ have been defined for certain $n \in \mathbb{N}$. Consider the closed subspace

$$T = (U \times \{0\}) \cup (A \times I)$$

of the product $U \times I$. Define $g\colon T \to A$ by

$$g(x,t) = \begin{cases} \Phi_n(x,s_n) & (x \in U,\ t = 0), \\ H(n)_{1-t}(x) & (x \in A,\ t \in I). \end{cases}$$

Observe that by $(4)_n$ g is well-defined and therefore continuous.

CLAIM 1: The composition $H(n+1)_1 \circ g\colon T \to A$ can be extended over a neighborhood of T in $U \times I$.

By assumption there is an open neighborhood E of A in C such that $H(n+1)_1$ can be extended over E. Since an open subspace of an **ANR** is an **ANR** (theorem 5.4.1), and C is an **ANR** by assumption, there is a neighborhood F of T in $U \times I$ such that g can be extended to a continuous function from F into E. Then F is clearly the required neighborhood of T.

By claim 1 and exercise 1.6.7, we can extend $H(n+1)_1 \circ g$ to a continuous function $G\colon U \times I \to A$. Now define Φ_{n+1} as follows:

$$\Phi_{n+1}(x,s) = \Phi_n(x,s) \qquad (x \in U,\ s \leq s_n),$$

$$\Phi_{n+1}(x,(1-\lambda)s_n + \lambda s_{n+1}) = \begin{cases} H(n+1)_{2\lambda}(\Phi_n(x,s_n)) & (x \in U,\ 0 \leq \lambda \leq \frac{1}{2}), \\ G(x,2\lambda-1) & (x \in U,\ \frac{1}{2} \leq \lambda \leq 1), \end{cases}$$

$$\Phi_{n+1}(x,s) = G(x,1) \qquad (x \in U,\ s \geq s_{n+1}).$$

It is easy to see that Φ_{n+1} is well-defined and therefore continuous. Also, it is easy to see that (2) and (4) hold for n+1. It remains to check (3). To this end, take arbitrary $a \in A$ and $s \in I$. By (2) we may assume that $s \geq s_n$. By the definition of Φ_{n+1}, both $\Phi_n(a,s)$ and $\Phi_{n+1}(a,s)$ belong to $H(n+1)((H(n)(\{a\} \times I)) \times I)$, which by the choice of the functions $H(n+1)$ and $H(n)$ has diameter less than $2^{-(n-1)}$.

Now define $K\colon A \times I \to A$ by

$$K(a,t) = \lim_{n\to\infty} \Phi_n(a,t).$$

Then K is well-defined by (2) and (4), and continuous by (3) and proposition 1.3.4. Observe that for every $a \in A$,

$$K(a,1) = a.$$

Similarly, define $H: U \times [0,1) \to A$ by

$$H(x,t) = \lim_{n\to\infty} \Phi_n(x,t).$$

Observe that H is also continuous.

We shall now construct a special Urysohn function $\lambda: U \to I$ which will be used in connection with the functions K and H to construct the desired retraction from U onto A.

Let d denote an arbitrary admissible metric on U and on $U \times [0,1)$ use the metric

$$\rho((u,s),(v,t)) = d(u,v) + |s - t|.$$

Define

$$V_0 = \{(u,t) \in U \times [0,1): \exists\ (v,s) \in A \times [0,1) \text{ such that } \rho((u,t),(v,s)) < 1 - t, \text{ and } d(H(u,t),H(v,s)) < 1 - t\}.$$

It is easy to see that V_0 is open and that $A \times [0,1) \subseteq V_0$. By compactness of the intervals $[s_n,s_{n+1}]$, we can find for each $n \in \mathbb{N}$ an open neighborhood U_n of A in U such that

$$U_n \times [s_n,s_{n+1}] \subseteq V_0.$$

We may assume that $\overline{U}_{n+1} \subseteq U_n$ for every n and also that

$$(5) \qquad \bigcap_{n=1}^{\infty} U_n = A.$$

(in fact, (5) follows automatically). By corollary 1.4.15 there exists for every $n \in \mathbb{N}$ a Urysohn function $\lambda_n: U \to I$ such that

$$\begin{cases} \lambda_n(U \setminus U_n) = 0, \\ \lambda_n(\overline{U}_{n+1}) = 1. \end{cases}$$

Now define λ: $U \to I$ by

$$\lambda(u) = \Sigma_{n=1}^{\infty} 2^{-n}\lambda_n(u).$$

Then λ is well-defined and continuous by proposition 1.3.4. Observe that for $u \in U\backslash U_1$, $\lambda(u) = 0$ and that for $u \in A$, $\lambda(u) = 1$.

CLAIM 2: For every $u \in U_1\backslash A$, $(u,\lambda(u)) \in V_0$.

Take an arbitrary $u \in U_1\backslash A$. By (5) there exists i such that $u \in U_i\backslash U_{i+1}$. Consequently,

$$\lambda_n(u) = \begin{cases} 1 & (n < i), \\ 0 & (n > i), \end{cases}$$

so that

$$\begin{aligned} \lambda(u) &= \tfrac{1}{2} + \cdots + \tfrac{1}{2^{i-1}} + \tfrac{1}{2^i}\lambda_i(u) = \\ &= 1 - \tfrac{1}{2^{i-1}} + \tfrac{1}{2^i}\lambda_i(u) = \\ &= s_i + \tfrac{1}{2^i}\lambda_i(u). \end{aligned}$$

We conclude that $\lambda(u) \in [s_i,s_{i+1}]$ and hence

$$(u,\lambda(u)) \in U_i \times [s_i,s_{i+1}] \subseteq V_0.$$

We now define a function r: $U \to A$ as follows:

$$r(u) = \begin{cases} H(u,\lambda(u)) & (u \in U\backslash A), \\ u & (u \in A). \end{cases}$$

Clearly H is well-defined since $\lambda(u) \neq 1$ for every $u \in U\backslash A$. We claim that r is continuous, and hence is a retraction. By continuity of H, r is continuous at all the points of $U\backslash A$. So it suffices to prove that for an arbitrary sequence $(u_n)_n$ of points in $U_1\backslash A$, converging to a point $a \in A$, we have that the sequence $(r(u_n))_n$ also converges to a. By claim 2 we conclude that for every n, $(u_n,\lambda(u_n)) \in V_0$ so that there is a point (a_n,t_n) in $A \times [0,1)$ such that

(6) $\rho((u_n,\lambda(u_n)),(a_n,t_n)) < 1 - \lambda(u_n)$, and

(7) $d(H(u_n,\lambda(u_n)),H(a_n,t_n)) < 1 - \lambda(u_n)$.

Since $u_n \to a$ $(n \to \infty)$ and since λ is continuous, we obtain $\lambda(u_n) \to \lambda(a) = 1$ $(n \to \infty)$. Consequently, (6), (7) and the definition of ρ yield

$$(8)\ (u_n,\lambda(u_n)) \to (a,1) \qquad (n \to \infty),$$
$$(9)\ \rho((u_n,\lambda(u_n)),(a_n,t_n)) \to 0 \qquad (n \to \infty),$$
$$(10)\ d(H(u_n,\lambda(u_n)),H(a_n,t_n)) \to 0 \qquad (n \to \infty).$$

By (8) and (9) it follows that $(a_n,t_n) \to (a,1)$ $(n \to \infty)$ and hence by the continuity of K that

$$(11) \qquad K(a_n,t_n) \to K(a,1) = a \qquad (n \to \infty).$$

Since $K(a_n,t_n) = H(a_n,t_n)$ for every n, (10) and (11) yield

$$(12) \qquad r(u_n) = H(u_n,\lambda(u_n)) \to a \qquad (n \to \infty).$$

We are done. □

We now come to the announced:

5.6.4. *Proof of theorem 5.6.1.*

First we prove the theorem for the **AR**-case, i.e. the case that X, Y and A are **AR**'s. We shall prove that $X \cup_f Y$ is an **AR**. Consider the topological sum $Z = X \oplus Y$, and the natural quotient map $p: Z \to S = X \cup_f Y$. For convenience, we identify Y and p(Y) and $X\backslash A$ and $p(X\backslash A)$, respectively. Let $\mathcal{U}$ be an open cover of S and let $\mathcal{W} = p^{-1}(\mathcal{U})$. By lemma 5.6.2 there is a $\mathcal{W}$-deformation $G: X \times I \to X$ having the following properties:

(1) for every $t \in I$ and $x \in A$, $G(x,t) = x$,

(2) there is an open neighborhood V of A in X such that $G_1(V) = A$.

Define $K: Z \times I \to Z$ by

$$\begin{cases} K(x,t) = G(x,t) & (x \in X,\ t \in I), \\ K(y,t) = y & (y \in Y,\ t \in I). \end{cases}$$

In addition, define $H: S \times I \to S$ by

$$H(x,t) = p \circ K_t(p^{-1}(x)).$$

By (1), H is single-valued, and by compactness of the spaces involved, H is continuous. It is clear that H is a $\mathcal{U}$-deformation. Observe that $H_1(p(V)) \subseteq f(A) \subseteq Y$.

Now assume that S is a closed subset of a convex set C in a normable linear space (lemma 1.2.3). Since C is an **AR** (theorem 1.5.1), by lemma 5.6.3 it suffices to prove that H_1 can be extended over C.

Consider the disjoint closed sets

$$B = p(X\backslash V) \text{ and } Y = p(Y)$$

in S. Since B and Y are closed in C, they have disjoint open neighborhoods in C, say B^* and Y^*, respectively (corollary 1.4.16). Put $E = C\backslash(B^* \cup Y^*)$. Then E is closed in C and $E \cap S$ is contained in $p(V\backslash A)$, which we identified with $V\backslash A$. Therefore $K_1 \mid E \cap S$ is a continuous function from $E \cap S$ into A, and since A is an **AR**, it can therefore be extended to a continuous function $\psi_0 \colon E \to A$. Now consider $B^* \cup E$ and observe that the function $\xi \colon (B^* \cap S) \cup E \to X$ defined by

$$\xi(x) = \begin{cases} K_1(x) & (x \in B^* \cap S), \\ \psi_0(x) & (x \in E), \end{cases}$$

is continuous. Since X is an **AR**, ξ can be extended to a continuous function $\psi_1 \colon B^* \cup E \to X$. Define $\psi_2 \colon B^* \cup E \to p(X)$ by $\psi_2 = p \circ \psi_1$. Observe that

$$\psi_2 \mid ((B^* \cup E) \cap p(X)) = H_1 \mid ((B^* \cup E) \cap p(X)),$$

and that

$$\psi_2(E) \subseteq p(A) \subseteq Y.$$

Now define $\psi_3 \colon E \cup (Y^* \cap S) \to Y$ by

$$\psi_3(x) = \begin{cases} \psi_2(x) & (x \in E), \\ H_1(x) & (x \in Y^* \cap S). \end{cases}$$

Since $H_1(p(V)) \subseteq f(A) \subseteq Y$, ψ_3 is well-defined and obviously continuous. Now since Y is an **AR**, we can extend ψ_3 to a continuous function $\psi_4 \colon E \cup Y^* \to Y$. The function $\psi_2 \cup \psi_4$ is a continuous extension of H_1 over C.

For the **ANR**-case, we use the just proved part of the theorem. So let, let X,Y and $A \subseteq X$ be compact **ANR**'s and let f: A → Y be continuous. We consider the compact **AR**'s $\Delta(X)$, $\Delta(Y)$ and $\Delta(A)$ (theorem 5.4.2). Define $\Delta(f)\colon \Delta(A) \to \Delta(Y)$ as follows:

$$\begin{cases} \Delta(f)(x,t) = (f(x),t) & (x \in A,\ 0 \le t < 1), \\ \Delta(f)(\infty) = \infty. \end{cases}$$

Then $\Delta(f)$ is clearly continuous and the adjunction space $\Delta(X) \cup_{\Delta(f)} \Delta(Y)$ is naturally homeomorphic to $\Delta(X \cup_f Y)$. By assumption it therefore follows that $\Delta(X \cup_f Y)$ is an **AR** and hence that $X \cup_f Y$ is an **ANR** (theorem 5.4.2). □

Important examples of adjunction spaces are the so-called mapping cylinders, which we shall define now. Let X and Y be compact spaces and let f: X → Y be continuous. It will be convenient to think of $X \times I$ and Y as being disjoint. The *mapping cylinder,* M(f), of f is the space we obtain from $(X \times I) \oplus Y$ by identifying for each $y \in f(X)$ the set $(f^{-1}(y) \times \{1\}) \cup \{y\}$ to a single point,

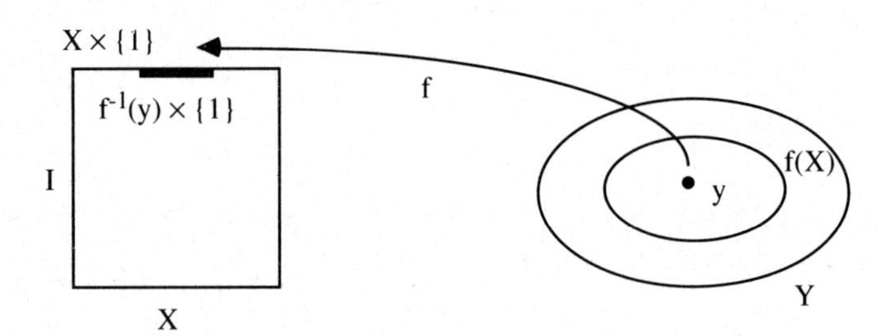

Figure 5.6.3.

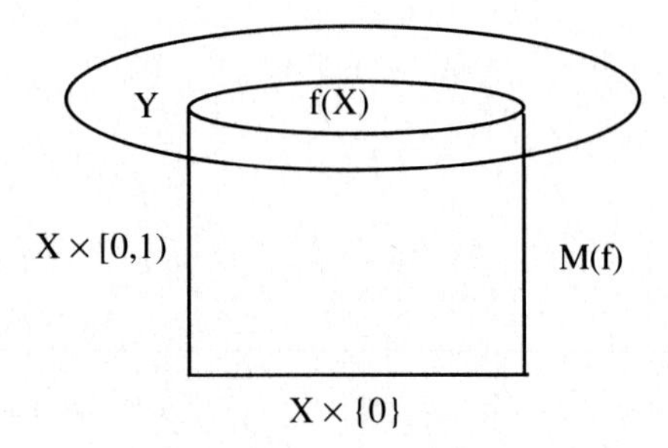

Figure 5.6.4.

i.e. $M(f) = (X \times I) \cup_g Y$, where $g: X \times \{1\} \to Y$ is defined by $g(x,1) = f(x)$. The resulting space after the identification looks something like the space in figure 5.6.4. It will be convenient to identify $X \times \{0\}$ and X and to think of M(f) as the space in figure 5.6.5. There is a natural retraction $c(f): M(f) \to Y$ which is formally defined as follows:

$$\begin{cases} c(f)(y) = y & (y \in Y), \\ c(f)(x,t) = f(x) & (x \in X,\ 0 \le t < 1). \end{cases}$$

It is easy to picture this mapping (see figure 5.6.6). Observe that c(f) is a retraction and has contractible point-inverses; it is called *the collapse to the base*. Mapping cylinders will play an important role in chapter 7.

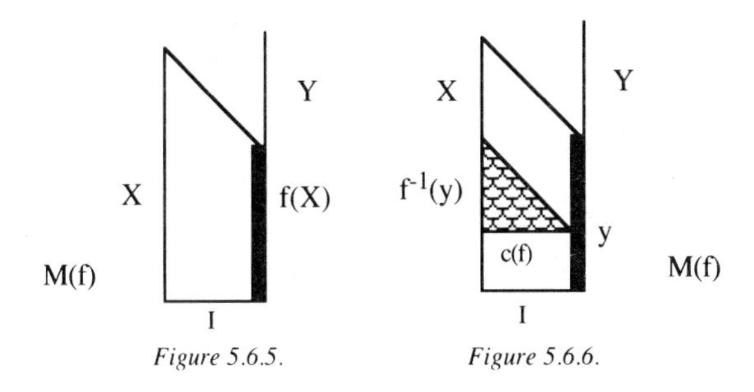

Figure 5.6.5. *Figure 5.6.6.*

Exercises for §5.6.

1. Let X be a space. Prove that the following statements are equivalent: (i) X is an **ANR**, (ii) for every open cover $\mathcal{U}$ of X there is an **ANR** Y that $\mathcal{U}$-dominates X, and (iii) for every open cover $\mathcal{U}$ of X there is a polytope P that $\mathcal{U}$-dominates X.

2. Let X be a compact space and let $f: X \to \{0\}$ be the constant function. Prove that the mapping cylinder M(f) of f is homeomorphic to $\Delta(X)$.

Notes and comments for chapter 5.

Most of the results in this chapter are well-known. Absolute (Neighborhood) Retracts were first defined by Borsuk [24], [25]. For a much more comprehensive study of the subject, see Borsuk [29] and Hu [74].

§1.

Theorems 5.1.1 and 5.1.8 are due to Hanner [69] and theorem 5.1.4 is due to Dugundji [55] and Lefschetz [94].

§2.

Theorem 5.2.1 is due to Dugundji [55] and Lefschetz [93]. Our proof of theorem 5.2.1 closely follows the exposition of Hu [74, chapter IV §4]. Theorem 5.2.12 is due to Toruńczyk [134].

§3.

Theorem 5.3.13 is due to Mazurkiewicz [99]. Theorem 5.3.14 is due to Wojdysławski [151].

§4.

Theorems 5.4.1 and 5.4.5 are due to Hanner [69]. Our use of cones simplifies the proof of theorem 5.4.5 slightly.

§5.

The results in this section are mainly due to Kuratowski [89].

§6.

Theorem 5.6.1 is due to Whitehead [150]. Lemma 5.6.3 is due to Hanner [69]. Our exposition closely follows Hu [74, Chapter IV theorem 5.3]. It is also possible to give a proof of Whitehead's Theorem via the characterization theorem 5.2.1; for details see Borsuk [29, Chapter V §9].

6. An Introduction to Infinite-Dimensional Topology

In this chapter we shall present some elementary results from infinite-dimensional topology. The results are elementary in the sense that no powerful apparatus is needed, but the proofs are not always easy. Our main results are that Hilbert space l^2 is homeomorphic to $\mathbb{R}^\infty$, and that $T \times Q$ and Q are homeomorphic, where T denotes the subspace $([0,1] \times \{0\}) \cup (\{\frac{1}{2}\} \times [0,1])$ of $[0,1] \times [0,1]$. These results are due to R.D. Anderson.

6.1. Constructing New Homeomorphisms From Old

In this section we present three methods to construct new homeomorphisms from old ones, namely,

(1) the Inductive Convergence Criterion,

(2) Bing's Shrinking Criterion, and

(3) by means of so-called isotopies.

Each method is illustrated with an application: from (1) we deduce the homogeneity of Q, (2) is used to prove that Q is homeomorphic to its own cone, and (3) is used in a result on extending homeomorphisms in the plane.

All three methods shall be used intensively in later sections.

A. The Inductive Convergence Criterion

We shall present a simple but useful technique enabling us to decide whether certain limits of homeomorphisms are again homeomorphisms. The following lemma will be helpful.

6.1.1. LEMMA: *Let* (X,d) *be a complete metric space and let* $(A_n)_n$ *be a sequence of subsets of* X. *Suppose that* $(x_n)_n$ *is a Cauchy sequence in* X *such that for every* n,

$$(1) \qquad d(x_{n+1},x_n) < 3^{-n}\cdot\min\{d(x_i,A_i): 1 \leq i \leq n\}.$$

Then $\lim_{n\to\infty} x_n \notin \bigcup_{n=1}^{\infty} A_n$.

PROOF: Take an arbitrary $n \in \mathbb{N}$. We shall prove that $x \notin A_n$, where $x = \lim_{n\to\infty} x_n$. Condition (1) implies that for every $m \in \mathbb{N}$,

$$d(x_{m+n},x_{(m-1)+n}) < 3^{-((m-1)+n)}\cdot d(x_n,A_n) \leq 3^{-m}\cdot d(x_n,A_n).$$

From this it follows that

$$d(x_{m+n},x_n) < \sum_{i=1}^{m} 3^{-i}\cdot d(x_n,A_n),$$

and as

$$\sum_{i=1}^{\infty} 3^{-i} = \tfrac{1}{2},$$

we obtain

$$d(x,x_n) = \lim_{m\to\infty} d(x_{m+n},x_n) \leq \sum_{i=1}^{\infty} 3^{-i}\cdot d(x_n,A_n) = \tfrac{1}{2}d(x_n,A_n),$$

so $x \notin A_n$. □

Observe that in the above lemma the distance between x_{n+1} and x_n "depends" only on the points $x_1,\cdots,x_n$. Hence if one wishes to choose the points $(x_n)_n$ inductively so that (1) is satisfied, at stage n+1 the choice of x_{n+1} is subject to n - hence finitely many - conditions.

Let X be a compact space and let $(h_n)_n$ be a sequence in $\mathcal{H}(X)$ (recall that $\mathcal{H}(X)$ is the autohomeomorphism group of X, cf. §1.3). It is clear that for each $n \in \mathbb{N}$ the function

$$f_n = h_n \circ \cdots \circ h_1$$

belongs to $\mathcal{H}(X)$. If $f = \lim_{n\to\infty} f_n$ exists then it will be denoted by

$$\lim_{n\to\infty} h_n \circ \cdots \circ h_1$$

and is called the *infinite left product* of the sequence $(h_n)_n$. We want to find conditions on the sequence $(h_n)_n$ which ensure that $\lim_{n\to\infty} h_n \circ \cdots \circ h_1$ exists and belongs to $\mathcal{H}(X)$. We adopt the notation of §1.3.

6.1.2. THEOREM ("The Inductive Convergence Criterion"): *Let* X *be a compact space and let* $(h_n)_n$ *be a sequence in* $\mathcal{H}(X)$ *such that for all* $n \in \mathbb{N}$,

(1) $d(h_{n+1}, 1_X) < 2^{-n}$,

(2) $d(h_{n+1}, 1_X) < 3^{-n} \cdot \min\{d(h_i \circ \cdots \circ h_1, \mathcal{G}_{1/i}(X,X)) : 1 \le i \le n\}$.

Then $h = \lim_{n\to\infty} h_n \circ \cdots \circ h_1$ *exists and is a homeomorphism of* X.

PROOF: For every n, put $f_n = h_n \circ \cdots \circ h_1$. Condition (1) and exercise 1.3.7 imply that for all n,

$$d(f_{n+1}, f_n) = d(h_{n+1} \circ \cdots \circ h_1, h_n \circ \cdots \circ h_1) = d(h_{n+1}, 1_X) < 2^{-n}. \tag{3}$$

We conclude that the sequence $(f_n)_n$ is d-Cauchy and consequently, $f = \lim_{n\to\infty} f_n$ exists (corollary 1.3.5). Since $f_n \in \mathcal{S}(X,X)$ for every n, proposition 1.3.7 implies that $f \in \mathcal{S}(X,X)$. Now by (2), (3), lemma 6.1.1 and lemma 1.3.9 we obtain

$$f \in \bigcap_{n=1}^{\infty} \mathcal{S}_{1/n}(X,X) = \mathcal{H}(X),$$

as required. □

The above result tells us that if the sequence $(d(h_n,1))_n$ converges rapidly to 0, then the infinite left product $\lim_{n\to\infty} h_n \circ \cdots \circ h_1$ is a homeomorphism. It turns out that we are not interested in the precise speed at which $(d(h_n,1))_n$ converges. We are always in the pleasant situation that while inductively defining the sequence $(h_n)_n$ we are able to choose the next homeomorphism so as to be "sufficiently close" to the identity. This simplifies life considerably.

Application 1: Topological homogeneity of the Hilbert cube.

In corollary 3.5.10 we showed that if $n \in \mathbb{N}$ then J^n is not homogeneous. As announced there, we shall now prove that Q *is* homogeneous. This is surprising. At first glance it seems that our finite-dimensional intuition predicts the truth, namely, that Q has a boundary and an interior. This finite-dimensional intuition however does not "work" in the infinite-dimensional Hilbert cube.

As in exercise 3.5.2, we put

$$s = \prod_{i=1}^{\infty} (-1,1)_i \text{ and } B(Q) = Q \setminus s,$$

respectively. We call s the *pseudo-interior* and B(Q) the *pseudo-boundary* of Q. It will be convenient for every i to let π_i: $Q \to [-1,1]_i$ denote the projection map $\pi_i(x) = x_i$.

6.1.3. LEMMA: *Suppose that* $x,y \in s \subseteq Q$. *Then there is a homeomorphism* $h \in \mathcal{H}(Q)$ *with* $h(x) = y$.

PROOF: Since for each i, $x_i,y_i \in (-1,1)_i$, we can find $h_i \in \mathcal{H}(J_i)$ such that $h_i(x_i) = y_i$. Take $h = h_1 \times h_2 \times \cdots$. □

We see that all points in the pseudo-interior are topologically equivalent. Consequently, if we show that every point in Q can be homeomorphed into s, then we have shown that Q is homogeneous. We need a preliminary lemma.

6.1.4. LEMMA: *Suppose that* $x \in Q$, *that* $m \in \mathbb{N}$, *and that* $\varepsilon > 0$. *Then there is an element* $h \in \mathcal{H}(Q)$ *such that*

(1) $d(h,1) < \varepsilon$,

(2) $h(x)_m \in (-1,1)_m$,

(3) h *does not affect the first* m-1 *coordinates of any point, i.e.* $h(y)_i = y_i$ *for all* $i \le m-1$ *and* $y \in Q$.

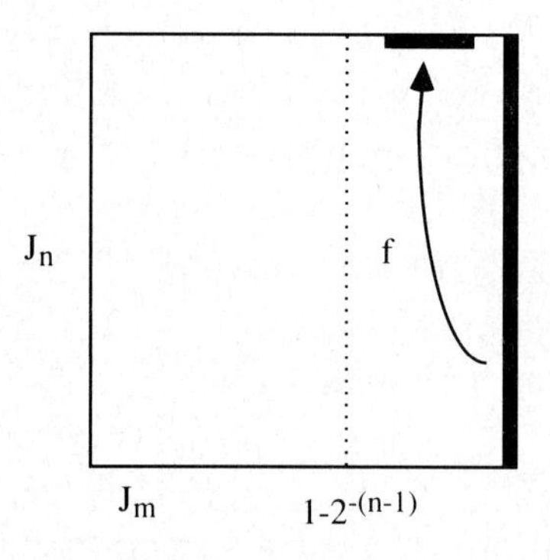

Figure 6.1.1.

PROOF: If $|x_m| \neq 1$, let h = 1. Therefore, without loss of generality assume that $x_m = 1$. Let $n > m$ be such that $2^{-(n-2)} < \varepsilon$. It is geometrically obvious that there is a homeomorphism ϕ: $J_m \times J_n \to J_m \times J_n$ such that

(4) $\phi(p,q) = (p,q)$ if $p \leq 1 - 2^{-(n-1)}$,

(5) $\phi(\{1\} \times J_n)$ is contained in $J_m \times \{1\}$ and projects within the open interval $(1-2^{-(n-1)},1)$.

For a moment we think of Q as $(J_m \times J_n) \times R$, where R denotes the product of all remaining factors. We define $h = \phi \times 1_R$. So h only affects the n-th and m-th coordinate of any point. To verify that h is as required, first observe that (2) and (3) are clearly satisfied. For (1), notice that for every $x \in Q$,

$$d(h(x),x) = 2^{-m} \cdot |h(x)_m - x_m| + 2^{-n} \cdot |h(x)_n - x_n|$$

$$\leq 1 \cdot 2^{-(n-1)} + 2^{-n} \cdot 2 = 2^{-(n-2)}$$

$$< \varepsilon.$$

From this it follows that $d(h,1_Q) < \varepsilon$. □

6.1.5. LEMMA: *Let* $x \in Q$. *There is an* $h \in \mathcal{H}(Q)$ *with* $h(x) \in s$.

PROOF: The homeomorphism h shall be of the form $\lim_{n\to\infty} h_n \circ \cdots \circ h_1$, where $(h_n)_n$ is an inductively constructed sequence of homeomorphisms of Q. At each stage of the construction, the next homeomorphism is constructed in accordance with the Inductive Convergence Criterion 6.1.2.

The sequence $(h_n)_n$ shall satisfy the following conditions:

$(1)_n$ h_n does not affect the first n-1 coordinates of any point,

$(2)_n$ $(h_n \circ \cdots \circ h_1(x))_i \in (-1,1)$ for every $i \leq n$, and

$(3)_n$ $d(h_n,1)$ is so small that the conditions (1) and (2) mentioned in the Inductive Convergence Criterion 6.1.2 are satisfied.

Apply lemma 6.1.4 to find a homeomorphism $f \in \mathcal{H}(Q)$ such that $f(x)_1 \in (-1,1)$ and define $h_1 = f$. Then h_1 is as required since $(1)_1$ and $(3)_1$ are empty conditions. Now suppose that h_i has been defined for every $i \leq n$. The Inductive Convergence Criterion 6.1.2 gives us a magic $\varepsilon > 0$ and tells us that we must choose the next homeomorphism ε-close to the identity. We are not interested at all in the precise value of ε. The only thing we need is that such ε exists. By lemma 6.1.4 there is a homeomorphism $h_{n+1} \in \mathcal{H}(Q)$ such that $d(h_{n+1},1) < \varepsilon$, h_{n+1} does not affect the first n coordinates of any point, and $(h_{n+1}(h_n \circ \cdots \circ h_1(x)))_{n+1} \in (-1,1)$. It is easily seen that

h_{n+1} is as required.

Now put $h = \lim_{n\to\infty} h_n \circ \cdots \circ h_1$. By construction, $h \in \mathcal{H}(Q)$. By continuity of the projections $\pi_m\colon Q \to J_m$ we find that for every m,

$$h(x)_m = (\lim_{n\to\infty} h_n \circ \cdots \circ h_1(x))_m = \lim_{n\geq m} (h_n \circ \cdots \circ h_1(x))_m =$$

$$= (h_m \circ \cdots \circ h_1(x))_m \in (-1,1),$$

i.e. $h(x) \in s$. □

We now come to the announced result:

6.1.6. THEOREM: Q *is homogeneous.*

PROOF: Take $x,y \in Q$. By lemma 6.1.5 we can find $h_1,h_2 \in \mathcal{H}(Q)$ with $h_1(x),h_2(y) \in s$. In addition, by lemma 6.1.3, we can find $f \in \mathcal{H}(Q)$ with $f(h_1(x)) = h_2(y)$. Then

$$g = h_2^{-1} \circ f \circ h_1 \in \mathcal{H}(Q)$$

and takes x onto y. □

B. Bing's Shrinking Criterion

Let X and Y be compact spaces. A continuous surjection $f\colon X \to Y$ is called a *near homeomorphism* provided that for any $\varepsilon > 0$ there is a homeomorphism $h\colon X \to Y$ such that $d(h,f) < \varepsilon$. It is easily seen that $f \in C(X,Y)$ is a near homeomorphism if and only if f belongs to the closure of $\mathcal{H}(X,Y)$ in $C(X,Y)$. Simple examples show that a near homeomorphism need not be a homeomorphism, see e.g. exercise 1.3.2. Near homeomorphisms play an important role in geometric topology.

Let X and Y be compact spaces. A continuous surjection $f\colon X \to Y$ is called *shrinkable* provided that for every $\varepsilon > 0$ there is a homeomorphism $h\colon X \to X$ satisfying

(1) $\mathrm{diam}(hf^{-1}(y)) < \varepsilon$ for every $y \in Y$ (i.e. $f \circ h^{-1}$ is an ε-map),
(2) $d(f,f \circ h) < \varepsilon$.

Shrinkability means that the point-inverses of f can be uniformly shrunk to small sets by a homeomorphism of X that, from the standpoint of the space Y, does not change f very much.

The reader should pause to get used to this definition.

Bing's Shrinking Criterion says that shrinkablity is the same as being a near homeomorphism. If the domain of the function under consideration has a rich supply of homeomorphisms, e.g. if it is a manifold, then verifying shrinkability is usually easier than verifying directly that the function in question is a near homeomorphism.

6.1.7. LEMMA: *Let* X *and* Y *be compact spaces and let* $f: X \to Y$ *be shrinkable. Then for each* $\varepsilon > 0$ *and for each* $\phi \in \mathcal{H}(X)$ *there exists* $\psi \in \mathcal{H}(X)$ *such that*

(1) $f \circ \psi$ *is an* ε*-map,*

(2) $d(f \circ \phi, f \circ \psi) < \varepsilon$.

PROOF: Let $\gamma > 0$ be such that for $A \subseteq X$, if $\text{diam}(A) < \gamma$ then $\text{diam}(\phi^{-1}(A)) < \varepsilon$ (use the fact that ϕ^{-1} is uniformly continuous). Since f is shrinkable, there is a homeomorphism $h \in \mathcal{H}(X)$ such that

(3) $d(f, f \circ h) < \varepsilon$,

(4) $\forall y \in Y$: $\text{diam}(hf^{-1}(y)) < \gamma$.

Since $d(f, f \circ h) < \varepsilon$, it follows that $d(fh^{-1}\phi, fhh^{-1}\phi) < \varepsilon$ (exercise 1.3.7), so that

$$d(f\phi, fh^{-1}\phi) < \varepsilon.$$

We claim that $fh^{-1}\phi$ is an ε-map. If so, then $\psi = h^{-1}\phi$ is as required. To this end, take an arbitrary $y \in Y$ and consider $(fh^{-1}\phi)^{-1}(y) = \phi^{-1}hf^{-1}(y)$. Since $\text{diam}(hf^{-1}(y)) < \gamma$ we see that $\text{diam}(\phi^{-1}hf^{-1}(y)) < \varepsilon$, and this is what we had to prove. □

6.1.8. THEOREM ("Bing's Shrinking Criterion"): *Let* X *and* Y *be compact spaces and let* $f: X \to Y$ *be a continuous surjection. Then* f *is a near homeomorphism if and only if* f *is shrinkable.*

PROOF: First assume that f is a near homeomorphism. Choose $\varepsilon > 0$ and let $g: X \to Y$ be a homeomorphism such that $d(g,f) < \varepsilon/2$. Find $\gamma > 0$ such that if $A \subseteq Y$ has diameter less than γ then $g^{-1}(A)$ has diameter less than ε (use the fact that g^{-1} is uniformly continuous). Let $p: X \to Y$ be a homeomorphism such that $d(p,f) < \min\{\gamma/2, \varepsilon/2\}$ and put $h = g^{-1} \circ p$. We claim that h is a shrinking homeomorphism. Since $d(f,p) < \varepsilon/2$ and

$$d(p,fh) = d(p,fg^{-1}p) = d(1,fg^{-1}) = d(gg^{-1},fg^{-1}) = d(g,f) < \varepsilon/2$$

(we used exercise 1.3.7 twice) it follows that

$$d(f,fh) \leq d(f,p) + d(p,fh) < \varepsilon/2 + \varepsilon/2 = \varepsilon.$$

Now take an arbitrary $y \in Y$. If $z \in p(f^{-1}(y))$ then since $d(p,f) < \gamma/2$, $d(y,z) < \gamma/2$. So, $\text{diam}(p(f^{-1}(y))) < \gamma$ from which it follows that $\text{diam}(g^{-1}p(f^{-1}(y))) < \varepsilon$. We conclude that $\text{diam}(h(f^{-1}(y))) < \varepsilon$.

Now assume that f is shrinkable. Let $\varepsilon > 0$. Let h_0 be the identity homeomorphism on X. By applying lemma 6.1.7 inductively, we find that there exists a sequence $(h_n)_{n\geq 1}$ in $\mathcal{H}(X)$ such that if we put

$$p_n = f \circ h_n$$

for each $n \geq 1$, then the following conditions are satisfied:

(1) p_n is a $1/n$-map,
(2) $d(p_{n+1},p_n) < 3^{-n}\cdot\varepsilon$ and $d(p_1,f) < \frac{\varepsilon}{2}$,
(3) $d(p_{n+1},p_n) < 3^{-n}\cdot\min\{d(p_i,\mathcal{G}_{1/i}(X,Y)): 1 \leq i \leq n\}$.

By (2), the sequence $(p_n)_n$ is Cauchy. So let $p = \lim_{n\to\infty} p_n$. Observe that by proposition 1.3.7, $p \in \mathcal{S}(X,Y)$. Again by (2), it follows that

$$d(f,p) < \frac{\varepsilon}{2} + \sum_{i=1}^{\infty} 3^{-i}\cdot\varepsilon = \frac{\varepsilon}{2} + \frac{\varepsilon}{2} = \varepsilon.$$

Also, (1),(3), lemma 6.1.1 and corollary 1.3.5 imply that

$$p \notin \bigcup_{n=1}^{\infty} \mathcal{G}_{1/n}(X,Y),$$

i.e. $p \in \mathcal{H}(X,Y)$ (lemma 1.3.9). We conclude that f is a near homeomorphism. □

6.1.9. COROLLARY: *Let* X *and* Y *be compact spaces and let* f: X → Y *be shrinkable. Then* X *and* Y *are homeomorphic.* □

Application 2: The cone over the Hilbert cube is the Hilbert cube.

6.1.10. LEMMA: *For every* $n \in \mathbb{N}$, $t \in [-1,1)$ *and* $\varepsilon > 0$ *there exists a homeomorphism* h:

$J^n \times J \to J^n \times J$ *such that*

(1) $h \mid J^n \times [-1,t] = 1$,

(2) *the diameter of* $h(J^n \times \{1\})$ *is less than* ε.

PROOF: The lemma is geometrically obvious. Alternatively, observe that the lemma is clear for $n = 1$ since h can then be found such as in the proof of lemma 6.1.4. Now proceed inductively. □

For the definition of the cone $\Delta(X)$ over a space X, see the exercises of §1.6. Lemma 6.1.10 now allows us to conclude that:

6.1.11. THEOREM: $\Delta(Q)$ *is homeomorphic to* Q.

PROOF: Observe that $\Delta(Q)$ is the space we get from $Q \times [0,1]$ by identifying $Q \times \{1\}$ to a single point ∞. Let $p: Q \times [0,1] \to \Delta(Q)$ be the decomposition map.

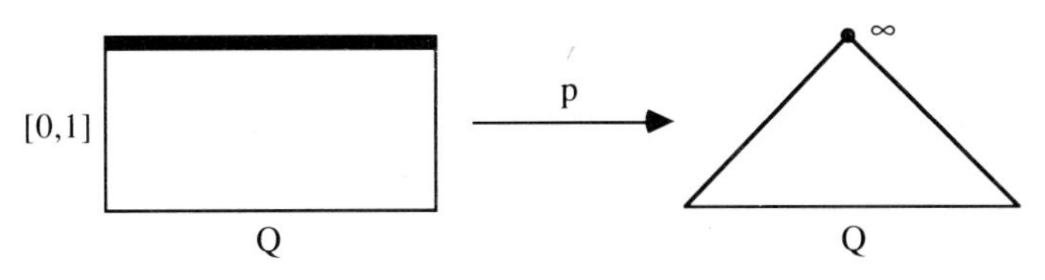

Figure 6.1.2.

We will show that p is shrinkable. From corollary 6.1.9 it then follows that

$$Q \approx Q \times [0,1] \approx \Delta(Q).$$

Let $\varepsilon > 0$ and let U be a neighborhood of $\infty \in \Delta(Q)$ having diameter less than ε. There exists $t \in (0,1)$ such that $(Q \times [t,1)) \cup \{\infty\} \subseteq U$. Find $n \in \mathbb{N}$ such that

$$\sum_{m=n}^{\infty} 2^{-m} < \tfrac{\varepsilon}{2},$$

and applying lemma 6.1.10 find a homeomorphism $h: J^n \times [0,1] \to J^n \times [0,1]$ such that

(1) $h \mid J^n \times [0,t] = 1$,

(2) $\operatorname{diam}(h(J^n \times \{1\})) < \tfrac{\varepsilon}{2}$.

Define $H \in \mathcal{H}(Q \times [0,1])$ by

$$H((x_1,\cdots,x_n,x_{n+1},\cdots),t) = (h(x_1,\cdots,x_n,t),x_{n+1},x_{n+2},\cdots)$$

(we make a few obvious identifications here).

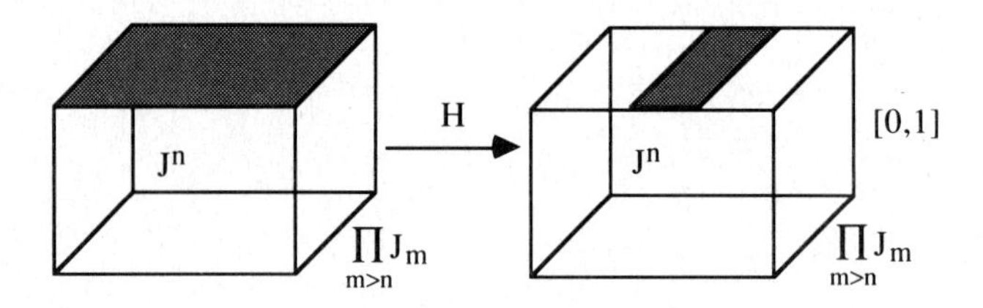

Figure 6.1.3.

By the special choice of n it follows easily that $\mathrm{diam}(H(Q \times \{1\})) < \varepsilon$. We claim that $d(p,p \circ H) < \varepsilon$. Take a point $(x,s) \in Q \times [0,1]$. If $s \leq t$ then $H(x,s) = (x,s)$, so $d(p(x,s),pH(x,s)) = 0$. In addition, if $s > t$ then $\{p(x,s),pH(x,s)\} \subseteq U$. Therefore, since $\mathrm{diam}(U) < \varepsilon$, we find that $d(p(x,s),pH(x,s)) < \varepsilon$. We conclude that p is shrinkable. □

Consider the point $\infty \in \Delta(Q)$. Basic closed neighborhoods of ∞ in $\Delta(Q)$ have the form

$$U_t = (Q \times [t,1)) \cup \{\infty\} \qquad (t \in (0,1)).$$

Observe that the boundary of U_t is equal to $Q \times \{t\}$. Consequently, theorem 6.1.11 implies that ∞ has arbitrarily small closed neighborhoods U such that

(1) U is homeomorphic to Q, and
(2) Bd(U) is homeomorphic to Q.

By the homogeneity of Q (theorem 6.1.6), *each* point of Q has arbitrarily small closed neighborhoods U satisfying (1) and (2). In particular, each point of Q has arbitrarily small closed neighborhoods with contractible boundaries.

This demonstrates a striking difference with the finite-dimensional situation and explains e.g. why homology theory was never of much use in infinite-dimensional topology.

We now present our last method for constructing new homeomorphisms from old ones.

C. Isotopies

Let X and Y be spaces. Recall that a homotopy from X to Y is a continuous function $H: X \times I \to Y$. For technical reasons, we call a continuous function $H: X \times K \to Y$, where K is any compact space, a *K-homotopy*. If $t \in K$ then the function $H_t: X \to Y$ defined by $H_t(x) = H(x,t)$ is called the t-th *level* of H. A *K-isotopy* from X to Y is a K-homotopy $H: X \times K \to Y$ each level of which is a homeomorphism from X onto Y. If $K \subseteq \mathbb{R}$ is a compact interval then a K-homotopy (resp. K-isotopy) is simply called a homotopy (resp. isotopy).

The proof of the following proposition is a triviality and is left as an exercise to the reader.

6.1.12. PROPOSITION: *For each* $n \in \mathbb{N}$ *let* $H^n: X_n \times K_n \to Y_n$ *be a* K_n*-isotopy. Then the function*

$$H: \prod_{n=1}^{\infty} X_n \times \prod_{n=1}^{\infty} K_n \to \prod_{n=1}^{\infty} Y_n,$$

defined by

$$H(x,t)_n = H^n(x_n,t_n) \qquad (n \in \mathbb{N}),$$

is a

$$\prod_{n=1}^{\infty} K_n \text{ - } isotopy. \square$$

We shall now present our last method for constructing new homeomorphisms from old ones. At first glance this method does not seem to be as important as the other two since direct appealing applications are hard to find. However, it turns out that this method is as fundamental as the preceding two.

6.1.13. THEOREM: *Let* K *be a compact space. If* $H: X \times K \to X$ *is a* K*-homotopy and* $\alpha: Y \to K$ *is continuous then so is the function* $f: X \times Y \to X \times Y$ *defined by*

$$f(x,y) = (H(x,\alpha(y)),y). \tag{1}$$

Moreover, if X *and* Y *are compact and* H *is a* K*-isotopy then* f *is a homeomorphism.*

Remark: Observe that the function f defined in (1) is "level preserving", i.e. f does not change the second coordinate of any point.

PROOF: That f is continuous is trivial. Assume next that X and Y are compact and that H is a

K-isotopy. Since f is "level preserving" and since each "level" of f is a homeomorphism from X onto X, it follows that f is surjective. By compactness we therefore only need to show that f is one-to-one (exercise 1.1.4). To this end, take distinct $(a,b), (a',b') \in X \times Y$. If $b \neq b'$ then clearly $f(a,b) \neq f(a',b')$. If $b = b'$ then $\alpha(b) = \alpha(b')$ and $a \neq a'$. Consequently, $H(a,\alpha(b)) \neq H(a',\alpha(b'))$ since $H_{\alpha(b)}$ is one-to-one. □

Application 3: Extending certain homeomorphisms.

Isotopies are very useful in situations where one wants to extend a given homeomorphism. This will be demonstrated in a simple situation here. The same ideas will be used to derive much more complicated results in §6.3.

Let $E,F \subseteq \mathbb{R}$ be compact subsets and let $f: E \to F$ be a homeomorphism. It is easy to see that it is not always possible to extend f to a homeomorphism $\bar{f}: \mathbb{R} \to \mathbb{R}$. For example, let $E = F = \{0,1,2\}$ and define f by $f(0) = 1$, $f(1) = 0$ and $f(2) = 2$.

Identify $\mathbb{R}$ and the x-axis in $\mathbb{R}^2$. Although in general f cannot be extended over $\mathbb{R}$, an easy application of exercise 1.4.6 yields that f can always be extended to a homeomorphism $\bar{f}: \mathbb{R}^2 \to \mathbb{R}^2$. We shall now present a different proof of this which much more than the solution of exercise 1.4.6 illustrates the technique of extending homeomorphisms that we are going to use in §6.3.

6.1.14. THEOREM: *Let* $E,F \subseteq \mathbb{R}$ *be compact and let* $f: E \to F$ *be a homeomorphism. Then* f *can be extended to a homeomorphism* $\bar{f}: \mathbb{R}^2 \to \mathbb{R}^2$.

PROOF: For technical reasons we shall prove the theorem for the interval $(-1,1)$ instead of $\mathbb{R}$. Since $(-1,1)$ and $\mathbb{R}$ are homeomorphic (exercise 1.1.8), we are allowed to do this.

So let $E,F \subseteq (-1,1)$ be compact and let $f: E \to F$ be a homeomorphism. Let $K \subseteq (-1,1)$ be a compact interval containing both E and F. For each $t \in K$ let $H_t: J \to J$ be the unique homeomorphism taking $[-1,t]$ linearly onto $[-1,0]$ and $[t,1]$ linearly onto $[0,1]$. It is easily seen that the function $H: J \times K \to J$ defined by $H(x,t) = H_t(x)$ is an isotopy.

Let $\Gamma \subseteq (-1,1)^2$ be the graph of the function f, i.e.

$$\Gamma = \{(x,f(x)): x \in E\}.$$

Our aim is to find a homeomorphism $h: J^2 \to J^2$ that takes Γ onto $\{0\} \times F$. We achieve this by applying theorem 6.1.13. Define a function $\phi: F \to K$ by $\phi(x) = f^{-1}(x)$. By the Tietze Extension Theorem 1.4.14, we can extend ϕ to a continuous function $\alpha: J \to K$. Now define $h: J^2 \to J^2$ by

$$h(x,y) = (H(x,\alpha(y)),y).$$

By theorem 6.1.13 it follows that h is a homeomorphism. Take $(x,f(x)) \in \Gamma$. Then

(1) $$h(x,f(x)) = (H(x,\alpha(f(x))),f(x)) = (H(x,x),f(x)) = (0,f(x)).$$

We conclude that h is as required. By precisely the same argumentation we can find a homeomorphism $g: J^2 \to J^2$ such that for every $(x,f(x)) \in \Gamma$,

(2) $$g(x,f(x)) = (x,0).$$

Now define $\xi: J^2 \to J^2$ by $\xi(x,y) = (y,x)$ and put $\bar{f} = \xi \circ h \circ g^{-1}$. Then $\bar{f}$ is a homeomorphism of J^2 and it is easily seen that $\bar{f}((-1,1)^2) = (-1,1)^2$ (alternatively, apply corollary 4.6.6). We claim that $\bar{f}$ extends f and is therefore as required. Take an arbitrary $x \in E$. Then

$$\bar{f}(x,0) = \xi h g^{-1}(x,0) = \xi h(x,f(x)) = \xi(0,f(x)) = (f(x),0)$$

(apply (1) and (2)). □

Exercises for §6.1.

1. Make the "it is geometrically obvious" part in the proof of lemma 6.1.4 precise by explicitly describing a homeomorphism ϕ as used there.

2. Show that Q is n-point order homogeneous for every n, i.e. that for all collections $\{x_1,\cdots,x_n\}$ and $\{y_1,\cdots,y_n\}$ of n distinct points in Q there is an $h \in \mathcal{H}(Q)$ with $h(x_i) = y_i$ for every $i \leq n$.

3. Prove that Q is strongly locally homogeneous, i.e. that there is an open base $\mathcal{U}$ for Q such that for all $U \in \mathcal{U}$ and $x,y \in U$ there is an $h \in \mathcal{H}(Q)$ such that $h(x) = y$ and $h \mid Q\backslash U = 1$.

4. Use theorem 6.1.2 and exercise 6.1.3 to show that Q is countable dense homogeneous, i.e. that for all countable dense subsets $D,E \subseteq Q$ there exists $h \in \mathcal{H}(Q)$ such that $h(D) = E$.

5. Make the "it is geometrically obvious" part in the proof of lemma 6.1.10 precise by explicitly describing a homeomorphism such as desired there.

6. Let X be a compact space. Let Z be the space we obtain from $\Delta(X) \times [0,1]$ by identifying $\Delta(X) \times \{1\}$ to a single point, and let $p: \Delta(X) \times [0,1] \to Z$ be the decomposition map. Prove that p is shrinkable and conclude

from this that $\Delta(\Delta(X)) \approx \Delta(X) \times [0,1]$.

7. Let X be a compact space and let $Y = \Delta(X) \times Q$. Prove that Y and $\Delta(Y)$ are homeomorphic.

8. Let X,Y and Z be compact spaces and let $f: X \to Y$ and $g: Y \to Z$ be near homeomorphisms. Prove that $g \circ f: X \to Z$ is a near homeomorphism.

9. Prove that Q is isotopically homogeneous, i.e. for every $x,y \in Q$ there exists an isotopy $H: Q \times I \to Q$ such that $H_0 = 1_Q$ and $H_1(x) = y$.

10. Give an example of a homogeneous continuum that is not isotopically homogeneous.

11. As usual, **C** denotes the Cantor middle-third set. Let A and B be closed and nowhere dense subsets (i.e. A and B have no interior). Prove that each homeomorphism $f: A \to B$ can be extended to a homeomorphism $\bar{f}: \mathbf{C} \to \mathbf{C}$ (Hint: Use the fact that **C** is zero-dimensional and apply lemma 1.4.12).

12. Let $\pi: Q \to \prod_{i=2}^{\infty} J_i$ be the projection. Prove that π is a near homeomorphism.

13. Let $D \subseteq Q$ be countable and dense. Prove that $Q \setminus D$ is homogeneous.

6.2. Z-Sets

Homeomorphism extension results are useful in infinite-dimensional topology. In this section we shall present a class of subsets of Q for which an important homeomorphism extension result shall be derived in section 6.4.

Let X be a space. A closed subset $A \subseteq X$ is called a *Z-set* in X provided that for every open cover $\mathcal{U}$ of X and every function $f \in C(Q,X)$ there is a function $g \in C(Q,X)$ such that

(1) f and g are $\mathcal{U}$-close, and
(2) $g(Q) \cap A = \emptyset$.

A *σZ-set* in X is a countable union of Z-sets. The collection of Z-sets and σZ-sets in X are denoted by $\mathcal{Z}(X)$ and $\mathcal{Z}_\sigma(X)$, respectively.

It will sometimes be convenient to have a "metric" translation of the concept of a Z-set.

6.2.1. LEMMA: *Let* X *be a space and let* $A \subseteq X$ *be closed. Then the following statements are* *equivalent:*

(a) $A \in \mathcal{Z}(X)$,

(b) $\forall \varepsilon > 0\ \forall f \in C(Q,X)\ \exists g \in C(Q,X)$ *such that* $d(f,g) < \varepsilon$ *and* $g(Q) \cap A = \emptyset$.

PROOF: Since (a) $\Rightarrow$ (b) is trivial, it suffices to establish the implication (b) $\Rightarrow$ (a). Let $\mathcal{U}$ be an open cover of X, and let $f: Q \to X$ be continuous. By compactness of f(Q), there exists $\delta > 0$ with the property that every $A \subseteq X$ with $\mathrm{diam}(A) < \delta$ and which moreover intersects f(Q), is contained in an element $U \in \mathcal{U}$ (lemma 1.1.1). By (b) there is a function $g \in C(Q,X)$ with $d(f,g) < \delta$ and $g(Q) \cap A = \emptyset$. We claim that f and g are $\mathcal{U}$-close. This is a triviality. Pick an arbitrary $x \in Q$. Then $d(f(x),g(x)) < \delta$ and since $f(x) \in f(Q)$, $\{f(x),g(x)\}$ is contained in an element $U \in \mathcal{U}$. □

We shall now derive some elementary properties of Z-sets.

6.2.2. LEMMA: *Let* X *be a space. Then*

(1) *If* $A \in \mathcal{Z}(X)$ *and* $B \subseteq A$ *is closed in* X *then* $B \in \mathcal{Z}(X)$.

(2) *If* $A \in \mathcal{Z}(X)$ *then* A *has empty interior in* X.

(3) *If* (X,d) *is complete,* $A \in \mathcal{Z}_\sigma(X)$ *and* A *is closed then* $A \in \mathcal{Z}(X)$; *in particular, finite unions of Z-sets are again Z-sets.*

(4) *If* (X,d) *is complete and* $A \in \mathcal{Z}_\sigma(X)$ *then* $\forall \varepsilon > 0\ \forall f \in C(Q,X)\ \exists g \in C(Q,X)$ *such that* $d(f,g) < \varepsilon$ *and* $g(Q) \cap A = \emptyset$.

(5) *If* $A \in \mathcal{Z}(X)$ *and* Y *is any space then* $A \times Y \in \mathcal{Z}(X \times Y)$.

(6) *If* $A \in \mathcal{Z}(X)$ *and* $h \in \mathcal{H}(X)$ *then* $h(A) \in \mathcal{Z}(X)$.

PROOF: (1), (2), (5) and (6) are trivial. For (4), write $A = \bigcup_{n=1}^{\infty} A_n$ with $A_n \in \mathcal{Z}(X)$ for every $n \in \mathbb{N}$. Using lemma 6.2.1, it is clearly possible to construct maps $f_n \in C(Q,X\backslash A_n)$, $n \in \mathbb{N}$, such that

(7) $d(f_1,f) < \frac{\varepsilon}{2}$ and $d(f_{n+1},f_n) < 3^{-n}.\frac{\varepsilon}{2}$,

(8) $d(f_{n+1},f_n) < 3^{-n}\cdot\min\{d(f_i(Q),A_i): 1 \le i \le n\}$.

By (7) and proposition 1.3.4, $F = \lim_{n\to\infty} f_n$ exists and is an element of C(Q,X). Take $x \in Q$ arbitrarily. Then again by (7) we find that

$$d(F(x),f(x)) = \lim_{n\to\infty} d(f_n(x),f(x)) \le \lim_{n\to\infty} \sum_{m=0}^{n-1} 3^{-m}.\frac{\varepsilon}{2} = \sum_{m=0}^{\infty} 3^{-m}.\frac{\varepsilon}{2} = \frac{3\varepsilon}{4} < \varepsilon.$$

From this we conclude that $d(F,f) < \varepsilon$. It now suffices to prove that $F(Q) \cap A = \emptyset$. This however is an immediate consequence of (8) and lemma 6.1.1. This proves (4).

Since (3) is a direct consequence of (4), we are done. □

We are interested particularly in Z-sets in the Hilbert cube Q. The following lemma provides us with "many" Z-sets in Q and provides a simple reformulation of the concept "Z-set" for the special case of the Hilbert cube.

6.2.3. LEMMA: *Let* $A \subseteq Q$ *be a closed set. Then:*

(1) A *is a Z-set iff* $\forall \varepsilon > 0\ \exists f \in C(Q,Q)$ *with* $d(f,1_Q) < \varepsilon$ *and* $f(Q) \cap A = \emptyset$.

(2) *If* $\pi_n(A) \neq [-1,1]_n$ *for infinitely many* $n \in \mathbb{N}$, *then* $A \in \mathcal{Z}(Q)$.

(3) *If* $\pi_n(A) \subseteq \{-1,1\}$ *for certain* $n \in \mathbb{N}$, *then* $A \in \mathcal{Z}(Q)$.

PROOF: For (1), let $A \in \mathcal{Z}(Q)$. Since $1 \in C(Q,Q)$, the definition of a Z-set immediately yields that $\forall \varepsilon > 0\ \exists f \in C(Q,Q)$ with $d(f,1_Q) < \varepsilon$ and $f(Q) \cap A = \emptyset$.

Now suppose that this last statement holds for a certain closed subset $A \subseteq Q$. Choose $\varepsilon > 0$ and $g \in C(Q,Q)$ arbitrarily. Find $f \in C(Q,Q)$ such that $d(f,1) < \varepsilon$ and $f(Q) \cap A = \emptyset$. Put $h = f \circ g$. Then clearly $h(Q) \cap A = \emptyset$, while moreover,

$$d(h,g) = d(f \circ g, g) \leq d(f,1) < \varepsilon$$

(exercise 1.3.7). We conclude that $A \in \mathcal{Z}(Q)$.

For (2), choose $\varepsilon > 0$ and find $n \in \mathbb{N}$ such that $2^{-(n-1)} < \varepsilon$ and $\pi_n(A) \neq [-1,1]_n$. Take an arbitrary $t \in [-1,1]_n \backslash \pi_n(A)$. Define $f\colon Q \to Q$ by

$$f(x_1,\cdots,x_{n-1},x_n,x_{n+1},\cdots) = (x_1,\cdots,x_{n-1},t,x_{n+1},\cdots).$$

Then clearly $f(Q) \cap A = \emptyset$ and $d(f,1) < \varepsilon$. So by (1), $A \in \mathcal{Z}(Q)$.

For (3), observe that $\{-1,1\} \in \mathcal{Z}([-1,1])$. Now apply lemma 6.2.2(1),(5). □

Observe that this lemma implies that each point of Q is a Z-set and also that every *endface* of Q, i.e. a set of the form

$$\pi_n^{-1}(\{1\}) \text{ or } \pi_n^{-1}(\{-1\}) \qquad (n \in \mathbb{N})$$

is a Z-set.

6.2.4. COROLLARY:

(1) $B(Q) \in \mathcal{Z}_\sigma(Q)$, *and*

(2) *if* $K \subseteq s$ *is compact then* $K \in \mathcal{Z}(Q)$. □

Exercises for §6.2.

1. Let $n \geq 1$ and $A \subseteq B^n = \{x \in \mathbb{R}^n : \Sigma_{i=1}^n x_i^2 \leq 1\}$ be closed. Prove that $A \in \mathcal{Z}(B^n)$ iff $A \subseteq S^{n-1}$.

A closed subset A of Q is called a *T-set* if for every open $U \subseteq Q$ such that U is homotopically trivial, the set $U \setminus A$ is again homotopically trivial.

2. Let $A \subseteq Q$ be a T-set. Prove that for each polyhedron P, for each $\varepsilon > 0$ and for each $f \in C(P,Q)$ there exists $g \in C(P,Q)$ such that $d(f,g) < \varepsilon$ and $g(P) \cap A = \emptyset$.

3. Let $A \subseteq Q$ be closed. Prove that A is a T-set iff A is a Z-set.

6.3. The Estimated Homeomorphism Extension Theorem for Compacta in s

The aim of this section is to prove that if $E,F \subseteq s$ are compact and if $f: E \to F$ is a homeomorphism such that $d(f,1_E) < \varepsilon$ then there is a homeomorphism $\bar{f}: Q \to Q$ that extends f and satisfies the same smallness condition, i.e. $d(\bar{f},1_Q) < \varepsilon$.

An element $h \in \mathcal{H}(Q)$ is called *boundary preserving* if $h(B(Q)) = B(Q)$, or equivalently, $h(s) = s$.

6.3.1. LEMMA: *For every compact subset* $K \subseteq s$ *such that* $K \neq \emptyset$ *there is a boundary preserving homeomorphism* $h: Q \to Q$ *such that* $\pi_1 h(K) = \{0\}$.

PROOF: Without loss of generality we may assume that

$$K = \Pi_{n=1}^{\infty}[a_n,b_n],$$

where $-1 < a_n \leq b_n < 1$ for $n \in \mathbb{N}$. For every $n > 1$, let $h_n: J_1 \times J_n \to J_1 \times J_n$ be a homeomorphism of the 2-cell $J_1 \times J_n$ having the following properties:

(1) h_1 does not change the J_1-coordinate of any point,
(2) every horizontal line intersects $h_n([a_1,b_1] \times [a_n,b_n])$ in a set of diameter at most $\frac{1}{n}$.

Here is a picture of h_n. As usual, it is geometrically obvious that h_n exists.

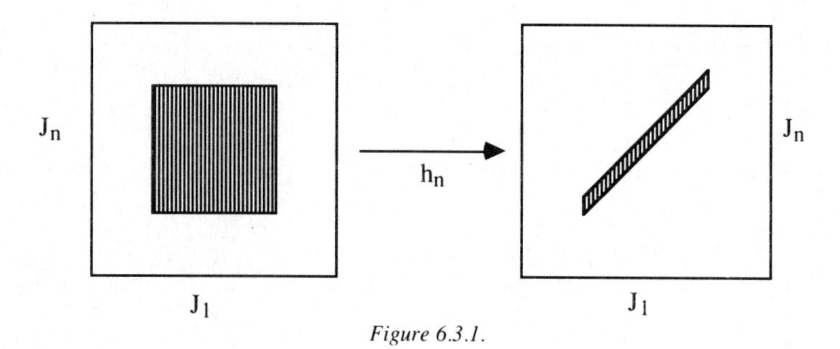

Figure 6.3.1.

Define f: Q → Q by

$$f(x_1,x_2,x_3,\cdots) = (x_1,y_2,y_3,\cdots),$$

where $(x_1,y_n) = h_n(x_1,x_n)$ for every $n > 1$. It is easy to verify that f is a homeomorphism and that f is boundary preserving (if necessary, apply exercise 2.1.2).

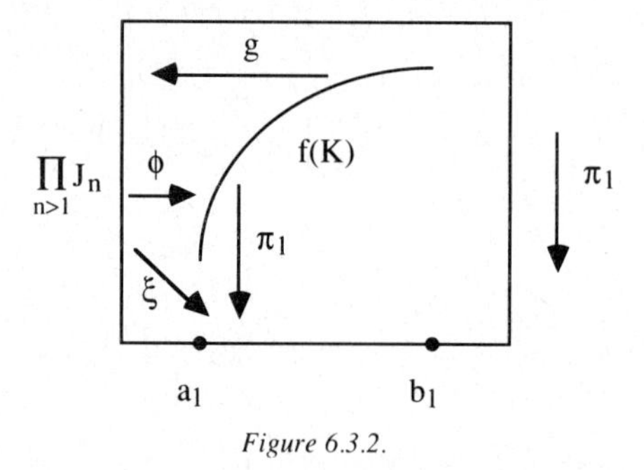

Figure 6.3.2.

Now think of Q as $J_1 \times \prod_{n=2}^{\infty} J_n$ and take two points in f(K) having the same second coordinates, i.e. points of the form (x,z) and (y,z). If $x \neq y$ then $|x - y| > \frac{1}{n}$ for certain $n \geq 2$. By the definition of f we have that the points (x,z_n) and (y,z_n) both belong to $h_n([a_1,b_1] \times [a_n,b_n])$. But this contradicts (2). So we conclude that the function $g\colon f(K) \to \prod_{n=2}^{\infty} J_n$ defined by $g(x,y) = y$ is one-to-one and by compactness is therefore an imbedding. Let $B = gf(K)$ and let $\phi\colon B \to f(K)$ be

the inverse of g. In addition, let $\xi: B \to [a_1,b_1]$ be ϕ followed by the projection π_1 onto J_1. We finish the proof by the same strategy as in the proof of theorem 6.1.14. By the Tietze Extension Theorem 1.4.14, we can extend ξ to a continuous function $\lambda: \prod_{n=2}^{\infty} J_n \to [a_1,b_1]$. For every $t \in [a_1,b_1]$ let H_t be the unique homeomorphism of J_1 taking $[-1,t]$ linearly onto $[-1,0]$ and $[t,1]$ linearly onto $[0,1]$. Then $H: J_1 \times [a_1,b_1] \to J_1$ defined by $H(x,t) = H_t(x)$ is an isotopy.
Now put

$$F(x,y) = (H(x,\lambda(y)),y).$$

Then F belongs to $\mathcal{H}(Q)$ by theorem 6.1.13. In addition, F has the property that for every $(x,y) \in f(K)$,

$$F(x,y) = (H(x,\lambda(y)),y) = (H(x,x),y) = (0,y).$$

Since F is clearly boundary preserving, we find that $h = F \circ f$ is as required. □

6.3.2. COROLLARY: *For every compact subset* $K \subseteq s$ *and* $\varepsilon > 0$ *there are an infinite* $N \subseteq \mathbb{N}$ *and a boundary preserving homeomorphism* $h: Q \to Q$ *such that*

(1) *the complement of* N *is infinite and* $\Sigma_{n\in N} 2^{-n} < \varepsilon$,

(2) $d(h,1_Q) < \varepsilon$,

(3) $\pi_n h(K) = \{0\}$ *for every* $n \in N$.

PROOF: Choose $n \in \mathbb{N}$ so large that

$$\sum_{m=n}^{\infty} 2^{-m} < \varepsilon.$$

Write $\mathbb{N}\setminus\{1,2,\cdots,n-1\}$ as the disjoint union of infinitely many infinite sets, say $C_1,C_2,\cdots$. For every $i \in \mathbb{N}$ let

$$p_i: Q \to \prod_{n\in C_i} J_n$$

be the projection and let

$$h_i: \prod_{n\in C_i} J_n \to \prod_{n\in C_i} J_n$$

be a homeomorphism with the following properties:

(1) h_i is boundary preserving (this has its obvious meaning),
(2) $h_i p_i(K)$ projects onto 0 in the first factor of the product $\prod_{n \in C_i} J_n$

(lemma 6.3.1). Define h: $Q \to Q$ by

$$h(x)_j = \begin{cases} x_j & (j \leq n-1), \\ (h_i p_i(x))_j & (j \in C_i). \end{cases}$$

It is clear that h is as required (observe that h is nothing but the product of the hi's and the identity on the first n-1 factors).

Finally, observe that the set $N = \{\min(C_i): i \in \mathbb{N}\}$ has infinite complement in $\mathbb{N}$ and $N \subseteq \{n, n+1, \cdots\}$ so $\Sigma_{m \in N} 2^{-m} < \varepsilon$. □

The following lemma is the key in deriving our main result. The strategy of the proof is similar to the one in theorem 6.1.14, but is more complicated.

It will be convenient to introduce some notation that shall be fixed throughout the remaining part of this section. Let A and B be complementary infinite subsets of $\mathbb{N}$. Put

$$Q_A = \{x \in Q: x_n = 0 \text{ if } n \notin A\} \text{ and } Q_B = \{x \in Q: x_n = 0 \text{ if } n \notin B\},$$

respectively.

It will be convenient to think of Q as $Q_A \times Q_B$. We will specify A and B later. Let $\delta > 0$ be such that

$$\sum_{n \in B} 2^{-n} < \delta/2.$$

The "origin" of Q, i.e. the point having all coordinates 0, shall be denoted by **0**, or, as we think of Q as $Q_A \times Q_B$, by $(\mathbf{0},\mathbf{0})$. Let X,Y and Z be compact subsets of s such that $X \cup Y \subseteq Q_A$ and $Z \subseteq Q_B$ and let p: $X \to Z$ and q: $Y \to Z$ be homeomorphisms such that

(*) $$d(q^{-1}p, 1_X) < \gamma$$

for certain $\gamma > 0$. Again we will specify p,q,γ,X,Y and Z later (see figure 6.3.3).

A homeomorphism f: $Q \to Q$ is called an *A-homeomorphism* (a *B-homeomorphism*) if for any

$x \in Q$ we have

$$f(x)_n = x_n$$

for all $n \in A$ ($n \in B$, respectively). So an A-homeomorphism is a "vertical action". Similarly, a B-homeomorphism acts "horizontally".

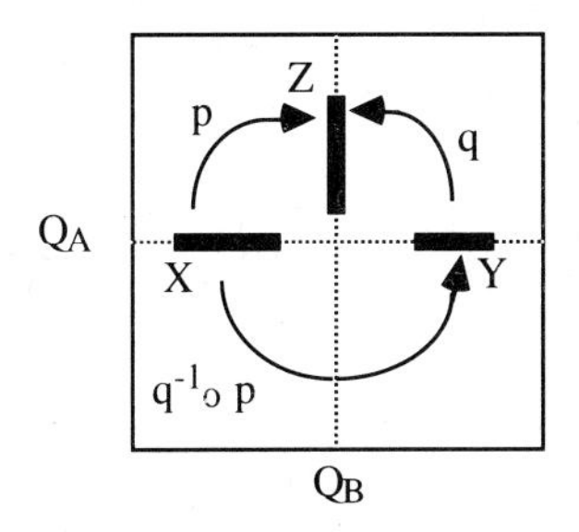

Figure 6.3.3.

6.3.3. LEMMA:

(1) *There is a boundary preserving* A-*homeomorphism* $h_1: Q \to Q$ *such that* $h_1(x,\mathbf{0}) = (x,p(x))$ *for every* $x \in X$.

(2) *There is a boundary preserving* A-*homeomorphism* $h_2: Q \to Q$ *such that* $h_2(y,\mathbf{0}) = (y,q(y))$ *for all* $y \in Y$.

(3) *There is a boundary preserving* B-*homeomorphism* $h_3: Q \to Q$ *such that* $h_3(x,p(x)) = (q^{-1}p(x),p(x))$ *for every* $x \in X$ *while moreover* $d(h_3,1_Q) < \gamma$.

The reader should pause to see what is going on. The homeomorphism h_1 takes X onto the "graph" G(p) of p. Similarly, h_2 takes Y onto the "graph" G(q) of q. Finally, h_3 is a "small" homeomorphism mapping G(p) onto G(q). So $h_2^{-1} \circ h_3 \circ h_1$ is a homeomorphism of Q extending the homeomorphism $q^{-1} \circ p: X \to Y$ (see figure 6.3.4).

PROOF: Since $X \cup Y \cup Z$ is compact we can find a sequence of intervals $[-r_n,r_n] \subseteq (-1,1)$, $n \in \mathbb{N}$, such that $X \cup Y \cup Z \subseteq \prod_{n=1}^{\infty}[-r_n,r_n]$. Put

$$K_A = \{x \in \prod_{n=1}^{\infty}[-r_n,r_n]: x_n = 0 \text{ if } n \notin A\}$$

and

$$K_B = \{x \in \prod_{n=1}^{\infty}[-r_n,r_n]: x_n = 0 \text{ if } n \notin B\},$$

respectively.

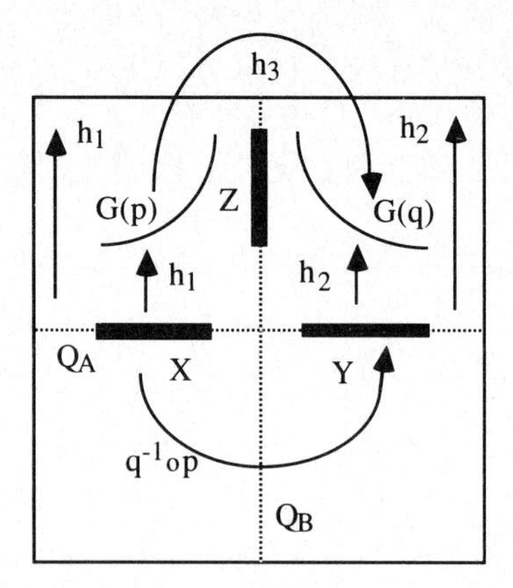

Figure 6.3.4.

Observe that both K_A and K_B are products of symmetric intervals, that $X \cup Y \subseteq K_A \subseteq Q_A$ and that $Z \subseteq K_B \subseteq Q_B$. By applying the Tietze Extension Theorem 1.4.14 to each factor of K_B separately, we can extend $p: X \to Z \subseteq K_B$ to a continuous function $\bar{p}: Q_A \to K_B$ (alternatively, apply corollary 1.5.5). For every $t \in (-1,1)$ let $\phi_t: J \to J$ be the unique homeomorphism taking the interval $[-1,0]$ linearly onto $[-1,t]$ and $[0,1]$ linearly onto $[t,1]$. For each $n \in B$ let the isotopy $H^n: J \times [-r_n,r_n] \to J$ be defined by $H^n(x,t) = \phi_t(x)$. The H^n's define a K_B-isotopy $H: Q_B \times K_B \to Q_B$ as follows:

$$H(y,x)_n = H^n(y_n,x_n) \qquad (n \in B)$$

(proposition 6.1.12) (we make an obvious identification here). Now define $h_1: Q_A \times Q_B \to Q_A \times Q_B$ by

$$h_1(x,y) = (x,H(y,\bar{p}(x))).$$

By theorem 6.1.13, h_1 is an A-homeomorphism. In addition, if $x \in X$ then

$$h_1(x,\mathbf{0}) = (x,H(\mathbf{0},\bar{p}(x))) = (x,H(\mathbf{0},p(x))) = (x,p(x))$$

(for the last equality, use the definition of the H^n's, $n \in B$). We conclude that h_1 is as required. The construction of h_2 is precisely the same.

The construction of h_3 is similar but slightly more complicated because of the smallness condition involved.

Consider $p^{-1}: Z \to X \subseteq K_A$ and $q^{-1}: Z \to Y \subseteq K_A$. By exercise 1.3.7 and (*),

$$d(p^{-1},q^{-1}) = d(p^{-1}p,q^{-1}p) = d(1_X,q^{-1}p) < \gamma.$$

CLAIM: There exist continuous extensions ξ,η: $Q_B \to K_A$ of p^{-1} and q^{-1}, respectively, such that $d(\xi,\eta) < \gamma$.

By applying the Tietze Extension Theorem 1.4.14 to each factor of K_A separately (alternatively, apply corollary 1.5.5), we can extend p^{-1} and q^{-1} to continuous functions f,g: $Q_B \to K_A$. Put

$$U = \{x \in Q_B: d(f(x),g(x)) < \gamma\}.$$

It is clear that U is an open neighborhood of Z in Q_B. Let α: $Q_B \to I$ be a Urysohn function such that $\alpha \mid Z \equiv 1$ and $\alpha \mid (Q_B \backslash U) \equiv 0$ (corollary 1.4.15). Now define ξ,η: $Q_B \to K_A$ by

$$\xi(x) = \alpha(x)\cdot f(x) \text{ and } \eta(x) = \alpha(x)\cdot g(x).$$

Observe that by the special choice of K_A these functions are well-defined (and obviously continuous). Now if $x \notin U$ then $\alpha(x) = 0$, so $\xi(x) = \eta(x)$. If $x \in U$ then

$$d(\xi(x),\eta(x)) = \sum_{n=1}^{\infty} 2^{-n}\alpha(x)\cdot|f(x)_n - g(x)_n| = \alpha(x)\cdot d(f(x),g(x)) < 1\cdot\gamma = \gamma.$$

From this we conclude that $d(\xi,\eta) < \gamma$. If $x \in Z$ then $\alpha(x) = 1$, so $\xi(x) = 1\cdot f(x) = f(x)$. We conclude that ξ extends f, and similarly, that η extends g. Consequently, ξ and η are as required.

For all $(x,y) \in (-1,1)^2$ let $\phi_{(x,y)}$: $J \to J$ be the unique homeomorphism taking $[-1,x]$ linearly onto $[-1,y]$ and $[x,1]$ linearly onto $[y,1]$. Observe the following easy:

CLAIM: If $(x,y) \in (-1,1)^2$ then $d(\phi_{(x,y)},1_J) = |x - y|$.

We now define a $(K_A \times K_A)$-isotopy F: $Q_A \times (K_A \times K_A) \to Q_A$ coordinatewise as follows:

$$F(q,x,y)_n = \phi_{(x_n,y_n)}(q_n) \qquad (n \in A),$$

cf. proposition 6.1.12. Now define h_3: $Q_A \times Q_B \to Q_A \times Q_B$ by

$$h_3(x,y) = (F_{(\xi(y),\eta(y))}(x),y).$$

Then h_3 is a B-homeomorphism (theorem 6.1.13). We shall prove that h_3 is as required.

Take an arbitrary $x \in X$. Then

$$h_3(x,p(x)) = (F_{(\xi p(x),\eta p(x))}(x),p(x)) = (F_{(x,q^{-1}p(x))}(x),p(x)) = (q^{-1}p(x),p(x))$$

(for the last equality, use the definition of F).

Finally, take an arbitrary $(x,y) \in Q_A \times Q_B$. Then

$$d(h_3(x,y),(x,y)) = \sum_{n\in A} 2^{-n}|(F_{(\xi(y),\eta(y))}(x))_n - x_n| = \sum_{n\in A} 2^{-n}|\phi_{(\xi(y)_n,\eta(y)_n)}(x_n) - x_n|$$

$$\leq \sum_{n\in A} 2^{-n}d(\phi_{(\xi(y)_n,\eta(y)_n)}, 1_J) = \sum_{n\in A} 2^{-n}|\xi(y)_n - \eta(y)_n|$$

$$\leq d(\xi,\eta)$$

$$< \gamma$$

(apply the claim). We conclude that $d(h_3,1) < \gamma$.□

We now come to the main result in this section.

6.3.4. THEOREM: *Let* $E,F \subseteq s$ *be compact and let* $f: E \to F$ *be a homeomorphism such that* $d(f,1_E) < \varepsilon$. *Then* f *can be extended to a boundary preserving homeomorphism* $\bar{f}: Q \to Q$ *such that* $d(\bar{f},1_Q) < \varepsilon$.

PROOF: Let $\varepsilon_1 = d(f,1_E)$ and put $\delta = (\varepsilon - \varepsilon_1)/6$ (this specifies δ). By corollary 6.3.2 we find a boundary preserving homeomorphism $g: Q \to Q$ and an infinite subset $B \subseteq \mathbb{N}$ such that

(1) $d(g,1) < \delta$,
(2) $\pi_n g(E \cup F) = \{0\}$ for every $n \in B$,
(3) $A = \mathbb{N}\setminus B$ is infinite and $\Sigma_{n\in B} 2^{-n} < \delta/2$

(this specifies A and B). Put $X = g(E)$, $Y = g(F)$ and $h = g \circ f \circ g^{-1}$. Observe that

$$d(h,1_X) < \varepsilon_1 + 2\delta.$$

Notice that $X \cup Y \subseteq Q_A$. Since $Q_B \approx Q$, we can find a topological copy Z of X in s such that $Z \subseteq Q_B$ (theorem 1.4.18) (Q_B is not a subset of s but a smaller infinite-dimensional "subcube" is). Let $p: X \to Z$ be any homeomorphism and let $q = p \circ (h)^{-1}$. Then clearly

$$d(q^{-1} \circ p, 1_X) = d(h, 1_X) < \varepsilon_1 + 2\delta.$$

Put $\gamma = \varepsilon_1 + 2\delta$. This specifies p,q and γ. Now let h_1,h_2 and h_3 be as in lemma 6.3.3. By (3) we have

$$d(h_1, 1) < \delta \text{ and } d(h_2, 1) < \delta.$$

Consequently, if we put

$$t = h_2^{-1} \circ h_3 \circ h_1,$$

then

$$d(t, 1) < \gamma + 2\delta = \varepsilon_1 + 4\delta.$$

Since t clearly extends h, we find that

$$\bar{f} = g^{-1} \circ t \circ g$$

extends f and that $d(\bar{f}, 1) < \varepsilon_1 + 4\delta + 2\delta = \varepsilon$. □

Exercises for §6.3.

1. Make the "geometrically obvious" part in the proof of lemma 6.3.1 precise by explicitly describing homeomorphisms h_n such as used there.

2. Let $\varepsilon > 0$. An isotopy $H: Q \times I \to Q$ is called an ε-isotopy provided that for every $x \in Q$ the diameter of the set $H(\{x\} \times I)$ is less than ε. Prove the following improvement of theorem 6.3.4. Let $E,F \subseteq s$ be compact and let $f: E \to F$ be a homeomorphism such that $d(f, 1_E) < \varepsilon$. Then there is an ε-isotopy $H: Q \times I \to Q$ such that $H_0 = 1$ and $H_1 \mid E = f$ (i.e. H_1 extends f).

6.4. The Estimated Homeomorphism Extension Theorem

The aim of this section is to prove that if $E,F \in \mathcal{Z}(Q)$ and if $f: E \to F$ is a homeomorphism such that $d(f,1_E) < \varepsilon$ then f can be extended to a homeomorphism $\bar{f}: Q \to Q$ such that $d(\bar{f},1_Q) < \varepsilon$. This result is known as the *estimated homeomorphism extension theorem* and is of fundamental importance in infinite-dimensional topology. The strategy of the proof is that we first push E and F into s by a small motion and then apply theorem 6.3.4.

We shall first fix some notation. As in §3.5, if $n \in \mathbb{N}$ and $\theta \in \{-1,1\}$ put

$$W_n(\theta) = \pi_n^{-1}(\{\theta\}),$$

i.e. $W_n(\theta)$ is an endface in the n-th coordinate direction. Throughout, let $K \subseteq s$ be a fixed compact subset. For each subset $F \subseteq \mathbb{N}$ let $p_F: Q \to \prod_{n\in F} J_n$ be the projection.

We will first show that endfaces can be pushed into s by a small movement. This is done by pushing endfaces into endfaces with larger and larger indexes.

6.4.1. LEMMA: *For each* $n \in \mathbb{N}$, $\theta \in \{-1,1\}$ *and* $\varepsilon > 0$ *there is a homeomorphism* $h: Q \to Q$ *and an* $m > n$ *such that*

(1) $h(W_n(\theta)) \cap \bigcup\{W_i(\mu): i < m$ *and* $\mu \in \{-1,1\}\} = \emptyset$,

(2) $h(W_n(\theta)) \subseteq W_m(1)$,

(3) $d(h,1) < \varepsilon$,

(4) $h \mid K = 1$.

PROOF: Let $\varepsilon > 0$ and choose $m > n$ such that $2^{-(m-2)} < \varepsilon/2$. We first push $W_n(\theta)$ into $W_m(1)$ and then away from the endfaces in the lower coordinate directions. Without loss of generality we assume that $\theta = 1$.

It is geometrically obvious that there is a homeomorphism $\phi: J_n \times J_m \to J_n \times J_m$ such that

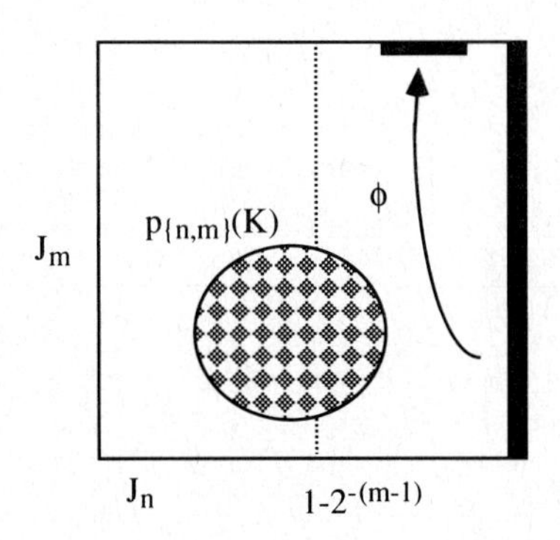

Figure 6.4.1.

(1) $\phi(\{1\} \times J_m) \subseteq J_n \times \{1\}$,
(2) $\phi \mid p_{\{n,m\}}(K) = 1$,
(3) if $x \leq 1 - 2^{-(m-1)}$ then $\phi(x,y) = (x,y)$ for every y.

Let $h_1 \in \mathcal{H}(Q)$ be the homeomorphism of Q that is defined by crossing ϕ with the identity on the other factors of Q.

CLAIM: (4) $d(h_1, 1_Q) \leq 2^{-(m-2)} < \frac{\varepsilon}{2}$,
(5) $h_1(W_n(\theta)) \subseteq W_m(1)$,
(6) $h \mid K = 1_K$.

Observe that (4) follows as in the proof of lemma 6.1.4. Since (5) and (6) are trivial, we are done.

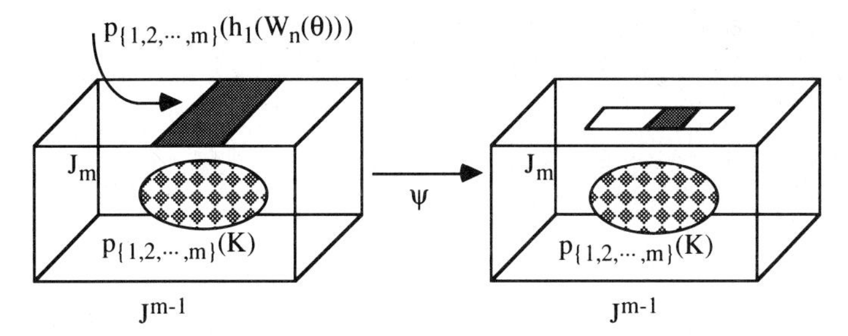

Figure 6.4.2.

By squeezing the (m-1)-cell J^{m-1} into its interior we can construct a homeomorphism $\psi: J^m \to J^m$ such that

(7) $\psi(J^{m-1} \times \{1\}) \subseteq \prod_{j=1}^{m-1}(-1,1)_j \times \{1\}_m$,
(8) $d(\psi, 1) < \frac{\varepsilon}{2}$,
(9) $\psi \mid p_{\{1,2,\cdots,m\}}(K) = 1$

(see figure 6.4.2).

Now let $h_2 \in \mathcal{H}(Q)$ be the homeomorphism that is defined by crossing ψ with the identity on the other factors of Q. It is clear that

(10) $h_2 \mid K = 1_K$,
(11) $d(h_2, 1) < \frac{\varepsilon}{2}$.

From (4),(5),(6),(7),(10) and (11) it follows easily that $h = h_2 \circ h_1$ is as required. □

6.4.2. COROLLARY: *For each* $n \in \mathbb{N}$, $\theta \in \{-1,1\}$ *and* $\varepsilon > 0$ *there is a homeomorphism* $h: Q \to Q$ *with*

(1) $h(W_n(\theta)) \subseteq s$,

(2) $d(h,1) < \varepsilon$,

(3) $h \mid K = 1$.

PROOF: Put $n_1 = n$. By lemma 6.4.1 we can push $W_n(\theta)$ into some other face $W_{n_2}(1)$ with $n_2 > n_1$, by a homeomorphism $h_1 \in \mathcal{H}(Q)$ which is as close to to the identity as we wish. Repeated applications of this yields a sequence $(h_i)_i \in \mathcal{H}(Q)$, each element of which is close enough to the identity in order to have

$$h = \lim_{i\to\infty} h_i \circ h_{i-1} \circ \cdots \circ h_1 \in \mathcal{H}(Q) \text{ and } d(h,1) < \varepsilon$$

(theorem 6.1.2). Moreover, each h_i can be taken to be the identity on K, to the effect that $h \mid K = 1_K$.

Finally, each h_i pushes the face $W_{n_i}(1)$ off the faces $W_j(\mu)$, $j < n_{i+1}$, $\mu \in \{-1,1\}$. For h_i close enough to the identity (as prescribed in lemma 6.1.1), we can keep the limit homeomorphism h "away from *all* faces of Q", that is $h(W_n(\theta)) \subseteq s$. □

We now know how to push an endface into s. We are going to use this result to push an arbitrary Z-set A away from a fixed endface, and then away from all endfaces, i.e. to push A into s.

6.4.3. PROPOSITION: *Let* X *be a compact space, let* $X_0 \subseteq X$ *be closed and let* $f: X \to s$ *be continuous such that* $f \mid X_0$ *is an imbedding. Then for every* $\varepsilon > 0$ *there is an imbedding* $g: X \to s$ *such that*

(1) $d(f,g) < \varepsilon$,

(2) $g \mid X_0 = f \mid X_0$.

PROOF: Let $\mathcal{B}$ be a countable basis for $X \backslash X_0$ consisting of compact sets (observe that $X \backslash X_0$ is locally compact). In addition, let $\{(F_i,G_i): i \in \mathbb{N}\}$ enumerate the set

$$\{(F,G) \in \mathcal{B} \times \mathcal{B}: F \cap G = \emptyset\}.$$

Choose $m \in \mathbb{N}$ such that

$$\sum_{n=m}^{\infty} 2^{-n} < \varepsilon.$$

Since $\pi_{m+i}f(X_0)$ is a compact subset of $(-1,1)_{m+i}$, for every $i \in \mathbb{N}$ we can apply the Tietze Extension Theorem 1.4.14 to construct a continuous function g_i: $X \to (-1,1)_{m+i}$ such that

(3) $g_i \mid X_0 = (\pi_{m+i} \circ f) \mid X_0$,
(4) $g_i(X_0) \cap (g_i(F_i) \cup g_i(G_i)) = \emptyset = g_i(F_i) \cap g_i(G_i)$.

(pick two distinct points $p,q \in (-1,1)_{m+i} \setminus \pi_{m+i}f(X_0)$ and extend the function that is constant p on F_i, constant q on G_i, and is equal to $\pi_{m+i} \circ f$ on X_0.) Define g: $X \to s$ by

$$g(x) = (f(x)_1, f(x)_2, \cdots, f(x)_m, g_1(x), g_2(x), \cdots).$$

It is clear that g is continuous, that $g \mid X_0 = f \mid X_0$ and that $d(f,g) < \varepsilon$. Take $x,y \in X$ such that $x \neq y$. If $x,y \in X_0$ then $g(x) \neq g(y)$ since $f \mid X_0$ is an imbedding. If $x \in X_0$ and $y \notin X_0$ then we can find an index $i \in \mathbb{N}$ such that $y \in F_i$. By (4) and the definition of g it now follows that $g(x)_{m+i} \neq g(y)_{m+i}$. Finally, if $x,y \notin X_0$ then we can find $i \in \mathbb{N}$ such that $x \in F_i$ and $y \in G_i$. As above it follows that $g(x)_{m+i} \neq g(y)_{m+i}$. We conclude that g is one-to-one and the compactness of X now yields that g is an imbedding (exercise 1.1.4). □

6.4.4. LEMMA: *Let* $A \in \mathcal{Z}(Q)$. *Then for each* $n \in \mathbb{N}$, $\theta \in \{-1,1\}$ *and* $\varepsilon > 0$ *there is a homeomorphism* h: $Q \to Q$ *with*

(1) $h(A) \cap W_n(\theta) = \emptyset$,
(2) $d(h,1) < \varepsilon$,
(3) $h \mid K = 1$.

PROOF: By corollary 6.4.2 we can find a homeomorphism $h_1 \in \mathcal{H}(Q)$ such that $d(h_1,1) < \varepsilon/3$, $h_1 \mid K = 1$ and $h_1(W_n(\theta)) \subseteq s$. Since $A \cup K \cup B(Q) \in \mathcal{Z}_\sigma(Q)$ (corollary 6.2.4), by lemma 6.2.2(4) there is a continuous function α: $Q \to Q\setminus(A \cup K \cup B(Q))$ such that $d(\alpha,1) < \varepsilon/3$. Consequently, $\alpha(Q) \subseteq s$ and $d(\alpha(Q), A \cup K) > 0$. By proposition 6.4.3 there is an imbedding β: $Q \to s$ with

(4) $\beta(Q) \cap (A \cup K) = \emptyset$,
(5) $d(\beta,\alpha) < \varepsilon/3$

(observe that we can achieve (4) since $d(\alpha(Q), A \cup K) > 0$). Notice that $d(\beta,1) < 2\varepsilon/3$. Put

$$\gamma = \beta \mid h_1(W_n(\theta)) \cup 1_K.$$

Since

$$K \cap (h_1(W_n(\theta)) \cup \beta h_1(W_n(\theta))) = \emptyset,$$

γ is a well-defined homeomorphism from $h_1(W_n(\theta)) \cup K$ onto $\beta h_1(W_n(\theta)) \cup K$ such that $d(\gamma,1) < 2\varepsilon/3$. By theorem 6.3.4, we can extend γ to a homeomorphism $G: Q \to Q$ such that $d(G,1) < 2\varepsilon/3$. Observe that

(6) $G \mid K = 1$,
(7) $Gh_1(W_n(\theta)) \cap A = \emptyset$.

Now put $h = (G \circ h_1)^{-1}$. Then h is as required since $d(h,1) < 2\varepsilon/3 + \varepsilon/3 = \varepsilon$.□

We can now prove the following:

6.4.5. THEOREM: *Let* $K \subseteq s$ *be compact and let* $A \in \mathcal{Z}(Q)$. *Then for every* $\varepsilon > 0$ *there is a homeomorphism* $h: Q \to Q$ *with*

(1) $d(h,1) < \varepsilon$,
(2) $h \mid K = 1_K$,
(3) $h(A) \subseteq s$.

PROOF: All we have to do is to apply lemma 6.4.4 inductively to free A from all endfaces $W_n(\theta)$ and keep K fixed. This has to be done with a little care so that once A is free from an endface, the limit homeomorphism does not carry it back to that endface. But this can be done again with the help of lemma 6.1.1 (cf. the proof of corollary 6.4.2).□

We can now prove some important results

6.4.6. THEOREM ("Homeomorphism Extension Theorem"): *Let* $E,F \in \mathcal{Z}(Q)$ *and let* $f: E \to F$ *be a homeomorphism such that* $d(f,1_E) < \varepsilon$. *Then* f *can be extended to a homeomorphism* $\bar{f}: Q \to Q$ *such that* $d(\bar{f},1) < \varepsilon$.

PROOF: Just apply theorems 6.3.4 and 6.4.5.□

Perhaps the reader feels that it is not worth the trouble in theorem 6.4.6 to prove that it is possible to extend "small" homeomorphisms to "small" homeomorphisms. At first glance it seems that the possibility of extending homeomorphisms is the most important fact and that the extra smallness condition is a technical curiosity. This is not true however. The possibility to extend "small" homeomorphisms to "small" homeomorphisms makes it possible to apply the inductive convergence criterion once again to create new homeomorphisms from old ones. This procedure turns out to be extremely powerful.

Observe that the metric d in theorem 6.4.6 is the standard "convex" metric on Q. It can be shown that the theorem is false if one replaces d by an equivalent "nonconvex" metric. For details see e.g. [13]. Of course, homeomorphisms between Z-sets can be extended, but problems arise with the smallness condition.

A Hilbert cube is a space homeomorphic to Q. Of course one would also like to have an estimated homeomorphism extension theorem for Hilbert cubes. We run into the same "smallness" problems here.

6.4.7. COROLLARY: *Let* (X,ρ) *be a metric space such that* X *is a Hilbert cube. Then for each* $\varepsilon > 0$ *there is* $\delta > 0$ *with the following property: if* $A,B \in \mathcal{Z}(X)$ *and* $f: A \to B$ *is a homeomorphism with* $\rho(f,1_A) < \delta$ *then there is a homeomorphism* $\bar{f}: X \to X$ *that extends* f *while moreover* $\rho(\bar{f},1_X) < \varepsilon$.

PROOF: Let $\phi: Q \to X$ be any homeomorphism. Now use theorem 6.4.6 and the fact that ϕ is uniformly continuous. □

Let X be a space. A continuous function $f: X \to Q$ is called a *Z-map* provided that $f(X) \in \mathcal{Z}(Q)$. If f is an imbedding as well as a Z-map then it is called a *Z-imbedding.*

The following result is a generalization of proposition 6.4.3. It will be used several times in the forthcoming.

6.4.8. THEOREM: *Let* X *be a compact space, let* $A \subseteq X$ *be closed and let* $f: X \to Q$ *be continuous such that* $f \mid A$ *is a Z-imbedding. Then for every* $\varepsilon > 0$ *there is a Z-imbedding* $g: X \to Q$ *such that* $d(f,g) < \varepsilon$ *and* $g \mid A = f \mid A$.

PROOF: Since $B(Q) \in \mathcal{Z}_\sigma(Q)$ (corollary 6.2.4), we can approximate f by a map $\bar{f}: X \to s$. By proposition 6.4.3 we can approximate $\bar{f}$ by an imbedding $\bar{\bar{f}}: X \to s$. Observe that $\bar{\bar{f}}(X) \in \mathcal{Z}(Q)$ (corollary 6.2.4). Then f(A) and $\bar{\bar{f}}(A)$ are homeomorphic Z-sets and we can now use theorem 6.4.6 to find a "small" homeomorphism $h \in \mathcal{H}(Q)$ extending the obvious homeomorphism between $\bar{\bar{f}}(A)$ and f(A). Then $g = h \circ \bar{\bar{f}}$ is as required. □

Exercises for §6.4.

1. Prove the following generalization of theorem 6.4.6: If $E,F \in \mathcal{Z}(Q)$ and if $f: E \to F$ is a homeomorphism such that $d(f,1) < \varepsilon$ then there is an ε-isotopy $H: Q \times I \to Q$ with $H_0 = 1_Q$ and $H_1 \mid E = f$, i.e. H_1 extends f.

2. Let X be a space. Suppose that $X = Q_0 \cup Q_1$, where $Q_0 \approx Q_1 \approx Q_0 \cap Q_1 \approx Q$, $Q_0 \cap Q_1 \in \mathcal{Z}(Q_0)$ and $Q_0 \cap Q_1 \in \mathcal{Z}(Q_1)$. Prove that $X \approx Q$.

6.5. Absorbers

The aim of this section is to characterize all elements $A \in \mathcal{Z}_\sigma(Q)$ having the property that for some homeomorphism $h: Q \to Q$ it is true that $h(A) = B(Q)$. As applications, we prove that an infinite product of intervals is homeomorphic to $\mathbb{R}^\infty$ iff infinitely many of them are not compact and that $\mathbb{R}^\infty \setminus E \approx \mathbb{R}^\infty$ for every σ-compact subset $E \subseteq \mathbb{R}^\infty$.

Let M^Q be a Hilbert cube. An element $A \in \mathcal{Z}_\sigma(M^Q)$ is called an *absorber* provided that for all $K,L \in \mathcal{Z}(M^Q)$ and $\varepsilon > 0$ there is a homeomorphism $h: M^Q \to M^Q$ such that

(1) $d(h,1) < \varepsilon$,
(2) $h \mid K = 1_K$,
(3) $h(L \setminus K) \subseteq A$.

Roughly speaking, A absorbs $L \setminus K$ by a small motion keeping K fixed.

An element $A \in \mathcal{Z}_\sigma(M^Q)$ is called a *skeletoid* provided that A can be written as

$$A = \bigcup_{n=1}^{\infty} A_n,$$

where $A_n \in \mathcal{Z}(M^Q)$ and $A_n \subseteq A_{n+1}$ for all $n \in \mathbb{N}$, while moreover the following absorption property holds: for every $\varepsilon > 0$, $n \in \mathbb{N}$, and $K \in \mathcal{Z}(M^Q)$ there are an $m \in \mathbb{N}$ and a homeomorphism $h: M^Q \to M^Q$ such that

(1) $d(h,1) < \varepsilon$,
(2) $h \mid A_n = 1$,
(3) $h(K) \subseteq A_m$.

Obviously, these two concepts are related. In abstract considerations it turns out that absorbers are very convenient to work with. On the other hand, if we wish to verify in a concrete situation that a certain set is an absorber we usually do that by first proving that it is a skeletoid, since this is often easy to verify, and then by applying the following result.

6.5.1. THEOREM: *Every skeletoid is an absorber.*

PROOF: Let $A \in \mathcal{Z}_\sigma(M^Q)$ be a skeletoid, say

$$A = \bigcup_{n=1}^{\infty} A_n,$$

where the A_n's are as above. Let K and L be Z-sets and let $\varepsilon > 0$. Let $L_0 \subseteq L_1 \subseteq \cdots$ be a sequence of closed subsets of $L \setminus K$ such that $L_0 = \emptyset$ and $L \setminus K = \bigcup_{i=0}^{\infty} L_i$. We shall inductively construct sequences $f_0, f_1, \cdots$ in $\mathcal{H}(M^Q)$ and $n(0) < n(1) < \cdots$ such that $f_0 = 1$ and $n(0) = 1$ while moreover the following statements hold:

(1) $d(f_i,1)$ is so small that the conditions mentioned in the Inductive Convergence Criterion 6.1.2 are satisfied,

(2) $d(f_i,1) < 3^{-i} \cdot \varepsilon$,

(3) $f_i \circ f_{i-1} \circ \cdots \circ f_0(L_i) \subseteq A_{n(i)}$,

(4) $f_i \mid K \cup A_{n(i-1)} = 1$.

This will establish the theorem since by (1) and (2), $f = \lim_{i \to \infty} f_i \circ \cdots \circ f_1$ exists, belongs to $\mathcal{H}(M^Q)$, and satisfies $d(f,1) < \varepsilon$, while moreover by (3) and (4), $f(L \setminus K) \subseteq A$ and $f \mid K = 1_K$.

It remains to perform the induction. Since f_0 and $n(0)$ are already defined, assume that f_j and $n(j)$ have been chosen up to $i \geq 0$. Find $\delta > 0$ such that if we choose f_{i+1} δ-close to 1 then (1) and (2) are automatically satisfied for $i+1$. Let $B = f_i \circ f_{i-1} \circ \cdots \circ f_0(L_{i+1})$ and observe that $B \in \mathcal{Z}(M^Q)$ (lemma 6.2.2(6)) and that $B \cap K = \emptyset$. Let $\gamma = d(B,K)$. Let $\xi > 0$ be such that every homeomorphism between Z-sets of M^Q that moves the points less than ξ can be extended to a homeomorphism of M^Q moving the points less than δ (corollary 6.4.7). Since A is a skeletoid, there are $m > n(i)$ and a homeomorphism $g: M^Q \to M^Q$ such that

(5) $d(g,1) < \min\{\xi, \gamma\}$,

(6) $g \mid A_{n(i)} = 1$,

(7) $g(B) \subseteq A_m$.

Observe that $g(B) \cap K = \emptyset$. It is clear that g is not quite yet as required since g might move K. However, clearly

$$(g \mid B) \cup 1_K \cup 1_{A_{n(i)}}$$

is a homeomorphism of $B \cup K \cup A_{n(i)}$ onto $g(B) \cup K \cup A_{n(i)}$. This homeomorphism can be extended to a homeomorphism $h: M^Q \to M^Q$ with $d(h,1) < \delta$. Now if we put $n(i+1) = m$ and $f_{i+1} = h$ then these choices are easily seen to be as required. □

We shall now present some fundamental properties of absorbers.

6.5.2. THEOREM:

(1) *If* A *is an absorber in* M^Q *and* $h: M^Q \to M^Q$ *is a homeomorphism then* $h(A)$ *is an absorber.*

(2) *If* A *is an absorber in* M^Q *and* $B \in \mathcal{Z}_\sigma(M^Q)$ *then* $A \cup B$ *is an absorber.*

(3) *If* A *and* B *are absorbers in* M^Q *then for each* $\varepsilon > 0$ *there is a homeomorphism* $h: M^Q \to M^Q$ *such that* $h(A) = B$ *and* $d(h,1) < \varepsilon$.

PROOF: For (1), observe that h is uniformly continuous. Notice that (2) is a triviality. Only (3) requires work. Write

$$A = \bigcup_{i=1}^{\infty} A_i \text{ and } B = \bigcup_{i=1}^{\infty} B_i,$$

where $A_i, B_i \in \mathcal{Z}(M^Q)$ for every $i \in \mathbb{N}$. We shall inductively construct a sequence $f_1, f_2, \cdots$ in $\mathcal{H}(M^Q)$ such that $f_1 = 1$ and

(4) $d(f_i,1)$ is small enough in order to apply theorem 6.1.2,

(5) $d(f_i,1) < 3^{-i} \cdot \varepsilon$,

(6) $B_i \subseteq f_i \circ g_{i-1}(A)$,

(7) $f_i \circ g_{i-1}(A_i) \subseteq B$,

(8) $f_i \mid \bigcup_{j=1}^{i-1}(g_{i-1}(A_j) \cup B_j) = 1$,

where $g_{i-1} = f_{i-1} \circ \cdots \circ f_1$.

Assume that $f_1, \cdots, f_i$ have been constructed. Find $\delta > 0$ such that if we choose f_{i+1} δ-close to 1 then (4) and (5) are satisfied for i+1. Observe that $g_i(A_{i+1}) \in \mathcal{Z}(M^Q)$ (lemma 6.2.2(6)). Also, by our induction hypothesis,

$$K = \bigcup_{j=1}^{i}(g_i(A_j) \cup B_j) \subseteq B.$$

Notice that $K \in \mathcal{Z}(M^Q)$ by lemma 6.2.2(3), (6). Since B is an absorber, there is a homeomorphism $\alpha: M^Q \to M^Q$ such that

(9) $d(\alpha,1) < \delta/2$,
(10) $\alpha \circ g_i(A_{i+1}) \subseteq B$ (i.e. $g_i(A_{i+1})$ is absorbed),
(11) $\alpha \mid K = 1$.

Since $\alpha \circ g_i$ belongs to $\mathcal{H}(M^Q)$, by (1) it follows that $\alpha \circ g_i(A)$ is an absorber. Notice that by (6) and (11) we have,

$$K' = K \cup \alpha \circ g_i(A_{i+1}) \subseteq \alpha \circ g_i(A).$$

Notice that $K' \cup B_{i+1} \in \mathcal{Z}(M^Q)$. Consequently, there is a homeomorphism $\beta: M^Q \to M^Q$ such that

(12) $d(\beta,1) < \delta/2$,
(13) $\beta(B_{i+1}) \subseteq \alpha \circ g_i(A)$ (i.e. B_{i+1} is absorbed),
(14) $\beta \mid K' = 1$.

Now define $f_{i+1} = \beta^{-1} \circ \alpha$. Then clearly $d(f_{i+1},1) < \delta$. Moreover, by (10) and (14),

$$f_{i+1} \circ g_i(A_{i+1}) = \beta^{-1} \circ \alpha(g_i(A_{i+1})) = \alpha \circ g_i(A_{i+1}) \subseteq B,$$

and by (13),

$$B_{i+1} \subseteq \beta^{-1}(\alpha \circ g_i(A)) = f_{i+1} \circ g_i(A).$$

Since clearly $f_{i+1} \mid K = 1$, we see that f_{i+1} is as required.

Now put $f = \lim_{i\to\infty} f_i \circ \cdots \circ f_1 = \lim_{i\to\infty} g_i$. Clearly $d(f,1) < \varepsilon$. By (8) and (7) we have,

$$(15) \qquad f(A) = \bigcup_{i=1}^{\infty} f(A_i) = \bigcup_{i=1}^{\infty} g_i(A_i) \subseteq B.$$

In addition, by (8) it follows that for every i,

$$\begin{aligned} f \circ g_i^{-1}(B_i) &= \lim_{n>i} (f_n \circ \cdots \circ f_{i+1} \circ g_i)(g_i^{-1})(B_i) \\ &= \lim_{n>i} (f_n \circ \cdots \circ f_i)(B_i) \end{aligned}$$

$$= B_i,$$

so that by (6),

$$B_i = f \circ g_i^{-1}(B_i) \subseteq f(A).$$

We conclude that

$$B = \bigcup_{i=1}^{\infty} B_i \subseteq f(A). \tag{16}$$

By (15) and (16), $f(A) = B$. We are done.□

A combination of (2) and (3) of the above theorem yields the following:

6.5.3. COROLLARY: *Let* $A,B \in \mathcal{Z}_\sigma(M^Q)$ *such that* A *is an absorber. Then there is a homeomorphism* $h: M^Q \to M^Q$ *with* $h(A \cup B) = A$.□

We now aim at proving that $B(Q)$ is an absorber in Q. Let $n \in \mathbb{N}$ and define

$$\Sigma_n = \prod_{i=1}^{\infty} [-1+2^{-n}, 1-2^{-n}]_i,$$

i.e. $\Sigma_n = \{x \in Q: (\forall i \in \mathbb{N})(|x_i| \le 1-2^{-n})\}$. Put

$$\Sigma = \bigcup_{n=1}^{\infty} \Sigma_n$$

and observe that Σ is a σ-compact subset of s.

6.5.4. PROPOSITION: Σ *is a skeletoid.*

PROOF: Observe that $\Sigma \in \mathcal{Z}_\sigma(Q)$ (corollary 6.2.4). Choose $K \in \mathcal{Z}(Q)$, $n \in \mathbb{N}$ and $\varepsilon > 0$ arbitrarily. By theorem 6.4.5 there is a homeomorphism $f \in \mathcal{H}(Q)$ such that $f(K) \subseteq s$, $d(f,1) < \varepsilon/2$ and $f \mid \Sigma_n = 1$. Choose $m \in \mathbb{N}$ so large that

$$\sum_{i=m}^{\infty} 2^{-i} < \frac{\varepsilon}{4}.$$

There obviously exists $k > n$ such that

$$\pi_i(f(K)) \subseteq [-1+2^{-k}, 1-2^{-k}]$$

for every $i < m$. For each $j \geq m$ let $h_j\colon [-1,1]_j \to [-1,1]_j$ be a homeomorphism having the following properties:

(1) $h_j([\min \pi_j(f(K)), \max \pi_j(f(K))]) \subseteq [-1+2^{-k}, 1-2^{-k}]$,

(2) $h_j \mid [-1+2^{-n}, 1-2^{-n}] = 1$

(It is a triviality that these homeomorphisms exist of course). Define $h\colon Q \to Q$ by

$$h(x_1, x_2, \cdots, x_m, x_{m+1}, \cdots) = (x_1, x_2, \cdots, x_{m-1}, h_m(x_m), h_{m+1}(x_{m+1}), \cdots)$$

(i.e. h is the product of the h_m's and the identity on the first $m-1$ factors of Q). Then h is a homeomorphism, $d(h,1) < \varepsilon/2$, $h \mid \Sigma_n = 1$ and $h(f(K)) \subseteq \Sigma_k$. We conclude that the homeomorphism $h \circ f$ and the natural number k show that Σ is a skeletoid. □

We are now in a position to prove the converse of theorem 6.5.1.

6.5.5. COROLLARY: *Every absorber in* Q *is a skeletoid.*

PROOF: Let A be an absorber. By proposition 6.5.4, Σ is a skeletoid and hence an absorber by theorem 6.5.1. Consequently, theorem 6.5.2(3) implies that there exists $h \in \mathcal{H}(Q)$ with $h(A) = \Sigma$. Now since Σ is a skeletoid, this easily implies that A is a skeletoid as well. □

Recall that for every $n \in \mathbb{N}$ and $\theta \in \{-1,1\}$, $W_n(\theta)$ denotes the endface $\pi_n^{-1}(\{\theta\})$ of Q.

6.5.6. LEMMA: *Let* $\mathcal{A}$ *be an infinite collection of endfaces of* Q. *Then there is a homeomorphism* $h \in \mathcal{H}(Q)$ *such that* $h(\Sigma) \subseteq \bigcup \mathcal{A}$.

PROOF: For every $A \in \mathcal{A}$ let $n(A) \in \mathbb{N}$ and $\theta(A) \in \{-1,1\}$ be such that $A = W_{n(A)}(\theta(A))$. Without loss of generality we assume that there exists an infinite subcollection $\mathcal{A}'$ of $\mathcal{A}$ such that for every $A \in \mathcal{A}'$, $\theta(A) = 1$. It is clear that there exists a partition $\{E_n\colon n \in \mathbb{N}\}$ of $\mathbb{N}$ into countably many infinite sets such that for every $n \in \mathbb{N}$, $\min(E_n) = n(A)$ for certain $A \in \mathcal{A}$. Factorize Q as

$$J^{E_1} \times J^{E_2} \times \cdots \times J^{E_n} \times \cdots\cdots,$$

and for every $n \in \mathbb{N}$ let

$$B(J^{E_n}) = \{x \in J^{E_n}: (\exists i \in E_n)(|x_i| = 1)\}.$$

By theorem 6.4.6, for every $n \in \mathbb{N}$ there exists a homeomorphism

$$h_n: J^{E_n} \to J^{E_n}$$

such that

$$h_n(\prod_{i \in E_n}[-1+2^{-n}, 1-2^{-n}]_i) \subseteq B(J^{E_n}).$$

Simply observe that

$$\prod_{i \in E_n}[-1+2^{-n}, 1-2^{-n}]_i \text{ and } \{x \in J^{E_n}: x_{\min(E_n)} = 1\}$$

are homeomorphic (to Q) and are both Z-sets of J^{E_n}.

It is easily seen that

$$h = h_1 \times h_2 \times \cdots \times h_n \times \cdots\cdots,$$

is as required. □

6.5.7. COROLLARY: *Let $\mathcal{A}$ be an infinite collection of endfaces of* Q. *Then $\cup\mathcal{A}$ is an absorber.*

PROOF: By lemma 6.5.6, the absorber Σ can be pushed into $\cup\mathcal{A}$ by a homeomorphism of Q. Now since $\cup\mathcal{A} \in \mathcal{Z}_\sigma(Q)$ (lemma 6.2.3(3)), the desired result is a direct application of theorem 6.5.2(2). □

We now come to the main result in this section.

6.5.8. THEOREM: *An element* $A \in \mathcal{Z}_\sigma(Q)$ *is an absorber if and only if there is a homeomor* *phism* $h: Q \to Q$ *such that* $h(A) = B(Q)$.

PROOF: Apply corollary 6.5.7 and theorem 6.5.2(3).□

We shall now present some other corollaries to the results obtained so far.

6.5.9. COROLLARY: *Let* A *be a* σ*-compact subset of* $\mathbb{R}^\infty$. *Then* $\mathbb{R}^\infty \setminus A$ *and* $\mathbb{R}^\infty$ *are homeomorphic.*

PROOF: We shall prove the corollary for s instead of $\mathbb{R}^\infty$. This is justified by exercise 1.1.8. By corollary 6.5.7, B(Q) is an absorber. In addition, $A \in \mathcal{Z}_\sigma(Q)$ by corollary 6.2.4. Consequently, $B(Q) \cup A$ is an absorber by theorem 6.5.2(2). Theorem 6.5.8 therefore implies the existence of an $h \in \mathcal{H}(Q)$ such that $h(B(Q)) = B(Q) \cup A$, i.e. $h(s) = s \setminus A$.□

6.5.10. COROLLARY: *For every n, let* A_n *be a nondegenerate interval in* $\mathbb{R}$. *Then*

$$\prod_{n=1}^{\infty} A_n \approx \mathbb{R}^\infty$$

iff infinitely many of the A_n *are not compact.*

PROOF: If only finitely many of the A_n are not compact then the product of the A_n's is locally compact and hence is certainly not homeomorphic to $\mathbb{R}^\infty$. So assume that infinitely many of the A_n's are not compact. Observe that every A_n is homeomorphic to $(-1,1)$, to $(-1,1]$, or to $[-1,1]$ (cf. exercises 1.1.8 and 1.1.9). Also, infinitely often, A_n is not homeomorphic to $[-1,1]$. Consequently, the product of the A_n's is homeomorphic to the complement in Q of a set consisting of infinitely many endfaces. Now apply corollary 6.5.7 and theorem 6.5.8.□

Exercises for §6.5.

1. Prove that $B(Q) \times Q$ and $B(Q) \times B(Q)$ are absorbers for $Q \times Q$.

2. Let $A \in \mathcal{Z}_\sigma(Q)$ be an absorber and let $B \in \mathcal{Z}(Q)$. Prove that $A \setminus B$ is an absorber (Hint: First observe that A is a skeletoid; let $A = \bigcup_{n=1}^{\infty} A_n$, where the A_n's are as in the definition of a skeletoid. For every n, put $B_n = \{x \in A_n : d(x,B) \geq 1/n\}$).

3. For each $i \in \mathbb{N}$ let $-1 < a_i < b_i < 1$. The set $\{x \in s\colon x_i \in [a_i,b_i]$ for all but finitely many $i \in \mathbb{N}\}$ is called the *basic core set structured on the core* $\prod_{i=1}^{\infty}[a_i,b_i]$. Prove that each basic core set is an absorber in Q.

4. Let A be a (relatively) closed subset of s and let B be its closure in Q. Prove that $A \in \mathcal{Z}(s)$ iff $B \in \mathcal{Z}(Q)$.

5. Let $A \in \mathcal{Z}_\sigma(\mathbb{R}^\infty)$. Prove that $\mathbb{R}^\infty \setminus A \approx \mathbb{R}^\infty$.

6.6. Hilbert Space is Homeomorphic to the Countable Infinite Product of Lines

The aim of this section is to prove that l^2 and $\mathbb{R}^\infty$ are homeomorphic. The strategy of the proof is roughly the following. We construct a compactification K of l^2 such that K is homeomorphic to Q while moreover the remainder $K \setminus l^2$ is an absorber in K. An application of §6.5 then yields the desired result. K is the so called *elliptical Hilbert cube*, i.e. the subspace

$$K = \{x \in Q\colon \sum_{i=1}^{\infty} x_i^2 \le 1\}$$

of Q. It is easy to see that K is an infinite-dimensional compact convex subset of Q. We shall see in chapter 8 that all such subsets of Q are homeomorphic to Q (this is Keller's Theorem). For the moment, knowing this for K is sufficient.

It will be convenient to introduce some notation. For every $n \in \{1,2,\cdots,\infty\}$ define

$$B^n = \{x \in J^n\colon \sum_{i=1}^{n} x_i^2 \le 1\}$$

(so $B^\infty = K$). Define a function $\Phi_n\colon J^n \to B^n$ ($n \in \{1,2,\cdots,\infty\}$) by

$$\Phi_n(x) = y \Leftrightarrow \begin{cases} y_1 = x_1, & \\ y_i = \sqrt{1 - \Sigma_{j=1}^{i-1} y_j^2} \cdot x_i & (i > 1). \end{cases}$$

6.6.1. LEMMA: Φ_n *is well-defined and continuous.*

PROOF: Let $n \in \{1,2,\cdots,\infty\}$ and let $x \in J^n$. By induction on i, we shall construct elements

$y_i \in J$ such that $y_1 = x_1$ and for $i > 1$,

$$y_i = \sqrt{1 - \Sigma_{j=1}^{i-1} y_j^2} \cdot x_i.$$

Let $i \geq 1$ and assume y_i to be defined. Then $y_1^2 = x_1^2 \leq 1$ and for $i > 1$,

$$y_i^2 = (1 - \Sigma_{j=1}^{i-1} y_j^2) x_i^2 \leq 1 - \Sigma_{j=1}^{i-1} y_j^2,$$

i.e.

$$\sum_{j=1}^{i} y_j^2 \leq 1,$$

so that

$$1 - \sum_{j=1}^{i} y_j^2 \geq 0.$$

From this we conclude that there is no problem with the construction of y_{i+1}.

Since there is at most one sequence $(y_i)_i$ as above, we conclude that Φ_n is well-defined. We shall next prove that all coordinate functions of Φ_n are continuous, i.e. that Φ_n is continuous. The first coordinate function of Φ_n is the projection onto the first factor of J^n, which is certainly continuous. The second coordinate function of Φ_n is a "formula function" in terms of the projection onto the second factor of J^n and the first coordinate function of Φ_n. So this function is continuous as well. By induction it is now easy to finish the job. □

Let $n < \infty$ and identify J^{n+1} and $J^n \times J$. Consider $\Phi_n \times 1_J$: $J^n \times J \to B^n \times J$, and define Ψ_n: $B^n \times J \to B^{n+1}$ by

$$\psi_n((x_1,\cdots,x_n),t) = (x_1,\cdots,x_n,\sqrt{1 - \Sigma_{j=1}^{n} x_j^2} \cdot t).$$

6.6.2. LEMMA: *Let* $n < \infty$. *Then* Ψ_n *is surjective and* $\Phi_{n+1} = \Psi_n \circ (\Phi_n \times 1)$.

PROOF: Take $(x_1,\cdots,x_{n+1}) \in B^{n+1}$. If

$$\sum_{i=1}^{n} x_i^2 = 1$$

then $x_{n+1} = 0$, so

$\Psi_n((x_1,\cdots,x_n),0) = (x_1,\cdots,x_{n+1})$.

Therefore, suppose that

$$\sum_{i=1}^{n} x_i^2 < 1 \text{ and put } t = \frac{x_{n+1}}{\sqrt{1 - \sum_{i=1}^{n} x_i^2}} .$$

Observe that

$$x_{n+1}^2 + \sum_{i=1}^{n} x_i^2 \leq 1$$

from which it follows that

$$t^2 = \frac{x_{n+1}^2}{1 - \sum_{i=1}^{n} x_i^2} \leq \frac{1 - \sum_{i=1}^{n} x_i^2}{1 - \sum_{i=1}^{n} x_i^2} = 1.$$

We conclude that $-1 \leq t \leq 1$. It is clear that

$$\Psi_n((x_1,\cdots,x_n),t) = (x_1,\cdots,x_{n+1}).$$

That $\Phi_{n+1} = \Psi_n \circ (\Phi_n \times 1)$ is a triviality (observe that we identified J^{n+1} and $J^n \times J$).□

6.6.3. COROLLARY: *Let* $n \leq \infty$. *Then* Φ_n *is surjective.*

PROOF: Suppose first that $n < \infty$. Since $\Phi_1 = 1$ the statement is clear for $n = 1$. Suppose that Φ_n is surjective. Since by lemma 6.6.2, $\Phi_{n+1} = \Psi_n \circ (\Phi_n \times 1)$ and ψ_n is surjective, we conclude that Φ_{n+1} is surjective.

Now suppose that $n = \infty$ and choose an arbitrary $(x_1,x_2,\cdots) \in B^\infty = K$. For each $m \in \mathbb{N}$ put

$$z^{(m)} = (x_1,\cdots,x_m,0,0,\cdots).$$

Applying the above, we can find a point $(y_1,\cdots,y_m) \in J^m$ such that

$$\Phi_m(y_1,\cdots,y_m) = (x_1,\cdots,x_m).$$

Clearly

$$\Phi_\infty(y_1,\cdots,y_m,0,0,\cdots) = z^{(m)}.$$

Now put $u^{(m)} = (y_1,\cdots,y_m,0,0,\cdots)$. Since Q is compact, there is a subsequence $(u^{(m_i)})_i$ of the sequence $(u^{(m)})_m$ and a point $u \in Q$ such that

$$u = \lim_{i\to\infty} u^{(m_i)}.$$

Then by continuity of Φ_∞ (lemma 6.6.1), we obtain

$$\Phi_\infty(u) = \lim_{i\to\infty} \Phi_\infty(u^{(m_i)}) = \lim_{i\to\infty} z^{(m_i)} = (x_1,x_2,\cdots),$$

which is as required. □

Again, let $n < \infty$ and consider $\Psi_n: B^n \times J \to B^{n+1}$. As usual, let

$$S^n = \{x \in \mathbb{R}^{n+1}: \|x\| = 1\}.$$

Observe that the collection of all nondegenerate point-inverses of Ψ_n is equal to the family

$$\{\{x\} \times J: x \in S^{n-1}\}.$$

So the case $n = 1$ is particularly simple: in J^2 the sets $\{1\} \times J$ and $\{-1\} \times J$ are both identified to a single point under Ψ_1.

The following lemma now comes as no surprise.

6.6.4. LEMMA: *Let* $n < \infty$. *Then* Ψ_n *is a near homeomorphism.*

PROOF: We consider the case $n = 1$ first. We shall show that Ψ_1 is shrinkable. By Bing's shrinking criterion 6.1.8 it then follows that Ψ_1 is a near homeomorphism.

Choose $\varepsilon > 0$. Find $\delta > 0$ such that if $A \subseteq J^2$ has diameter less than δ then $\Psi_1(A)$ has diameter less than $\varepsilon/2$.

Precisely as in the proof of lemma 6.1.4 we can construct a homeomorphism $h: J^2 \to J^2$ such that

(1) $h \mid [-1+\delta, 1-\delta] \times J = 1$,

(2) $\operatorname{diam} h(\{-1\} \times J) < \varepsilon$ and $\operatorname{diam} h(\{1\} \times J) < \varepsilon$.

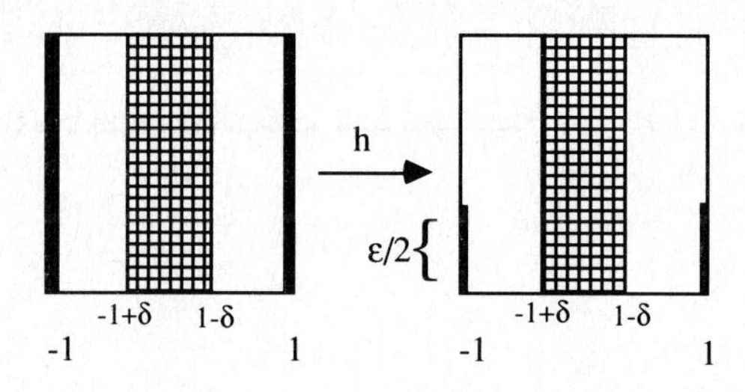

Figure 6.6.1.

Of course, h is the desired "shrinking homeomorphism". Just observe that from (2) it follows that for every y,

$$\operatorname{diam}(h(\Psi_1^{-1}(y))) < \varepsilon.$$

We shall show that $d(\Psi_1, \Psi_1 \circ h) < \varepsilon$. If $x \in [-1+\delta, 1-\delta] \times J$ then $\Psi_1(x) = \Psi_1(h(x))$, so $d(\Psi_1(x), \Psi_1(h(x))) = 0$. Now choose e.g. $x \in [-1, -1+\delta) \times J$. There is a point $p \in \{-1\} \times J$ such that $d(p,x) < \delta$. There is also a point $p' \in \{-1\} \times J$ such that $d(p', h(x)) < \delta$. Since $\Psi_1(p) = \Psi_1(p')$ it follows that

$$d(\Psi_1(x), \Psi_1(h(x))) \leq d(\Psi_1(x), \Psi_1(p)) + d(\Psi_1(p'), \Psi_1(h(x)))$$

$$< \varepsilon/2 + \varepsilon/2 = \varepsilon.$$

We shall proceed a little faster now. The above proof can easily be generalized for all $n < \infty$. Alternatively, we can use the $n = 1$ case to derive the general case. Let $n \geq 2$. Choose $\delta > 0$ such that if $A \subseteq B^n \times J$ and $\operatorname{diam}(A) < \delta$ then $\operatorname{diam}(\Psi_n(A)) < \varepsilon/2$. Let h be the homeomorphism we just constructed for the $n = 1$ case. We shall assume that $\delta < 1$. Define $H: B^n \times J \to B^n \times J$ by

$$H(x,t) = \begin{cases} (x,t) & (x = 0), \\ \left(h(\|x\|,t)_1 \cdot \dfrac{x}{\|x\|}, h(\|x\|,t)_2\right) & (x \neq 0). \end{cases}$$

where $h(\|x\|,t) = (h(\|x\|,t)_1, h(\|x\|,t)_2)$. By observing that the function

$$\|\cdot\| \times 1: [0, \frac{x}{\|x\|}] \times J \to [0,1] \times J$$

is an isometry, it follows easily that H is the desired "shrinking" homeomorphism. □

6.6.5. THEOREM: *Let* $n \leq \infty$. *Then* $\Phi_n: J^n \to B^n$ *is a near homeomorphism.*

PROOF: Assume first that $n < \infty$. Since $\Phi_1 = 1$, for $n = 1$ there is no problem. Suppose therefore that $n \geq 1$ and that Φ_n is a near homeomorphism. Since by lemma 6.6.2,

$$\Phi_{n+1} = \Psi_n \circ (\Phi_n \times 1)$$

we find by lemma 6.6.4 that Φ_{n+1} is the composition of two near homeomorphisms, and is therefore a near homeomorphism itself (see exercise 6.1.8).

We shall now prove that Φ_∞ is a near homeomorphism. For each $n \in \mathbb{N}$ we identify

(a) J^n and $J^n \times \{(0,0,\cdots)\} \subseteq J^\infty$, and

(b) B^n and $B^n \times \{(0,0,\cdots)\} \subseteq B^\infty$.

Let $\rho_n: J^\infty \to J^n$ and $r_n: B^\infty \to B^n$ be the projections, i.e.

$$\rho_n(x_1,\cdots,x_n,x_{n+1},\cdots) = (x_1,\cdots,x_n,0,0,\cdots),$$

and

$$r_n(x_1,\cdots,x_n,x_{n+1},\cdots) = (x_1,\cdots,x_n,0,0,\cdots),$$

respectively. It is a triviality that the diagram

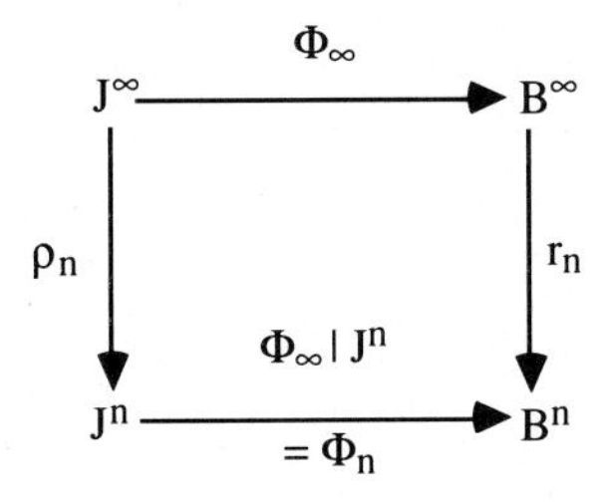

commutes.

We claim that for every $x \in B^\infty$ there exists $y \in B^n$ such that

$$\Phi_\infty^{-1}(x) \subseteq \rho_n^{-1}\Phi_n^{-1}(y).$$

Choose $x \in B^\infty$ and observe that since the diagram commutes, the point $y = r_n(x)$ is as required.

Choose $\varepsilon > 0$ and let $\delta > 0$ be such that if $A \subseteq J^\infty$ and $\text{diam}(A) < \delta$ then $\text{diam}(\Phi_\infty(A)) < \varepsilon/3$. Choose $n \in \mathbb{N}$ so large that each set of the form $\rho_n^{-1}(x)$, $x \in J^n$, has diameter less than $\min\{\delta, \varepsilon/3\}$. Since Φ_n is shrinkable (because it is a near homeomorphism), there is a homeomorphism $h: J^n \to J^n$ satisfying

(2) $\forall y \in B^n$: $\text{diam}(h\Phi_n^{-1}(y)) < \min\{\delta, \frac{\varepsilon}{3}\}$,

(3) $d(\Phi_n, \Phi_n \circ h) < \frac{\varepsilon}{3}$.

Define $H: J^\infty \to J^\infty$ by crossing h with the identity on the other factors of Q.

We shall prove that H is the desired "shrinking" homeomorphism. We shall first prove that for every $x \in B^\infty$, $\text{diam}(H\Phi_\infty^{-1}(x) < \varepsilon$. By the above there exists a point $y \in B^n$ such that $\Phi_\infty^{-1}(x) \subseteq \rho_n^{-1}\Phi_n^{-1}(y)$. Consequently,

$$H\Phi_\infty^{-1}(x) \subseteq \rho_n^{-1} h\Phi_n^{-1}(y).$$

Since

$$\text{diam}(h\Phi_n^{-1}(y)) < \tfrac{\varepsilon}{3} \text{ and for every } z \in J_n,\ \text{diam}(\rho_n^{-1}(z)) < \tfrac{\varepsilon}{3},$$

it follows easily that

$$\text{diam}(H\Phi_\infty^{-1}(x)) \leq \text{diam}(\rho_n^{-1} h\Phi_n^{-1}(y)) < \tfrac{\varepsilon}{3} + \tfrac{\varepsilon}{3} = \tfrac{2\varepsilon}{3} < \varepsilon.$$

We next claim that $d(\Phi_\infty, \Phi_\infty \circ H) < \varepsilon$. Factorize Q as $J^n \times Q_n$, where Q_n denotes the product of all other factors. Take an arbitrary $(x,p) \in Q$. Then $d((x,p),(x,\mathbf{0})) < \delta$ (here $\mathbf{0}$ is the "origin" of Q_n, i.e. the point having all coordinates equal to 0), so

$$d(\Phi_\infty(x,p), \Phi_\infty(x,\mathbf{0})) = d(\Phi_\infty(x,p), (\Phi_n(x),0)) < \tfrac{\varepsilon}{3}.$$

Since $d(\Phi_n(x), \Phi_n h(x)) < \varepsilon/3$ we conclude that

$$d(\Phi_\infty(x,p), (\Phi_n h(x),0)) = d(\Phi_\infty(x,p), \Phi_\infty H(x,\mathbf{0})) < \tfrac{2\varepsilon}{3}.$$

Also $d(H(x,p), H(x,\mathbf{0})) = d((h(x),p),(h(x),\mathbf{0})) < \delta$, so that

$$d(\Phi_\infty(x,p),\Phi_\infty(H(x,p))) < \tfrac{2\varepsilon}{3} + \tfrac{\varepsilon}{3} = \varepsilon.$$

By theorem 6.1.8 it follows that Φ_∞ is a near homeomorphism.□

6.6.6. COROLLARY: $K \approx Q$.□

We shall now construct the announced absorber for K. For every $0 < \varepsilon < 1$ put

$$K(\varepsilon) = \{x \in Q: \sum_{i=1}^{\infty} x_i^2 \le 1 - \varepsilon\}.$$

We shall prove that

$$B = \{x \in K: \sum_{i=1}^{\infty} x_i^2 < 1\} = \bigcup_{n=2}^{\infty} K(\tfrac{1}{n})$$

is an absorber. After this has been established, the proof that l^2 and $\mathbb{R}^\infty$ are homeomorphic is almost completed.

6.6.7. LEMMA: *If* $0 < \varepsilon < 1$ *then* $K(\varepsilon) \subseteq K$, $K(\varepsilon) \in \mathcal{Z}(K)$ *and* $K(\varepsilon) \approx K \approx Q$.

PROOF: That $K(\varepsilon) \subseteq K$ is trivial. Define $\phi: K \to K(\varepsilon)$ by

$$\phi(x) = \sqrt{1-\varepsilon} \cdot x.$$

It is clear that ϕ is a homeomorphism. Since by corollary 6.6.6, $K \approx Q$, we conclude that $K(\varepsilon) \approx Q$.

Let $\delta > 0$. We shall construct a continuous function $f: K \to K \backslash K(\varepsilon)$ such that $d(f,1) < \delta$. It then follows by lemma 6.2.3 that $K(\varepsilon) \in \mathcal{Z}(K)$. Choose $n \in \mathbb{N}$ so large that

$$\sum_{i=n}^{\infty} 2^{-i} < \tfrac{\delta}{2}.$$

Define $f: K \to K$ by

$$f(x_1,x_2,\cdots) = (x_1,x_2,\cdots,x_{n-1}, \sqrt{(1 - \Sigma_{i=1}^{n-1} x_i^2)(1 - \tfrac{\varepsilon}{2})}, 0,0,0,\cdots).$$

Then f is well-defined, continuous and $d(f,1) < \delta$. Choose $x \in K$ arbitrarily and let $y = f(x)$. Then

$$\sum_{i=1}^{\infty} y_i^2 = x_1^2 + x_2^2 + \cdots + x_{n-1}^2 + (1-\Sigma_{i=1}^{n-1} x_i^2)(1 - \tfrac{\varepsilon}{2})$$

$$\geq (x_1^2 + x_2^2 + \cdots + x_{n-1}^2)(1 - \tfrac{\varepsilon}{2}) + (1-\Sigma_{i=1}^{n-1} x_i^2)(1 - \tfrac{\varepsilon}{2})$$

$$= 1 - \tfrac{\varepsilon}{2} > 1 - \varepsilon.$$

We conclude that $f(K) \cap K(\varepsilon) = \emptyset$.□

For each $n \geq 2$ let $\phi_n: K \to K(1/n)$ be the homeomorphism defined in the previous proof, i.e.

$$\phi_n(x) = \sqrt{1 - \tfrac{1}{n}} \cdot x.$$

6.6.8. LEMMA: $d(\phi_n, 1) < \frac{1}{n}$.

PROOF: Take $x \in K$. Then

$$d(\phi_n(x), x) = d(\sqrt{1-\tfrac{1}{n}} \cdot x, x) = (1 - \sqrt{1-\tfrac{1}{n}})\sum_{i=1}^{\infty} 2^{-i}|x_i| \leq 1 - \sqrt{1-\tfrac{1}{n}} < \tfrac{1}{n}$$

since $|x_i| \leq 1$ for every $i \in \mathbb{N}$ and $\sum_{i=1}^{\infty} 2^{-i} = 1$.□

We are now in a position to prove that B is an absorber.

6.6.9. THEOREM: B *is an absorber.*

PROOF: We shall prove that B is a skeletoid and then apply theorem 6.5.1. Naturally, we shall prove that the sequence $(K(1/n))_{n \geq 2}$ does the job for us. To this end, let $n \geq 2$, $\varepsilon > 0$, and $Z \in \mathcal{Z}(K)$. Since K is a Hilbert cube by corollary 6.6.6, corollary 6.4.7 implies that there exists $\varepsilon_1 > 0$ such that "ε_1-small" homeomorphisms between Z-sets in K can be extended to "ε-small" homeomorphisms of K. By another application of corollary 6.4.7, there exists $\varepsilon_2 > 0$ such that "ε_2-small" homeomorphisms between Z-sets in K can be extended to "$\varepsilon_1/2$-small" homeomorphisms of K. Choose $m > n$ such that $1/m < \min\{\varepsilon_1/2, \varepsilon_2\}$. Since $m > n$, lemma 6.6.7 implies that $K(1/n) \in \mathcal{Z}(K(1/m))$ (the reader should have a little imagination here). Consequently,

$$\phi_m^{-1}(K(\tfrac{1}{n})) \in \mathcal{Z}(K).$$

Let

$$\xi = \phi_m^{-1} \mid K(\tfrac{1}{n}): K(\tfrac{1}{n}) \to \phi_m^{-1}(K(\tfrac{1}{n})).$$

Since by lemmas 6.6.7 and 6.6.8, $K(1/n) \in \mathcal{Z}(K)$ and

$$d(\xi,1) \leq d(\phi_m^{-1},1) = d(\phi_m,1) \leq \tfrac{1}{m} < \varepsilon_2,$$

there is a homeomorphism $f: K \to K$ that extends ξ while moreover $d(f,1) < \varepsilon_1/2$. Define

$$\eta: K(\tfrac{1}{n}) \cup Z \to K$$

by

$$\eta(x) = \phi_m f(x).$$

Then by exercise 1.3.7, $d(\eta,1) \leq \varepsilon_1/2 + 1/m < \varepsilon_1/2 + \varepsilon_1/2 = \varepsilon_1$. It is clear that $\eta(K(\frac{1}{n}) \cup Z) \subseteq K(\frac{1}{m})$ and that η is an imbedding. Since $K(\frac{1}{n}) \cup Z \in \mathcal{Z}(K)$ and $d(\eta,1) < \varepsilon_1$ there is a homeomorphism $\gamma: K \to K$ that extends η while moreover $d(\gamma,1_K) < \varepsilon$. Then $\gamma(Z) = \eta(Z) \subseteq K(\frac{1}{m})$ and if $x \in K(\frac{1}{n})$ then

$$\gamma(x) = \eta(x) = \phi_m f(x) = \phi_m \xi(x) = \phi_m \phi_m^{-1}(x) = x,$$

i.e. $\gamma \mid K(\frac{1}{n}) = 1$.□

6.6.10. COROLLARY: *The subspace* $\{x \in K: \sum_{i=1}^{\infty} x_i^2 = 1\}$ *of* K *is homeomorphic to* $\mathbb{R}^\infty$.

PROOF: The complement of an absorber in a Hilbert cube is homeomorphic to $\mathbb{R}^\infty$ by corollary 6.5.8.□

We now come to the main result in this section.

6.6.11. THEOREM ("The Anderson Theorem"): l^2 *is homeomorphic to* $\mathbb{R}^\infty$.

PROOF: By corollary 6.6.10 we have

$$E = \{x \in K: \sum_{i=1}^{\infty} x_i^2 = 1\} \approx \mathbb{R}^\infty.$$

In addition, by lemma 1.2.6, $E \approx S = \{x \in l^2 : \|x\| = 1\}$. So we conclude that $S \approx \mathbb{R}^\infty$.

Let $v = (-1,0,0,\cdots) \in S$. We shall prove that $l^2 \approx S\backslash\{v\}$. Since $S \approx \mathbb{R}^\infty$, by corollary 6.5.9 it then follows that

$$l^2 \approx S\backslash\{v\} \approx S \approx \mathbb{R}^\infty,$$

which is as required.

Let $l_0^2 = \{x \in l^2 : x_1 = 0\}$ and define $g: l^2 \to l_0^2$ by

$$g(x_1,x_2,\cdots) = (0,x_1,x_2,\cdots).$$

Then g is an isometry, so $l^2 \approx l_0^2$ and all we need to show is that $S\backslash\{v\} \approx l_0^2$.

Define $h: S\backslash\{v\} \to l_0^2$

$$h(x) = \begin{cases} (0,x_2,x_3,\cdots) & (x_1 \geq 0), \\ \dfrac{(0,x_2,x_3,\cdots)}{\|(0,x_2,x_3,\cdots)\|^2} & (x_1 \leq 0). \end{cases}$$

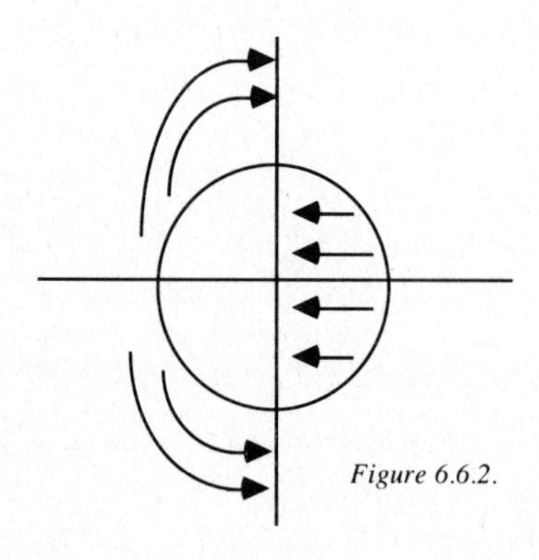

Figure 6.6.2.

It is a triviality to verify that h is a homeomorphism. □

Exercises for §6.6.

1. Let $n \leq \infty$ and let Φ_n: $J^n \to B^n$ be the function defined in this section. Prove that for all $x,y \in J^n$: $\Phi_n(x) = \Phi_n(y) \Leftrightarrow (x = y)$ or $((\exists m)(x_i = y_i$ for all $i < m$ and $x_m = y_m \in \{-1,1\}))$. Use this to describe explicitly the non-degenerate point-inverses of Φ_3.

2. Prove that $l^2 \approx l^2 \times [0,1) \approx l^2 \times I \approx l^2 \times Q$.

3. Prove that l^2 is homeomorphic to its own cone.

A continuous function f: $X \to Y$ is called *monotone* provided that every point-inverse $f^{-1}(y)$ is connected.

4. Let $A,B \in \mathcal{Z}_\sigma(Q)$ be such that $Q\backslash A \approx Q\backslash B$. Prove that there are a compact space Z and monotone surjections f,g: $Z \to Q$ such that $f^{-1}(A) = g^{-1}(B)$ (Hint: let h: $Q\backslash A \to Q\backslash B$ be a homeomorphism and let Z be the closure of the graph of h in $Q \times Q$).

5. Prove that B(Q) is path-connected. Use this and exercises 6.6.4 and 4.6.1 to prove that if E is any countable subset of Q then $Q\backslash E$ is not homeomorphic to l^2.

6.7. Inverse Limits

Sometimes it is possible to "approximate" a complicated space "arbitrarily closely" by less complicated objects. In this section we shall formally define what we mean by this and we shall derive an "approximation theorem". The results in this section are used in §6.8 to prove that "the letter T" crossed by Q is homeomorphic to Q.

An *inverse sequence* is a sequence of pairs $(X_n,f_n)_{n\geq 1}$ of spaces X_n and continuous functions f_n: $X_{n+1} \to X_n$. The spaces X_n are called *coordinate spaces* and the mappings f_n are called *bonding maps*. The *inverse limit* of the inverse sequence $(X_n,f_n)_n$, denoted by

$$\varprojlim (X_n,f_n)_n,$$

is defined to be the following subspace of the product of the X_n:

$$\{x \in \prod_{n=1}^{\infty} X_n : (\forall n \in \mathbb{N})(f_n(x_{n+1}) = x_n)\}.$$

It will sometimes be convenient to denote the inverse limit of the sequence $(X_n,f_n)_n$ by X_∞.

For every $n \in \mathbb{N}$, the restriction to $\lim\limits_{\leftarrow} (X_n,f_n)_n$ of the projection onto the n-th factor of the product $\prod_{n=1}^{\infty} X_n$ shall be denoted by $f_{\infty,n}$. By definition it follows that for every n, $f_n \circ f_{\infty,n+1} = f_{\infty,n}$. The functions $f_{\infty,n}$ are called the *projections* of the inverse sequence.

6.7.1. LEMMA: (1) *The inverse limit of a sequence of compact spaces is compact.*

(2) *Let* $(X_n,f_n)_n$ *be an inverse sequence with surjective bonding maps. Then the projections of* $(X_n,f_n)_n$ *are also surjective.*

PROOF: For (1), it is easy to see that the inverse limit of the sequence $(X_n,f_n)_n$ is a closed subspace of the product $\prod_{n=1}^{\infty} X_n$. So the compactness of the inverse limit follows by the compactness of the coordinate spaces. For (2), fix $n \in \mathbb{N}$ and take an arbitrary $x \in X_n$. Since f_n is surjective, there exists $x_{n+1} \in X_{n+1}$ with $f_n(x_{n+1}) = x$. Similarly, there exists $x_{n+2} \in X_{n+2}$ with $f_{n+1}(x_{n+2}) = x_{n+1}$, etc. Define $x_n = x$ and $x_{n-1} = f_{n-1}(x_n)$, $x_{n-2} = f_{n-2}(x_{n-1})$, etc. The sequence $(x_n)_n \in \prod_{n=1}^{\infty} X_n$ belongs to $\lim\limits_{\leftarrow} (X_n,f_n)_n$ and since clearly $f_{\infty,n}((x_n)_n) = x_n = x$, we are done. □

The following proposition can be used to "recognize" inverse limits.

6.7.2. PROPOSITION: *Let* X *be a compact space, let* $(X_n)_n$ *be an increasing sequence of closed subsets of* X, *and for* $n \in \mathbb{N}$, *let* $f_n\colon X_{n+1} \to X_n$ *be continuous and surjective. Suppose that the following conditions hold:*

(1) $\bigcup_{n=1}^{\infty} X_n$ *is dense in* X,

(2) *for every* $n \in \mathbb{N}$, $d(f_n, 1_{X_{n+1}}) \le 2^{-n}$,

(3) *for every* $n \in \mathbb{N}$ *and* $x,y \in X_{n+1}$, $d(f_n(x),f_n(y)) \le d(x,y)$.

Then X *and* $\lim\limits_{\leftarrow} (X_n,f_n)$ *are homeomorphic.*

PROOF: Let X_∞ denote the inverse limit of $(X_n,f_n)_n$. For every n, consider the function $f_{\infty,n}\colon X_\infty \to X_n \subseteq X$.

CLAIM 1: The sequence $(f_{\infty,n})_n$ is Cauchy in the function space $C(X_\infty,X)$. In fact, if $n \in \mathbb{N}$ then $d(f_{\infty,n},f_{\infty,n+1}) \le 2^{-n}$,

Take arbitrary $x \in X_\infty$ and $n \in \mathbb{N}$. Then by (2),

$$d(f_{\infty,n}(x),f_{\infty,n+1}(x)) = d(f_n(x_{n+1}),x_{n+1}) \le 2^{-n}.$$

This implies that

$$d(f_{\infty,n},f_{\infty,n+1}) \leq 2^{-n}.$$

By proposition 1.3.4, the function $F = \lim_{n\to\infty} f_{\infty,n}: X_\infty \to X$ is well-defined and continuous. We claim that F is a homeomorphism.

CLAIM 2: F is surjective.

By compactness of X_∞ (lemma 6.7.1(1)) it suffices to prove that $F(X_\infty)$ is dense. To this end, take arbitrary $x \in X$ and $\varepsilon > 0$. There exists $n \in \mathbb{N}$ such that $2^{-n} < \varepsilon/4$. By (1) and the fact that the sequence of the X_n's increases, there also exists $m \geq n$ such that $X_m \cap B(x,\varepsilon/2) \neq \emptyset$, say $y \in X_m \cap B(x,\varepsilon/2)$. By lemma 6.7.1(2) there exists $p \in X_\infty$ such that $f_{\infty,m}(p) = y$. By claim 1 it follows that

$$d(F(p),y) \leq \sum_{i=m}^{\infty} 2^{-i} = 2^{-(m-1)} < \frac{\varepsilon}{2},$$

i.e. $d(F(p),x) < \varepsilon$. We conclude that $F(X_\infty) \cap B(x,\varepsilon) \neq \emptyset$.

CLAIM 3: F is one-to-one.

Take distinct $x,y \in X_\infty$. There exists i with $x_i \neq y_i$. Put $\delta = d(x_i,y_i)$. There exists $k > i$ such that $d(F(x),x_k) < \delta/2$ and $d(F(y),y_k) < \delta/2$. By (3) it clearly follows that

$$d(x_k,y_k) \geq d(x_i,y_i) = \delta.$$

Consequently, $F(x) \neq F(y)$, which is as required.

We conclude that F is a homeomorphism (exercise 1.1.4). □

The above proposition enables us to give easy examples of inverse systems and their limits. For each $n \in \mathbb{N}$ let $J_n = [-1 + 2^{-n}, 1 - 2^{-n}]$ and define $f_n: J_{n+1} \to J_n$ by

$$f_n(x) = \begin{cases} x & (-1+2^{-n} \leq x \leq 1-2^{-n}), \\ -1 + 2^{-n} & (-1+2^{-(n+1)} \leq x \leq -1+2^{-n}), \\ 1 - 2^{-n} & (1-2^{-n} \leq x \leq 1-2^{-(n+1)}). \end{cases}$$

By proposition 6.7.2,

$$\lim_{\leftarrow} (J_n,f_n)_n \approx [-1,1].$$

Before we are in a position to formulate and prove the desired "approximation theorem" for inverse systems, we have to fix some notation first. Let $(X_n,f_n)_n$ be an inverse system with inverse limit X_∞. It is sometimes useful to have an explicit admissible metric for X_∞. For every $n \in \mathbb{N}$ let d_n be an admissible metric for X_n which is bounded by 1 (exercise 1.1.7). The formula

$$d(x,y) = \sum_{n=1}^{\infty} 2^{-n} d_n(x_n,y_n)$$

defines an admissible metric for the product of the X_n, and hence also for X_∞. Observe that d is also bounded by 1.

6.7.3. LEMMA: *Let* $(X_n,f_n)_n$ *be an inverse system with inverse limit* X_∞. *In addition, for every* n *let* d_n *be an admissible metric for* X_n *which is bounded by* 1. *Then with respect to the metric* d *for* X_∞ *defined above, we have that every* $f_{\infty,n}\colon X_\infty \to X_n$ *is a* $2^{-(n-1)}$*-mapping.*

PROOF: This is easy. Fix $n \in \mathbb{N}$. Let $p,q \in X_\infty$ be such that $f_{\infty,n}(p) = f_{\infty,n}(q)$. Observe that $p_i = q_i$ for every $i \le n$. Consequently,

$$d(p,q) = \sum_{i=1}^{\infty} 2^{-i} d_i(p_i,q_i) = \sum_{i=n+1}^{\infty} 2^{-i} d_i(p_i,q_i) \le \sum_{i=n+1}^{\infty} 2^{-i} = 2^{-n}.$$

We conclude that $f_{\infty,n}$ is a $2^{-(n-1)}$-mapping. □

6.7.4. THEOREM ("Brown's Approximation Theorem"): *Let* $(X_n,f_n)_n$ *be an inverse sequence consisting of compact spaces with inverse limit* X_∞. *If each* f_n *is a near homeomorphism, then so is each* $f_{\infty,n}$. *In particular, if each* f_n *is a near homeomorphism then* X_∞ *is homeomorphic to* X_1.

PROOF: We shall prove that $f_{\infty,1}\colon X_\infty \to X_1$ is a near homeomorphism. For each n, let d_n be an admissible metric for X_n which is bounded by 1 (exercise 1.1.7) and let d be the metric for X_∞ introduced above. Let $\varepsilon > 0$. Inductively, we shall construct for every $n \in \mathbb{N}$ a homeomorphism $h_n\colon X_{n+1} \to X_n$ such that the functions $g_n\colon X_\infty \to X_1$ defined by $g_n = h_1 \circ \cdots \circ h_n \circ f_{\infty,n+1}$ have the following properties (we adopt the notation introduced in §1.3):

(1) $d_1(g_1,f_{\infty,1}) < \varepsilon/2$,

(2) $d_1(g_n,g_{n+1}) < 3^{-n}\cdot\varepsilon$,

(3) $d_1(g_n,g_{n+1}) < 3^{-n}\cdot\min\{d_1(g_i,\mathcal{G}_{1/i}(X_\infty,X_1)): 1 \le i \le n\}$.

Since f_1 is a near homeomorphism and $f_{\infty,1} = f_1 \circ f_{\infty,2}$, it is clear that h_1 can be constructed so that condition (1) is met. Suppose that the homeomorphisms h_i are defined for $1 \le i \le n$. Observe that for every $1 \le i \le n$, the function $g_i \in \mathcal{S}(X_\infty,X_1)\backslash\mathcal{G}_{1/i}(X_\infty,X_1)$ (lemmas 6.7.1(2) and 6.7.3). Let

$$\delta = 3^{-n}\cdot \min\{\varepsilon, \min\{d_1(g_i,\mathcal{G}_{1/i}(X_\infty,X_1)): 1 \le i \le n\}\}$$

and notice that $\delta > 0$. Since

$$g_n = h_1 \circ \cdots \circ h_n \circ f_{\infty,n+1} = h_1 \circ \cdots \circ h_n \circ f_{n+1} \circ f_{\infty,n+2}$$

and f_{n+1} is a near homeomorphism, we can find a homeomorphism $h_{n+1}: X_{n+2} \to X_{n+1}$ such that

$$d(h_1 \circ \cdots \circ h_n \circ f_{n+1} \circ f_{\infty,n+2}, h_1 \circ \cdots \circ h_n \circ h_{n+1} \circ f_{\infty,n+2}) < \delta.$$

(use the fact that $h_1 \circ \cdots \circ h_n$ is uniformly continuous). It is clear that h_{n+1} is as required, which completes the inductive construction of the h_n.

By (2) it follows that the sequence $(g_n)_n$ is d_1-Cauchy and consequently, $g = \lim_{n\to\infty} g_n$ exists. Since $g_n \in \mathcal{S}(X_\infty,X_1)$ for every n, proposition 1.3.7 implies that $g \in \mathcal{S}(X_\infty,X_1)$. Observe that (1) and (2) imply that $d_1(f_{\infty,1},g) < \varepsilon$. Also, (3), lemma 6.1.1 and lemma 1.3.9 imply that

$$g \in \bigcap_{n=1}^{\infty} \mathcal{S}_{1/n}(X_\infty,X_1) = \mathcal{H}(X_\infty,X_1),$$

as required. □

6.7.5. COROLLARY: *Let* $(Q_n,f_n)_n$ *be an inverse sequence consisting of Hilbert cubes. If each* f_n *is a near homeomorphism then*

$$\varprojlim (Q_n,f_n)_n$$

is a Hilbert cube.

Exercises for §6.7.

1. Let X and Y be compact spaces and let $f: X \to Y$ and $g: Y \to X$ be continuous functions such that both $g \circ f: X \to X$ and $f \circ g: Y \to Y$ are near homeomorphisms. Prove that X and Y are homeomorphic.

2. Let X be a compact space. Prove that $\dim X \leq n < \infty$ if and only if X is homeomorphic to the inverse limit of an inverse sequence of at most n-dimensional polyhedra (this is the "Freudenthal Expansion Theorem") (hint: use exercise 3.6.6).

3. Explicitly describe an inverse sequence of finite spaces the inverse limit of which is homeomorphic to the Cantor set **C**.

4. Prove that the circle S^1 is not homeomorphic to the inverse limit of an inverse sequence of copies of the unit interval [0,1].

5. Let X be the inverse limit of an inverse sequence of copies of the unit interval [0,1]. Prove that X has the fixed-point property.

6. Show that in proposition 6.7.2 the assumption that the functions f_n are surjective is superfluous.

6.8. Hilbert Cube Factors

A *Hilbert cube factor* is a space X such that $X \times Q$ is homeomorphic to Q. By proposition 1.5.4, a necessary condition for a space X to be a Hilbert cube factor is that X be a compact **AR**. Interestingly, this condition is also sufficient (this is the Edwards Theorem). For details see §7.8. The aim of this section is to present an elementary proof of the fact that "the letter T" is a Hilbert cube factor.

6.8.1. LEMMA: *Let* X *be a space. The following statements are equivalent:*

(1) X *is a Hilbert cube factor,*

(2) *there is a space* Y *such that* $X \times Y$ *is homeomorphic to* Q.

PROOF: It is clear that we only need to show that (2) implies (1). For this implication, simply observe that

$$Q \approx Q^\infty \approx (X \times Y)^\infty \approx X^\infty \times Y^\infty \approx X \times X^\infty \times Y^\infty \approx X \times Q. \square$$

The following simple result will gives us examples of "nontrivial" Hilbert cube factors, cf. exercise 6.4.2.

6.8.2. LEMMA: *Let* X *be a space that can be written as* $Q_A \cup Q_B$ *such that both* Q_A *and* Q_B *are Hilbert cubes. If* $Q_A \cap Q_B$ *is a Hilbert cube which is a Z-set both in* Q_A *and in* Q_B *then* X *is a Hilbert cube.*

PROOF: Let f: $Q_A \cap Q_B \to \{0\} \times \prod_{i=2}^{\infty}[-1,1]_i$ be a homeomorphism. The Homeomorphism Extension Theorem 6.4.6 implies that f can be extended to a homeomorphism

$$f_A: Q_A \to [-1,0] \times \prod_{i=2}^{\infty}[-1,1]_i.$$

Similarly, f can be extended to a homeomorphism

$$f_B: Q_B \to [0,1] \times \prod_{i=2}^{\infty}[-1,1]_i.$$

Then $\bar{f} = f_A \cup f_B$ is a homeomorphism from X onto Q.□

6.8.3. COROLLARY: *Let* X *be a space that can be written as* $A \cup B$ *such that both* A *and* B *are Hilbert cube factors. If* $A \cap B$ *is a Hilbert cube factor which is a Z-set both in* A *and in* B *then* X *is a Hilbert cube factor.*

PROOF: Observe that $X \times Q$ can be written as $Q_A \cup Q_B$, where $Q_A = A \times Q$ and $Q_B = B \times Q$, respectively. By assumption, Q_A, Q_B and $Q_A \cap Q_B$ are Hilbert cubes and lemma 6.2.2(5) gives us that $Q_A \cap Q_B$ is a Z-set both in Q_A as well as in Q_B. Now apply lemma 6.8.2.□

This corollary implies that spaces of the form

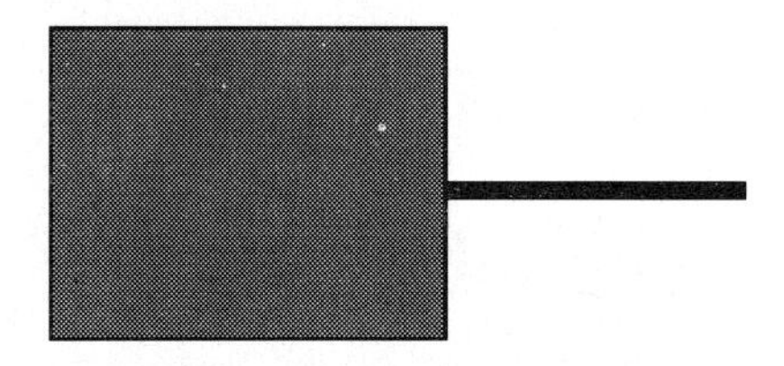

Figure 6.8.1.

are Hilbert cube factors, but *not* that "the letter T" is one. This is substantially more complicated to prove. The main result in this section is that every space X that can be written as $A \cup B$ such that A, B and $A \cap B$ are Hilbert cube factors with $A \cap B$ a Z-set in A, is a Hilbert cube factor. This is a nontrivial improvement of corollary 6.8.3, since from this result it follows that "the letter T" is a Hilbert cube factor.

6.8.4. LEMMA: *Let* M^Q *be a Hilbert cube. In addition, let* A *be a subspace of* M^Q *and let* i: $A \to Q \times \{0\}$ *be a homeomorphism. Then* i *can be extended to an imbedding* $\bar{i}$: $M^Q \to Q \times [-1,0]$.

PROOF: Let h: $M^Q \to Q \times \{-1\}$ be a homeomorphism. Observe that $Q \times \{0\} \in \mathcal{Z}(Q \times [-1,0])$ (lemma 6.2.2(5)). By theorem 6.4.6 there is a homeomorphism h': $Q \times [-1,0] \to Q \times [-1,0]$ which extends the homeomorphism $i \circ h^{-1}$: $h(A) \to i(A)$. Then $\bar{i} = h' \circ h$ is an imbedding of M^Q into $Q \times [-1,0]$ and obviously extends i. □

We now come to the main result in this section.

6.8.5. THEOREM: *Let* X *be a space that can be written as* $A \cup B$ *such that both* A *and* B *are Hilbert cube factors. If* $A \cap B$ *is a Hilbert cube factor which is a Z-set in* A *then* X *is a Hilbert cube factor.*

PROOF: The strategy of the proof will be to represent $X \times Q \times Q$ as the inverse limit of an inverse sequence of Hilbert cubes the bonding maps of which are near homeomorphisms. An appeal to corollary 6.7.5 then yields the desired result.

Put $X_1 = A \times Q$ and $X_2 = B \times Q$, respectively. Then X_1 and X_2 are both Hilbert cubes, and $X_1 \cap X_2$ is a Hilbert cube that is a Z-set in X_1 (lemma 6.2.2(5)). Consequently, theorem 6.4.6 implies that we may identify the pairs $(X_1, X_1 \cap X_2)$ and $(Q \times [0,1], Q \times \{0\})$. By lemma 6.8.4 we may also assume that X_2 is contained in $Q \times [-1,0]$. After these "identifications" we arrived at the following situation:

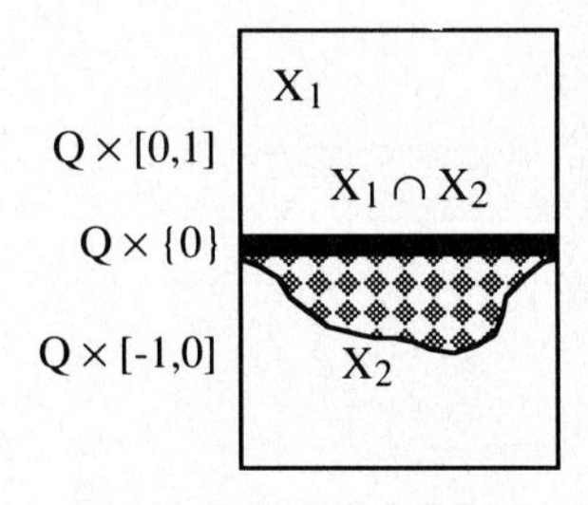

Figure 6.8.2.

For every i let

$$P_i = \{x \in Q: \text{for every } j > i,\ x_j \le 0\},$$

and let Y_i be the subspace

$$(X_2 \times Q) \cup (Q \times [0,1] \times P_i)$$

of the space $Q \times [-1,1] \times Q$. Observe that $Y_1 \subseteq Y_2 \subseteq \cdots$. We endow $Q \times [-1,1] \times Q$ with the admissible metric

$$\rho((x_1,t_1,y_1),(x_2,t_2,y_2)) = d(x_1,x_2) + |t_1 - t_2| + d(y_1,y_2).$$

For $i > 1$, define a retraction $r_i: Y_i \to Y_{i-1}$ as follows:

$$\begin{cases} r_i \mid Y_{i-1} = 1, \\ r_i(x,t,y) = (x,0,(y_1,y_2,\cdots,y_{i-1},y_i-2^i t,y_{i+1},y_{i+2},\cdots)) & (t \ge 0,\ y \in P_i,\ 2^i t \le y_i), \\ r_i(x,t,y) = (x,t-2^{-i}y_i,(y_1,y_2,\cdots,y_{i-1},0,y_{i+1},y_{i+2},\cdots)) & (t \ge 0,\ y \in P_i,\ 2^i t \ge y_i). \end{cases}$$

Observe that r_i affects two coordinate directions only. Here is a picture of what r_i does there.

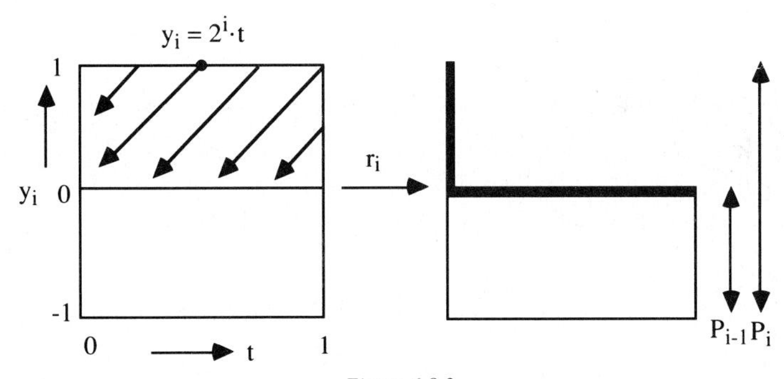

Figure 6.8.3.

CLAIM 1: For every $i > 1$, $\rho(r_i, 1_{Y_i}) \le 2^{-(i-1)}$.

Take an arbitrary point $(x,t,y) \in Y_i \setminus Y_{i-1}$. Consider e.g. the case that $2^i t \le y_i$. Then

$$|0 - t| \le |2^{-i} y_i| = 2^{-i}|y_i| \le 2^{-i}.$$

Also,

$$d((y_1,y_2,\cdots),(y_1,y_2,\cdots,y_{i-1},y_i-2^it,y_{i+1},y_{i+2},\cdots)) = 2^{-i}|y_i - (y_i - 2^it)| \le 2^{-i}|y_i| \le 2^{-i}.$$

From this we find that $\rho(r_i(x,t,y),(x,t,y)) \le 2^{-(i-1)}$, which is as required.

The case that $2^it > y_i$ is similar.

CLAIM 2: If $(x,t,y),(a,s,b) \in Y_i$ and $i > 1$ then $\rho(r_i(x,t,y),r_i(a,s,b)) \le \rho((x,t,y),(a,s,b))$.

Take arbitrary $(x,t,y),(a,s,b) \in Y_i$.

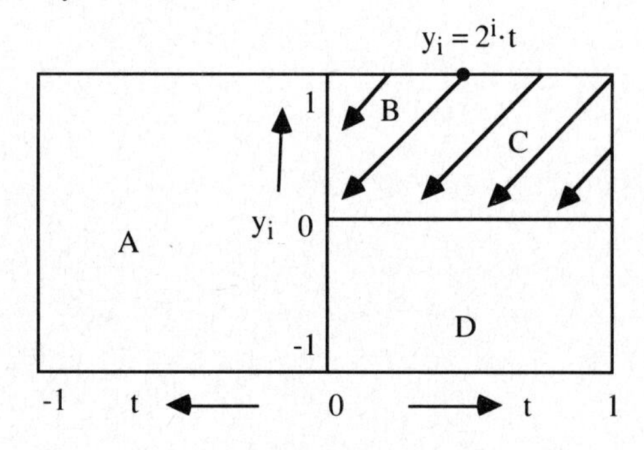

Figure 6.8.4.

There are several cases to consider. If (t,y_i) and (s,b_i) both belong to $A \cup D$ then $\rho(r_i(x,t,y),r_i(a,s,b)) = \rho((x,t,y),(a,s,b))$. So assume e.g. that $(t,y_i) \in B$ and $(s,b_i) \in A$. Then

$$\xi = \rho(r_i(x,t,y),r_i(a,s,b)) - \rho((x,t,y),(a,s,b)) = |s| + 2^{-i}|(y_i - 2^it) - b_i| - |t - s| -2^{-i}|y_i - b_i| =$$

$$= |s| + 2^{-i}|(y_i - b_i) - 2^it| - |t - s| -2^{-i}|y_i - b_i|.$$

Put $p = 2^{-i}(y_i - b_i)$. Then since $s \le 0$ and $t \ge 0$, we obtain

$$\xi = |s| + |p - t| - |t - s| - |p| = -s + |p - t| - t + s - |p| = |p - t| - |p| - t \le 0,$$

which is as required. Suppose next that e.g. $(x,t,y),(a,s,b) \in B$. Then

$$\rho(r_i(x,t,y),r_i(a,s,b)) - \rho((x,t,y),(a,s,b)) = 2^{-i}|(y_i-2^it) - (b_i-2^is)| - |t - s| - 2^{-i}|y_i - b_i| =$$

$$= |2^{-i}(y_i - b_i) + (s - t)| - |t - s| - 2^{-i}|y_i - b_i| \leq 0,$$

which is again as required.

The other cases can be proved similarly.

We next prove that for every i, Y_i is a Hilbert cube. First observe that $P_i \in \mathcal{Z}(Q)$ (lemma 6.2.3(2)), from which it follows that $Q \times \{0\} \times P_i \in \mathcal{Z}(X_2 \times Q)$ (lemma 6.2.2(1) and (5)). By the Homeomorphism Extension Theorem 6.4.6 there exists a homeomorphism $\phi_i: X_2 \times Q \to Q \times [-1,0] \times P_i$ such that $\phi_i | Q \times \{0\} \times P_i = 1$. Then $\psi_i: Y_i \to Q \times [-1,1] \times P_i$ defined by

$$\psi_i = \phi_i \cup 1_{Q \cup [0,1] \times P_i}$$

is a homeomorphism, which implies that Y_i is a Hilbert cube.

Now for every $i > 1$, define $s_i: Q \times [-1,1] \times P_i \to Q \times [-1,1] \times P_i$ by

$$s_i = \psi_i \circ r_i \circ \psi_i^{-1}.$$

Then $s_i(Q \times [-1,1] \times P_i) = (Q \times [-1,0] \times P_i) \cup (Q \times [0,1] \times P_{i-1})$. Moreover, s_i has the following properties:

$$\begin{cases} s_i | Q \times [-1,0] \times P_i = 1, \\ s_i | Q \times [0,1] \times P_i = r_i | Q \times [0,1] \times P_i. \end{cases}$$

So we can think of s_i as the product of the function in the figures 6.8.3 and 6.8.4 crossed with the identity on the other factors of the product $Q \times [-1,0] \times P_i$.

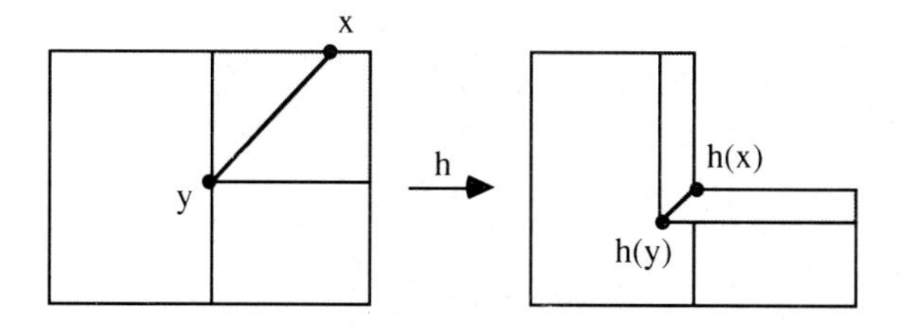

Figure 6.8.5.

CLAIM 3: Let $i > 1$. Then $r_i: Y_i \to Y_{i-1}$ is a near homeomorphism.

Since ψ_i is a homeomorphism, exercise 6.1.8 implies that it suffices to prove that s_i is a near homeomorphism. This is a triviality however. For each $\varepsilon > 0$ it is geometrically obvious that there exists a homeomorphism $h: J^2 \to ([-1,0] \times J) \cup ([0,1] \times [-1,0])$ such that $d(h,\eta_i) < \varepsilon$, where η_i is the function in figures 6.8.3 and 6.8.4. Then h crossed with the identity on the other factors of the product $Q \times [-1,0] \times P_i$ is a homeomorphism "close to" s_i (alternatively, prove that s_i is shrinkable).

By proposition 6.7.2 we conclude that $X \times Q \times Q$ is homeomorphic to the inverse limit of an inverse sequence of Hilbert cubes with bonding maps that are all near homeomorphisms. An application of corollary 6.7.5 therefore yields that $X \times Q \times Q$ is a Hilbert cube.□

6.8.6. COROLLARY: *"The letter* T*" is a Hilbert cube factor.*□

Exercises for §6.8.

Let X be a space. A space Y is called an *X-manifold* if Y admits an open cover by sets homeomorphic to open subsets of X.

1. Let P be a polytope. Prove that $P \times Q$ is a Q-manifold.

2. Let P be a 1-dimensional polyhedron. Prove that $P \times Q$ is a Hilbert cube if and only if P is contractible.

Notes and comments for chapter 6.

§1.

The Inductive Convergence Criterion 6.1.2 is due to Fort [64] and was later rediscovered by Anderson [8]. Theorem 6.1.6 is due to Keller [81]. Bing's Shrinking Criterion 6.1.8 is due to Bing [23]. As was observed by many mathematicians, theorem 6.1.11 follows directly from Keller's Theorem [81]. Exercise 6.1.6 is due to Schori [123]; the proof suggested by us seems to be new. In addition, exercise 6.1.11 is due to Ryll-Nardzewski.

§2.

The concept of a Z-set is due to Anderson [8].

§3.

Theorem 6.3.4 is due to Anderson [8] and Barit [17]. The technique used in the proof is motivated by work of

Klee [82].

§4.

The results in this section are basically due to Anderson [8].

§5.

Skeletoids were introduced independently by Anderson [11] and Bessaga and Pełczyński [20]. Absorbers were first defined by Toruńczyk (unpublished) and later by West [147]. Theorem 6.5.2 is due to Toruńczyk (unpublished) and West [147]. Theorem 6.5.8 is due to Anderson [11]. Corollaries 6.5.9 and 6.5.10 are also due to Anderson [9]. Exercise 6.5.5 is due to Anderson [10].

§6.

Theorem 6.6.10 is due to Anderson [8]. The proof presented here is due to Bessaga and Pełczyński [19] (see also [21]). Exercise 6.6.4 is due to Anderson, Curtis and van Mill [13].

§7.

Theorem 6.7.4 is due to Brown [38]. Exercise 6.7.2 is due to Freudenthal [65].

§8.

Lemma 6.8.2 is due to Anderson (unpublished). Theorem 6.8.5 is due to West [146]. The proof presented here is due to Wong and Kroonenberg [153]. Corollary 6.8.6 is due to Anderson [6]. Exercise 6.8.1 is due to West [146]. Interestingly, the converse to exercise 6.8.1 is also true, i.e. for every Q-manifold M there exists a polytope P such that M and $P \times Q$ are homeomorphic, see Chapman [46], [45].

7. Cell-like Maps and Q-manifolds

In the process of finding a usable topological characterization of the Hilbert cube, maps of a certain type play a prominent role. These are the so-called *cell-like maps*. The aim of this chapter is to study them in some detail. In addition, we shall present some basic facts concerning Hilbert cube manifolds.

Our main results are "The Approximation Theorem" and "The Characterization Theorem", proved in sections 7.5 and 7.8, respectively. These results are due to H. Toruńczyk. As a consequence we obtain that for every compact **ANR** X, $X \times Q$ is a Q-manifold (this is Edward's Theorem). Applications of our results are presented in chapter 8.

7.1. Cell-like Maps and Fine Homotopy Equivalences

Let X be a space. A subset $A \subseteq X$ is said to be *contractible* in X provided that there is a homotopy $H: A \times I \to X$ such that $H_0 = 1_A$ and H_1 is constant. So a contractible space is contractible in itself.

A compactum X has *trivial shape* if for each imbedding f of X into an **ANR** Z, f(X) is contractible in any of its neighborhoods. It is clear that each contractible compactum has trivial shape (the familiar sin(1/x)-continuum (exercise 1.5.4) is an example of a trivial shape continuum which is not contractible). So in particular, if X is a compact **AR** then X has trivial shape. It is easy to see that each compactum with trivial shape is connected. As a consequence, S^0 does not have trivial shape. It is natural to wonder whether there is a connected compactum not having trivial shape. A natural candidate for such a space is any n-sphere for $n \geq 1$. Since S^n is an **ANR** that is not contractible in itself (corollary 3.5.6), it cannot have trivial shape.

7.1.1. THEOREM: *Let* X *be a compact space and let* Y *be an* **ANR** *containing* X. *The following statements are equivalent:*

(a) X *has trivial shape,*

(b) *if* U *is a neighborhood of* X *in* Y *then* X *is contractible in* U.

PROOF: The implication (a) $\Rightarrow$ (b) is a triviality. For (b) $\Rightarrow$ (a), let Z be an **ANR** containing X and let U be an open neighborhood of X in Z. Since U is an **ANR** (theorem 5.4.1), there is an open neighborhood V of X in Y such that the identity $1: X \to X$ can be extended to a continuous function $f: V \to U$. By assumption, there is a homotopy $H: X \times I \to V$ such that $H_0 = 1_X$ and H_1 is constant. Now define $F: X \times I \to U$ by

$$F(x,t) = f(H(x,t)).$$

Then F is clearly a homotopy such that $F_0 = 1_X$ and F_1 is constant.□

Since by corollary 1.5.5 Q is an **AR**, we immediately obtain the following

7.1.2. COROLLARY: *Let* $X \subseteq Q$ *be compact. The following statements are equivalent:*

(a) X *has trivial shape,*

(b) *if* U *is a neighborhood of* X *in* Q *then* X *is contractible in* U.□

We conclude that Q is a test space for detecting compacta having trivial shape.

Let X and Y be spaces. A continuous surjection $f: X \to Y$ is called *proper* if $f^{-1}(K)$ is compact for every compact subset $K \subseteq Y$. It is easy to see that f is proper if and only if f is *perfect*, i.e. f is closed and the point-inverses of f are compact. A proper map $f: X \to Y$ is called *cell-like* provided that $f^{-1}(y)$ has trivial shape for every $y \in Y$. Cell-like maps are very important since they often turn out to be near homeomorphisms if their domain and range are nice.

7.1.3. PROPOSITION: *Let* X *be an* **ANR**, *let* $K \subseteq X$ *be closed and let* $U \subseteq X$ *be a neighborhood of* K *such that* K *is contractible in* U. *Then there is a neighborhood* V *of* K *such that* $V \subseteq U$ *and* V *is contractible in* U.

PROOF: We may assume that U is open in X. Since then U is an **ANR** (theorem 5.4.1), we may even assume that U = X.

Let $H: K \times I \to X$ be a homotopy such that $H_0 = 1$ and H_1 is a constant function, say with value c. Put $Z = (X \times \{0,1\}) \cup (K \times I) \subseteq X \times I$ and define $f: Z \to X$ by

$$\begin{cases} f(x,0) = x & (x \in X), \\ f(x,t) = H(x,t) & (x \in K, t \in I), \\ f(x,1) = c & (x \in X). \end{cases}$$

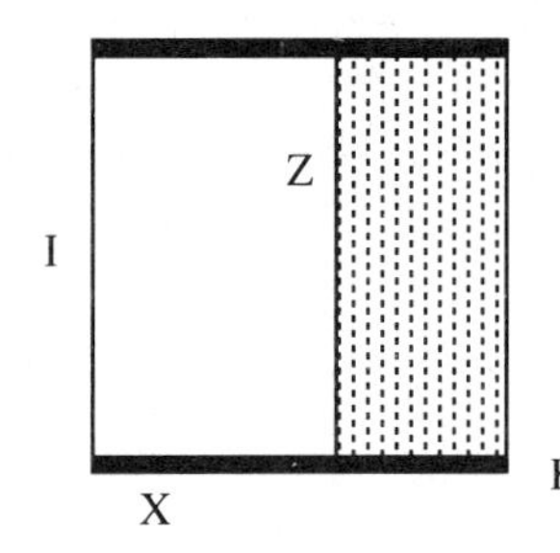

Figure 7.1.1.

Then f is clearly continuous and since X is an **ANR** we can find a neighborhood W of Z in $X \times I$ such that f can be extended continuously to a map g: $W \to X$. By compactness of I, we can find a neighborhood V of K in X such that $V \times I \subseteq W$ (cf. the proof of lemma 1.6.2). Then V is as required for let $F = g \mid (V \times I)$. Then F is a homotopy such that $F_0 = 1$ and F_1 is constant.□

7.1.4. COROLLARY: *Let* X *be an* **ANR** *and let* f: $X \to Y$ *be cell-like. Then for every* $y \in Y$ *and for every neighborhood* U *of* y *in* Y *there is a neighborhood* V *of* y *in* Y *contained in* U *such that* $f^{-1}(V)$ *is contractible in* $f^{-1}(U)$.

PROOF: By proposition 7.1.3 there is an open neighborhood W of $f^{-1}(y)$ such that W is contractible in $f^{-1}(U)$. Since f is a closed map, $V = Y \backslash f(X \backslash W)$ is a neighborhood of y in Y. Since clearly $f^{-1}(V) \subseteq W$, we are done (cf. exercise 1.1.10).□

7.1.5. PROPOSITION: *Let* X *be a compact* **ANR** *and let* f: $X \to Y$ *be cell-like. Moreover, let* K *be a polyhedron, let* $L \subseteq K$ *be a subpolyhedron, and let* ϕ: $K \to Y$ *and* ψ': $L \to X$ *be continuous mappings such that* $f \circ \psi' = \phi \mid L$. *Then for each* $\varepsilon > 0$ *there is a continuous function* ψ: $K \to X$ *extending* ψ' *such that* $d(f \circ \psi, \phi) < \varepsilon$.

PROOF: Suppose that n = dim K. By corollary 7.1.4 and lemma 5.1.7 we can find a collection $\mathcal{U}_0 < \mathcal{U}_1 < \cdots < \mathcal{U}_n$ of finite open covers of Y such that

(1) mesh($\mathcal{U}_n$) < ε,

(2) if $0 \le i \le n-1$ then for each $U \in \mathcal{U}_i$ there is $V \in \mathcal{U}_{i+1}$ such that $f^{-1}(St(U,\mathcal{U}_i))$ is contractible in $f^{-1}(V)$.

Let $\mathcal{T}$ be a triangulation of K such that for every $\sigma \in \mathcal{T}$ there exists a $U \in \mathcal{U}_0$ such that $\phi(\sigma) \subseteq U$ (theorem 3.6.12). For $0 \leq i \leq n$, put $K(i) = L \cup |\mathcal{T}^{(i)}|$. Observe that $K(n) = K$. We shall now inductively construct functions $\psi_i\colon K(i) \to X$ having the following properties:

(3) ψ_0 extends ψ' and if $i \leq n-1$ then ψ_{i+1} extends ψ_i,
(4) for each simplex $\sigma \in \mathcal{T}^{(i)}$ there exists $U \in \mathcal{U}_i$ containing $\phi(\sigma) \cup f(\psi_i(\sigma))$.

For every vertex $v \in \mathcal{T}^{(0)}$ that does not belong to L let $\psi_0(v)$ be an arbitrary point from $f^{-1}(\phi(v))$ (here we use that f is surjective). Define $\psi_0(x) = \psi'(x)$ for every $x \in L$. Then ψ_0 is clearly as required. Now suppose that for certain $0 \leq i \leq n-1$ the function ψ_i has been constructed. Take an arbitrary $\sigma \in \mathcal{T}^{(i+1)}$. We shall define a continuous function $\rho_\sigma\colon \sigma \to X$. If $\sigma \subseteq K(i)$ then define $\rho_\sigma = \psi_i \mid \sigma$. So suppose that σ is not a subset of K(i). Then the geometrical interior of σ misses K(i). By construction there exists $U_0 \in \mathcal{U}_0$ containing $\phi(\sigma)$. Since $\mathcal{U}_0 < \mathcal{U}_i$, there also exists $U_1 \in \mathcal{U}_i$ containing $\phi(\sigma)$. Let τ be any proper face of σ. By (4) there is $U_2 \in \mathcal{U}_i$ containing $\phi(\tau) \cup f(\psi_i(\tau))$. Since $\phi(\tau) \subseteq \phi(\sigma)$ we find that $U_2 \cap U_1 \neq \emptyset$, and consequently, $U_2 \subseteq St(U_1,\mathcal{U}_i)$. Since τ was arbitrary we find that

$$\phi(\sigma) \cup f(\psi_i(\partial\sigma)) \subseteq St(U_1,\mathcal{U}_i).$$

By (2) we there exists an element $V \in \mathcal{U}_{i+1}$ such that $f^{-1}(St(U_1,\mathcal{U}_i))$ is contractible in $f^{-1}(V)$. Let $\rho_\sigma\colon \sigma \to f^{-1}(V)$ be an arbitrary continuous extension of the function $\psi_i \mid \partial\sigma$ (such a function exists since $\partial\sigma \approx S^i$, $\sigma \approx B^{i+1}$ and $f^{-1}(St(U_1,\mathcal{U}_i))$ is contractible in $f^{-1}(V)$). Now let ψ_{i+1} be the union of all the ρ_σ's, $\sigma \in \mathcal{T}^{(i+1)}$. It is easily seen that ψ_{i+1} is as required (lemma 3.6.6). This completes the induction.

Now put $\psi = \psi_n$. Then by (1), (3) and (4) we conclude that ψ is the desired extension. □

Let X and Y be spaces. A continuous function $f\colon X \to Y$ is called a *homotopy equivalence* if there is a continuous function $g\colon Y \to X$ such that $g \circ f \simeq 1_X$ and $f \circ g \simeq 1_Y$. If such an f exists then we say that X and Y have the same *homotopy type*. A continuous function $f\colon X \to Y$ is called a *fine homotopy equivalence* if for any open cover $\mathcal{U}$ of Y there is a continuous function $g\colon Y \to X$ such that

(1) $f \circ g \underset{\mathcal{U}}{\simeq} 1_Y$, and
(2) $g \circ f \underset{f^{-1}(\mathcal{U})}{\simeq} 1_X$

(for the definition of a $\mathcal{V}$-homotopy, see §5.1). Roughly speaking, a fine homotopy equivalence is a homotopy equivalence "with control". So each fine homotopy equivalence is a homotopy equivalence; the converse is not true.

Let X be a space with open cover $\mathcal{U}$. As usual, $St^1(\mathcal{U}) = St(\mathcal{U})$ and $St^{n+1}(\mathcal{U}) = St(St^n(\mathcal{U}))$, $n \in \mathbb{N}$. An open cover $\mathcal{V}$ of X is called a *St^n-refinement* of $\mathcal{U}$ if $St^n(\mathcal{V}) < \mathcal{U}$. It follows by lemma 5.1.7 that $\mathcal{U}$ has St^n-refinements for every $n \in \mathbb{N}$.

We now come to the main result in this section.

7.1.6. THEOREM ("Haver's Theorem"): *Let* X *and* Y *be compact* **ANR**'s *and let* $f: X \to Y$ *be a continuous surjection. The following statements are equivalent:*

(a) f *is cell-like,*

(b) f *is a fine homotopy equivalence.*

PROOF: *We prove (b) ⇒ (a).* Pick an arbitrary $y \in Y$ and let V be a neighborhood of $f^{-1}(y)$. We have to show that $f^{-1}(y)$ is contractible in V (theorem 7.1.1). Since f is a closed map, there exists a neighborhood U_0 of y such that $f^{-1}(U_0) \subseteq V$ (exercise 1.1.10). Let $U_1 = Y\setminus\{y\}$. Then $\mathcal{U} = \{U_0,U_1\}$ is an open cover of Y and by assumption there is a continuous function $g: Y \to X$ such that, among other things,

$$g \circ f \underset{f^{-1}(\mathcal{U})}{\simeq} 1_X.$$

Let $H: X \times I \to X$ be a homotopy such that is limited by $f^{-1}(\mathcal{U})$ while moreover $H_0 = 1_X$ and $H_1 = g \circ f$. If $x \in f^{-1}(y)$ then $x \notin f^{-1}(U_1)$ so

$$H(\{x\} \times I) \subseteq f^{-1}(U_0) \subseteq V.$$

Since $H_0 = 1$ and $H_1(x) = g \circ f(x) = g(y)$ for $x \in f^{-1}(y)$, we see that H contracts $f^{-1}(y)$ to the point g(y) in V.

We prove (a) ⇒ (b).

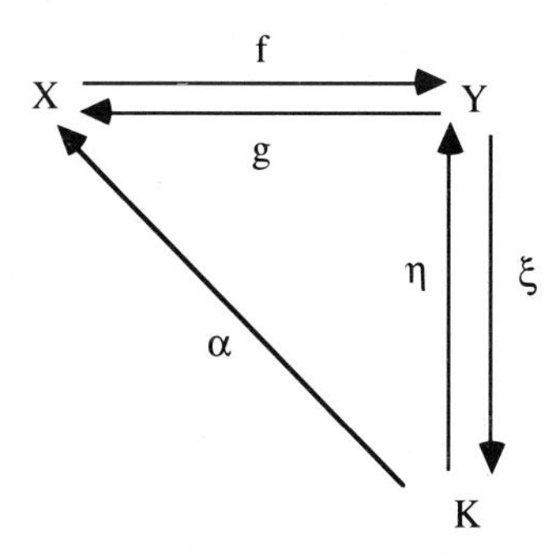

Let $\mathcal{U}$ be an open cover of Y and let $\mathcal{U}_1$ be a St^2-refinement of $\mathcal{U}$. In addition, let $\mathcal{U}_2$ be a star-refinement of $\mathcal{U}_1$. Find $\delta > 0$ such that any two δ-close maps from any space into Y are homotopic by a homotopy that is limited by $\mathcal{U}_2$ (theorem 5.1.1; observe that Y is compact so any open cover of Y has a Lebesgue number (lemma 1.1.1)). Let K be a polyhedron for which there exist continuous functions $\xi\colon Y \to K$ and $\eta\colon K \to Y$ such that $\eta \circ \xi$ and 1_Y are $\mathcal{U}_2$-homotopic (theorem 5.1.8). By proposition 7.1.5 there is a continuous function $\alpha\colon K \to X$ with $d(f \circ \alpha, \eta) < \delta$. Now define $g\colon Y \to X$ by $g = \alpha \circ \xi$. Observe that $d(fg, \eta\xi) = d(f\alpha\xi, \eta\xi) \le d(f\alpha, \eta) < \delta$ (exercise 1.3.7). So fg and $\eta\xi$ are $\mathcal{U}_2$-homotopic. Since $\eta\xi$ and 1_Y are also $\mathcal{U}_2$-homotopic, we conclude that fg and 1_Y are $\mathcal{U}_1$-homotopic, so they are also $\mathcal{U}$-homotopic. Let $h\colon Y \times I \to Y$ be a homotopy that is limited by $\mathcal{U}_1$ while moreover $h_0 = 1_Y$ and $h_1 = fg$.

We shall show that gf and 1_X are $f^{-1}(\mathcal{U})$-homotopic. Let $\mathcal{W}$ be a common open refinement of $f^{-1}(\mathcal{U}_1)$ and $(gf)^{-1}(f^{-1}(\mathcal{U}_1))$. In addition, let L be a polyhedron for which there exist continuous functions $\beta\colon X \to L$ and $\gamma\colon L \to X$ such that $\gamma \circ \beta$ is $\mathcal{W}$-homotopic to 1_X (theorem 5.1.8). Let $H\colon X \times I \to X$ be a homotopy that is limited by $\mathcal{W}$ while moreover $H_0 = \gamma \circ \beta$ and $H_1 = 1_X$. Now define $F\colon L \times I \to Y$ by $F(x,t) = h_t(f\gamma(x))$. Observe that $F(x,0) = f\gamma(x)$ and $F(x,1) = fgf\gamma(x)$, for every $x \in L$. Define $G'\colon L \times \{0,1\} \to X$ by $G'(x,0) = \gamma(x)$ and $G'(x,1) = gf\gamma(x)$. The diagram

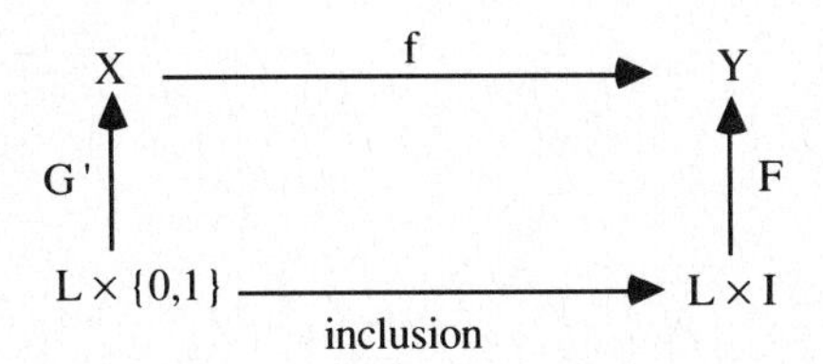

clearly commutes so by proposition 7.1.5 there is a continuous function $G\colon L \times I \to X$ such that $G \mid L \times \{0,1\} = G'$ and such that F and fG are $\mathcal{U}_1$-close, i.e. for every $x \in L \times I$ there exists $U \in \mathcal{U}_1$ containing both $F(x)$ and $fG(x)$ (observe that $L \times \{0,1\}$ and $L \times I$ are polyhedra and that there is a triangulation of $L \times I$ such that $L \times \{0,1\}$ is a subpolyhedron of $L \times I$ (exercise 3.6.1)). Define $\psi\colon X \times I \to X$ by

$$\psi(x,t) = G(\beta(x),t).$$

Observe that for every $x \in X$,

$$\psi(x,0) = G(\beta(x),0) = G'(\beta(x),0) = \gamma\beta(x),$$

and

$$\psi(x,1) = G(\beta(x),1) = G'(\beta(x),1) = gf\gamma\beta(x).$$

Recall that H is a homotopy joining $\gamma\beta$ and 1_X and is limited by $f^{-1}(\mathcal{U}_1)$. Also, H is limited by $(gf)^{-1}(f^{-1}(\mathcal{U}_1))$, so that the homotopy $gfH\colon X \times I \to X$ is limited by $f^{-1}(\mathcal{U}_1)$ and joins the functions $gf\gamma\beta$ and gf. We therefore have the following situation:

$$(1)\qquad gf \underset{f^{-1}(\mathcal{U}_1)}{\overset{gfH}{\sim}} gf\gamma\beta \overset{\psi}{\sim} \gamma\beta \underset{f^{-1}(\mathcal{U}_1)}{\overset{H}{\sim}} 1_X.$$

CLAIM: ψ is limited by $St(f^{-1}(\mathcal{U}_1))$.

Since $\mathcal{U}_1$ is a St^2-refinement of $\mathcal{U}$ this implies that gf and 1_X are $f^{-1}(\mathcal{U})$-homotopic, which is as desired.

It clearly suffices to prove that the composition $f\psi$: $X \times I \to Y$ is limited by $St(\mathcal{U}_1)$. Take $x \in X$ and fix an arbitrary $t \in I$. Since h is limited by $\mathcal{U}_1$ there exists an element $U \in \mathcal{U}_1$ such that

$$h(f\gamma\beta(x) \times I) \subseteq U.$$

Also, $f\psi(x,t) = fG(\beta(x),t)$ and since fG and F are $\mathcal{U}_1$-close, there exists $U_1 \in \mathcal{U}_1$ containing the points

$$f\psi(x,t) \text{ and } F(\beta(x),t) = h_t(f\gamma\beta(x)).$$

Observe that $U \cap U_1 \neq \emptyset$ since they both contain $h_t(f\gamma\beta(x))$. Consequently,

$$f\psi(\{x\} \times I) \subseteq St(U,\mathcal{U}_1)) \in St(\mathcal{U}_1).\square$$

7.1.7. COROLLARY: *Let* X *and* Y *be compact* **ANR**'s *and let* $f: X \to Y$ *be cell-like. In addition, let* $A \subseteq Y$ *be closed. If* U *is a neighborhood of* A *in* Y *such that* A *is contractible in* U *then* $f^{-1}(A)$ *is contractible in* $f^{-1}(U)$.

PROOF: Put $B = f^{-1}(A)$. Let $H: A \times I \to U$ be a homotopy that contracts A to a point. Let $\varepsilon > 0$ be such that the open ball about $A' = H(A \times I)$ with radius ε is contained in U. Let $\mathcal{U}$ be the cover of Y consisting of all open sets with diameter less than ε. By theorem 7.1.6 there is a continuous function $g: Y \to X$ such that

(1) $f \circ g \underset{\mathcal{U}}{\sim} 1_Y$, and

(2) $g \circ f \underset{f^{-1}(\mathcal{U})}{\sim} 1_X$.

By (2) there consequently is an $f^{-1}(\mathcal{U})$-homotopy $S: B \times I \to X$ such that $S_0 = 1_B$ and $S_1 = (g \circ f) \mid B$. Observe that $S(B \times I) \subseteq f^{-1}(U)$. Define $T: B \times I \to X$ to be the composition

$g \circ H \circ (f \times 1_I)$.

Then T is a homotopy connecting the function $(g \circ f) \mid B$ and a constant function. In addition, $T(B \times I) \subseteq g(A')$. So it suffices to prove that $g(A') \subseteq f^{-1}(U)$. This is easy. Take a point $a' \in A'$. Then by (1), $d(a', fg(a')) < \varepsilon$, so $fg(a') \in U$, from which it follows that $g(a') \in f^{-1}(U)$.□

7.1.8. COROLLARY: *Let* X *and* Y *be compact* **ANR**'s *and let* $f: X \to Y$ *be cell-like. If* $K \subseteq Y$ *has trivial shape then* $f^{-1}(K)$ *has trivial shape.*

PROOF: Apply corollary 7.1.7 and theorem 7.1.1.□

7.1.9. COROLLARY: *Any finite composition of cell-like maps between compact* **ANR**'s *is cell-like.*□

7.1.10. *Remark:* It is possible to give an example of two compacta X and Y such that Y is an **ANR**, and a cell-like map $f: X \to Y$ such that f is not a fine homotopy equivalence; this is easy. It is also possible to construct two compacta X and Y such that X is an **ANR** and a cell-like map $f: X \to Y$ that is not a fine homotopy equivalence, see Taylor [133]; this is quite complicated. We conclude that the restriction to **ANR**'s in theorem 7.1.6 is essential.

7.2. Z-Sets in ANR's

The aim of this section is, among other things, to derive a characterization of Z-sets in an arbitrary **ANR** similar to lemma 6.2.3(1).

Let X be a space. We define $\mathcal{K}(X)$ to be the collection of all spaces Y having the following property:

> For every $A \in \mathcal{Z}(X)$, for every open cover $\mathcal{V}$ of X and for every continuous function $f: Y \to X$ there exists a continuous function $g: Y \to X \setminus A$ such that f and g are $\mathcal{V}$-close.

So $Q \in \mathcal{K}(X)$. We shall prove in this section that for each **ANR** X, $\mathcal{K}(X)$ is the class of all spaces.

7.2.1. LEMMA: *Let* X *be an* **ANR** *and let* $A \in \mathcal{Z}(X)$. *If* $Y \in \mathcal{K}(X)$, $B \subseteq Y$ *is closed and* $f: Y \to X$ *is continuous such that* $\overline{f(B)} \cap A = \emptyset$, *then for every open cover* $\mathcal{U}$ *of* X *there is a continuous function* $g: Y \to X$ *such that*

(1) f *and* g *and* $\mathcal{U}$*-close,*

(2) $g \mid B = f \mid B$,

(3) $g(Y) \cap A = \emptyset$.

PROOF: Let $\mathcal{T}$ be the open cover $\{X\setminus A, X\setminus\overline{f(B)}\}$ of X and let $\mathcal{V}$ be a common open refinement of the covers $\mathcal{U}$ and $\mathcal{T}$. Finally, let $\mathcal{W}$ be an open refinement of $\mathcal{V}$ such that any two $\mathcal{W}$-close maps from any space into X are $\mathcal{V}$-homotopic (theorem 5.1.1). Since $A \in \mathcal{Z}(X)$ there exists a continuous function $h: Y \to X$ such that h and f are $\mathcal{W}$-close and $h(Y) \cap A = \emptyset$. Then h and f are $\mathcal{V}$-homotopic, so there is a homotopy $H: Y \times I \to X$ which is limited by $\mathcal{V}$ and which moreover connects h and f, i.e. $H_0 = h$ and $H_1 = f$. Since $\mathcal{V} < \mathcal{T}$, $H(B \times I) \cap A = \emptyset$. By continuity there is a neighborhood V of B in Y such that $H(V \times I) \cap A = \emptyset$. Let $\alpha: Y \to I$ be a Urysohn function (corollary 1.4.15) such that $\alpha(B) = 1$ and $\alpha(Y\setminus V) = 0$. Define $g: Y \to X$ by

$$g(x) = H_{\alpha(x)}(x).$$

Then g is clearly continuous and we claim that g is as required. First we shall show that $g(Y) \cap A = \emptyset$. If $x \in Y\setminus V$ then $\alpha(x) = 0$, so $g(x) = H_0(x) = h(x) \notin A$. Also, if $x \in V$ then $g(x) \in H(V \times I)$ and since $H(V \times I)$ misses A, we also conclude that $g(x) \notin A$. Secondly, observe that $g \mid B = f$ is a triviality. Finally, that f and g are $\mathcal{U}$-close follows easily from the fact that H is limited by $\mathcal{V}$ and the fact that $H_1 = f$. □

This lemma is used in the proof of the following

7.2.2. PROPOSITION: *Let* X *be an* **ANR**, *let* $U \subseteq X$ *be open and let* $A \in \mathcal{Z}(U)$ *be such that* A *is closed in* X. *Then for every polyhedron* K, *for every open cover* $\mathcal{U}$ *of* X *and for every continuous function* $f: K \to X$ *there is a continuous function* $g: K \to X$ *such that*

(1) f *and* g *are* $\mathcal{U}$*-close,*

(2) $g(K) \cap A = \emptyset$.

PROOF: Since A has empty interior in U (lemma 6.2.2(2)), A has empty interior in X. So the proposition trivially holds for zero-dimensional polyhedra (= finite spaces). Suppose that the proposition is true for all polyhedra of dimension $n \geq 0$ and let a polyhedron K of dimension n+1 and let a continuous function $f: K \to X$ be given. Let $\mathcal{F}$ be an open cover of X that refines both $\mathcal{U}$ and the cover $\{U, X\setminus A\}$ of X. By lemma 5.1.7 there are open covers $\mathcal{D}$ and $\mathcal{E}$ of X such that

$$\mathcal{D} \overset{*}{<} \mathcal{E} \overset{*}{<} \mathcal{F}.$$

Since every singleton in X has trivial shape (being contractible), for every $x \in X$ there is a neighborhood C_x of x which is contractible in an element of $\mathcal{D}$ (proposition 7.1.3). Let $\mathcal{C}$ be the cover $\{C_x : x \in X\}$ and, again by lemma 5.1.7, let $\mathcal{B}$ be an open cover of X such that $\mathcal{B} \overset{*}{<} \mathcal{C}$. In addition, let $\mathcal{A}$ be an open cover of X such that $\mathcal{A} \overset{*}{<} \mathcal{B}$. Observe that the following holds:

(*) $$\mathcal{A} \overset{*}{<} \mathcal{B} \overset{*}{<} \mathcal{C} < \mathcal{D} \overset{*}{<} \mathcal{E} \overset{*}{<} \mathcal{F}.$$

By compactness of K, the cover $f^{-1}(\mathcal{A})$ of K has a Lebesgue number (lemma 1.1.1). Consequently there exists a triangulation $\mathcal{T}$ of K each element of which is contained in an element of the cover $f^{-1}(\mathcal{A})$ (theorem 3.6.12). By our inductive hypothesis there exists a continuous function $h: |\mathcal{T}^{(n)}| \to X$ such that $h(|\mathcal{T}^{(n)}|) \cap A = \emptyset$, while moreover $f \mid |\mathcal{T}^{(n)}|$ and h are $\mathcal{A}$-close.

CLAIM 1: For every $\sigma \in \mathcal{T}^{(n)}$ there exists $B \in \mathcal{B}$ such that $h(\sigma) \cup f(\sigma) \subseteq B$.

This is easy. Let $\vec{A} \in \mathcal{A}$ be such that $f(\sigma) \subseteq \vec{A}$ and take an arbitrary $p \in \sigma$. Then the set $\{f(p),h(p)\}$ is contained in an element of $\mathcal{A}$. We conclude that $h(p) \in St(\vec{A},\mathcal{A})$. Consequently, since p was arbitrary, $h(\sigma) \cup f(\sigma) \subseteq St(\vec{A},\mathcal{A})$ and since $\mathcal{A}$ is a star-refinement of $\mathcal{B}$, the existence of the required B follows.

CLAIM 2: For every $\sigma \in \mathcal{T}^{(n+1)}$ there exists $C \in \mathcal{C}$ such that $h(\partial\sigma) \cup f(\partial\sigma) \subseteq C$.

It is easy to see that the claim is true for $n = 0$. Therefore, without loss of generality assume that $n \geq 1$. By claim 1, for every n-dimensional face τ of σ, there exists an element $B_\tau \in \mathcal{B}$ containing $h(\tau) \cup f(\tau)$. Since the n-dimensional faces of σ pairwise intersect, this implies that for every n-dimensional face τ of σ we have $h(\partial\sigma) \cup f(\partial\sigma) \subseteq St(B_\tau,\mathcal{B})$. Since $\mathcal{B} \overset{*}{<} \mathcal{C}$, we are done.

Now let $\sigma \in \mathcal{T}^{(n+1)}$. By claim 2 there exists $D_\sigma \in \mathcal{D}$ such that $h(\partial\sigma)$ is contractible in D_σ. Therefore, the function $h \mid \partial\sigma: \partial\sigma \to X$ can be extended to a map $\xi_\sigma: \sigma \to D_\sigma$. The union of the ξ_σ's defines a continuous function $\xi: K \to X$ that extends h (lemma 3.6.6).

CLAIM 3: For every $\sigma \in \mathcal{T}$, $f(\sigma) \cup \xi(\sigma)$ is contained in an element of $\mathcal{E}$.

First assume that $\sigma \in \mathcal{T}^{(n)}$. Then claim 1, the fact that ξ extends h and the fact that $\mathcal{B} < \mathcal{E}$ give us what we want. Suppose therefore that $\sigma \in \mathcal{T}^{(n+1)}$. By claim 2 there exists $C \in \mathcal{C}$ such that $h(\partial\sigma) \cup f(\partial\sigma) \subseteq C$. In addition, by construction there exists $\vec{A} \in \mathcal{A}$ such that $f(\sigma) \subseteq \vec{A}$. Also, there exists $D \in \mathcal{D}$ with $\xi(\sigma) \subseteq D$. From this and (*) it follows easily that $f(\sigma) \cup \xi(\sigma)$ is contained in an element of $\mathcal{E}$.

Again, take an arbitrary $\sigma \in \mathcal{T}^{(n+1)}$. By claim 3, $f(\sigma) \cup \xi(\sigma)$ is contained in an element of $\mathcal{E}$. Since $\mathcal{E} < \mathcal{F}$, this element is either contained in U or in X\A. In the second case, put $g_\sigma = \xi_\sigma$. In the first case, since an open subset of an **ANR** is an **ANR** (theorem 5.4.1), lemma 7.2.1 gives us a continuous function $g_\sigma: \sigma \to U$ such that

(1) $g_\sigma \mid \partial\sigma = h$,
(2) $g_\sigma(\sigma) \cap A = \emptyset$, and
(3) g_σ and ξ_σ are $\mathcal{E}$-close

(observe that $\sigma \approx J^{n+1}$ and hence belongs to $\mathcal{K}(U)$ since $Q \in \mathcal{K}(U)$). The union of the g_σ's defines a continuous function $g: K \to X\backslash A$ (lemma 3.6.6). Finally, claim 3 and (3) easily imply that f and g are $\mathcal{U}$-close. □

7.2.3. COROLLARY: *Let* X *be an* **ANR**. *Then* $\mathcal{K}(X)$ *contains all polyhedra.* □

7.2.4. PROPOSITION: *Let* X *be an* **ANR**, *let* $U \subseteq X$ *be open and let* $A \in \mathcal{Z}(U)$ *be such that* A *is closed in* X. *Then for every polytope* K, *for every open cover* $\mathcal{U}$ *of* X *and for every continuous function* $f: K \to X$ *there is a continuous function* $g: K \to X$ *such that*

(1) f *and* g *are* $\mathcal{U}$-close,
(2) $g(K) \cap A = \emptyset$.

PROOF: By theorem 1.4.19 there exists an admissible metric d on X such that the collection of all open d-balls with radius 1 refines $\mathcal{U}$. It is clear that there exists a sequence of subpolyhedra (i.e. subpolytopes having finitely many simplexes only) $K_1 \subseteq K_2 \subseteq \cdots$ of K such that

$$K = \bigcup_{n=1}^{\infty} K_n.$$

Finally, since X is an **ANR**, theorem 5.1.1 implies that there exist open covers $\mathcal{U}_0, \mathcal{U}_1, \mathcal{U}_2, \cdots$ of X such that

(1) $\mathcal{U}_0 = \mathcal{U}$ and for $n \geq 1$, $\text{mesh}(\mathcal{U}_n) < 2^{-n}$,
(2) for every $n \geq 1$, for every space Y, any two $\mathcal{U}_n$-close maps $f,g: Y \to X$ are $\mathcal{U}_{n-1}$-homotopic.

By induction on $n \geq 0$ we shall construct a continuous function $f_n: K \to X$ having the following properties:

(3) $f_0 = f$ and for $n \geq 1$, $f_n \mid K_{n-1} = f_{n-1} \mid K_{n-1}$,
(4) for $n \geq 1$, f_n and f_{n-1} are $\mathcal{U}_n$-close,
(5) for $n \geq 1$, $f_n(K_n) \cap A = \emptyset$.

Assume that for some $n \geq 1$, the functions $f_0,\cdots,f_{n-1}$ have been defined. By corollary 7.2.3, $K_n \in \mathcal{K}(X)$, so by lemma 7.2.1 there exists a continuous function $\gamma: K_n \to X$ such that

(6) $\gamma \mid K_{n-1} = f_{n-1} \mid K_{n-1}$,
(7) γ and $f_{n-1} \mid K_n$ are $\mathcal{U}_{n+1}$-close,
(8) $\gamma(K_n) \cap A = \emptyset$.

Then by (7) and (2), γ and $f_{n-1} \mid K_n$ are $\mathcal{U}_n$-homotopic. Since $f_{n-1} \mid K_n$ can be extended to the function $f_{n-1}: K \to X$, theorem 5.1.3 implies that γ can be extended to a continuous function $f_n: K \to X$ such that f_{n-1} and f_n are $\mathcal{U}_n$-close. Then f_n is clearly as required. This completes the induction.

Now define $g: K \to X$ by

$$g(x) = \lim_{n\to\infty} f_n(x).$$

Observe that by (3) the function g is well-defined, and that by (3) and (5) we have $g(K) \cap A = \emptyset$. We shall prove that g is continuous. To this end, let σ be one of the simplexes of K. Since σ has only finitely many vertices, there has to be an index n such that $\sigma \subseteq K_n$. Again by (3) this gives us that $g \mid \sigma = f_n \mid \sigma$. Consequently, the continuity of the function f_n implies that $g \mid \sigma$ is continuous. Now apply lemma 3.6.6. Finally, we shall prove that g and f are $\mathcal{U}$-close. To this end, take an arbitrary $x \in K$, say $x \in K_n$. Then by (3), $g(x) = f_n(x)$. Consequently, (4) and (1) imply that

$$d(f(x),g(x)) = d(f_0(x),f_n(x)) \leq \Sigma_{i=0}^{n-1} d(f_i(x),f_{i+1}(x)) \leq \Sigma_{i=0}^{n-1} 2^{-(i+1)} < 1.$$

Consequently, by the special choice of d, we are done. □

We now come to the main result in this section.

7.2.5. THEOREM: *Let* X *be an* **ANR** *and let* $A \subseteq X$ *be closed. The following statements are equivalent:*

(a) $A \in \mathcal{Z}(X)$,
(b) *for every neighborhood* U *of* A *in* X, $A \in \mathcal{Z}(U)$,
(c) *for some neighborhood* U *of* A *in* X, $A \in \mathcal{Z}(U)$,

(d) *for every open cover* $\mathcal{U}$ *of* X *there exists a continuous function* $f: X \to X \setminus A$ *such that* f *and* 1_X *are* $\mathcal{U}$*-close* .

PROOF: That (a) ⇒ (b) ⇒ (c) is trivial and the proof of (d) ⇒ (a) is similar to the proof of lemma 6.2.3(1). It therefore suffices to prove that (c) ⇒ (d). Without loss of generality we assume that U is open. Let $\mathcal{V}$ be a star-refinement of $\mathcal{U}$ (lemma 5.1.7). By theorem 5.1.8 there exists a polytope K and continuous functions $\xi: X \to K$ and $\eta: K \to X$ such that $\eta \circ \xi$ and 1_X are $\mathcal{V}$-close. By proposition 7.2.4 there is a continuous function $h: K \to X$ such that h and η are $\mathcal{V}$-close while moreover $h(K) \cap A = \emptyset$. Now put $f = h \circ \xi$. Then clearly $f(X) \cap A = \emptyset$. We claim that f and 1_X are $\mathcal{U}$-close. To this end pick an arbitrary $x \in X$. There exist $V_1 \in \mathcal{V}$ and $V_2 \in \mathcal{V}$ such that $\{x,\eta(\xi(x))\} \subseteq V_1$ and $\{\eta(\xi(x)),h(\xi(x))\} = \{\eta(\xi(x)),f(x)\} \subseteq V_2$. Now since $\mathcal{V}$ is a star-refinement of $\mathcal{U}$, we are done.□

7.2.6. COROLLARY: *Let* X *be an* **ANR**. *Then* $\mathcal{K}(X)$ *is the class of all spaces.*

PROOF: This is easy. Let $\mathcal{U}$ be an open cover of X, let Y be any space, let $A \in \mathcal{Z}(X)$ and let $f: Y \to X$ be continuous. By theorem 7.2.5(d) there exists a continuous function $\xi: X \to X \setminus A$ such that ξ and 1_X are $\mathcal{U}$-close. Define $g: Y \to X$ by $g = \xi \circ f$. Clearly, g and f are $\mathcal{U}$-close and $g(Y) \cap A = \emptyset$.□

We now aim at a generalization of proposition 7.2.4 for topologically complete spaces that shall be important in future considerations.

7.2.7. LEMMA: *Let* X *be a topologically complete* **ANR** *and let* $A \in \mathcal{Z}_\sigma(X)$. *If* K *is a compact space and* K_0 *is closed in* K *then for every open cover* $\mathcal{U}$ *of* X *and every continuous function* $f: K \to X$ *there is a continuous function* $g: K \to X$ *such that*

(1) f *and* g *are* $\mathcal{U}$*-close,*

(2) $f \mid K_0 = g \mid K_0$,

(3) $g(K \setminus K_0) \cap A = \emptyset$.

PROOF: Let $A = \bigcup_{n=1}^{\infty} A_n$, where $A_n \in \mathcal{Z}(X)$ for every n and let $K \setminus K_0 = \bigcup_{n=1}^{\infty} K_n$, where each K_n is compact.

Define

$$\mathcal{A} = \{g \in C(K,X): g \mid K_0 = f \mid K_0\}.$$

Then $\mathcal{A}$ is clearly a closed subspace of the topologically complete space C(K,X) (corollary 1.3.5). Consequently, $\mathcal{A}$ is topologically complete itself. For all $n,m \in \mathbb{N}$ define

$$\mathcal{A}_{n,m} = \{g \in \mathcal{A}: g(K_n) \cap A_m = \emptyset\}.$$

By compactness of K_n it follows easily that $\mathcal{A}_{n,m}$ is an open subspace of $\mathcal{A}$. We claim that $\mathcal{A}_{n,m}$ is also dense. To this end, take an arbitrary $g \in \mathcal{A}$. Then g can be approximated by maps of the form

$$K \xrightarrow{g} X \xrightarrow{\xi} X \backslash A_m,$$

where $d(\xi,1)$ is arbitrarily small (theorem 7.2.5). Since X is an **ANR** we can choose a suitable ξ such that g and $\xi \circ g$ are homotopic by a small homotopy H (theorem 5.1.1). Let $\alpha: K \to I$ be a Urysohn function (corollary 1.4.15) such that $\alpha(K_0) = 0$ and $\alpha(K_n) = 1$. Define $g': K \to X$ by

$$g'(x) = H_{\alpha(x)}(x).$$

Then $d(g,g')$ is small, $g' \in \mathcal{A}$ and $g'(K_n) \cap A_m = \emptyset$.

By the Baire category theorem it now follows that

$$\mathcal{B} = \bigcap_{n=1}^{\infty} \bigcap_{m=1}^{\infty} \mathcal{A}_{n,m}$$

is dense in $\mathcal{A}$.

By lemma 1.1.1 there is $\varepsilon > 0$ such that every subset B of X with diameter less than ε and which moreover intersects f(K) is contained in an element of $\mathcal{U}$. It is easy to see that any function g in $\mathcal{B}$ having distance less than ε from f is as required. □

7.2.8. COROLLARY: *Let* X *be a topologically complete* **ANR** *and let* $A \in \mathcal{Z}_\sigma(X)$. *If* K *is a* σ*-compact space then for every open cover* $\mathcal{U}$ *of* X *and every continuous function* $f: K \to X$ *there is a continuous function* $g: K \to X$ *such that*

(1) f and g are $\mathcal{U}$-close,

(2) $g(K) \cap A = \emptyset$.

PROOF: We argue such as in the proof of proposition 7.2.4; for completeness sake we shall give all details. By theorem 1.4.19 there exists an admissible metric d on X such that the collection of all open d-balls with radius 1 refines $\mathcal{U}$. Let $K_1 \subseteq K_2 \subseteq \cdots$ be compact subsets of K such that

$$K = \bigcup_{n=1}^{\infty} K_n.$$

Since X is an **ANR**, theorem 5.1.1 implies that there exist open covers $\mathcal{U}_0$, $\mathcal{U}_1$, $\mathcal{U}_2$, $\cdots$ of X such that

(1) $\mathcal{U}_0 = \mathcal{U}$ and for $n \geq 1$, $\text{mesh}(\mathcal{U}_n) < 2^{-n}$,
(2) for every $n \geq 1$, for every space Y, any two $\mathcal{U}_n$-close maps $f,g: Y \to X$ are $\mathcal{U}_{n-1}$-homotopic.

By induction on $n \geq 0$ we shall construct a continuous function $f_n: K \to X$ having the following properties:

(3) $f_0 = f$ and for $n \geq 1$, $f_n \mid K_{n-1} = f_{n-1} \mid K_{n-1}$,
(4) for $n \geq 1$, f_n and f_{n-1} are $\mathcal{U}_n$-close,
(5) for $n \geq 1$, $f_n(K_n) \cap A = \emptyset$.

Assume that for some $n \geq 1$, the functions $f_0,\cdots,f_{n-1}$ have been defined. By lemma 7.2.7 there exists a continuous function $\gamma: K_n \to X$ such that

(6) $\gamma \mid K_{n-1} = f_{n-1} \mid K_{n-1}$,
(7) γ and $f_{n-1} \mid K_n$ are $\mathcal{U}_{n+1}$-close,
(8) $\gamma(K_n) \cap A = \emptyset$.

Then by (7) and (2), γ and $f_{n-1} \mid K_n$ are $\mathcal{U}_n$-homotopic. Since $f_{n-1} \mid K_n$ can be extended to the function $f_{n-1}: K \to X$, theorem 5.1.3 implies that γ can be extended to a continuous function $f_n: K \to X$ such that f_{n-1} and f_n are $\mathcal{U}_n$-close. Then f_n is clearly as required. This completes the induction.

Now define $g: K \to X$ by

$$g(x) = \lim_{n\to\infty} f_n(x).$$

Observe that by (3) the function g is well-defined, and that by (3) and (5) we have $g(K) \cap A = \emptyset$. Also, g is continuous by (1), (4) and proposition 1.3.4. Finally, by (1), (6) and (7), $d(f(x),g(x)) < 1$ for every x.□

7.2.9. COROLLARY: *Let* X *be a topologically complete* **ANR** *and let* A *be a* σZ*-set in* X. *Then for every open cover* $\mathcal{U}$ *of* X *there is a continuous function* $f: X \to X \setminus A$ *such that* 1_X *and* f *are* $\mathcal{U}$*-close.*

PROOF: Let $\mathcal{V}$ be a star-refinement of $\mathcal{U}$ (lemma 5.1.7). By theorem 5.1.8 there exist a polytope K and continuous functions $\xi: X \to K$ and $\eta: K \to X$ such that $\eta \circ \xi$ and 1_X are $\mathcal{V}$-close. Since K is σ-compact, by corollary 7.2.8 there is a continuous function $\rho: K \to X \backslash A$ such that η and ρ are $\mathcal{V}$-close. It is clear that $f = \rho \circ \xi: X \to X$ is as required.□

We finish this section by deriving a few simple results on cell-like maps that shall be important later.

7.2.10. PROPOSITION: *Let* X *and* Y *be compact* **ANR**'s *and let* $f: X \to Y$ *be cell-like. Then for each* $A \in \mathcal{Z}(Y)$, *each neighborhood* V *of* $f^{-1}(A)$ *and each open covering* $\mathcal{U}$ *of* Y *there is a homotopy* $\alpha: X \times I \to X$ *having the following properties:*

(1) α *is limited by* $f^{-1}(\mathcal{U})$,

(2) $\alpha_0 = 1_X$ *and* $\alpha_1(X) \cap f^{-1}(A) = \emptyset$,

(3) *every* α_t $(t \in I)$ *restricts to the identity on* $X \backslash V$.

PROOF: Let $\mathcal{E}$ be a finite open St^2-refinement of $\mathcal{U}$ (lemma 5.1.7) and let $\xi: Y \to Y \backslash A$ be continuous such that ξ and 1_Y are $\mathcal{E}$-homotopic; such a function ξ exists by theorems 7.2.5 and 5.1.1. Let $H: Y \times I \to Y$ be a homotopy that is limited by $\mathcal{E}$ such that $H_0 = 1_Y$ and $H_1 = \xi$. In addition, let $\mathcal{F}$ be a finite open refinement of $\mathcal{E}$ having the following property:

(1) if $F \in \mathcal{F}$ and $F \cap \xi(Y) \neq \emptyset$ then $F \cap A = \emptyset$.

Since f is a fine homotopy equivalence (theorem 7.1.6) there is a continuous function $g: Y \to X$ such that

(2) $g \circ f \underset{f^{-1}(\mathcal{F})}{\simeq} 1_X$.

Let $L: X \times I \to X$ be a homotopy that is limited by $f^{-1}(\mathcal{F})$ such that $L_0 = 1_X$ and $L_1 = g \circ f$. Let p be the composition $g \circ \xi \circ f$. We shall prove that $p(X) \cap f^{-1}(A) = \emptyset$. Take an arbitrary $x \in X$. Then $\xi f(x) \notin A$ and since f is surjective, there is a point $y \in X$ such that $f(y) = \xi f(x)$. By (2) there is an element $F \in \mathcal{F}$ such that $\{y, gf(y)\} \subseteq f^{-1}(F)$. Then $f(y) = \xi f(x) \in F$, so that by (1), $F \cap A = \emptyset$. So $gf(y) = g\xi f(x) = p(x) \in f^{-1}(F)$ and $f^{-1}(F) \cap f^{-1}(A) = \emptyset$. We conclude that $p(x) \notin f^{-1}(A)$. Next we shall prove that p and 1_X are $f^{-1}(\mathcal{U})$-homotopic. Define a homotopy $G: X \times I \to X$ by

(3) $G(x,t) = gH(f(x),t)$.

Then $G_0 = g \circ f$ and $G_1 = g \circ \xi \circ f = p$. Now take an arbitrary $x \in X$. There exists an element $E \in \mathcal{E}$ such that $H(\{f(x)\} \times I) \subseteq E$. Take an arbitrary point $f(z) \in H(\{f(x)\} \times I)$. There is an element $E_0 \in \mathcal{E}$ such that $\{z, gf(z)\} \subseteq f^{-1}(E_0)$. Since $z \in f^{-1}(E)$ we conclude that the point $gf(z)$ belongs to $f^{-1}(St(E,\mathcal{E}))$. Since $f(z)$ was chosen arbitrarily, this proves that

$$(4)\ gH(\{f(x)\} \times I) \subseteq f^{-1}(St(E,\mathcal{E})).$$

By (2) there is an element $E_1 \in \mathcal{E}$ such that

$$(5)\ L(\{x\} \times I) \subseteq f^{-1}(E_1).$$

Since $gf(x) \in f^{-1}(St(E,\mathcal{E})) \cap f^{-1}(E_1)$ we conclude from (4) and (5) that

$$(6)\ G(\{x\} \times I) \cup L(\{x\} \times I) \text{ is contained in an element of } f^{-1}(St^2(\mathcal{E})).$$

Since $\mathcal{E}$ is a St^2-refinement of $\mathcal{U}$ this shows that p and 1_X are $f^{-1}(\mathcal{U})$-homotopic.

Now assume that V is a neighborhood of $f^{-1}(A)$. Since f is a closed map, there is an open neighborhood W of A such that $f^{-1}(W) \subseteq V$ (exercise 1.1.10). It is clear that we may assume without loss of generality that our open cover $\mathcal{U}$ has the additional property that every $U \in \mathcal{U}$ which intersects A is contained in W. By the above there is a homotopy $\Phi: X \times I \to X$ such that Φ is limited by $f^{-1}(\mathcal{U})$, while moreover $\Phi_0 = 1_X$ and $\Phi_1(X) \cap f^{-1}(A) = \emptyset$. Put

$$G = \{x \in X: \Phi(\{x\} \times I) \cap f^{-1}(A) \neq \emptyset\}.$$

It is easy to see that G is closed in X. Also, observe that $f^{-1}(A) \subseteq G$. We claim that $S = X \backslash f^{-1}(W)$ misses G. That is easy. Take an arbitrary point $x \in S$. Then $\Phi(\{x\} \times I)$ is contained in an element of $f^{-1}(\mathcal{U})$, say $f^{-1}(U_0)$. Since every $U \in \mathcal{U}$ which intersects A is contained in W, U_0 misses A.

Now let $\gamma: X \to I$ be a Urysohn function such that $\gamma(S) = 0$ and $\gamma(G) = 1$. Define $\alpha: X \times I \to X$ by

$$\alpha(x,t) = \Phi(x, t\cdot\gamma(x)).$$

A straightforward check shows that α is limited by $f^{-1}(\mathcal{U})$ and that $\alpha_0 = 1_X$. We claim that $\alpha_1(X) \cap f^{-1}(A) = \emptyset$. If $x \notin G$ then $\alpha_1(x) \notin f^{-1}(A)$ since $\alpha_1(x) \in \Phi(\{x\} \times I)$. In addition, if $x \in G$ then $\alpha_1(x) = \Phi_1(x)$ which by the properties of Φ does also not belong to $f^{-1}(A)$. By construction, every α_t restricts to the identity on $X \backslash V$. □

7.2.11. COROLLARY: *Let* X *and* Y *be compact* **ANR**'s *and let* $f\colon X \to Y$ *be cell-like. In addition, let* $A \in \mathcal{Z}(Y)$. *If for every* $a \in A$, $f^{-1}(a)$ *is a single point, then* $f^{-1}(A) \in \mathcal{Z}(X)$. □

7.2.12. PROPOSITION: *Let* X *and* Y *be compact* **ANR**'s *and let* $f\colon X \to Y$ *be cell-like. If* $A \subseteq Y$ *is such that* $f^{-1}(A) \in \mathcal{Z}(X)$ *then* $A \in \mathcal{Z}(Y)$.

PROOF: Let $\mathcal{U}$ be an open cover of Y and let $\mathcal{V}$ be an open star-refinement of $\mathcal{U}$ (lemma 5.1.7). By theorem 7.1.6 there exists a continuous function $g\colon Y \to X$ such that (among other things) $f \circ g$ and 1_Y are $\mathcal{V}$-close. Since $f^{-1}(A) \in \mathcal{Z}(X)$, by theorem 7.2.5 there exists a continuous function $\xi\colon X \to X \backslash f^{-1}(A)$ such that 1_X and ξ are $f^{-1}(\mathcal{V})$-close. Now define $\eta\colon Y \to Y$ by $\eta = f \circ \xi \circ g$. Since $\mathcal{V}$ is a star-refinement of $\mathcal{U}$, it is easy to see that 1_Y and η are $\mathcal{U}$-close. In addition, clearly $\eta(Y) \cap A = \emptyset$. By theorem 7.2.5(d) we therefore conclude that $A \in \mathcal{Z}(Y)$. □

7.3. The Disjoint-Cells Property

Let X be a space. We say that X has the *disjoint-cells property* provided that for every $n \in \mathbb{N}$, every continuous function $f\colon I^n \times \{0,1\} \to X$ is approximable (arbitrarily closely) by maps sending $I^n \times \{0\}$ and $I^n \times \{1\}$ to disjoint sets. Formally, for every n, every continuous function $f\colon I^n \times \{0,1\} \to X$ and every $\varepsilon > 0$ there is a continuous function $g\colon I^n \times \{0,1\} \to X$ such that

(1) $d(f,g) < \varepsilon$, and
(2) $g(I^n \times \{0\}) \cap g(I^n \times \{1\}) = \emptyset$

(by using the same technique as in lemma 6.2.1 it is possible to translate this into an "open cover statement"; this is left as an exercise to the reader). This property will play a crucial role later on in the process of topologically characterizing the Hilbert cube Q and spaces modeled on Q. The aim of this section is to derive some reformulations of the disjoint-cells property.

We shall verify that Q has the disjoint-cells property. To this end, let $\varepsilon > 0$ and let a continuous $f\colon I^n \times \{0,1\} \to Q$ be given. There exists $m \in \mathbb{N}$ such that $2^{-(m-1)} < \varepsilon$. Define $g\colon I^n \times \{0,1\} \to Q$ as follows:

$$\begin{cases} g(x,0) = (f(x,0)_1,\cdots,f(x,0)_{m-1},-1,f(x,0)_{m+1},\cdots), \\ g(x,1) = (f(x,1)_1,\cdots,f(x,1)_{m-1},1,f(x,1)_{m+1},\cdots). \end{cases}$$

An easy check shows that g satisfies (1) and (2). We shall prove in §7.8 that Q is the only compact **AR** with the disjoint cells property.

A space X is said to have the *Z-approximation property* provided that for every $n \in \mathbb{N}$, every continuous function $f: I^n \to X$ is approximable (arbitrarily closely) by Z-maps.

7.3.1. LEMMA: *A space X has the disjoint-cells property if and only if every continuous function* $f: Q \times \{0,1\} \to X$ *can be approximated by maps sending* $Q \times \{0\}$ *and* $Q \times \{1\}$ *to disjoint sets.*

PROOF: Suppose that X has the disjoint-cells property. Since every continuous function f: $Q \to X$ can be approximated by functions of the form

$$Q \xrightarrow{\pi} J^n \xrightarrow{f \mid J^n} X,$$

where J^n is identified with $J^n \times \{(0,0,\cdots)\} \subseteq Q$ and $\pi: Q \to J^n$ is the projection, we conclude that X has the "disjoint Hilbert cube property". The remaining part of the proof is left as an exercise to the reader. □

This simple lemma is used in the proof of the following:

7.3.2. PROPOSITION: *Let X be a topologically complete space. The following statements are equivalent:*

(a) *X has the disjoint-cells property,*

(b) $C(Q,X)$ *contains a countable dense set consisting entirely of Z-maps,*

(c) $C(Q,X)$ *contains a countable dense set consisting entirely of Z-maps having pairwise disjoint images,*

(d) *every continuous function* $f: Q \to X$ *is approximable by Z-maps,*

(e) *X has the Z-approximation property.*

PROOF: We shall first prove that (b) ⇒ (a). We wish to apply lemma 7.3.1. To this end, let f: $Q \times \{0,1\} \to X$ be continuous. By assumption, we can approximate $f \mid Q \times \{0\}$ by a map the range A of which is a Z-set in X. By the definition of Z-set, we can now approximate $f \mid Q \times \{1\}$ by a map the range of which misses A.

We next prove (a) ⇒ (c). By proposition 1.3.3 the function space $C(Q,X)$ has a countable dense set, say F. Let $\{f_i: i \in \mathbb{N}\}$ be a listing of F such that each $f \in F$ is listed infinitely often. Observe that $C(Q,X)$ is topologically complete by corollary 1.3.5. Define

$$Z = \prod_{i=1}^{\infty} C(Q,X)_i,$$

i.e. the product of countably-infinitely many copies of $C(Q,X)$, and put

$$Y = \{(g_1,g_2,\cdots) \in Z: d(f_i,g_i) < \tfrac{1}{i} \text{ for every } i \in \mathbb{N}\}.$$

Observe that Z is topologically complete. Since Y is the intersection of countably many open subsets of Z it follows that Y is topologically complete as well (theorem 4.7.4). For each $n \in \mathbb{N}$ put

$$U_n = \{(g_1,g_2,\cdots) \in Y: \text{for every } 1 \le i < j \le n,\ g_i(Q) \cap g_j(Q) = \emptyset\}.$$

It is easy to see that each U_n is open in Y. Moreover, the disjoint Hilbert cube property for X clearly implies that each U_n is dense in Y. By the Baire category theorem we can therefore find a point

$$(g_1,g_2,\cdots) \in \bigcap_{n=1}^{\infty} U_n.$$

Observe that $d(f_i,g_i) < \frac{1}{i}$ for every i and that the collection

$$\{g_i(Q): i \in \mathbb{N}\}$$

is pairwise disjoint. Since each $f \in F$ appears in the listing infinitely often, the set $\{g_i: i \in \mathbb{N}\}$ is dense in $C(Q,X)$. Moreover, each g_i can be approximated by the functions g_j with $j \neq i$. Since the collection $\{g_i(Q): i \in \mathbb{N}\}$ is pairwise disjoint, we find that each g_i is a Z-map.

Since (c) $\Rightarrow$ (b) $\Rightarrow$ (d) are trivialities, we shall proceed by establishing the implication (d) $\Rightarrow$ (e). Let $n \in \mathbb{N}$ and let $f: J^n \to X$ be continuous. Identify J^n and $J^n \times \{(0,0,\cdots)\} \subseteq Q$ and let $\pi: Q \to J^n$ be the projection. By (d), the composition

$$Q \xrightarrow{\pi} J^n \xrightarrow{f} X$$

can be approximated by Z-maps. This implies that f can be approximated by Z-maps.

We shall finally prove that (e) $\Rightarrow$ (a). To this end, let $n \in \mathbb{N}$ and let $f,g: I^n \to X$ be continuous. Then f can be approximated by a Z-map $\bar{f}: I^n \to X$. Consequently, g can be approximated by a map $\bar{g}: I^n \to X \setminus \bar{f}(I^n)$. □

7.3.3. COROLLARY: *Let* X *be a topologically complete* **ANR**. *Then the following statements are equivalent:*

(a) X *has the disjoint-cells property,*

(b) *if* K *is any compact space then any continuous function* $f: K \to X$ *can be approximated by continuous functions* $g,h: K \to X$ *having the property that* $g(K) \cap h(K) = \emptyset$.

PROOF: Since $I^n \times \{0,1\}$ is compact for every n, the implication (b) $\Rightarrow$ (a) is a triviality. We shall now prove that (a) $\Rightarrow$ (b). By proposition 7.3.2, there is a countable dense set $F \subseteq C(Q,X)$ consisting entirely of Z-maps. By lemma 7.2.7 we can approximate f by a continuous function $g: K \to X$ having the property that

$$g(K) \cap \bigcup_{\xi \in F} \xi(Q) = \emptyset.$$

Since F is dense, g is a Z-map. By theorem 7.2.5 we can approximate g by a function h of the form

$$K \overset{g}{\to} X \overset{\eta}{\to} X \backslash g(K),$$

with $d(\eta,1)$ arbitrarily small. It is clear that g and h are as required. □

7.3.4. COROLLARY: *Let* X *be a topologically complete* **ANR** *with the disjoint-cells property. If* K *is a compact space and if* $E,F,G \subseteq K$ *are pairwise disjoint closed subsets of* K *then any continuous map* $f: K \to X$ *such that* $f \mid G$ *is a Z-map can be approximated by continuous functions* $g: K \to X$ *such that*

(1) $g(E)$, $g(F)$ *and* $g(G)$ *are pairwise disjoint,*

(2) $g \mid G = f \mid G$.

PROOF: Let $\mathcal{U}$ be an open cover of X. In addition, let $\mathcal{V}$ be an open refinement of $\mathcal{U}$ such that for every space Y, any two $\mathcal{V}$-close maps $f,g: Y \to X$ are $\mathcal{U}$-homotopic (theorem 5.1.1). By compactness of K, corollaries 7.3.3 and 7.2.6 imply the existence of two continuous functions $\xi,\eta: K \to X \backslash f(G)$ having the following properties:

(3) $\xi(K) \cap \eta(K) = \emptyset$,

(4) both ξ and η are $\mathcal{V}$-close to f.

There consequently exist $\mathcal{U}$-homotopies $H,S: K \times I \to X$ such that $H_0 = S_0 = f$, $H_1 = \xi$ and $S_1 = \eta$. Let $\alpha: K \to [-1,1]$ be a continuous function with $\alpha(E) = -1$, $\alpha(G) = 0$ and $\alpha(F) = 1$ (corollary 1.4.14). In addition, let $K_0 = \alpha^{-1}([-1,0])$ and $K_1 = \alpha^{-1}([0,1])$ and define the function $g: K \to X$ as follows:

$$g(x) = \begin{cases} H_{-\alpha(x)}(x) & (x \in K_0), \\ S_{\alpha(x)}(x) & (x \in K_1). \end{cases}$$

An easy check shows that g is $\mathcal{U}$-close to f and satisfies (1) and (2).□

We now come to the announced reformulation of the disjoint-cells property.

7.3.5. THEOREM: *Let* X *be a topologically complete* **ANR**. *The following statements are equivalent:*

(a) *for every* n, *every continuous function* $f\colon I^n \times \{0,1\} \to X$ *is approximable by maps sending* $I^n \times \{0\}$ *and* $I^n \times \{1\}$ *to disjoint sets, i.e.* X *has the disjoint-cells property,*

(b) *every continuous function* $f\colon Q \times \{0,1\} \to X$ *is approximable by maps sending* $Q \times \{0\}$ *and* $Q \times \{1\}$ *to disjoint sets,*

(c) *for every* $n \in \mathbb{N}$, *every continuous function* $f\colon I^n \to X$ *is approximable by Z-maps, i.e.* X *has the Z-approximation property,*

(d) *every continuous function* $f\colon Q \to X$ *is approximable by Z-maps,*

(e) *for every compact space* K, *every continuous function* $f\colon K \to X$ *is approximable by Z-imbeddings,*

(f) *for every compact space* K, *the function space* C(K,X) *contains a countable dense set of Z-imbeddings having pairwise disjoint images,*

(g) *for every compact space* K *and closed subset* $K_0 \subseteq K$, *and for every* $Z \in \mathcal{Z}_\sigma(X)$, *every continuous function* $f\colon K \to X$ *such that* $f \mid K_0$ *is a Z-imbedding, is approximable by a Z-imbedding* $g\colon K \to X$ *such that* $f \mid K_0 = g \mid K_0$ *and* $f(K \setminus K_0) \subseteq X \setminus Z$.

Remark: The equivalences (a) ⇔ (e) and (a) ⇔ (g) are of particular importance.

PROOF: (a) ⇔ (b) is lemma 7.3.1, and the equivalences (a) ⇔ (c) ⇔ (d) are stated in proposition 7.3.2.

We shall prove that (e) ⇒ (a). To this end, let a continuous function $f\colon I^n \times \{0,1\} \to X$ be given. By (e) we find that f can be approximated by an imbedding. However, if $\phi\colon I^n \times \{0,1\} \to X$ is an imbedding then clearly $\phi(I^n \times \{0\}) \cap \phi(I^n \times \{1\}) = \emptyset$. So we are done.

Since the empty set is clearly a Z-set, (g) implies (e). In addition, the implication (f) ⇒ (e) is a triviality.

We shall now prove that (d) ⇒ (g). Let $F \subseteq C(Q,X)$ be a countable dense set consisting of Z-maps (proposition 7.3.2). Define

$$\mathcal{A} = \{g \in C(K,X)\colon f \mid K_0 = g \mid K_0\}.$$

Then $\mathcal{A}$ is clearly closed in the topologically complete space C(K,X) (corollary 1.3.5) so that $\mathcal{A}$ is topologically complete itself. Let $\mathcal{B}$ be a countable open basis for the topology of $K\backslash K_0$ such that each $B \in \mathcal{B}$ has compact closure in $K\backslash K_0$. If $A,B \in \mathcal{B}$ and $\overline{A} \cap \overline{B} = \emptyset$ then

$$\{g \in \mathcal{A}: g(\overline{A}),\ g(\overline{B}) \text{ and } g(K_0) \text{ are pairwise disjoint}\}$$

is clearly open in $\mathcal{A}$. This set is also dense in $\mathcal{A}$ by corollary 7.3.4. We conclude that the collection

$$\mathcal{E} = \{g \in \mathcal{A}: (\forall A,B \in \mathcal{B})(\overline{A} \cap \overline{B} = \emptyset \Rightarrow g(\overline{A}),\ g(\overline{B}) \text{ and } g(K_0) \text{ are pairwise disjoint}\}$$

is a dense G_δ in $\mathcal{A}$. In addition, since $K\backslash K_0$ is a countable union of compacta, the collection

$$\mathcal{F} = \{g \in \mathcal{A}: g(K\backslash K_0) \cap (\bigcup_{\xi \in F} \xi(Q) \cup Z) = \emptyset\}$$

is also a dense G_δ in $\mathcal{A}$ by lemma 7.2.7. We conclude that $\mathcal{G} = \mathcal{E} \cap \mathcal{F}$ is a dense G_δ in $\mathcal{A}$. It is easy to see that every $g \in \mathcal{G}$ close to f has the required properties. So we are done.

It therefore remains to establish the implication (e) ⇒ (f). Again by proposition 1.3.3, there is a countable dense subset $F \subseteq C(K,X)$. Enumerate F as $\{f_n: n \in \mathbb{N}\}$ such that each $f \in F$ occurs infinitely often in the listing. It clearly suffices to construct by induction on n a Z-imbedding $\alpha_n: K \to X$ such that $d(\alpha_n,f_n) < 2^{-n}$ while moreover $\alpha_n(K) \cap \bigcup_{i=1}^{n-1}\alpha_i(K) = \emptyset$. For $n = 1$, apply (e) to obtain a Z-imbedding $\alpha_1: K \to X$ such that $d(\alpha_1,f_1) < 2^{-1}$. Now suppose that the imbeddings $\alpha_1,\cdots,\alpha_{n-1}$ have been constructed for some $n > 1$. Since $Z = \bigcup_{i=1}^{n-1}\alpha_i(K)$ is a Z-set in X (lemma 6.2.2(3)), there exists by corollary 7.2.6 a continuous function $g_n: K \to X$ such that $g_n(K) \cap Z = \emptyset$, while moreover $d(f_n,g_n) < 2^{-(n+1)}$. Put $\varepsilon = d(g_n(K),Z)$. Now again by applying (e) we obtain a Z-imbedding $\alpha_n: K \to X$ with $d(\alpha_n,g_n) < \min\{2^{-(n+1)},\varepsilon\}$. It is clear that α_n is as required. □

We shall now present some classes of spaces with the disjoint-cells property. Recall that a Q-manifold is a space that admits an open cover by sets homeomorphic to open subsets of Q, cf. exercise 6.8.1.

7.3.6. THEOREM: (1) *If* X *has the disjoint-cells property then for any space* Y, X × Y *has the disjoint-cells property.*

(2) *Every* Q*-manifold is an* **ANR** *and has the disjoint-cells property.*

PROOF: The proof of (1) is an elementary exercise that is left to the reader. For (2), Let M be a Q-manifold. Since open subsets of Q are **ANR**'s (theorem 5.4.1), M admits an open cover by

ANR's and is therefore an **ANR** itself (theorem 5.4.5). We shall show that every continuous function $f: K \to M$, where K is any polyhedron, is approximable by Z-maps. An appeal to theorem 7.3.5 then gives us that M has the disjoint-cells property.

Since every singleton in Q is a Z-set in Q (lemma 6.2.3(2)), an easy application of theorem 7.2.5 gives us that every singleton in M is a Z-set in M. So our claim is certainly true for zero-dimensional (= finite) polyhedra. Suppose that the claim is true for all n-dimensional polyhedra, $n \geq 0$, and let an (n+1)-dimensional polyhedron K, a continuous function $f: K \to M$ and $\varepsilon > 0$ be given. Let $\mathcal{V}$ be an open cover of M each element of which has diameter less than $\varepsilon/3$. Since M is a Q-manifold, there exists a refinement $\mathcal{U}$ of $\mathcal{V}$ each element of which is homeomorphic to an open subset of Q. Since Q has an open base consisting of contractible open sets (each "standard" open subcube is contractible), without loss of generality we assume that each element of $\mathcal{U}$ is contractible. By compactness of f(K) there exists $\delta > 0$ such that if a subset A of M has diameter less than δ and intersects f(K) then A is contained in an element of $\mathcal{U}$ (lemma 1.1.1). Without loss of generality assume that $\delta < \varepsilon$. There is a triangulation $\mathcal{T}$ of K such that $\operatorname{diam}(f(\sigma)) < \delta/3$ for every $\sigma \in \mathcal{T}$ (theorem 3.6.12). For convenience, let K^n denote $|\mathcal{T}^{(n)}|$. By our inductive assumption, there is a continuous function $h: K^n \to M$ such that $h(K^n) \in \mathcal{Z}(M)$ and $d(h, f \mid K^n) < \delta/6$. Let $\sigma \in \mathcal{T}$ be an arbitrary (n+1)-simplex. Since the faces of σ pairwise intersect if $n \geq 1$, it is clear that in that case

$$(*) \qquad \operatorname{diam}(h(\partial\sigma)) < \tfrac{2\delta}{3}.$$

If $n = 0$ then (*) also follows because $\operatorname{diam}(f(\sigma)) < \delta/3$ and $d(h, f \mid K^n) < \delta/6$. Let p be any vertex of σ. Then $d(f(p), h(p)) < \delta/3$ and since h(p) belongs to $h(\partial\sigma)$, we conclude that the diameter of $h(\partial\sigma) \cup \{f(p)\}$ is smaller than δ. By the special choice of $\mathcal{U}$ there consequently exists $U \in \mathcal{U}$ containing $h(\partial\sigma) \cup \{f(p)\}$. Since U is contractible, the function $h \mid \partial\sigma$ can be extended to a continuous function $\xi_\sigma: \sigma \to U$. We shall prove that

$$d(f \mid \sigma, \xi_\sigma) < \tfrac{2\varepsilon}{3}.$$

This is easy. Simply observe that $\operatorname{diam}(U) < \varepsilon/3$, that $\operatorname{diam}(f(\sigma)) < \varepsilon/3$ and that $f(p) \in U \cap f(\sigma)$. The union of the ξ_σ's and $h \mid K^n$ defines a continuous function $\xi: K \to M$ having the property that

$$d(f, \xi) < \tfrac{2\varepsilon}{3}.$$

This function is not as required yet, so we shall adjust it a little further. Again, let $\sigma \in \mathcal{T}$ be an arbitrary (n+1)-simplex. By construction, there exists $U \in \mathcal{U}$ containing $\xi(\sigma)$. By abuse of notation we assume that U *is* an open subset of Q. Since $\xi(\partial\sigma) \in \mathcal{Z}(M)$ it follows that it belongs to

$\mathcal{Z}(U)$ and hence to $\mathcal{Z}(Q)$ (we used theorem 7.2.5 twice). By theorem 6.4.8 there is a Z-map $\psi_\sigma: \sigma \to U$ such that

(1) $\psi_\sigma \mid \partial\sigma = h$,
(2) $d(\psi_\sigma, \xi_\sigma) < \frac{\varepsilon}{3}$.

The union of the ψ_σ's and $h \mid K^n$ defines a continuous function $\psi: K \to M$ having the property that $d(\psi,\xi) < \varepsilon/3$. Arguing as above yields that for every $\sigma \in \mathcal{T}$, $\psi(\sigma) \in \mathcal{Z}(M)$. From this we conclude that $\psi(K) \in \mathcal{Z}(M)$ (lemma 6.2.2(3)). Finally,

$$d(f,\psi) \le d(f,\xi) + d(\xi,\psi) < \frac{2\varepsilon}{3} + \frac{\varepsilon}{3} = \varepsilon,$$

so ψ is the required "Z-approximation" to f. □

7.3.7. COROLLARY: *For every space* X, $X \times Q$ *has the disjoint-cells property.* □

7.4. Z-Sets in Q-manifolds

In this section we shail present some basic facts concerning Z-sets in Q-manifolds such as the Z-set Unknotting Theorem 7.4.9.

Let X be a space. A subset $Y \subseteq X$ is called *locally collared in* X if for every $y \in Y$ there is an open neighborhood U of y in Y and an open imbedding $\phi: U \times [0,1) \to X$ such that $\phi(u,0) = u$ for every $u \in U$. If there exists an open imbedding $\phi: Y \times [0,1) \to X$ such that $\phi(y,0) = y$ for every $y \in Y$ then we say that Y is *collared* in X and we call $\phi(Y \times [0,1))$ a *collared neighborhood* of Y.

7.4.1. THEOREM ("Brown's Theorem"): *Let* X *be a space and let* $Y \subseteq X$ *be closed. If* Y *is locally collared in* X *then* Y *is collared in* X.

Let X and Y be as in the theorem and put $X^+ = (X \times \{0\}) \cup (Y \times [-\frac{1}{2},0]) \subseteq X \times [-1,0]$ (see figure 7.4.1). We shall construct a homeomorphism $g: X \to X^+$ such that $g(Y) = Y \times \{-\frac{1}{2}\}$. Since $Y \times \{-\frac{1}{2}\}$ is obviously collared in X^+, we then find that Y is collared in X. We need to derive a simple lemma first.

Let $f: Y \to [-\frac{1}{2},0]$ be continuous. Define

$$X(f) = (X \times \{0\}) \cup \{(x,t): x \in Y \text{ and } f(x) \le t\}.$$

7.4.2. LEMMA: *Let* $f: Y \to [-\frac{1}{2},0]$ *be continuous. In addition, let* $U \subseteq Y$ *be open in* Y *and let* $\phi: U \times [0,1) \to X$ *be an open imbedding such that* $\phi(u,0) = u$ *for every* $u \in U$. *Then for each open subset* V *in* Y *such that* $V \subseteq \overline{V} \subseteq U$ *there exist a continuous function* $g: Y \to [-\frac{1}{2},0]$ *and a homeomorphism* $F: X(f) \to X(g)$ *such that:*

(1) *if* $x \in \overline{V}$ *then* $g(x) = -\frac{1}{2}$, *and if* $x \notin U$ *then* $f(x) = g(x)$,

(2) *if* $f(x) = -\frac{1}{2}$ *for certain* $x \in Y$ *then* $f(x) = g(x)$, *and*

(3) *if* $x \in Y$ *then* $F(x,f(x)) = (x,g(x))$.

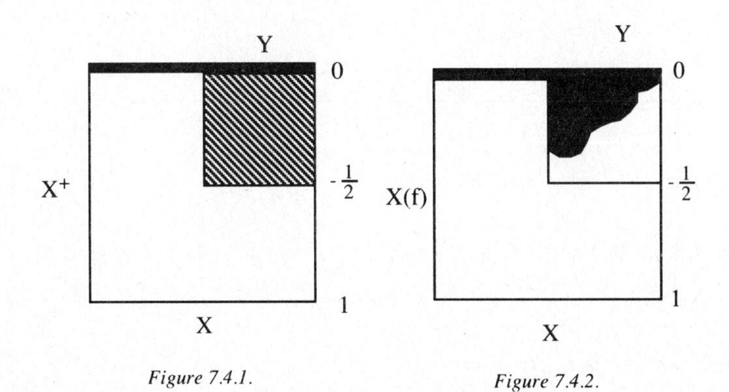

Figure 7.4.1. *Figure 7.4.2.*

PROOF: Let $W \subseteq Y$ be open such that $V \subseteq \overline{V} \subseteq W \subseteq \overline{W} \subseteq U$ and let $\lambda: Y \to I$ be a Urysohn function (corollary 1.4.15) such that $\lambda | (Y \backslash W) \equiv 0$ and $\lambda | \overline{V} \equiv 1$. Define $g: Y \to [-\frac{1}{2},0]$ by

$$g(x) = (1 - \lambda(x))f(x) + \frac{-\lambda(x)}{2}.$$

It is clear that g is well-defined, continuous, and satisfies (1) and (2). It remains to construct the homeomorphism $F: X(f) \to X(g)$ satisfying (3). By abuse of notation we assume that $U \times [0,1)$ *is* an open subset of X with $(U \times [0,1)) \cap Y = U$, where $(u,0)$ and u are identified for each $u \in U$. For each $x \in Y$ let $s_x: [-\frac{1}{2},\frac{1}{2}] \to [-\frac{1}{2},\frac{1}{2}]$ be the unique order preserving homeomorphism taking $[f(x),\frac{1}{2}]$ linearly onto $[g(x),\frac{1}{2}]$. Now define $F: X(f) \to X(g)$ as follows:

$$F(x,t) = \begin{cases} (x,s_x(t)) & ((x,t) \in (\overline{W} \times [-\frac{1}{2},\frac{1}{2}]) \cap X(f)), \\ (x,t) & ((x,t) \notin \overline{W} \times [-\frac{1}{2},\frac{1}{2}]). \end{cases}$$

The reader should draw a picture to see what is going on and to find that F is clearly as required, cf. the proof of theorem 6.1.13.□

We are now prepared to give the proof of theorem 7.4.1.

7.4.3. *Proof of Theorem 7.4.1:* Let $\mathcal{U} = \{U_i: i \in \mathbb{N}\}$ be an open cover of Y such that for every i there exists an open imbedding $\phi_i: U_i \times [0,1) \to X$ such that $\phi_i(u,0) = u$ for every $u \in U_i$. By lemma 1.4.1 we may assume without loss of generality that $\mathcal{U}$ is locally finite. By proposition 4.3.3, for every i there exists a closed set F_i in Y such that $F_i \subseteq U_i$ while moreover the family $\{F_i: i \in \mathbb{N}\}$ covers Y. Finally, by corollary 1.4.16 for every i there exists an open subset V_i in Y with

(1) $F_i \subseteq V_i \subseteq \overline{V}_i \subseteq U_i$.

Observe that the family $\mathcal{V} = \{V_i: i \in \mathbb{N}\}$ covers Y. By induction on $i \geq 0$, we shall now inductively construct continuous functions $f_i: Y \to [-\frac{1}{2},0]$ and homeomorphisms $g_i: X \to X(f_i)$ such that

(2) f_0 is the constant function with value 0,
(3) $g_0(x) = (x,0)$ for every $x \in X$,
(4) $f_i(x) = -\frac{1}{2}$ if $x \in \overline{V}_1 \cup \cdots \cup \overline{V}_i$,
(5) $g_i(x) = (x,f_i(x))$ for every $x \in Y$.

That the functions f_i and g_i can be constructed follows easily from lemma 7.4.2. Define $g: X \to X^+$ by

(6) $g(x) = \lim_{i\to\infty} g_i(x)$.

Since the collection $\overline{\mathcal{V}}$ is locally finite, for each $x \in X$ the sequence $(g_i(x))_i$ is eventually constant on a neighborhood of x. We conclude that g is a well-defined homeomorphism such that $g(Y) = Y \times \{-\frac{1}{2}\}$. As remarked above, this proves the theorem. □

Theorem 7.4.1 has the following corollaries.

7.4.4. COROLLARY: (1) *If* M *and* N *are Q-manifolds and* $M \subseteq N$ *is a Z-set, then* M *is collared in* N.

(2) *If* M *is a Q-manifold then* $M \times [0,1)$ *is homeomorphic to an open subset of* Q.

PROOF: For (1), choose an arbitrary $x \in M$ and let $U \subseteq N$ be an open neighborhood of x in N that is homeomorphic to an open subset of Q. Let V be a closed neighborhood of x in M such that $V \approx Q$ and $V \subseteq U$. Since $M \in \mathcal{Z}(N)$ it follows that $V \in \mathcal{Z}(U)$ (use that U is open). By abuse of notation we assume that U *is* an open subset of Q. Theorem 7.2.5 now implies that $V \in \mathcal{Z}(Q)$.

Since V is homeomorphic to the endface $W = \{1\} \times J \times J \times \cdots \subseteq Q$ and since W is obviously collared in Q, theorem 6.4.6 implies that V is collared in Q. By compactness of V this implies that V is collared in U and hence in N. We conclude that M is locally collared in N, hence collared by theorem 7.4.1.

For (2), let γM be a compactification of M. We may assume that γM is a subset of an endface of Q (theorem 1.4.18). Observe that $Z = \gamma M \setminus M$ is closed in Q. We conclude that $M = \gamma M \cap (Q \setminus Z)$ is a Z-set in the Q-manifold $Q \setminus Z$. By (1), M is collared in $Q \setminus Z$, from which the desired result follows immediately. □

7.4.5. THEOREM: *Let* M *be a compact Q-manifold. If* $A \in \mathcal{Z}(M)$ *then there is a neighborhood of* A *which is homeomorphic to an open subset of* Q.

PROOF: By theorem 7.3.6(2), M is an **ANR** and has the disjoint-cells property. Consequently, theorem 7.3.5(g) implies that the identity $1: M \to M$ can be approximated by a Z-imbedding $f: M \to M$ such that $f \mid A = 1_A$. By corollary 7.4.4, f(M) is collared in M and $f(M) \times [0,1)$ is homeomorphic to an open subset of Q. Let $\phi: f(M) \times [0,1) \to M$ be an open imbedding such that $\phi(x,0) = x$ for every $x \in f(M)$. Since $f \mid A = 1_A$, we conclude that the open neighborhood $U = \phi(f(M) \times [0,1))$ of A is as required. □

We shall use theorem 7.4.5 to derive a homeomorphism extension result for compact Q-manifolds. We have to be careful since the topological sum of $S^1 \times Q$ and Q is clearly a compact Q-manifold that is not homogeneous. So a homeomorphism extension result as simple as in the case of the Hilbert cube will not work; some homotopy conditions have to be met.

7.4.6. LEMMA: *Let* $A,B \in \mathcal{Z}(Q)$. *Then for every neighborhood* V *of* B *and for every open cover* $\mathcal{U}$ *of* V *there is an isotopy* $H: Q \times I \to Q$ *such that*

(1) $H_0 = 1_Q$ *and* $H_1(B) \cap A = \emptyset$,

(2) $H_t \mid Q \setminus V = 1$ *for every* $t \in I$,

(3) $H \mid V \times I \to V$ *is limited by* $\mathcal{U}$.

PROOF: Without loss of generality we assume that $\mathcal{U}$ consists of subsets of V. Factorize Q as $J \times \vec{Q}$, where $\vec{Q}$ denotes $\Pi_{i=2}^{\infty} J_i$. By theorem 6.4.6 there exist subsets $A', B' \subseteq \vec{Q}$ and a homeomorphism $f: Q \to Q$ such that $f(A) = \{0\} \times A'$ and $f(B) = \{0\} \times B'$. By compactness, it is clear that there are a neighborhood O of B' in $\vec{Q}$ and a $t > 0$ such that for every $x \in O$ there is a $U \in \mathcal{U}$ with

$$[-2t,2t] \times \{x\} \subseteq f(U).$$

Let $\alpha: \vec{Q} \to J$ be a Urysohn function (corollary 1.4.15) such that $\alpha \mid B' \equiv 1$ and $\alpha \mid \vec{Q} \setminus O \equiv 0$. In addition, let $F: J \times I \to J$ be an isotopy having the following properties:

(1) $F_s \mid J \setminus [-2t,2t] = 1$ for every $s \in I$,

(2) $F_0 = 1$ and $F_1(0) = t$.

Define $G: Q \times I \to Q$ by

$$G((x,y),s) = (F_{s \cdot \alpha(y)}(x),y)$$

(recall that we identified Q and $J \times \vec{Q}$). Then G is clearly an isotopy (theorem 6.1.13). Also, $G_0 = 1$ and $G_1(f(B)) \cap f(A) = \emptyset$.

Choose $((p,q),s) \in Q \times I$ arbitrarily. If $p \notin [-2t,2t]$ then $G((p,q),s) = (p,q)$ (use (1)). In addition, if $q \notin O$ then

$$G((p,q),s) = (F_{s \cdot \alpha(q)}(p),q) = (F_0(p),q) = (p,q)$$

(use (2)). Finally, if $(p,q) \in [-2t,2t] \times O$ then

$$G((p,q) \times I) \subseteq [-2t,2t] \times \{q\} \subseteq f(U)$$

for certain $U \in \mathcal{U}$. From this we conclude that

$G_s \mid Q \setminus f(V) = 1$ for every $s \in I$, and

$G \mid f(V) \times I: f(V) \times I \to f(V)$ is limited by $f(\mathcal{U})$.

Now define $H: Q \times I \to Q$ by

$$H_s = f^{-1} \circ G_s \circ f \qquad (s \in I).$$

It is clear that H is as required. □

7.4.7. LEMMA: *Let* $V \subseteq Q$ *be open and let* $\mathcal{U}$ *be an open cover of* V. *If* A *is compact and if* $F: A \times I \to V$ *is a homotopy such that* F_0 *and* F_1 *are Z-imbeddings such that* $F_0(A) \cap F_1(A) = \emptyset$ *then there is an isotopy* $H: Q \times I \to Q$ *such that*

(1) $H_0 = 1$,

(2) $H_1 \circ F_0 = F_1$,

(3) $H_t \mid Q \setminus V = 1$ *for every* $t \in I$.

Moreover, if F *is limited by* $\mathcal{U}$ *then we may construct* H *in such a way that* $H \mid V \times I \to V$ *is also limited by* $\mathcal{U}$.

PROOF: Without loss of generality we assume that $\mathcal{U}$ consists of subsets of V. For technical reasons we assume that F is defined on $A \times [-\frac{1}{2},\frac{1}{2}]$. By theorem 7.3.5 we can replace F by a Z-imbedding $G: A \times [-\frac{1}{2},\frac{1}{2}] \to Q$ such that

(1) $G_{-\frac{1}{2}} = F_{-\frac{1}{2}}$, $G_{\frac{1}{2}} = F_{\frac{1}{2}}$ and $G(A \times [-\frac{1}{2},\frac{1}{2}]) \subseteq V$,

(2) G is limited by $\mathcal{U}$.

Factorize Q as $J \times \vec{Q}$, where $\vec{Q}$ denotes $\prod_{i=2}^{\infty} J_i$. We may assume that $A \subseteq \vec{Q}$ and that $A \in \mathcal{Z}(\vec{Q})$ (just put A in some endface of $\vec{Q}$). Then, by lemma 6.2.2(5), $A \times [-\frac{1}{2},\frac{1}{2}] \in \mathcal{Z}(Q)$. By theorem 6.4.6 there exists a homeomorphism $f: Q \to Q$ such that

$$f \circ G = 1_{A \times [-\frac{1}{2},\frac{1}{2}]},$$

i.e. f extends the homeomorphism $G^{-1}: G(A \times [-\frac{1}{2},\frac{1}{2}]) \to A \times [-\frac{1}{2},\frac{1}{2}]$. By compactness, it is clear that we can find a neighborhood O of A in $\vec{Q}$ and a point $t \in (\frac{1}{2},1)$ such that for every $x \in O$ there exists $U \in \mathcal{U}$ with

$$[-t,t] \times \{x\} \subseteq f(U).$$

Just as in the proof of lemma 7.4.6 we can now construct an isotopy $S: Q \times [-\frac{1}{2},\frac{1}{2}] \to Q$ having the following properties:

(3) $S_{-\frac{1}{2}} = 1$, $S_{\frac{1}{2}}(-\frac{1}{2},a) = (\frac{1}{2},a)$ for every $a \in A$,

(4) $S_\lambda \mid Q \backslash f(V) = 1$ for every $\lambda \in [-\frac{1}{2},\frac{1}{2}]$, and

(5) $S \mid f(V) \times [-\frac{1}{2},\frac{1}{2}]: f(V) \times [-\frac{1}{2},\frac{1}{2}] \to f(V)$ is limited by $f(\mathcal{U})$.

Now define $H: Q \times [-\frac{1}{2},\frac{1}{2}] \to Q$ by

$$H_\lambda = f^{-1} \circ S_\lambda \circ f \qquad (\lambda \in [-\tfrac{1}{2},\tfrac{1}{2}]).$$

It is clear that H is as required. □

7.4.8. COROLLARY: *Let $V \subseteq Q$ be open and let $\mathcal{U}$ be an open cover of V. If A is compact and if $F: A \times I \to V$ is a homotopy such that F_0 and F_1 are Z-imbeddings then there is an isotopy $H: Q \times I \to Q$ such that*

(1) $H_0 = 1$,

(2) $H_1 \circ F_0 = F_1$,

(3) $H_t \mid Q \backslash V = 1$ *for every* $t \in I$.

Moreover, if F is limited by $\mathcal{U}$ then we can construct H in such a way that $H \mid V \times I \to V$ is also limited by $\mathcal{U}$.

PROOF: Without loss of generality we assume that $\mathcal{U}$ consists of subsets of V. Let F be limited by $\mathcal{U}$.

CLAIM: There exists $\varepsilon > 0$ such that for every $a \in A$, $B(F(\{a\} \times I),\varepsilon)$ is contained in an element of $\mathcal{U}$.

To the contrary, assume that such $\varepsilon > 0$ does not exist. Then for every $n \in \mathbb{N}$ there is a point $a_n \in A$ such that $B(F(\{a_n\} \times I),1/n)$ is not contained in an element of $\mathcal{U}$. Since A is compact, without loss of generality we assume that the sequence $(a_n)_n$ converges to $a \in A$. Pick an element $U \in \mathcal{U}$ such that $F(\{a\} \times I) \subseteq U$ and let $\delta = d(F(\{a\} \times I),Q \backslash U)$. There exists $n \in \mathbb{N}$ so large that

$$\tfrac{1}{n} < \tfrac{1}{2}\delta \text{ and } F(\{a_n\} \times I) \subseteq B(F(\{a_n\} \times I),\tfrac{1}{2}\delta).$$

But this implies that $B(F(\{a_n\} \times I),1/n) \subseteq U$, which is a contradiction.

Let $\mathcal{W} = \{B(F(\{a\} \times I),\frac{1}{2}\varepsilon): a \in A\}$ and put $V^* = \cup\mathcal{W}$. By lemma 7.4.6 there is an isotopy $S: Q \times I \to Q$ such that

(1) $S_0 = 1$ and $S_1(F_0(A)) \cap F_1(A) = \emptyset$,

(2) $S_t \mid Q \backslash V^* = 1$ for every $t \in I$,

(3) for every $x \in Q$, $\operatorname{diam}(S\{x\} \times I)) < \frac{1}{2}\varepsilon$.

Define $\bar{F}: A \times I \to Q$ by

$$\bar{F}(a,t) = \begin{cases} S_{1-2t}(F_0(a)) & (0 \le t \le \frac{1}{2}), \\ F_{2t-1}(a) & (\frac{1}{2} \le t \le 1). \end{cases}$$

Observe that $\bar{F}_0(A) \cap \bar{F}_1(A) = \emptyset$ and that for every $a \in A$,

$$(4)\ \bar{F}(\{a\} \times I) \subseteq B(F(\{a\} \times I), \tfrac{1}{2}\varepsilon).$$

By lemma 7.4.7 there is an isotopy $T: Q \times I \to Q$ such that

(5) $T_0 = 1$ and $T_1 \circ \bar{F}_0 = \bar{F}_1$,
(6) $T_t \mid Q \setminus V^* = 1$ for every $t \in I$,
(7) $T \mid V^* \times I \to V^*$ is limited by $\mathcal{W}$.

Now define $H: Q \times I \to Q$ by

$$H_t = \begin{cases} S_{2t} & (0 \leq t \leq \frac{1}{2}), \\ T_{2t-1} \circ S_1 & (\frac{1}{2} \leq t \leq 1). \end{cases}$$

Clearly, $H_0 = 1$. Also,

$$H_1 \circ F_0 = T_1 \circ S_1 \circ F_0 = T_1 \circ \bar{F}_0 = \bar{F}_1 = F_1.$$

By (2) and (6), $H_t \mid Q \setminus V^* = 1$ for every $t \in I$ so that $H_t \mid Q \setminus V = 1$ for every $t \in I$. It therefore suffices to prove that H is limited by $\mathcal{U}$. Take an arbitrary $x \in V^*$. Then

$$H(\{x\} \times I) = S(\{x\} \times I) \cup T(\{S_1(x)\} \times I).$$

By (7) there exists $a \in A$ such that

$$T(\{S_1(x)\} \times I) \subseteq B(F(\{a\} \times I), \tfrac{1}{2}\varepsilon).$$

In addition, $\mathrm{diam}(S(\{x\} \times I) < \frac{1}{2}\varepsilon$ by (3). Since

$$S_1(x) \in S(\{x\} \times I) \cap T(\{S_1(x)\} \times I),$$

we therefore obtain

$$H(\{x\} \times I) \subseteq B(F(\{a\} \times I), \varepsilon).$$

By the claim this is as required. □

We now come to the main result in this section.

7.4.9. THEOREM ("The Z-set Unknotting Theorem"): *Let* M *be a compact Q-manifold and let* $\mathcal{U}$ *be an open cover of* M. *If* A *is compact and if* $F: A \times I \to M$ *is a homotopy that is limited by* $\mathcal{U}$ *while moreover* F_0 *and* F_1 *are Z-imbeddings, then for every neighborhood* E *of* $F(A \times I)$ *there is an isotopy* $H: M \times I \to M$ *such that*

(1) $H_0 = 1$ *and* $H_1 \circ F_0 = F_1$,

(2) H *is limited by* $\mathcal{U}$,

(3) $H_t \mid M \setminus E = 1$ for every $t \in I$.

PROOF: By theorem 7.3.6(2) and theorem 7.3.5 we can assume without loss of generality that $F(A \times I) \in \mathcal{Z}(M)$. By theorem 7.4.5 there is a neighborhood V of $F(A \times I)$ that is homeomorphic to an open subset of Q. Let W be an open neighborhood of $F(A \times I)$ such that $W \subseteq \overline{W} \subseteq V \cap E$ and $\overline{W}$ is compact. By abuse of notation we assume that V *is* an open subset of Q. Then $F(A \times I) \in \mathcal{Z}(Q)$ by theorem 7.2.5. By corollary 7.4.8 there is an isotopy $G: \overline{W} \times I \to \overline{W}$ such that

(1) $G_0 = 1$ and $G_1 \circ F_0 = F_1$,

(2) $G_t \mid \overline{W} \setminus W = 1$ for every $t \in I$,

(3) G is limited by $\{U \cap \overline{W}: U \in \mathcal{U}\}$.

Now define $H: M \times I \to M$ by

$$H(x,t) = \begin{cases} G(x,t) & (x \in \overline{W}), \\ x & (x \in M \setminus \overline{W}). \end{cases}$$

It is clear that H is as required. □

7.5. Toruńczyk's Approximation Theorem and Applications

In this section we shall prove The Approximation Theorem. As applications we shall prove that each compact Q-manifold M is stable in the sense that $M \times Q \approx M$, and that a contractible compact Q-manifold is homeomorphic to Q. Further applications can be found in §7.8 and chapter 8.

The proof of the Approximation Theorem will be given in several steps.

7.5.1. LEMMA: *Let* X *be a compact space such that* $X \times Q \approx X$. *Then the projection* $\pi: X \times J \to X$ *is a near homeomorphism.*

PROOF: Since $X \times Q \approx X$, all we have to do is to show that the projection

$$\bar{\pi}: (X \times Q) \times J \to X \times Q$$

is a near homeomorphism. For each $n \in \mathbb{N}$ define $f_n: (X \times Q) \times J \to X \times Q$ by

$$f_n((x,(q_1,q_2,\cdots)),t) = (x,(q_1,q_2,\cdots,q_{n-1},t,q_n,q_{n+1},\cdots)).$$

It is easy to see that each f_n is a homeomorphism and that

$$\lim_{n\to\infty} f_n = \bar{\pi},$$

i.e. $\bar{\pi}$ is a near homeomorphism. □

Assume that X and Y are compact spaces and that $f: X \to Y$ is continuous. As in §5.6, M(f) denotes the mapping cylinder of f and $c(f): M(f) \to Y$ is the collapse to the base. Observe that if f is surjective then the natural map $\pi_f: X \times I \to M(f)$ defined by

$$\pi_f(x,t) = \begin{cases} (x,t) & (0 \le t < 1), \\ f(x) & (t = 1), \end{cases}$$

is continuous.

7.5.2. LEMMA: *Let* M *be a compact Q-manifold, let* X *be an* **ANR** *and let* $f: M \to X$ *be cell-like. Then* $\pi_f: M \times I \to M(f)$ *is a near homeomorphism.*

PROOF: Let $\varepsilon > 0$. We shall produce a homeomorphism $h: M \times I \to M \times I$ such that

(1) $d(\pi_f \circ h, \pi_f) < \varepsilon$,
(2) $\forall x \in X$: $\operatorname{diam} h(f^{-1}(x) \times \{1\}) < \varepsilon$.

Since $\pi_f \mid M \times [0,1)$ is one-to-one, theorem 6.1.8 then implies that π_f is a near homeomorphism.

For convenience we shall identify M and $M \times \{1\}$. Consequently, $\pi_f \mid M = f$. Let $\mathcal{U}$ be the cover of M(f) consisting of all open sets of diameter less than ε and let $\mathcal{U}'$ be a finite open star-refinement of $\mathcal{U}$(lemma 5.1.7). Put $\mathcal{V} = \{U \cap X: U \in \mathcal{U}'\}$. Since f is cell-like and M and X are **ANR**'s (theorem 7.3.6), by theorem 7.1.6 there is a continuous function $g: X \to M$ such that

(3) $g \circ f \underset{f^{-1}(\mathcal{V})}{\sim} 1_M$ and $f \circ g \underset{\mathcal{V}}{\sim} 1_X$.

Since M is a Q-manifold, by theorems 7.3.6(2) and 7.3.5 we can approximate $g \circ f$ by imbeddings. Since $\text{diam}((g \circ f)(f^{-1}(x))) = 0$ for every $x \in X$ and since "close" maps into an **ANR** are homotopic by a "small" homotopy (theorem 5.1.1), by compactness of M there is an imbedding $\xi: M \to M$ such that

(4) $\xi \underset{f^{-1}(\mathcal{V})}{\sim} g \circ f$,

(5) $\forall x \in X: \text{diam}(\xi f^{-1}(x)) < \varepsilon$.

Then ξ and 1_M are homotopic Z-imbeddings of M into the Q-manifold $M \times I$. By (3) and (4) these imbeddings are homotopic by a homotopy that is limited by $\pi_f^{-1}(\mathcal{U})$. By theorem 7.4.9 we can therefore extend ξ to a homeomorphism $h: M \times I \to M \times I$ such that

(6) h and $1_{M\times I}$ are $\pi_f^{-1}(\mathcal{U})$-close,

which directly gives us that

(7) $d(\pi_f \circ h, \pi_f) < \varepsilon$.

Also, since h extends ξ, (5) implies that

(8) $\forall x \in X: \text{diam}(hf^{-1}(x)) < \varepsilon$.

We are done. □

7.5.3. COROLLARY: *Let* M *be a compact Q-manifold, let* X *be an* **ANR** *and let* $f: M \to X$ *be cell-like. Then* M(f) *is a Q-manifold.* □

We now come to the basic tool in the proof of the Approximation Theorem.

7.5.4. PROPOSITION: *Let* M *be a compact Q-manifold, let* X *be an* **ANR** *and let* $f: M \to X$ *be cell-like. Then for every* $Y \in \mathcal{Z}(X)$ *and* $\varepsilon > 0$ *there is a near homeomorphism* $g: M(f) \to M(f)$ *such that*

(1) $d(c(f) \circ g, c(f)) < \varepsilon$,

(2) $g \mid g^{-1}(Y) = c(f) \mid c(f)^{-1}(Y)$.

PROOF: Let Z be the space we obtain from M(f) by identifying each set of the form $c(f)^{-1}(y)$, $y \in Y$, to a single point and let $q: M(f) \to Z$ be the quotient map (Z is something like a reduced mapping cylinder).

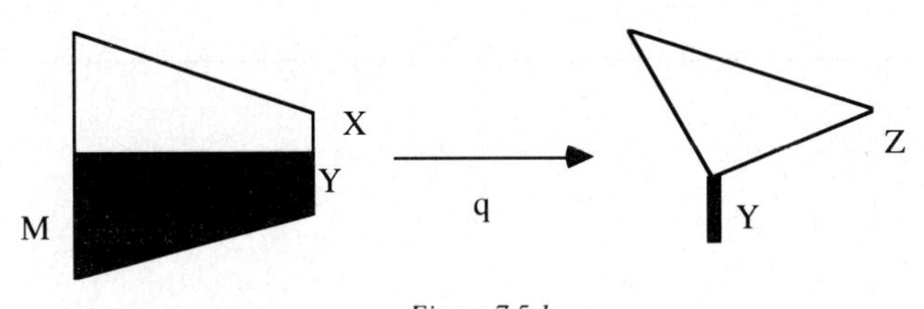

Figure 7.5.1.

We shall construct for every $\delta > 0$ a homeomorphism $h: M(f) \to M(f)$ such that

(1) $d(q \circ h, q) < \delta$,
(2) $h \mid X = 1_X$,
(3) $\forall y \in Y$: $\mathrm{diam}(h(c(f)^{-1}(y))) < \delta$.

By (1) and (3) and theorem 6.1.8 it then follows that q is a near homeomorphism. We shall need (2) and the fact that q is a near homeomorphism to construct the desired near homeomorphism g.

We remind the reader that we think of M and X as being subspaces of M(f).

Let $\delta > 0$ and let $\mathcal{U}$ be a finite open cover of Z such that each $U \in \mathcal{U}$ has diameter less than δ. For each $y \in Y$ choose an open neighborhood W_y of y in X such that $c(f)^{-1}(W_y) \subseteq q^{-1}(U)$ for some $U \in \mathcal{U}$. Let $\mathcal{W} = \{W_y : y \in Y\} \cup \{X \setminus Y\}$ and let $\mathcal{V}$ be a finite open cover of X such that

(4) $\mathrm{St}(\mathcal{V}) < \mathcal{W}$,
(5) $\mathrm{mesh}(\mathrm{St}(\mathcal{V})) < \frac{1}{2}\delta$ (here diameter means the diameter in M(f))

(lemma 5.1.7). By proposition 7.2.10 there is a homotopy $\alpha: M \times I \to M$ such that

(6) α is limited by $f^{-1}(\mathcal{V}) = \{c(f)^{-1}(V) \cap M : V \in \mathcal{V}\}$,
(7) $\alpha_1(M) \cap f^{-1}(Y) = \emptyset = \alpha_1(M) \cap c(f)^{-1}(Y)$.

By theorems 7.3.6 and 7.3.5 we may assume that α_1 is an imbedding. Since $M \times [0,\frac{1}{2}]$ is a compact Q-manifold and since $M \times \{0\}$ and $M \times \{\frac{1}{2}\}$ are clearly Z-sets in $M \times [0,\frac{1}{2}]$, by (6) and theorem 7.4.9 there exists a homeomorphism $\underline{h}_1: M \times [0,\frac{1}{2}] \to M \times [0,\frac{1}{2}]$ such that

(8) $\underline{h}_1(\alpha_1(M) \times \{0\}) = M \times \{0\}$,
(9) $\underline{h}_1 \mid M \times \{\frac{1}{2}\} = 1$,

(10) $\underline{h}_1$ and $1_{M\times[0,\frac{1}{2}]}$ are $\mathcal{E}$-homotopic, where $\mathcal{E}$ denotes the cover

$$\{c(f)^{-1}(V) \cap (M \times [0,\tfrac{1}{2}]) \colon V \in \mathcal{V}\}.$$

Let $h_1\colon M(f) \to M(f)$ be the union of $\underline{h}_1$ and the identity on $M(f)\backslash(M \times [0,\frac{1}{2}))$. Then h_1 is clearly a homeomorphism having the following properties:

(11) $h_1(\alpha_1(M)) = M$,
(12) h_1 and $1_{M(f)}$ are $c(f)^{-1}(\mathcal{V})$-close,
(13) h_1 restricts to the identity on $M(f)\backslash(M \times [0,\frac{1}{2}))$.

Observe that by (7) and (8), $h_1(c(f)^{-1}(Y)) \cap M = \emptyset$.

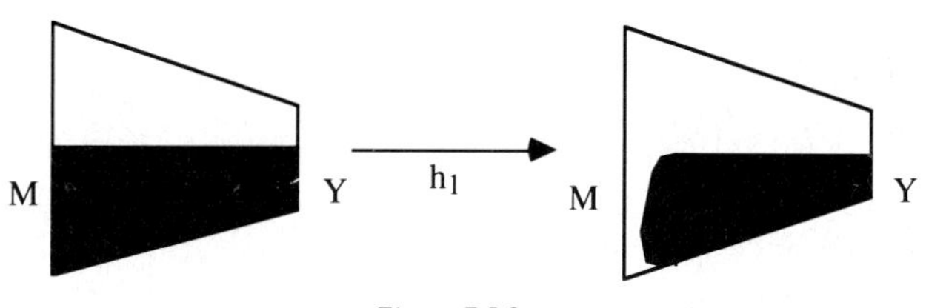

Figure 7.5.2.

Let $O = St(Y,\mathcal{V})$. Observe that O is an open neighborhood of Y in X.

CLAIM 1: For every $y \in Y$ we have $c(f)^{-1}(y) \cup h_1(c(f)^{-1}(y)) \subseteq c(f)^{-1}(St(y,\mathcal{V})) \subseteq c(f)^{-1}(O)$.

Take an arbitrary $y \in Y$ and pick $z \in c(f)^{-1}(y)$. By (12) there is $V \in \mathcal{V}$ such that

$$\{z,h_1(z)\} \subseteq c(f)^{-1}(V).$$

So $y \in V$ from which it follows that $\{z,h_1(z)\} \subseteq c(f)^{-1}(St(y,\mathcal{V}))$. We are done.

By (5), each set of the form $St(y,\mathcal{V})$ has diameter less than δ in M(f). By compactness we can find a point $t \in (\frac{1}{2},1)$ such that the diameter of each set of the form

$$(14) \qquad St(y,\mathcal{V}) \cup (f^{-1}(St(y,\mathcal{V})) \times [t,1)) \qquad (y \in Y)$$

is less than δ (see figure 7.5.3). There is a point $b \in (0,\frac{1}{2})$ such that $(M \times [0,b]) \cap h_1(c(f)^{-1}(Y)) = \emptyset$ (see figure 7.5.4.). Let $A = c(f)(h_1(c(f)^{-1}(Y)))$. By claim 1, A is a compact subset of O. Let

β: M $\rightarrow$ I be a Urysohn function such that $\beta \mid f^{-1}(A) \equiv 1$ and $\beta \mid (M \backslash f^{-1}(O)) \equiv 0$. In addition, let S: I $\times$ I $\rightarrow$ I be an isotopy having the following properties:

(15) $S_0 = 1_I$, $S_1(b) = t$,
(16) $S_i \mid [\frac{1}{2}(1+t),1] = 1_I$ for every $i \in I$.

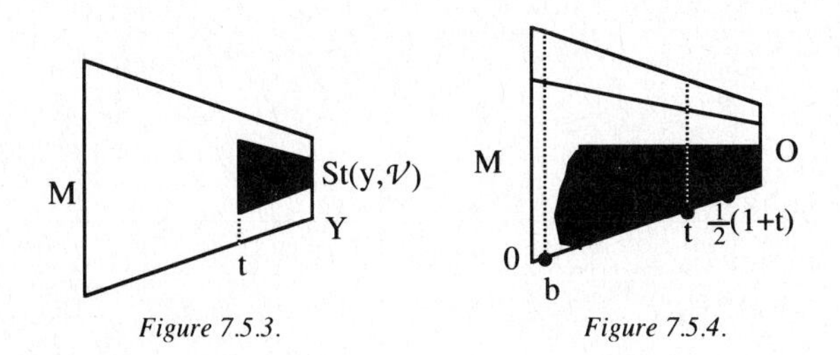

Figure 7.5.3. *Figure 7.5.4.*

Define ϕ: M(f) $\rightarrow$ M(f) by

$$\begin{cases} \phi(x) = x & (x \in X), \\ \phi(p,i) = (p, S_{\beta(p)}(i)) & ((p,i) \in M(f)\backslash X). \end{cases}$$

The reader should have no trouble verifying that ϕ is a homeomorphism, see the proof of theorem 6.1.13.

CLAIM 2: If $y \in Y$ then $\text{diam}(\phi h_1(c(f)^{-1}(y))) < \delta$.

Choose an arbitrary $y \in Y$. By claim 1, $h_1(c(f)^{-1}(y)) \subseteq c(f)^{-1}(\text{St}(y,\mathcal{V}))$. From the definition of ϕ it follows easily that ϕ maps $c(f)^{-1}(\text{St}(y,\mathcal{V}))$ onto itself. Since β is 1 on $f^{-1}(A)$, we find by (15) that

$$\phi h_1(c(f)^{-1}(y)) \subseteq \text{St}(y,\mathcal{V}) \cup (f^{-1}(\text{St}(y,\mathcal{V})) \times [t,1)). \quad (17)$$

Since by (14) the right-hand side has diameter less than δ, we are done.

Now define h: M(f) $\rightarrow$ M(f) by $h = h_1^{-1} \circ \phi \circ h_1$. We claim that h satisfies (1), (2) and (3). That $h \mid X = 1_X$ is clear since $h_1 \mid X = \phi \mid X = 1_X$ (by (13) and the definition of ϕ). Since $t > \frac{1}{2}$, h_1 is the identity on $\phi h_1(c(f)^{-1}(Y))$ (by (17) and (13)). So by claim 2, h satisfies (3). It therefore remains to check that h satisfies (1).

To this end, take an arbitrary $x \in M(f)$. If $h_1(x) \notin c(f)^{-1}(O)$ then by the definition of ϕ we have $\phi h_1(x) = h_1(x)$, so that

$$h(x) = h_1^{-1}\phi h_1(x) = x, \text{ i.e. } d(q(x),qh(x)) = 0.$$

So suppose that $h_1(x) \in c(f)^{-1}(O)$. There exists $y \in Y$ such that

$$h_1(x) \in c(f)^{-1}(St(y,\mathcal{V})).$$

By (12) there exists $V_0 \in \mathcal{V}$ such that $\{x,h_1(x)\} \subseteq c(f)^{-1}(V_0)$. Consequently,

$$c(f)h_1(x) \in V_0 \cap St(y,\mathcal{V}),$$

so there exists $V \in \mathcal{V}$ such that

(19) $$y \in V \text{ and } c(f)h_1(x) \in V \cap V_0.$$

By (12) there exists an element $V_1 \in \mathcal{V}$ such that

$$\{\phi h_1(x),h_1^{-1}\phi h_1(x)\} \subseteq c(f)^{-1}(V_1).$$

By the definition of ϕ it follows that $\phi h_1(x) \in c(f)^{-1}c(f)h_1(x)$. We conclude that $c(f)h_1(x) \in V_1$. Consequently,

$$x \in c(f)^{-1}(V_0),\ \ h(x) \in c(f)^{-1}(V_1),$$

and by (19)

$$V \cap V_0 \cap V_1 \neq \emptyset.$$

Therefore, $\{x,h(x)\} \subseteq c(f)^{-1}(St(V,\mathcal{V}))$ and $St(V,\mathcal{V}) \cap Y \neq \emptyset$ because $y \in V$. By (4) there is an element $W \in \mathcal{W}$ such that $St(V,\mathcal{V}) \subseteq W$. Since $St(V,\mathcal{V}) \cap Y \neq \emptyset$, $W = W_{y'}$ for certain $y' \in Y$. By the choice of $W_{y'}$ there is an element $U \in \mathcal{U}$ such that $c(f)^{-1}(W) \subseteq q^{-1}(U)$. We conclude that

$$\{x,h(x)\} \subseteq c(f)^{-1}(St(V,\mathcal{V})) \subseteq c(f)^{-1}(W) \subseteq q^{-1}(U).$$

From this we conclude that $d(q(x),qh(x)) < \delta$, and since x was chosen arbitrarily, we are done.

We shall now produce the desired near homeomorphism $g: M(f) \to M(f)$. As remarked above, by (1) and (3), q is shrinkable, so it is a near homeomorphism by theorem 6.1.8. The near homeomorphism g shall have the form $\gamma^{-1} \circ q$, where γ: M(f) → Z is a carefully chosen homeomorphism approximating q.

From the proof of Bing's Shrinking Criterion 6.1.8, it follows that q can be approximated by homeomorphisms of the form

$$(20)\ \lim_{n\to\infty} q \circ f_n^{-1} \circ f_{n-1}^{-1} \circ \cdots \circ f_1^{-1},$$

where each f_n: M(f) → M(f) is a "shrinking" homeomorphism of q. By (1), (2) and (3) we can choose each f_n to be the identity on X. So from (20) we conclude that there is a homeomorphism γ: M(f) → Z such that

(21) $d(q,\gamma)$ is as small as we please,

(22) for every $x \in X$, $q(x) = \gamma(x)$.

Observe that there is a continuous function τ: Z → X such that $c(f) = \tau \circ q$. Given $\varepsilon > 0$ find $\delta > 0$ such that if $A \subseteq Z$ has diameter less than δ then $\tau(A)$ has diameter less than ε. Now choose γ: M(f) → Z such as in (22) and satisfying $d(q,\gamma) < \delta$. An easy check shows that $g = \gamma^{-1} \circ q$ is as required. □

7.5.5. PROPOSITION: *Let* M *be a compact Q-manifold, let* X *be an* **ANR** *with the disjoint cells property, and let* g: M → X *be cell-like. If proj*: M × Q → M *denotes the projection, then the composition*

$$M \times Q \xrightarrow{proj} M \xrightarrow{g} X$$

is a near homeomorphism. Consequently, X *is homeomorphic to the Q-manifold* M × Q.

PROOF: Put N = M × Q and $f = g \circ proj$. Then N is a compact Q-manifold and f is cell-like by corollary 7.1.9. Let p: N × I → N be the projection. The diagram

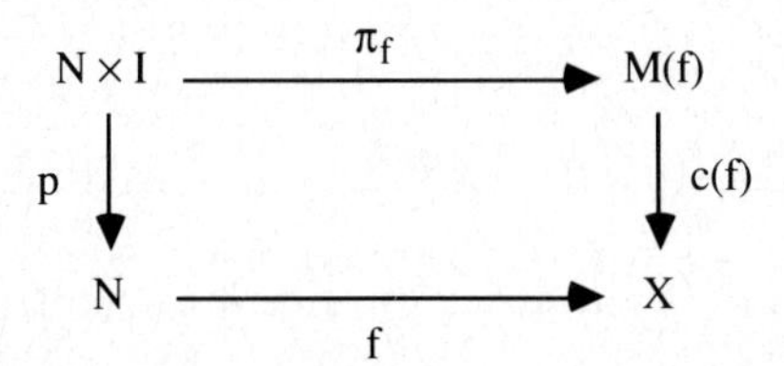

clearly commutes. Since N × Q ≈ N, p is a near homeomorphism by lemma 7.5.1. Also, π_f is a near homeomorphism by lemma 7.5.2. An easy "diagram chase" now establishes that f is a near homeomorphism if and only if c(f) is a near homeomorphism. We shall prove that c(f) is a near homeomorphism.

Choose $\varepsilon > 0$. We shall construct a near homeomorphism $h: M(f) \to M(f)$ such that

(1) $d(c(f) \circ h, c(f)) < \varepsilon$,

(2) $\forall x \in X$: $\mathrm{diam}(h(c(f)^{-1}(x))) < \varepsilon$.

Each homeomorphism $\phi: M(f) \to M(f)$ closely approximating h satisfies the conditions of the Bing Shrinking Criterion, so from theorem 6.1.8 we then conclude that c(f) is a near homeomorphism.

The near homeomorphism h will have the form $\tau \circ \mu^{-1} \circ g$, where g is a near homeomorphism such as in proposition 7.5.4 and τ and μ are certain auxiliary homeomorphisms of M(f) to be constructed below.

Let $\mathcal{U} = \{U \subseteq X: U \text{ is open and } \mathrm{diam}(U) < \frac{1}{3}\varepsilon\}$. Here diam(U) means the diameter of U as a subset of M(f) (we remind the reader again that we consider X to be a subset of M(f)). It is clear that we can find a point $\lambda \in (0,1)$ such that for every $U \in \mathcal{U}$ we have

(3) $\mathrm{diam}(c(f)^{-1}(U) \setminus (N \times [0,\lambda])) < \frac{1}{3}\varepsilon$.

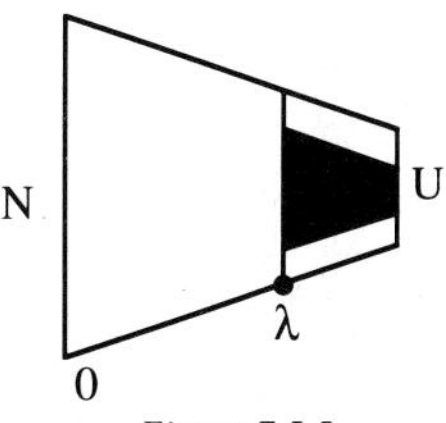

Figure 7.5.5.

Choose finitely many points $0 = t_0 < t_1 < \cdots < t_n$ in the interval [0,1) such that

(4) $\lambda \leq t_{n-1}$

and a finite open cover $\mathcal{V}$ of N such that for every $V \in \mathcal{V}$ and $1 \leq i \leq n-1$ we have

(5) $\mathrm{diam}(V \times (t_{i-1}, t_{i+1})) < \varepsilon$ and $\mathrm{diam}(V \times [0, t_1)) < \varepsilon$.

(see figure 7.5.6).

By lemma 1.1.1 there exists $\delta > 0$ such that every $A \subseteq M(f)$ of diameter less than δ which intersects $N \times [0,\lambda]$ is contained in an element of the collection

$$\{V \times (t_{i-1},t_{i+1}): V \in \mathcal{V}, 1 \le i \le n-1\} \cup \{V \times [0,t_1): V \in \mathcal{V}\}.$$

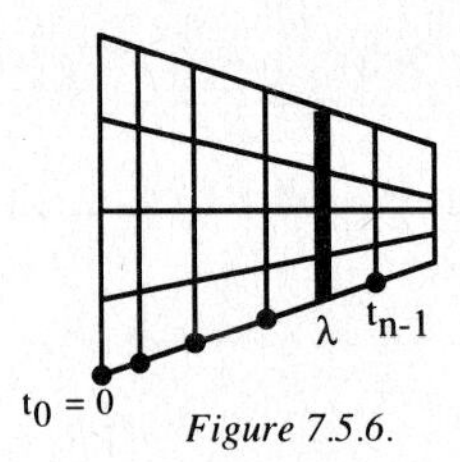

Figure 7.5.6.

Since X is a compact **ANR** with the disjoint cells property, theorems 5.1.1 and 7.3.5 imply that there is a homotopy $H: N \times I \to X$ that is limited by $\mathcal{U}$ while moreover $H_0 = f$ and H_1 is a Z-imbedding. Define a homotopy $F: N \times I \to M(f)$ by

$$F(x,t) = \begin{cases} \pi_f(x,2t) & (t \in [0,\tfrac{1}{2}]), \\ H(x,2t-1) & (t \in [\tfrac{1}{2},1]). \end{cases}$$

Observe that F is continuous, that $F_0: N \to M(f)$ is the inclusion, that $\xi = F_1: N \to M(f)$ is an imbedding and that F is limited by $c(f)^{-1}(\mathcal{U})$. By corollary 7.5.3 it follows that M(f) is a compact Q-manifold. Since $N \times \{1\} \in \mathcal{Z}(N \times [0,1])$, it follows from proposition 7.2.12 that $X \in \mathcal{Z}(M(f))$. Also, $N \times \{0\} \in \mathcal{Z}(M(f))$, so by theorem 7.4.9 there is an isotopy $G: M(f) \times I \to M(f)$ such that

(6) G is limited by $c(f)^{-1}(\mathcal{U})$,
(7) $G_0 = 1$ and $G_1 | N = \xi$.

Let $\mu = G_1$. Observe that

(8) $d(c(f) \circ \mu, c(f)) = d(c(f) \circ \mu^{-1}, c(f)) < \frac{1}{3}\varepsilon$,
(9) $\mu(N) = \xi(N)$.

Since $\xi(N) \in \mathcal{Z}(X)$, by proposition 7.5.4 there is a near homeomorphism $g: M(f) \to M(f)$ such that

(10) $d(c(f) \circ g, c(f)) < \frac{1}{6}\varepsilon$,
(11) $g | g^{-1}(\xi(N)) = c(f) | c(f)^{-1}(\xi(N))$.

CLAIM 1: For each $x \in X$, if $\mu^{-1}g(c(f)^{-1}(x)) \cap N \neq \emptyset$ then $\mu^{-1}g(c(f)^{-1}(x))$ is a single point.

If $\mu^{-1}g(c(f)^{-1}(x)) \cap N \neq \emptyset$ then $g(c(f)^{-1}(x)) \cap \mu(N) = g(c(f)^{-1}(x)) \cap \xi(N) \neq \emptyset$ so that

$$c(f)^{-1}(x) \cap g^{-1}(\xi(N)) \neq \emptyset.$$

By (11) this implies that $x \in \xi(N)$ so that $g(c(f)^{-1}(x))$ is a single point. Since μ is a homeomorphism, we are done.

Define

$$A = \bigcup\{\mu^{-1}g(c(f)^{-1}(x)) : x \in X \text{ and } \operatorname{diam}(\mu^{-1}g(c(f)^{-1}(x))) \geq \delta\}.$$

Then A is obviously closed and, by claim 1, $A \cap N = \emptyset$. So we can find a "level" $s_0 \in (0,1)$ such that $A \cap (N \times [0,s_0]) = \emptyset$. Without loss of generality we assume that

$$(12)\ s_0 \leq \lambda.$$

Now put

$$A' = \bigcup\{\mu^{-1}g(c(f)^{-1}(x)) : x \in X \text{ and } \mu^{-1}g(c(f)^{-1}(x)) \cap \pi_f(N \times [s_0,1]) \neq \emptyset\}.$$

It is again clear that A' is closed, so by another application of claim 1 we can find a "level" $s_1 \in (0,s_0)$ such that $A' \cap (N \times [0,s_1]) = \emptyset$. Continuing in this way we get finitely many points

$$s_0 > s_1 > \cdots > s_{n-1} > s_n = 0$$

(remember that n is not arbitrary) in [0,1) such that

(13) if $\operatorname{diam}(\mu^{-1}g(c(f)^{-1}(x))) \geq \delta$ then $\mu^{-1}g(c(f)^{-1}(x)) \cap (N \times [0,s_0]) = \emptyset$,
(14) each $\mu^{-1}g(c(f)^{-1}(x))$ intersects at most one of the "levels" $\{N \times \{s_i\} : 0 \leq i \leq n\}$.

Now let η: $[0,1) \to [0,1)$ be a homeomorphism having the following properties:

$$(15)\ \eta([s_{i+1},s_i]) = [t_{n-(i+1)},t_{n-i}] \qquad (0 \leq i \leq n-1).$$

Define a homeomorphism τ: $M(f) \to M(f)$ as follows:

$$(16)\ \begin{cases} \tau(x) = x & (x \in X), \\ \tau(p,u) = (p,\eta(u)) & (p \in N,\ u \in [0,1)). \end{cases}$$

Observe that τ is indeed a homeomorphism and that for every $x \in M(f)$,

$$(17)\ c(f)(x) = c(f)(\tau(x)).$$

CLAIM 2: $\forall x \in X$: $\operatorname{diam}(\tau\mu^{-1}g(c(f)^{-1}(x))) < \varepsilon$.

Take an arbitrary $p \in X$. By (10) there exists an element $U \in \mathcal{U}$ containing p such that

$$g(c(f)^{-1}(p)) \subseteq c(f)^{-1}(U).$$

Since by (6) the isotopy $G: M(f) \times I \to M(f)$ is limited by $c(f)^{-1}(\mathcal{U})$, the following statement holds:

$$(18)\ \forall p \in X\ \exists\ U \in \mathcal{U} \text{ containing p such that } \mu^{-1}g(c(f)^{-1}(p)) \subseteq c(f)^{-1}(St(U,\mathcal{U})).$$

Now take an arbitrary $x \in X$ and without loss of generality assume that $\mu^{-1}g(c(f)^{-1}(x))$ is not a single point.

CASE 1: $\mu^{-1}g(c(f)^{-1}(x)) \cap (N \times [0,s_1]) = \emptyset$.

It follows by (4), (18) and the definition of τ that

$$(19)\ \tau\mu^{-1}g(c(f)^{-1}(x)) \subseteq \bigcup\{c(f)^{-1}(U')\backslash(N \times [0,\lambda]): U' \in \mathcal{U} \text{ and } U' \cap U \neq \emptyset\}.$$

By (3) we therefore obtain $\operatorname{diam}(\tau\mu^{-1}g(c(f)^{-1}(x))) < \varepsilon$, which is as desired.

CASE 2: $\mu^{-1}g(c(f)^{-1}(x)) \cap (N \times [0,s_1]) \neq \emptyset$.

Observe that by (13), $\operatorname{diam}(\mu^{-1}g(c(f)^{-1}(x))) < \delta$. By (14) and claim 1 there is an element $i \in \{1,2,\cdots,n-1\}$ such that

$$(20)\ \mu^{-1}g(c(f)^{-1}(x)) \subseteq N \times (s_{i+1},s_{i-1}).$$

In addition, by (12) and by the special choice of δ, there are an element $V \in \mathcal{V}$ and say an index $j \in \{1,2,\cdots,n-1\}$ such that

$$(21)\ \mu^{-1}g(c(f)^{-1}(x)) \subseteq V \times (t_{j-1},t_{j+1}).$$

By (20), (21) and the definition of τ it therefore follows that

$$(22)\ \tau\mu^{-1}g(c(f)^{-1}(x)) \subseteq V \times (t_{n-(i+1)}, t_{n-(i-1)}).$$

By (5), $\operatorname{diam}(V \times [t_{n-(i+1)}, t_{n-(i-1)}]) < \varepsilon$, from which the desired result obviously follows.

CLAIM 3: $d(c(f) \circ \tau \circ \mu^{-1} \circ g, c(f)) < \varepsilon$.

Take an arbitrary $x \in M(f)$ and put $y = c(f)(x)$. By (18) there is an element $U \in \mathcal{U}$ containing y such that

$$\mu^{-1}g(c(f)^{-1}(y)) \subseteq c(f)^{-1}(St(U, \mathcal{U})).$$

By (17) we get

$$\tau\mu^{-1}g(c(f)^{-1}(y)) \subseteq c(f)^{-1}(St(U, \mathcal{U})).$$

Since $\operatorname{diam}(St(U, \mathcal{U})) < \varepsilon$, we are done.

We conclude that the near homeomorphism $h = \tau \circ \mu^{-1} \circ g$ is as required.□

We can now derive several interesting results.

7.5.6. THEOREM ("Stability of Q-manifolds"): *Let* M *be a compact Q-manifold. Then* $M \times Q$ *and* M *are homeomorphic. Moreover, the projection proj:* $M \times Q \to M$ *is a near homeomorphism.*

PROOF: Observe that M is an **ANR** with the disjoint cells property (theorem 7.3.6). The identity $1: M \to M$ is obviously cell-like. So by proposition 7.5.5 we conclude that the composition

$$M \times Q \xrightarrow{proj} M \xrightarrow{1} M$$

is a near homeomorphism.□

7.5.7. THEOREM ("The Approximation Theorem"): *Let* M *be a compact Q-manifold, let* X *be an* **ANR** *and let* $f: M \to X$ *be cell-like. The following statements are equivalent:*

(a) f *is a near homeomorphism,*

(b) X *has the disjoint-cells property.*

PROOF: The implication (a) $\Rightarrow$ (b) is obvious since each Q-manifold has the disjoint-cells property (theorem 7.3.6). For (b) $\Rightarrow$ (a), observe that by proposition 7.5.5 the composition

$$M \times Q \xrightarrow{proj} M \xrightarrow{f} X$$

is a near homeomorphism. Since by theorem 7.5.6 *proj:* $M \times Q \to M$ is a near homeomorphism, it follows easily that f is a near homeomorphism. □

7.5.8. THEOREM ("Classification"): *Every compact contractible Q-manifold is homeomorphic to* Q.

PROOF: Let M be a compact contractible Q-manifold. Consider the projection

$$\pi: M \times Q \to Q.$$

Since M is contractible, π is cell-like. Since Q satisfies the disjoint cells property, $M \times Q$ and Q are homeomorphic by theorem 7.5.7. Moreover, $M \times Q \approx M$ by theorem 7.5.6. We conclude that $M \approx Q$. □

7.5.9. *Remark:* In view of theorem 7.5.7 it is natural to ask which compact **ANR**'s are the cell-like image of a Q-manifold. That turns out to be the case for *every* compact **ANR** (this is The Miller-West Theorem). In the forthcoming sections we shall present a proof of this fact and combine it with earlier results to derive The Characterization Theorem.

7.6. Cell-like Maps I

In this section we shall prove a technical proposition on cell-like maps that shall be used in the forthcoming sections to prove The Characterization Theorem.

Let M be a compact Q-manifold and let r: $M \to A$ be a retraction. Define $r_1: M \times [0,2] \to A$ by

$$r_1(m,t) = r(m),$$

i.e. r_1 is the projection followed by r. Now consider the topological sum of the spaces M(r) and $M \times [1,2]$. By identifying every $a \in A \subseteq M(r)$ with the point $(a,1) \in M \times [1,2]$ we obtain the space in figure 7.6.1 which we shall denote by $M(r) \cup (M \times [1,2])$. The collapse to base on M(r) and the identity on the product $M \times [1,2]$ form a natural retraction from $M(r) \cup (M \times [1,2])$ onto $M \times [1,2]$. This retraction followed by the map

$$(m,t) \rightarrow r(m) \qquad (m \in M, t \in [1,2])$$

shall be denoted by r_2.

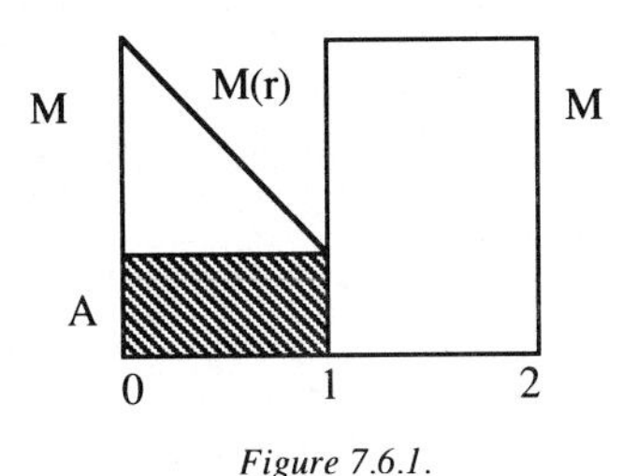

Figure 7.6.1.

A continuous function $f: X \rightarrow X$ is said to be *supported on* a set $A \subseteq X$ if f restricts to the identity on $X \backslash A$.

7.6.1. PROPOSITION: *For each* $\varepsilon > 0$ *there is a cell-like map* $\beta: M \times [0,2] \rightarrow M(r) \cup (M \times [1,2])$ *having the following properties:*

(1) *for every* $a \in A$ *and* $m \in M$, $\beta^{-1}(a,0) = \{(a,0)\}$ *and* $\beta(m,2) = (m,2)$,

(2) $d(r_2 \circ \beta, r_1) < \varepsilon$.

PROOF: We shall first construct a cell-like map β which satisfies (1) and then later indicate which changes have to be made in order for (2) to hold.

The map β shall be the composition of four auxiliary maps, namely, u,v,w, and ρ.

Step 1: The construction of u.

Since M is a Q-manifold, the projection π from $M \times [0,2] \times Q$ onto $M \times [0,2]$ is a near homeomorphism (theorem 7.5.6). Let $\mathbf{0}$ be the origin of Q, i.e. the points all coordinates of which are 0. A homeomorphism ϕ closely approximating π maps $M \times \{0\} \times \{\mathbf{0}\}$ close to $M \times \{0\}$. So we can use the Z-set Unknotting Theorem 7.4.9 to adjust ϕ so that it has the additional property that $\phi(m,0,\mathbf{0}) = (m,0)$ for every $m \in M$. Similarly, by the same technique, the adjusted ϕ can be adjusted further to have the additional property that $\phi(m,2,\mathbf{0}) = (m,2)$ for every $m \in M$ (both adjustments can also be made simultaneously of course). So we let $u: M \times [0,2] \rightarrow M \times [0,2] \times Q$ be a homeomorphism such that $u^{-1}: M \times [0,2] \times Q \rightarrow M \times [0,2]$ is a homeomorphism closely approximating π and having the additional property that $u^{-1}(m,\xi,\mathbf{0}) = (m,\xi)$ for every $m \in M$ and $\xi \in \{0,2\}$.

Step 2: The construction of v.

$v: M \times [0,2] \times Q \to M \times [1,2] \times Q$ is a homeomorphism defined by the formula

$$v(m,t,q) = (m,1+\tfrac{1}{2}t,q).$$

Step 3: The construction of w.

Let $\phi: M \to Q$ be a map such that $\phi(M\backslash A) \subseteq Q\backslash\{\mathbf{0}\}$, $\phi(A) = \{\mathbf{0}\}$ and $\phi \mid M\backslash A$ is one-to-one. Such a map can be obtained as follows. First, consider the quotient space M/A. This space can be imbedded in Q by theorem 1.4.18. Now, by the homogeneity of Q (theorem 6.1.6), we can easily adjust the imbedding such that the point $\{A\}$ in M/A corresponds to the point $\mathbf{0}$ in Q. Let ϕ be the quotient map $M \to M/A$ followed by this imbedding.

Define a function $\theta: M \to M \times [1,2] \times Q$ by

$$(*) \qquad \theta(m) = (r(m),1,\phi(m)).$$

Observe that for every $a \in A$,

$$\theta(a) = (a,1,\mathbf{0})$$

and also that $\theta(M)$ is a copy of the graph of the function $r \times \phi$.

CLAIM 1: θ is a Z-imbedding.

Since $\{1\} \in \mathcal{Z}([1,2])$, it is clear that θ is a Z-map. Consequently, it suffices to show that θ is one-to-one. That follows easily since r is a retraction and ϕ is one-to-one on $M\backslash A$.

Now let X be the mapping cylinder $M(\theta) = (M \times [0,1]) \cup_\theta (M \times [1,2] \times Q)$; recall that X is the space we obtain from the disjoint topological sum of the spaces

$$M \times [0,1] \text{ and } M \times [1,2] \times Q.$$

by identifying for each $m \in M$ the point $(m,1)$ in $M \times [0,1]$ and the point $\theta(m)$ in $M \times [1,2] \times Q$ to a single point.

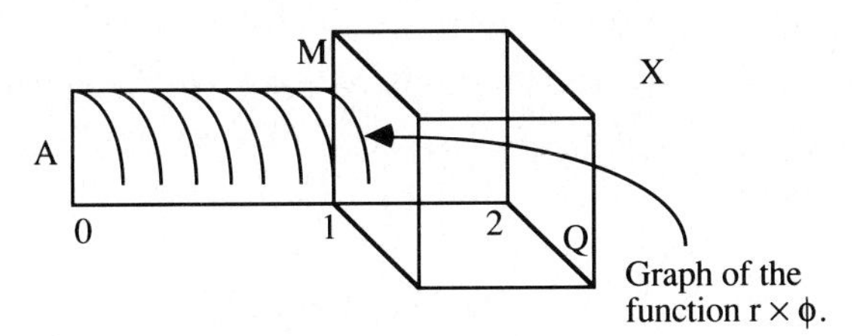

Figure 7.6.2.

By claim 1 and corollary 7.4.4(1) there is an open imbedding Ψ: $M \times [1,2) \to M \times [1,2] \times Q$ such that $\Psi_1 = \theta$. Without loss of generality we assume that $\Psi(M \times [1,2))$ misses $M \times \{2\} \times Q$. Observe that X contains a "canonical" *open* homeomorph of $M \times [0,2)$.

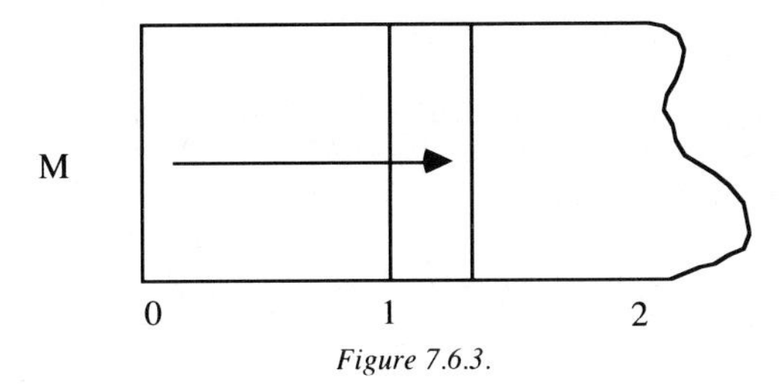

Figure 7.6.3.

For a small $\delta > 0$, consider a homeomorphism of the interval [0,2) onto [1,2) which is supported on $[0,1+\delta]$. It is a triviality to use this homeomorphism to construct a homeomorphism η: $X \to M \times [1,2] \times Q$ having the following properties:

(4) η^{-1} is the identity outside $\Psi(M \times [1,1+\delta])$,

(5) for every $m \in M$, $\eta(m,0) = (r(m),1,\phi(m))$,

Now let w be η^{-1}.

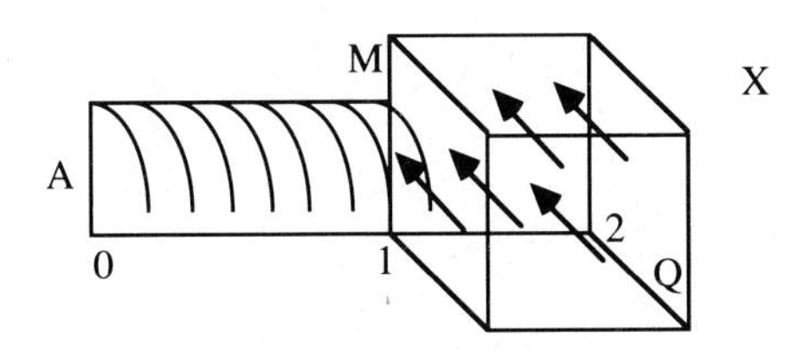

Figure 7.6.4.

Step 4: The construction of ρ.

Consider the space X. If in X we "project" $M \times [1,2] \times Q$ onto $M \times [1,2]$ then, by the definition of θ (see (*)), we obtain the space $M(r) \cup (M \times [1,2])$ (precisely because of this, we did not take for θ an arbitrary Z-imbedding). Let ρ be this "projection". Observe that ρ is cell-like (in fact, each fiber is contractible) and that in a sense we "collapsed out the Q-factor".

Now let $\beta = \rho \circ w \circ v \circ u$. We claim that β is as required. Take an arbitrary $a \in A$. Then

$$\beta^{-1}(a,0) = u^{-1}v^{-1}w^{-1}\rho^{-1}(a,0) = u^{-1}v^{-1}w^{-1}(a,0) = u^{-1}v^{-1}\eta(a,0) = u^{-1}v^{-1}(a,1,\mathbf{0})$$

$$= u^{-1}(a,0,\mathbf{0}) = \{(a,0)\}.$$

Also, take an arbitrary $m \in M$. Then

$$\beta(m,2) = \rho w v u(m,2) = \rho w v(m,2,\mathbf{0}) = \rho w(m,2,\mathbf{0}) = \rho(m,2,\mathbf{0}) = (m,2).$$

We are done.

We shall now indicate which changes have to be made in order for (2) to hold. To this end, let $\varepsilon > 0$. Recall that $u: M \times [0,2] \to M \times [0,2] \times Q$ is a homeomorphism such that its inverse u^{-1}: $M \times [0,2] \times Q \to M \times [0,2]$ is as close to the projection $\pi: M \times [0,2] \times Q \to M \times [0,2]$ as we want. Therefore, by compactness of all spaces under consideration, we can choose u^{-1} so close to π that

(6) $d(r_1, r_1 \circ \pi \circ u) < \frac{1}{2}\varepsilon$.

Let $\bar{r}_1: M \times [1,2] \to A$ be the function

$$\bar{r}_1(m,t) = r(m).$$

It will be convenient to let π also denote the projection $M \times [1,2] \times Q \to M \times [1,2]$. Observe that

(7) $r_1 \circ \pi = \bar{r}_1 \circ \pi \circ v$.

Let $\Psi: M \times [1,2) \to M \times [1,2] \times Q$ be the open imbedding constructed above such that $\Psi \mid M \times \{1\} = \theta$. Now choose $0 < \delta < 1$ so small that

(8) $\forall m \in M: \operatorname{diam}(\bar{r}_1 \pi \Psi(\{m\} \times [1,1+\delta])) < \frac{1}{2}\varepsilon$.

As above, consider a homeomorphism of the interval $[0,2)$ onto $[1,2)$ which is supported on $[0,1+\delta]$ and maps 1 onto $1+\delta/2$. Use this homeomorphism to construct the homeomorphism $\eta: X \to M \times [1,2] \times Q$. Observe that η is nothing but a "horizontal push".

It will be convenient to consider $M \times [1,2] \times Q$ to be a subspace of X. Observe that after this "identification",

$$\bar{r}_1 \circ \pi = r_2 \circ \rho \mid (M \times [1,2] \times Q).$$

We claim that the above δ "works". To this end, take an arbitrary point $(m,s) \in M \times [0,1]$. By (6) and (7), for $vu((m,s)) = (\overline{m},\overline{s},q) \in M \times [1,2] \times Q$ we have

$$(9)\ d(r_1(m,s),\bar{r}_1 \circ \pi(\overline{m},\overline{s},q)) = d(r(m),r(\overline{m})) < \tfrac{1}{2}\varepsilon.$$

First assume that $(\overline{m},\overline{s},q) \notin \Psi(M \times [1,1+\delta))$. Then $r_2\beta(m,s) = r_2\rho\, w(\overline{m},\overline{s},q) = r_2\rho(\overline{m},\overline{s},q) = r_2(\overline{m},\overline{s}) = r(\overline{m})$, so by (9) there is nothing to prove. Next assume that for an element

$$(*)\qquad (\overline{\overline{m}},\overline{\overline{s}}) \in M \times [1,1+\delta),\ \Psi(\overline{\overline{m}},\overline{\overline{s}}) = (\overline{m},\overline{s},q).$$

We consider two cases:

CASE 1: $\overline{\overline{s}} < 1+\delta/2$.

Consider the points $\Psi(\overline{\overline{m}},1) = \theta(\overline{\overline{m}},1)$ and $\Psi(\overline{\overline{m}},1+\delta/2)$. Observe that $w\Psi(\overline{\overline{m}},1+\delta/2) = \theta(\overline{\overline{m}},1)$, so that

$$(10)\ \bar{r}_1\pi\Psi(\overline{\overline{m}},1) = r_2\rho\, w\Psi(\overline{\overline{m}},1+\delta/2).$$

By (8) and (9) it follows that

$$\begin{aligned} d(\bar{r}_1\pi\Psi(\overline{\overline{m}},1),r_1(m,s)) &\le d(\bar{r}_1\pi\Psi(\overline{\overline{m}},1),\bar{r}_1\pi\Psi(\overline{\overline{m}},\overline{\overline{s}})) + d(\bar{r}_1\pi\Psi(\overline{\overline{m}},\overline{\overline{s}}),r_1(m,s)) \\ &< \tfrac{1}{2}\varepsilon + \tfrac{1}{2}\varepsilon \\ &= \varepsilon. \end{aligned}$$

In view of (10) it therefore suffices to prove that

$$r_2\beta(m,s) = r_2\rho\, w\Psi(\overline{\overline{m}},1+\delta/2).$$

To this end, take $t \in [0,1)$ such that $w(\overline{m},\overline{s},q) = (\overline{\overline{m}},t)$. Then,

$$r_2\beta(m,s) = r_2\rho\, w(\overline{m},\overline{s},q) = r_2\rho(\overline{\overline{m}},t) = r_2(\overline{\overline{m}},t) = r(\overline{\overline{m}}),$$

and

$$w\Psi(\overline{\overline{m}},1+\delta/2) = \theta(\overline{\overline{m}},1) = (r(\overline{\overline{m}}),1,\phi(\overline{\overline{m}})),$$

so that

$$r_2\rho\, w\Psi(\overline{\overline{m}},1+\delta/2) = r(r(\overline{\overline{m}})) = r(\overline{\overline{m}}).$$

We are done.

CASE 2: $\overline{\overline{s}} \geq 1+\delta/2$.

Take $t \in [1,1+\delta)$ such that $w(\overline{m},\overline{s},q) = \Psi(\overline{\overline{m}},t)$. Observe that

$$\bar{r}_1\pi\Psi(\overline{\overline{m}},t) = r_2\rho\,\Psi(\overline{\overline{m}},t) = r_2\rho\, w(\overline{m},\overline{s},q) = r_2\beta(m,s).$$

By (8) and (9) we therefore obtain

$$\begin{aligned} d(r_2\beta(m,s),r_1(m,s)) &= d(\bar{r}_1\pi\Psi(\overline{\overline{m}},t),r_1(m,s)) \\ &\leq d(\bar{r}_1\pi\Psi(\overline{\overline{m}},t),\bar{r}_1\circ\pi(\overline{m},\overline{s},q)) + d(\bar{r}_1\circ\pi(\overline{m},\overline{s},q),r_1(m,s)) \\ &\leq d(\bar{r}_1\pi\Psi(\overline{\overline{m}},t),\bar{r}_1\pi\Psi(\overline{\overline{m}},\overline{\overline{s}})) + \tfrac{1}{2}\varepsilon \\ &< \tfrac{1}{2}\varepsilon + \tfrac{1}{2}\varepsilon = \varepsilon, \end{aligned}$$

as required. □

7.7. Cell-like Maps II

Throughout this section, $A \subseteq Q$ is a compact **AR** and $r: Q \to A$ is a retraction. We consider the mapping cylinder $M(r) = (Q \times [0,1)) \cup A$ of r. Observe that we consider A to be a subspace of M(r).

Our aim is to prove the following result.

7.7.1. THEOREM: *Let* $r: Q \to A$ *be a retraction. Then there is a cell-like map* $\psi: M(r) \to A \times [0,1]$ *such that for every* $a \in A$, $\psi(a) = (a,1)$.

We shall need this in §7.8 to present a proof of Toruńczyk's Q-manifold Characterization Theorem.

The strategy of the proof is roughly speaking the following. Consider the mapping cylinder M(r).

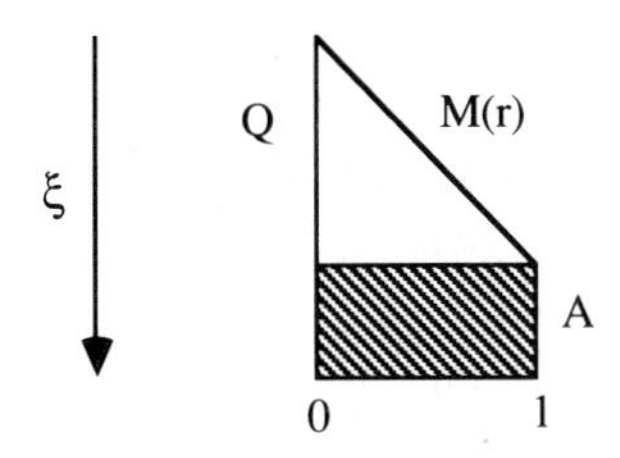

Figure 7.7.1.

Observe that M(r) contains a canonical homeomorph of $A \times [0,1]$. It will be convenient to identify $A \times [0,1]$ with this homeomorph.

Also, observe that there is a vertical retraction ξ: $M(r) \to A \times [0,1]$ which is induced by r. It is easy to see that if r is not cell-like then ξ is not cell-like. By making suitable "subdivisions" of M(r) into smaller and smaller "mapping cylinders", we obtain the desired cell-like map as the limit of an inductively chosen sequence of maps that are "more and more" cell-like.

For each $n \geq 0$ consider the space X_n which is pictured in figure 7.7.2.

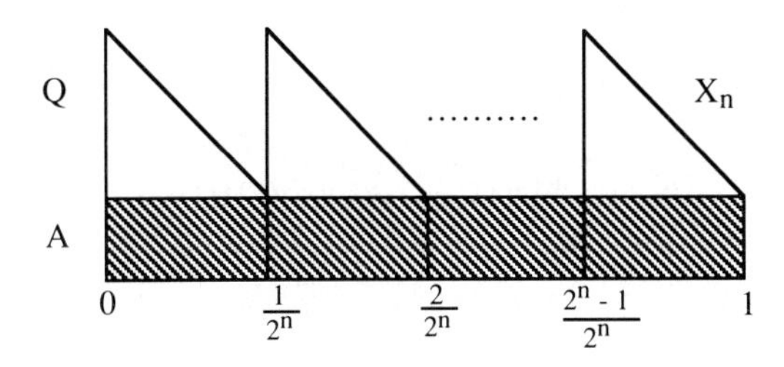

Figure 7.7.2.

X_n is a union of 2^n copies of M(r) which are sewn together along A and $A \times \{0\}$ (so $X_0 = M(r)$). Observe that the "horizontal diameter" of each of the mapping cylinders is 2^{-n}. In addition, note that $A \times [0,1] \subseteq X_n$ and that there is a (vertical) retraction s_n: $X_n \to A \times [0,1]$ which is "induced" by r; if $x \in X_n$, say x = "(p,t)" for $p \in Q$ and $t \in [0,1]$, then

$$s_n(x) = (r(p),t).$$

For each $1 \leq m \leq 2^n$ let $X_{n,m}$ be the m-th building block" of X_n, i.e. the space

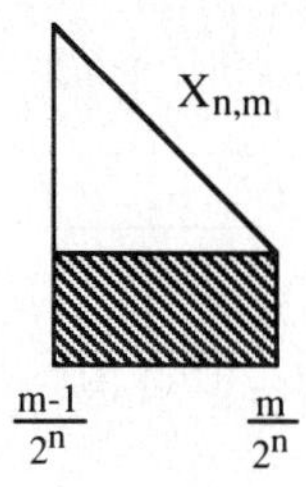

Figure 7.7.3.

It is clear that the spaces X_n are compact. In addition, they are **AR**'s by theorem 5.6.1 and 1.5.9.

We shall now construct a cell-like map $f_n: X_n \to X_{n+1}$. We think of X_n as the space in figure 7.7.4.

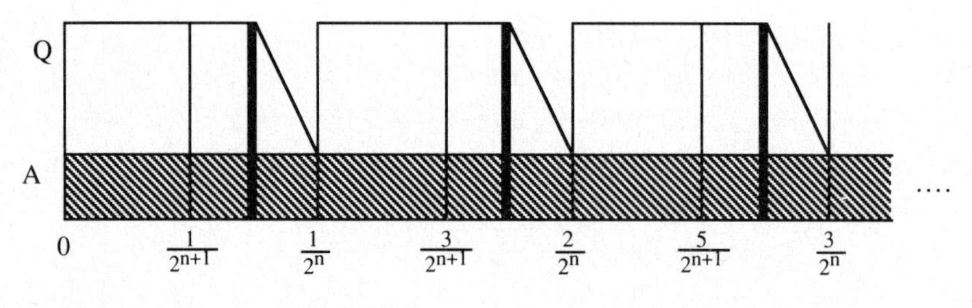

Figure 7.7.4.

By proposition 7.6.1 there is a cell-like map (see figure 7.7.5)

$$\eta: Q \times [0,\tfrac{1}{2}(2^{-n}+2^{-(n+1)})] \to X_{n+1,1} \cup (Q \times [2^{-(n+1)},\tfrac{1}{2}(2^{-n}+2^{-(n+1)})])$$

such that for each $a \in A$ and $x \in Q$,

$$(1)\ \eta^{-1}(a,0) = \{(a,0)\} \text{ and } \eta(x,\tfrac{1}{2}(2^{-n}+2^{-(n+1)})) = (x,\tfrac{1}{2}(2^{-n}+2^{-(n+1)})),$$

and

$$(2)\ d(r_2 \circ \eta, r_1) < 2^{-n}$$

(we adopt the notation in proposition 7.6.1).

The union of this map and the identity on $(Q \times [\frac{1}{2}(2^{-n}+2^{-(n+1)}),2^{-n})) \cup (A \times \{2^{-n}\})$ will be denoted by η^*. Now by applying the same map η^* on all the mapping cylinders that form X_n, we obtain for every $n \geq 0$ a map $f_n: X_n \to X_{n+1}$. It is easy to see that this function is cell-like. Also,

(3) f_n restricts to the identity on $A \times \{\frac{m}{2^n}\}$ for every $0 \le m \le 2^{-n}$.

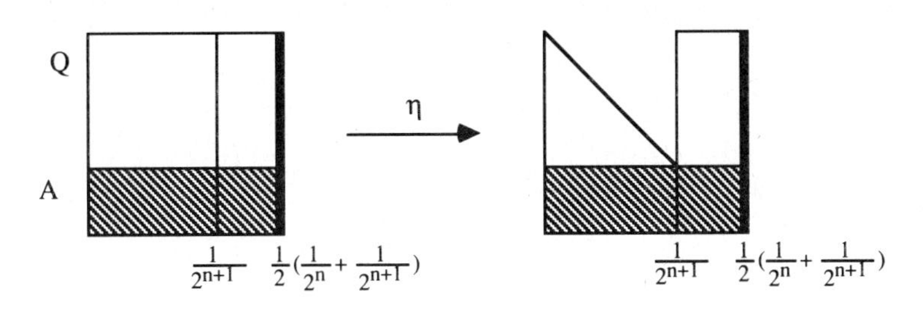

Figure 7.7.5.

Let d be an admissible metric on A. On $A \times [0,1]$ we use the admissible metric

$$\rho((x,t),(y,s)) = \max\{d(x,y),|t - s|\}.$$

CLAIM 1: For every $n \ge 0$, $\rho(s_{n+1} \circ f_n, s_n) \le 2^{-n}$.

This follows by (2) and by the fact that the "horizontal diameter" of every building block of X_n is equal to 2^{-n}.

We conclude that for every $m \ge 0$ the sequence

$$(s_{m+n+1} \circ f_{m+n} \circ \cdots \circ f_{m+1} \circ f_m)_n$$

is Cauchy in the function space $C(X_m, A \times [0,1])$, so it converges to a continuous surjection α_m: $X_m \to A \times [0,1]$ (propositions 1.3.4 and 1.3.7).

Our aim is to prove that the function α_0 is cell-like. Before we are able to do that, we need one more technical result.

CLAIM 2: For every $n \ge 0$, $\rho(s_n, \alpha_n) \le \frac{2}{2^n}$.

By claim 1 and the definition of α_n it follows that for $n \ge 0$,

$$\rho(s_n,\alpha_n) \le \sum_{m=0}^{\infty} \frac{1}{2^{n+m}} = \frac{1}{2^n}\sum_{m=0}^{\infty} 2^{-m} = \frac{2}{2^n}.$$

We shall now prove that α_0: $X_0 \to A \times [0,1]$ is cell-like. To this end, take an arbitrary point $(a,t) \in A \times [0,1]$. For $\gamma > 0$ let $B_\rho((a,t),\gamma)$ denote the open ball about (a,t) of ρ-radius γ.

Take $\gamma > 0$ and let $U = B_\rho((a,t),\gamma)$. Since A is an **AR** there is $\varepsilon > 0$ such that $B(a,\varepsilon)$ is contractible in $B(a,\gamma)$ (exercise 1.6.2). Now take $n \in \mathbb{N}$ so large that

(3) $\frac{9}{2^n} < \varepsilon.$

Consider $B = \alpha_n^{-1}(a,t)$. We shall prove that B is contractible in $\alpha_n^{-1}(U)$. To this end, take arbitrary $p,q \in B$. Then by claim 2,

$$\rho(s_n(p),s_n(q)) \leq \rho(s_n(p),\alpha_n(p)) + \rho(\alpha_n(p),\alpha_n(q)) + \rho(\alpha_n(q),s_n(q))$$

$$\leq \frac{2}{2^n} + 0 + \frac{2}{2^n} = \frac{4}{2^n}.$$

Consequently, there is a collection of at most 5 consecutive "building blocks" of X_n the union of which contains B. For covenience, let us assume that B is contained in the union of the first 5 "building blocks" of X_n (it will be clear from the construction later that this does not violate generality); let this union be denoted by Z.

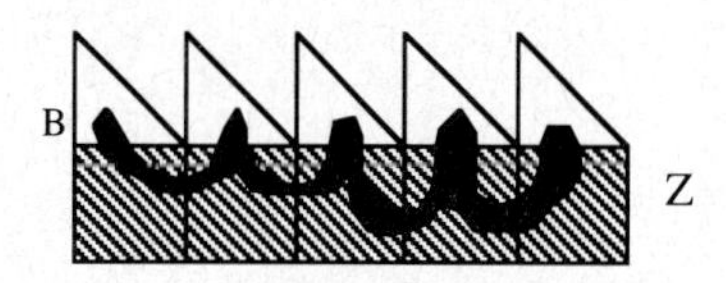

Figure 7.7.6.

Now by pushing B horizontally along the "[0,1]-axis" of each of the mapping cylinders (see figure 7.7.7), we obtain a homotopy $H: B \times I \to Z$ such that

(4) $H_0 = 1_B$,

(5) $H_1(B) \subseteq A \times \{\frac{5}{2^n}\}$.

CLAIM 3: $H(B \times I) \subseteq \alpha_n^{-1}(B_\rho((a,t),\varepsilon))$.

Let us see what happens to a point (x,λ) that belongs to e.g. $X_{n,1} \cap B$. The homotopy H pushes (x,λ) along the set $\{x\} \times [\lambda,\frac{1}{2^n})$ to the point $(r(x),\frac{1}{2^n})$ (see figure 7.7.8). Then H pushes $(r(x),\frac{1}{2^n})$ horizontally to the point $(r(x),\frac{5}{2^n})$. Observe that the second part of the route takes place in $A \times [0,1]$. Also observe that s_n maps the arc $[(x,\lambda),(r(x),\frac{1}{2^n})]$ onto the arc $[(r(x),\lambda),(r(x),\frac{1}{2^n})]$. Consequently, $s_n H(\{(x,\lambda)\} \times I) = [(r(x),\lambda),(r(x),\frac{5}{2^n})]$ and has therefore ρ-diameter at most $\frac{5}{2^n}$. By claim 2 we therefore obtain that

$$\operatorname{diam}_\rho(\alpha_n H(\{(x,\lambda)\} \times I)) \leq \frac{2}{2^n} + \frac{2}{2^n} + \frac{5}{2^n} = \frac{9}{2^n}.$$

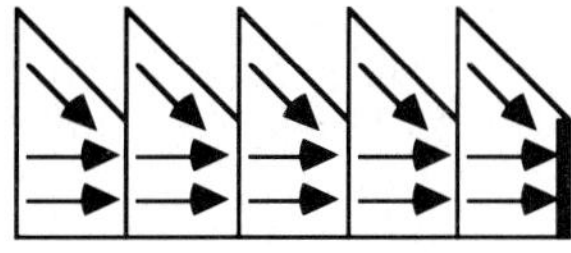

Figure 7.7.7.

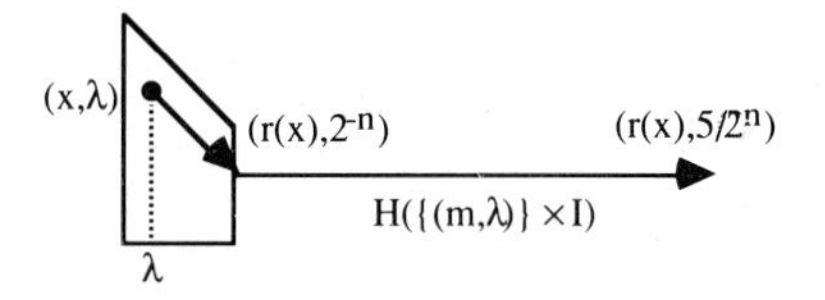

Figure 7.7.8.

Since $\alpha_n(x,\lambda) = (a,t)$ we conclude by (3) that

$$\alpha_n H(\{(x,\lambda)\} \times I) \subseteq D_\rho((a,t),\tfrac{9}{2^n}) \subseteq B_\rho((a,t),\varepsilon),$$

as required.

Now consider $B_1 = H_1(B)$. We want to contract B_1 to a point in $\alpha_n^{-1}(U) = \alpha_n^{-1}(B_\rho((a,t),\gamma))$. Observe that α_n restricts to the identity on $A \times \{\frac{5}{2^n}\}$ from which it follows by claim 3 that

$$B_1 = \alpha_n(B_1) \subseteq B_\rho((a,t),\varepsilon) \cap (A \times \{\tfrac{5}{2^n}\}) = B(a,\varepsilon) \times \{\tfrac{5}{2^n}\}.$$

Consequently, since $B(a,\varepsilon)$ is contractible in $B(a,\gamma)$, B_1 can be contracted to a point in

$$E = \alpha_n(E) = B(a,\gamma) \times \{\tfrac{5}{2^n}\}.$$

Since $E \subseteq \alpha_n^{-1}(E)$ we therefore conclude that B_1 can be contracted to a point in

$$\alpha_n^{-1}(E) \subseteq \alpha_n^{-1}(U),$$

as required.

We conclude that B can be contracted to a point in $\alpha_n^{-1}(U)$. Since $\alpha_0\colon X_0 \to A \times [0,1]$ is equal to the composition

$$\alpha_n \circ f_{n-1} \circ \cdots \circ f_0,$$

and $f_{n-1} \circ \cdots \circ f_0$ is cell-like (corollary 7.1.9), it follows that the set $\alpha_0^{-1}(x,t)$ is contractible in $\alpha_0^{-1}(U)$ (corollary 7.1.7). By theorem 7.1.1 it now follows that α_0 is cell-like.

Observe that by construction, for every $a \in A$, $\alpha_0(a) = (a,1)$. □

7.8. The Characterization Theorem

The aim of this section is to present a proof of Toruńczyk's Q-manifold Characterization Theorem. We first prove Edward's Theorem that for every compact **ANR** X, $X \times Q$ is a Q-manifold (this result immediately implies the Miller-West Theorem that every compact **ANR** is the cell-like image of a compact Q-manifold). Our strategy is to prove Edward's Theorem for compact **AR**'s first and then use the established **AR**-case to derive the **ANR**-case.

7.8.1. THEOREM ("Edward's Theorem"): *Let* X *be a compact* **ANR**. *Then* $X \times Q$ *is a Q-manifold.*

PROOF: Let A be a compact **AR** and let $r: Q \to A$ be a retraction. As in §7.6 consider the topological sum of the spaces M(r) and $Q \times [1,2]$. By identifying every $a \in A \subseteq M(r)$ with the point $(a,1) \in M \times [1,2]$ we obtain the following space S (cf. figure 7.6.1):

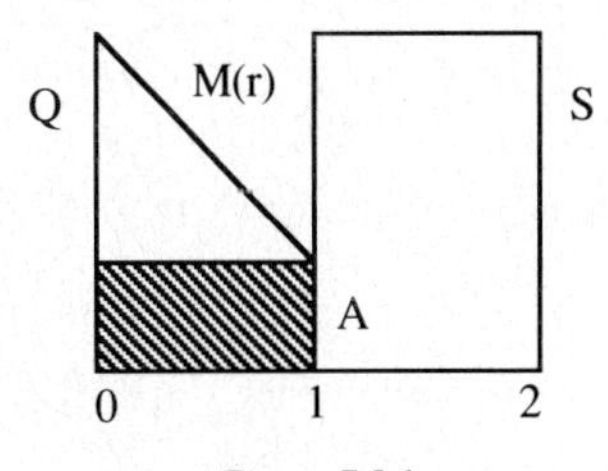

Figure 7.8.1.

By lemma 7.6.2 there is a cell-like map $\beta: Q \times [0,2] \to S = M(r) \cup (Q \times [1,2])$. Observe that by theorems 5.6.1 and 1.5.9, S is an **AR**.

Consider the subspace $T = (A \times [0,1]) \cup (Q \times [1,2])$ of S.

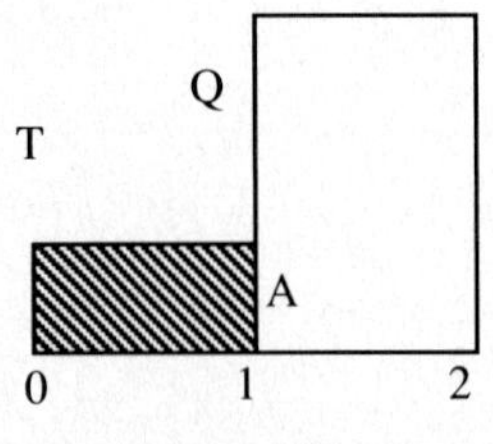

Figure 7.8.2.

Observe that by theorem 1.5.9, T is an **AR**. By theorem 7.7.1 there is a cell-like map γ: $M(r) \to A \times [0,1]$ such that for every $a \in A$, $\gamma(a) = (a,1)$. Now define a function $\bar{\gamma}$: $M(r) \cup (Q \times [1,2]) \to T$ by

$$\bar{\gamma} = \gamma \cup 1_{Q\times[1,2]}.$$

Then $\bar{\gamma}$ is well-defined because for every $a \in A$, $\gamma(a) = (a,1)$. In addition, since γ is cell-like, so is $\bar{\gamma}$. Since any finite composition of cell-like maps between compact **ANR**'s is again cell-like (corollary 7.1.9), the composition

$$\xi = \bar{\gamma} \circ \beta\colon Q \times [0,2] \to T$$

is cell-like. Consequently, the map

$$\xi \times 1_Q\colon Q \times [0,2] \times Q \to T \times Q$$

is cell-like as well. Since T is a compact **AR** and $T \times Q$ has the disjoint-cells property (theorem 7.3.6), theorem 7.5.7 yields that $T \times Q \approx Q$. Consequently, $A \times [0,1) \times Q$ is a Q-manifold being an open subspace of $T \times Q$. From this we conclude that $A \times [0,1] \times Q$ is a Q-manifold being the union of two open subspaces each homeomorphic to $A \times [0,1) \times Q$. We conclude that $A \times Q$ is a Q-manifold. This concludes the proof of the **AR**-case.

Now let X be an arbitrary compact **ANR**. By theorem 5.4.2, the cone $\Delta(X)$ is a compact **AR**. Consequently, by the above, $\Delta(X) \times Q$ is a Q-manifold. Since $X \times [0,1) \times Q$ is an open subspace of $\Delta(X) \times Q$, we obtain as above that

$$X \times [0,1] \times Q \approx X \times Q$$

is a Q-manifold. □

7.8.2. COROLLARY ("The Miller-West Theorem"): *Every compact* **ANR** *is the cell-like image of a Q-manifold.* □

These results now yield the following interesting

7.8.3. THEOREM ("The Characterization Theorem"): *Let* X *be a compact space. The following statements are equivalent:*

(a) X *is a Q-manifold,*

(b) X *is an* **ANR** *with the disjoint-cells property.*

PROOF: For (a) $\Rightarrow$ (b), see theorem 7.3.6.

For (b) $\Rightarrow$ (a), observe that by the above, $X \times Q$ is a Q-manifold. So the projection $\pi: X \times Q \to X$ is a near homeomorphism by theorem 7.5.7. We conclude that X is a Q-manifold. □

7.8.4. COROLLARY: *A space* X *is homeomorphic to the Hilbert cube* Q *if and only if it is a compact* **AR** *with the disjoint-cells property.*

PROOF: That the conditions on X are necessary is clear. So assume that X is a compact **AR** with the disjoint-cells property. By theorem 7.8.3 and corollary 1.6.7, X is a contractible compact Q-manifold, and hence a Hilbert cube by theorem 7.5.8. □

For applications of The Characterization Theorem, see chapter 8.

Notes and comments for chapter 7.

§1.

Theorem 7.1.6 is due to Haver [72].

§3.

The results in this section are basically due to Toruńczyk [136].

§4.

Theorem 7.4.1 is due to Brown [39]; the proof presented here however is due to Connely [47]. Theorem 7.4.5 is due to Chapman [45; corollary 17.3]. The Z-set Unknotting Theorem 7.4.9 is due to Anderson and Chapman [12]; the proof presented here was taken from Chapman [45].

§5.

Theorem 7.5.7 is due to Toruńczyk [136]. The proof presented here is due to Edwards [57]. Toruńczyk's original proof of theorem 7.5.7 used the result obtained earlier by Edwards (see [45, chapter XIV]) that every locally compact **ANR** is a Q-manifold factor. For an alternate proof of theorem 7.5.7 see Walsh [145]. Theorem 7.5.6 is due to Anderson and Schori [15]. Finally, theorem 7.5.8 is due to Chapman [44].

§6 and §7.

Theorem 7.7.1 is due to Miller [107] and West [148]. The proof presented here is a simplification of the proof due to Chapman and West presented in [45, chapter XIII].

§8.

As observed above, theorem 7.8.1 is due to Edwards. This result in combination with Chapman's Triangulation Theorem [45, 46] implies the important result that every compact **ANR** has the homotopy type of a polyhedron: this is due to West [148]. Theorem 7.8.3 is due to Toruńczyk [136].

8. Applications

In this chapter we shall present some applications of the results obtained in chapter 7. They will illustrate the power of the Q-manifold characterization theorem.

8.1. Infinite Products

In §7.8 we proved that for every compact **ANR** X, $X \times Q$ is a Q-manifold. It is natural to ask which infinite products are compact Q-manifolds. The answer to this question is given in the following

8.1.1 THEOREM: *For every* $n \in \mathbb{N}$ *let* X_n *be a non-degenerate compact space. The following statements are equivalent:*

(a) $\prod_{n=1}^{\infty} X_n$ *is a Q-manifold,*

(b) *each* X_n *is an* **ANR** *and there is an* $n \in \mathbb{N}$ *such that* X_m *is an* **AR** *for every* $m \geq n$.

PROOF: Since every Q-manifold is an **ANR** (theorem 7.3.6(2)), the implication (a) $\Rightarrow$ (b) follows from theorem 1.5.8. For (b) $\Rightarrow$ (a), again use theorem 1.5.8 to conclude that $X = \prod_{n=1}^{\infty} X_n$ is an **ANR**. By theorem 7.8.3 it suffices to show that X has the disjoint-cells property. This is a triviality however. Given $\varepsilon > 0$ and continuous functions $f, g: I^n \to X$, find a large index $N \in \mathbb{N}$ and two distinct points $a, b \in X_N$ such that the functions $f', g': I^n \to X$ defined by

$$f'(u) = (f(u)_1, \cdots, f(u)_{N-1}, a, f(u)_{N+1}, \cdots\cdots),$$

and

$$g'(u) = (g(u)_1,\cdots,g(u)_{N-1},b,g(u)_{N+1},\cdots\cdots),$$

are ε-close to f and g, respectively.□

8.1.2. COROLLARY: *For every* $n \in \mathbb{N}$ *let* X_n *be a non-degenerate compact space. The following statements are equivalent:*

(a) $\prod_{n=1}^{\infty} X_n$ *and* Q *are homeomorphic,*

(b) *each* X_n *is an* **AR**.

PROOF: The proof of (a) ⇒ (b) follows from proposition 1.5.4. For (b) ⇒ (a), use theorem 8.1.1 to conclude that $X = \prod_{n=1}^{\infty} X_n$ is a Q-manifold. Consequently, since X is obviously contractible (corollary 1.6.7), X is a contractible compact Q-manifold, and hence a Hilbert cube by theorem 7.5.8.□

8.2. Keller's Theorem

In this section we shall prove "the first non-trivial result in infinite-dimensional topology", namely, Keller's Theorem that every compact convex and infinite-dimensional subspace of Hilbert space l^2 is homeomorphic to Q.

8.2.1. LEMMA: *Let* K *be a non-empty closed convex subset of* l^2 *and let* $x \in l^2$. *Then there is a unique point* $y \in K$ *with* $d(x,K) = \|x - y\| = d(x,y)$.

PROOF: We shall first prove that for some $y \in K$, $d(x,K) = d(x,y)$. Let $\delta = d(x,K)$.

Choose an arbitrary sequence $(y_n)_n$ in K such that

$$\lim_{n\to\infty} \|y_n - x\| = \delta. \tag{1}$$

We shall prove that $(y_n)_n$ is a Cauchy sequence. To this end, let $\varepsilon > 0$. By the parallelogram law (see page 9) it follows that for all $n,m \in \mathbb{N}$,

$$2\|y_m - x\|^2 + 2\|y_n - x\|^2 = \|(y_m + y_n) - 2x\|^2 + \|y_m - y_n\|^2. \tag{2}$$

Since K is convex, $\frac{1}{2}(y_m + y_n) \in K$ for all n and m, so that by the choice of δ,

$$\|(y_m + y_n) - 2x\| = 2\|\tfrac{1}{2}(y_m + y_n) - x\| \geq 2\delta. \tag{3}$$

Observe that by (1) there exists $N \in \mathbb{N}$ such that for every $n \geq N$,

$$(4) \qquad \|y_n - x\| < \sqrt{\delta^2 + \frac{\varepsilon^2}{4}}\,.$$

It follows from (2), (3) and (4) that for all $m,n \geq N$,

$$\|y_m - y_n\|^2 = 2\|y_m - x\|^2 + 2\|y_n - x\|^2 - \|(y_m + y_n) - 2x\|^2 <$$

$$< 2(\delta^2 + \frac{\varepsilon^2}{4}) + 2(\delta^2 + \frac{\varepsilon^2}{4}) - 4\delta^2 = \varepsilon^2.$$

We conclude that $(y_n)_n$ is Cauchy and hence, by the completeness of l^2 (lemma 1.2.5), converges.

Now let y be the limit of this sequence. Since K is closed, $y \in K$ and by (1),

$$d(x,y) = \lim_{n\to\infty} d(x,y_n) = \delta = d(x,K).$$

Consequently, a y such as in the formulation of the lemma exists. We shall next prove that y is unique. Let $y' \in K$ be such that $d(x,K) = \|x - y'\|$. For $n \in \mathbb{N}$ put $h_n = y$ for even n and $h_n = y'$ for odd n. Then

$$\lim_{n\to\infty} \|x - h_n\| = \lim_{n\to\infty} \delta = \delta,$$

so by what we just proved, the sequence $(h_n)_n$ is Cauchy. But this is possible only when $y = y'$. □

8.2.2. COROLLARY: *Let* K *be a non-empty closed and convex subset of* l^2. *Then there is a retraction* $r\colon l^2 \to K$ *such that for every* $x \in l^2$, $d(x,K) = \|x - r(x)\|$.

PROOF: For every $x \in l^2$ let $r(x) \in K$ be such that $d(x,K) = \|x - r(x)\|$; by lemma 8.2.1, r is well-defined. It is clear that r restricts to the identity on K. Also, r is continuous, for let $(x_n)_n$ be a sequence in l^2 converging to a point $x \in l^2$. Then by lemma 8.2.1 for every $n \in \mathbb{N}$ we have

$$d(x,K) = \|x - r(x)\| \leq \|x - r(x_n)\| \leq \|x - x_n\| + \|x_n - r(x_n)\| = \|x - x_n\| + d(x_n,K).$$

Therefore, since $\|x - x_n\| \to 0$, $n \to \infty$, and $d(x_n,K) \to d(x,K) = \|x - r(x)\|$, $n \to \infty$, we conclude that

$$\lim_{n\to\infty} \|x - r(x_n)\| = d(x,K).$$

From the proof of lemma 8.2.1 it now follows that the sequence $(r(x_n))_n$ converges to $r(x)$, as required. □

8.2.3. LEMMA: *Let* $F \subseteq l^2$ *be finite. If* $n = \dim \operatorname{conv}(F)$ *then* $\operatorname{conv}(F)$ *is homeomorphic to the n-cube* I^n.

PROOF: We assume without loss of generality that all finite subsets F of l^2 under consideration contain the origin $\underline{0} = (0,0,\cdots)$. By induction on the algebraic dimension dim lin(F) of the linear hull lin(F) of the set F we shall prove that conv(F) has nonempty interior in lin(F). If dim lin(F) = 0 then there is nothing to prove. So assume the lemma to be true for all finite subsets of l^2 the linear hulls of which have algebraic dimension at most n, $n \geq 0$. Assume that $F = \{x_0,\cdots,x_m\}$ is a subset of l^2 such that $x_0 = \underline{0}$ and dim lin(F) = n+1. Since $\dim \operatorname{lin}(\{x_0\}) = 0$ there is an index $i \leq m$ such that $\dim \operatorname{lin}(\{x_0,\cdots,x_i\}) = n$. Put $F_0 = \{x_0,\cdots,x_i\}$. By assumption, $\operatorname{conv}(F_0)$ has nonempty interior in $\operatorname{lin}(F_0)$. Since dim lin(F) = n+1, there is an index $j > i$, say $j = i+1$, such that $x_{i+1} \notin \operatorname{lin}(F_0)$. Observe that $\dim \operatorname{lin}\{x_0,\cdots,x_{i+1}\} = n+1 = \dim \operatorname{lin}(F)$, so $\operatorname{lin}(\{x_0,\cdots,x_{i+1}\}) = \operatorname{lin}(F)$. From figure 8.2.1 it is geometrically obvious that $\operatorname{conv}(\{x_0,\cdots,x_{i+1}\})$ has nonempty interior in $\operatorname{lin}(\{x_0,\cdots,x_{i+1}\})$.

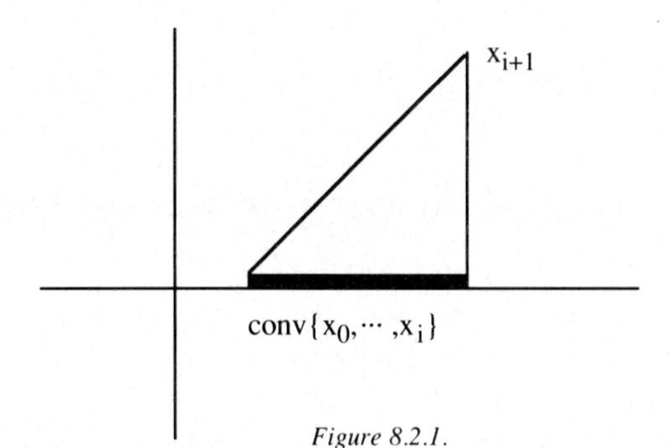

Figure 8.2.1.

Since $\operatorname{lin}(F) = \operatorname{lin}(F_0)$ and $\operatorname{conv}(\{x_0,\cdots,x_{i+1}\}) \subseteq \operatorname{conv}(F)$, we are done.

By exercise 3.5.8(c) it now follows that for every finite $F \subseteq l^2$ there is an $n \in \mathbb{N}$ such that conv(F) is homeomorphic to I^n. Since $\dim I^n = n$ (theorem 4.3.10), n is equal to dim conv(F). □

We now come to the main result in this section.

8.2.4. THEOREM ("Keller's Theorem"): *Every compact convex infinite-dimensional subspace of* l^2 *is homeomorphic to* Q.

PROOF: Let K be a compact convex, infinite-dimensional subspace of l^2. By theorem 1.5.1, K is an **AR** so by corollary 7.8.4, it suffices to prove that K has the disjoint-cells property.

Let $D = \{x_n: n \in \mathbb{N}\}$ be a countable dense subset of K, and for every n put

$$K_n = \text{conv}(\{x_1,\cdots,x_n\}).$$

CLAIM: The set $\{\dim K_n: n \in \mathbb{N}\}$ is unbounded.

Suppose that this is not the case. Then there is an integer $m \in \mathbb{N}$ such that $\dim K_n \leq m$ for every n. Since $\dim K \geq m+1$, there is an essential family $\{(A_i,B_i): i \leq m+1\}$ of pairs of disjoint closed subsets of K. By corollary 8.2.2 and the fact that D is dense there are an integer $N \in \mathbb{N}$ and a retraction $r: K \to K_N$ such that for every $i \leq m+1$,

$$r(A_i) \cap r(B_i) = \emptyset.$$

Since $\dim K_N \leq m$, there are partitions S_i $(i \leq m+1)$ in K_N between $r(A_i)$ and $r(B_i)$ such that

$$\bigcap_{i=1}^{m+1} S_i = \emptyset.$$

It follows that the sets $r^{-1}(S_i)$ $(i \leq m+1)$ are partitions between A_i and B_i in K having empty intersection; contradiction.

Now let $\varepsilon > 0$, let X be an at most n-dimensional compactum and let $f: X \to K$ be continuous. We shall prove that f can be approximated by imbeddings; since $I^n \times \{0,1\}$ is n-dimensional this implies that K has the disjoint-cells property. By corollary 8.2.2 there are an $m \in N$ and a retraction $r: K \to K_m$ such that

(1) $\qquad d(r,1_K) < \frac{1}{3}\varepsilon,$

(2) $\qquad \dim K_m \geq 2n+1.$

Put $g = r \circ f$. Observe that

(3) $\qquad d(f,g) < \frac{1}{3}\varepsilon.$

By lemma 8.2.3, K_m is homeomorphic to a cube I^k, with $k \geq 2n+1$. Since the geometrical boundary ∂I^k is a Z-set in I^k, there is a map $h: X \to I^k \backslash \partial I^k$ such that

(4) $d(g,h) < \frac{1}{3}\varepsilon$

(corollary 7.2.6). Now since $I^k \backslash \partial I^k$ is homeomorphic to $\mathbb{R}^k$, dim $X \leq n$ and $k \geq 2n+1$, theorem 4.4.4 implies that there is an imbedding $\xi: X \to I^k \backslash \partial I^k$ such that

(5) $d(\xi,h) < \frac{1}{3}\varepsilon.$

By (3), (4) and (5) it follows that $d(f,\xi) < \varepsilon$. We are done.□

8.3. Cone Characterization of the Hilbert cube

In theorem 6.1.11 we proved that the Hilbert cube Q is homeomorphic to its own cone. It turns out that Q is the *only* compact **AR** with this property.

We first derive the following simple lemma, cf. exercise 6.1.6.

8.3.1. LEMMA: *Let* X *be a compact space. Then* $\Delta(\Delta(X))$ *is homeomorphic to* $\Delta(X) \times I$.

PROOF: Observe that $\Delta(\Delta(X))$ is the space we obtain from $\Delta(X) \times I$ by identifying the set $\Delta(X) \times \{1\}$ to a single point. Let $\pi: \Delta(X) \times I \to \Delta(\Delta(X))$ be the quotient map. We shall prove that π is shrinkable, and hence a near homeomorphism (theorem 6.1.8). Observe that π has only one nondegenerate point-inverse. Consider a homeomorphism ϕ of I^2 onto itself that is supported on a small neighborhood of $I \times \{1\}$, shrinks $I \times \{1\}$ to a small set and restricts to the identity on $\{1\} \times I$. This homeomorphism induces a homeomorphism f of $\Delta(X) \times I$ by "crossing" it by the identity on the "factor" X. It is clear that ϕ serves as the desired "shrinking homeomorphism" of π.□

We now come to the main result in this section.

8.3.2. THEOREM: *For a space* X *the following statements are equivalent:*

(1) X *is homeomorphic to the Hilbert cube* Q,

(2) X *is a compact* **AR** *and is homeomorphic to its own cone* $\Delta(X)$.

PROOF: For (1) ⇒ (2), see theorem 6.1.11. For (2) ⇒ (1), observe that for every n, $X \times I^n \approx X$ (lemma 8.3.1). We shall prove that X has the disjoint-cells property. To this end, let Y be an n-dimensional space, let $\varepsilon > 0$ and let $f: Y \to X$ be continuous. We think of X as $X \times I^{2n+1}$ and we let $\pi_1: X \times I^{2n+1} \to X$ and $\pi_2: X \times I^{2n+1} \to I^{2n+1}$ denote the projections (the reader has to interpret "we think of" properly of course). As in the proof of theorem 8.2.4 it follows that the function $\pi_2 \circ f: Y \to I^{2n+1}$ can be approximated by an imbedding, say ξ. Then the function $g: Y \to X \times I^{2n+1}$ defined by

$$g(y) = (\pi_1(f(y)),\xi(y))$$

is an imbedding "close" to f. □

8.4. The Curtis-Schori-West Hyperspace Theorem

Let X be a Peano continuum and let 2^X denote its hyperspace, cf. §4.7. In §5.3 we proved that 2^X is an **AR** if and only if X is a Peano continuum (theorem 5.3.14). In this section we shall prove that 2^X is in fact homeomorphic to the Hilbert cube (this is the Curtis-Schori-West Hyperspace Theorem). As in the previous sections, all we need to prove is that 2^X has the disjoint-cells property.

Again, let X be a Peano continuum. Recall that for every closed subspace A of X we can identify 2^A and the subspace $\{B \in 2^X: B \subseteq A\}$ of 2^X, see the remarks following corollary 5.3.3. For every $n \in \mathbb{N}$, put

$$\mathcal{F}_n(X) = \{A \in 2^X: |A| \le n\}.$$

Since the function $f: X^n \to 2^X$ defined by $f(x_1,\cdots,x_n) = \{x_1,\cdots,x_n\}$ is continuous (lemma 5.3.4), it follows that each $\mathcal{F}_n(X)$ is a Peano continuum (exercise 5.3.1). Observe that by lemma 5.3.9(2),

$$\mathcal{F}_\infty(X) = \bigcup_{n=1}^\infty \mathcal{F}_n(X)$$

is dense in 2^X.

8.4.1. LEMMA: *For each* $n \in \mathbb{N}$, *there exists a map* $r: B^{n+1} \to \mathcal{F}_3(S^n)$ *such that for every* $x \in S^n$, $r(x) = \{x\}$.

PROOF: First consider the case n = 1. Coordinatize S^1 as $[0,2\pi]$, with 0 and 2π identified. For $\theta \in [0,2\pi]$, set $t_\theta = \pi - |\theta - \pi|$. Define a homotopy $H: S^1 \times [0,\pi] \to \mathcal{F}_2(S^1)$ by

$$H(\theta,t) = \begin{cases} \{\theta - t, \theta + t\} & (0 \le t \le t_\theta), \\ \{\theta - t_\theta, \theta + t_\theta\} & (t_\theta \le t \le \pi). \end{cases}$$

Observe that $H_0(\theta) = \{\theta\}$ for every $\theta \in S^1$. In addition, define a homotopy $K: S^1 \times [0,\pi] \to \mathcal{F}_3(S^1)$ by

$$K(\theta,t) = \{t,2\pi-t\} \cup (\{\theta-t_\theta,\theta+t_\theta\} \cap [t,2\pi-t]).$$

It is easy to see that the homotopy H followed by K provides a homotopy from S^1 into $\mathcal{F}_3(S^1)$ connecting the inclusion map $\theta \to \{\theta\}$ and the constant map $\theta \to \{\pi\}$.

Now we proceed by induction on n. Assume that for certain $n \geq 2$ there exists a map $f: B^n \to \mathcal{F}_3(S^{n-1})$ such that $f(x) = \{x\}$ for every $x \in S^{n-1}$. We identify B^{n+1} and the two-point compactification of $B^n \times (-1,1)$. Now define a map $g: B^{n+1} \to \mathcal{F}_3(S^n)$ as follows:

$$\begin{cases} g(x,t) = \{(p,t): p \in f(x)\} & ((x,t) \in B^n \times (-1,1)), \\ g(+\infty) = \{+\infty\}, \\ g(-\infty) = \{-\infty\}. \end{cases}$$

An easy check shows that g is as required. □

This lemma will be used in the proof of the following

8.4.2. PROPOSITION: *Let* $P = |\mathcal{T}|$ *be an n-dimensional polyhedron and let* $K = |\mathcal{K}|$ *be the 1-skeleton of* P. *There is a map* $\xi: P \to \mathcal{F}_{3n-1}(K)$ *such that if* τ *is the carrier of* $x \in P$ *then* $\xi(x)$ *is contained in the 1-skeleton of* τ.

PROOF: We induct on the dimension of P. If $\dim P \leq 1$ then define $\sigma: P \to \mathcal{F}_1(K)$ by $\sigma(x) = \{x\}$. Now assume the lemma to be true for all polyhedra of dimension at most n, $n \geq 1$. Let P be (n+1)-dimensional and let $Q = |\mathcal{T}^{(n)}|$. By assumption, there is a map $\rho: Q \to \mathcal{F}_{3n-1}(K)$ such that if τ is the carrier of $x \in Q$ then $\rho(x)$ is contained in the 1-skeleton of τ. Now let σ be an arbitrary (n+1)-dimensional simplex in $\mathcal{T}$. By lemma 8.4.1 there is a continuous function $f_\sigma: \sigma \to \mathcal{F}_3(\partial\sigma)$ such that for every $x \in \partial\sigma$, $f_\sigma(x) = \{x\}$. Now define $g_\sigma: \sigma \to 2^K$ as follows:

$$g_\sigma(x) = \bigcup\{\rho(y): y \in f_\sigma(x)\}.$$

Then g_σ is continuous by corollary 5.3.7. Also, g_σ extends $\rho \mid \partial\sigma$ because if $x \in \partial\sigma$ then

$$g_\sigma(x) = \bigcup\{\rho(y): y \in f_\sigma(x)\} = \bigcup\{\rho(y): y \in \{x\}\} = \rho(x).$$

CLAIM 1: $g_\sigma(\sigma)$ is a collection of subsets of the 1-skeleton of σ.

Since $g_\sigma \mid \partial\sigma = \rho \mid \partial\sigma$, by our inductive hypothesis, it suffices to prove that for an arbitrary $x \in \sigma\backslash\partial\sigma$ we have that $g_\sigma(x)$ is a subset of the 1-skeleton of σ. Take $z \in g_\sigma(x)$. There exists a point $y \in f_\sigma(x)$ such that $z \in \rho(y)$. Observe that $f_\sigma(x) \subseteq \partial\sigma$. So the carrier τ of y is a proper face of σ and is therefore contained in Q. Consequently, by our inductive assumption, $\rho(y)$ is a subset of the 1-skeleton of τ, which is contained in the 1-skeleton of σ. Consequently, z belongs to the 1-skeleton of σ.

Now let ξ be the union of the functions g_σ.

CLAIM 2: For every $x \in P$, $\xi(x) \in \mathcal{F}_{3^n}(K)$.

This is easy. Let σ be the carrier of x. If $x \in Q$ then we are done because ξ extends ρ. Now assume that $x \notin Q$. Then σ is (n+1)-dimensional. Since $f_\sigma(x)$ has cardinality at most 3 and for every $y \in f_\sigma(x)$, $\rho(y)$ has cardinality at most 3^{n-1}, by the definition of ξ we conclude that

$$|\xi(x)| \leq 3 \cdot 3^{n-1} = 3^n,$$

as required. □

A *graph* Γ is an at most 1-dimensional polyhedron. A subcollection $\mathcal{E}$ of $\mathcal{F}_\infty(X)$ is called an *expansion hyperspace* of X if for every $E \in \mathcal{E}$ and $F \in \mathcal{F}_\infty(X)$ we have:

$$\text{if } E \subseteq F \text{ then } F \in \mathcal{E}.$$

Observe that $\mathcal{F}_\infty(X)$ is an expansion hyperspace.

We now come to the following:

8.4.3. PROPOSITION: *Let* X *be a Peano continuum and let* $\mathcal{E}$ be *an expansion hyperspace of* X *which is dense in* 2^X. *Then for every* $\varepsilon > 0$ *there is a map* $f: 2^X \to \mathcal{E}$ *with* $d_H(f, 1_{2^X}) < \varepsilon$.

PROOF: To begin with, let us establish the following claim:

CLAIM 1: Let $\delta > 0$ and let Γ be a graph. Then for every continuous function $\phi: \Gamma \to 2^X$ there is a continuous function $\psi: \Gamma \to \mathcal{E}$ such that for every $t \in \Gamma$, $d_H(\phi(t), \psi(t)) < \delta$.

There exists $\gamma > 0$ such that $\gamma \leq \delta$ while moreover every pair of points $x,y \in X$ with $d(x,y) < \gamma$ can be joined by a path having diameter less than $\frac{1}{6}\delta$ (construct an open cover of X by "small" connected open sets and apply lemma 1.1.1 and theorem 5.3.13). For every vertex $v \in \Gamma$ pick a point $\psi(v) \in \mathcal{E}$ such that $d_H(\psi(v),\phi(v)) < \frac{1}{3}\gamma$. Now fix a 1-dimensional simplex σ of Γ. By abuse of notation it will be convenient to denote σ by $[v,w]$, where v and w the vertices of σ. There is a partition $v = a_1 < a_2 < \cdots < a_n = w$ of $[v,w]$ such that for every $i \leq n-1$ we have

$$\text{diam}(\phi([a_i,a_{i+1}])) < \tfrac{1}{3}\gamma.$$

Since $\mathcal{E}$ is dense in 2^X, for every $2 \leq i \leq n-1$ there is a point $F_i \in \mathcal{E}$ such that $d_H(F_i,\phi(a_i)) < \frac{1}{3}\gamma$; for notational convenience put $F_1 = \psi(v)$ and $F_n = \psi(w)$. Now observe that for every i,

$$d_H(F_i,F_{i+1}) < \gamma.$$

Fix $i \leq n-1$. Without loss of generality assume that $|F_i| \geq |F_{i+1}|$. There is a surjective function $s\colon F_i \to F_{i+1}$ such that for every $x \in F_i$, $d(x,s(x)) < \gamma$. So by construction there is for every $x \in F_i$ a path $P_x\colon I \to X$ connecting x and s(x) and having diameter less than $\frac{1}{6}\delta$. Now define a path κ_i connecting F_i and $F_i \cup F_{i+1}$ by

$$\kappa_i(t) = F_i \cup \{P_x(t)\colon x \in F_i\}.$$

Then $\kappa_i(I)$ has clearly diameter less than $\frac{1}{6}\delta$. Also, $\kappa_i(I) \subseteq \mathcal{E}$. Similarly, define a path λ_i connecting F_{i+1} and $F_{i+1} \cup F_i$ by

$$\lambda_i(t) = F_{i+1} \cup \{P_x(1-t)\colon x \in F_i\}.$$

As above, λ_i has diameter less than $\frac{1}{6}\delta$ and is contained in $\mathcal{E}$. We conclude that F_i and F_{i+1} can be joined by a path ψ_i in $\mathcal{E}$ of diameter less than $\frac{1}{3}\delta$.

The union of the ψ_i's defines a path $\psi_\sigma\colon I \to \mathcal{E}$. Take an arbitrary $t \in [a_i,a_{i+1}]$, $i \leq n-1$. Observe that

$$\text{diam}(\phi([a_i,a_{i+1}]) \cup \psi_i([a_i,a_{i+1}])) < \tfrac{1}{3}\gamma + \tfrac{1}{3}\delta + \tfrac{1}{3}\gamma \leq \delta.$$

Consequently, $d_H(\phi(t),\psi_\sigma(t)) < \delta$.

From this we see that the union of the ψ_σ's is as required.

Now let $\varepsilon > 0$. Since 2^X is an **AR** (theorem 5.3.14), there are a polyhedron P and maps $\alpha: 2^X \to P$ and $\beta: P \to 2^X$ such that

$$(1) \qquad d(\beta \circ \alpha, 1_{2^X}) < \tfrac{1}{2}\varepsilon$$

(theorem 5.1.8). Let $n = \dim P$. There is a triangulation $\mathcal{T}$ for P such that for every $\sigma \in \mathcal{T}$,

$$(2) \qquad \operatorname{diam}(\beta(\sigma)) < \tfrac{1}{8}\varepsilon$$

(theorem 3.6.12). Let Γ denote the 1-skeleton of P. By claim 1 there is a map $\eta: \Gamma \to \mathcal{E}$ such that

$$(3) \qquad d(\eta, \beta \mid \Gamma) < \tfrac{1}{8}\varepsilon.$$

In addition, by proposition 8.4.2 there is a function $\xi: P \to \mathcal{F}_{3n-1}(\Gamma)$ such that if τ is the carrier of $x \in P$ then $\xi(x)$ is contained in the 1-skeleton of τ. Observe that for every $x \in P$, $\xi(x)$ is a finite subset of Γ; consequently, $\eta(\xi(x))$ is a finite subset of $\mathcal{E}$ and hence is a finite collection of finite subsets of X. Now define $\Phi: P \to \mathcal{E}$ by

$$\Phi(x) = \bigcup \eta \circ \xi(x).$$

By the above, Φ is well-defined, and by corollary 5.3.7, Φ is continuous.

CLAIM 2: $d(\Phi, \beta) < \tfrac{1}{2}\varepsilon$.

Take an arbitrary $x \in P$, let τ be the carier of x, and let v be a vertex of τ. By (2) and (3) it follows easily that

$$(4) \qquad \eta(\tau \cap \Gamma) \cup \beta(\tau) \subseteq B(\beta(v), \tfrac{1}{4}\varepsilon).$$

Now consider the collection $\mathcal{S} = \eta \circ \xi(x)$. By (4), for every $S \in \mathcal{S}$ it follows that $d_H(\beta(v), S) < \tfrac{1}{4}\varepsilon$. This easily implies that $d_H(\beta(v), \Phi(x)) = d_H(\beta(v), \cup\mathcal{S}) < \tfrac{1}{4}\varepsilon$. By (4) it therefore follows that

$$d_H(\Phi(x), \beta(x)) < \tfrac{1}{2}\varepsilon,$$

as required.

Now define $f: 2^X \to \mathcal{E}$ by $f = \alpha \circ \Phi$. By (1) and claim 2, we are done. □

8.4.4. COROLLARY: Let X *be a Peano continuum and let* $\mathcal{K}$ *be a compact subset of* $\mathcal{F}_\infty(X)$. *Then* $\mathcal{K}$ *is a Z-set in* 2^X.

PROOF: Define

$$\mathcal{E} = \{F \in \mathcal{F}_\infty(X): F \text{ is not contained in an element of } \mathcal{K}\}.$$

Observe that $\mathcal{E}$ is an expansion hyperspace of X and that $\mathcal{E} \cap \mathcal{K} = \emptyset$. We claim that $\mathcal{E}$ is dense in 2^X. Since $\mathcal{F}_\infty(X)$ is dense in 2^X it suffices to prove that for arbitrary $F \in \mathcal{F}_\infty(X)$ and $\varepsilon > 0$ there exists an element $E \in \mathcal{E}$ with $d_H(F,E) < \varepsilon$. So let $F \in \mathcal{F}_\infty(X)$ and $\varepsilon > 0$ be given. Pick a point $x \in F$ and a sequence $(x_n)_n$ in $X \setminus \{x\}$ converging to x, with $d(x_n,x) < \varepsilon$ for each n. Define

$$F_n = F \cup \{x_1,\cdots,x_n\} \qquad (n \in \mathbb{N}).$$

Observe that for each n, $F_n \in \mathcal{F}_\infty(X)$ and $d_H(F,F_n) < \varepsilon$. We claim that for some n, $F_n \in \mathcal{E}$. If not, then there exists a sequence $(K_n)_n$ in $\mathcal{K}$ with $F_n \subseteq K_n$ for each n. By compactness of $\mathcal{K}$ we assume without loss of generality that $K_n \to K$, $n \to \infty$.

CLAIM: $\{x_1,x_2,x_3,\cdots\} \subseteq K$.

Suppose that for some n, $x_n \notin K$. Since $K_n \to K$, $n \to \infty$, there is an $m \geq n$ such that $K_m \subseteq K \setminus \{x_n\}$. This is a contradiction because $x_n \in F_m \subseteq K_m \subseteq K \setminus \{x_n\}$.

From the claim we now conclude that K is infinite, contradicting the fact that $K \in \mathcal{F}_\infty(X)$.

We conclude that $\mathcal{E}$ is dense in 2^X. By proposition 8.4.3 there are arbitrarily small maps $2^X \to \mathcal{E}$. Since $\mathcal{E} \cap \mathcal{K} = \emptyset$ this implies that $\mathcal{K} \in \mathcal{Z}(2^X)$, as required. □

8.4.5. THEOREM ("The Curtis-Schori-West Hyperspace Theorem"): *Let* X *be compact space. The following statements are equivalent:*

(a) X *is a Peano continuum,*

(b) 2^X *is homeomorphic to the Hilbert cube* Q.

PROOF: As observed above, it suffices to prove that for a Peano continuum X, 2^X has the disjoint-cells property or, equivalently, the Z-approximation property (theorem 7.3.5). Since 2^X admits arbitrarily small maps into $\mathcal{F}_\infty(X)$ (proposition 8.4.3) and since every compact subset of $\mathcal{F}_\infty(X)$ is a Z-set in 2^X (corollary 8.4.4), this is obvious. □

Notes and comments for chapter 8.

§1.

The results in this section are due to West [146] and Edwards [45; chapter 14].

§2.

Keller's Theorem is due to Keller [81].

§3.

Lemma 8.3.1 is due to Schori [123]; the proof seems to be new. Theorem 8.3.2 is due independently to Lay and Walsh [92], Toruńczyk [135] and Wong [152].

§4.

Wojdysławski [151] proved that the hyperspace of every Peano continuum is an **AR** and conjectured that in fact $2^X \approx Q$ if and only if X is a Peano continuum. In [125], Schori and West verified this conjecture in the case X = [0,1]. Their result was used in [124] to prove that $2^\Gamma \approx Q$ for every connected graph. Based on these results of Schori and West, in [49] Curtis and Schori verified Wojdysławski's conjecture. The proof that 2^X has the disjoint-cells property presented in this section is implicit in Curtis and To Nhu [48]. Using the existence of convex metrics on Peano continua it is possible to give different proofs of this fact (Toruńczyk [137] and Curtis (private communication)).

What Next ?

In this book we touched upon several branches of general and geometric topology. For example, general topology (for further information see e.g. [59]), dimension theory (see e.g. [80], [60]), shape theory and algebraic topology (see e.g. [98], [130] and [111]) and manifold topology (see e.g. [50], [21] and [45]). Much emphasis in chapter 7 was on the so-called cell-like maps. There are numerous interesting problems about them, the most prominent whether the cell-like image of a finite-dimensional compactum is again finite-dimensional (this question was solved recently in the negative by A.N. Dranishnikov, *On a problem of P.S. Alexandrov,* Matem. Sbornik 135 (1988) 551-557). For information about cell-like maps see e.g. [5] and [53]. Let us also notice that Toruńczyk's Q-manifold Characterization Theorem can also be derived for locally compact spaces [136] and that a similar result holds for spaces modeled on Hilbert space l^2 [137] (see also [31]) and on certain countable dimensional linear spaces [108]. For open problems about infinite-dimensional topology see [67] and for "finite-dimensional" results that are "infinite-dimensional" in spirit, see [22].

BIBLIOGRAPHY

1. Alexandrov, P., *Une définition des nombres de Betti pour un ensemble fermé quelconque,* C.R. Acad. Paris 184 (1927) 317-319.
2. Alexandrov, P., *Dimensionstheorie. Ein Beitrag zur Geometrie der abgeschlossenen Mengen,* Math. Ann. 106 (1932) 161-238.
3. Alexandrov, P. and P. Urysohn, *Über nuldimensionale Punktmengen,* Math. Ann. 98 (1928) 89-106.
4. Alexandrov, P. and P. Urysohn, *Mémoire sur les espaces topologiques compacts,* Verh. Akad. Wetensch. Amsterdam 14 (1929).
5. Ancel, F.D., *The role of countable dimensionality in the theory of cell-like relations,* Trans. Am. Math. Soc. 287 (1985) 1-40.
6. Anderson, R.D., *The Hilbert cube as a product of dendrons,* Notices Am. Math. Soc. 11 (1964) p. 572.
7. Anderson, R.D., *Hilbert space is homeomorphic to the countable infinite product of lines,* Bull. Am. Math. Soc. 72 (1966) 515-519.
8. Anderson, R.D., *On topological infinite deficiency,* Mich. Math. J. 14 (1967) 365-383.
9. Anderson, R.D., *Topological properties of the Hilbert cube and the infinite product of open intervals,* Trans. Am. Math. Soc. 126 (1967) 200-216.
10. Anderson, R.D., *Strongly negligible sets in Fréchet manifolds,* Bull. Am. Math. Soc. 75 (1969) 64-67.
11. Anderson, R.D., *On sigma-compact subsets of infinite-dimensional manifolds,* unpublished manuscript.
12. Anderson, R.D. and T.A. Chapman, *Extending homeomorphisms to Hilbert cube manifolds,* Pacific J. Math. 38 (1971) 281-293.
13. Anderson, R.D., D.W. Curtis and J. van Mill, *A fake topological Hilbert space,* Trans. Am. Math. Soc. 272 (1982) 311-321.
14. Anderson, R.D. and J.E. Keisler, *An example in dimension theory,* Proc. Am. Math. Soc. 18 (1967) 709-713.
15. Anderson, R.D. and R. Schori, *Factors of infinite-dimensional manifolds,* Trans. Am. Math. Soc. 142 (1969) 315-330.
16. Banach S., *Théorie des opérations linéaires,* PWN Warszawa, 1932.
17. Barit, B., *Small extensions of small homeomorphisms*, Notices Am. Math. Soc. 16 (1969) p. 295.
18. Bartle, R.G. and L.M. Graves, *Mappings between function spaces,* Trans. Am. Math. Soc. 72 (1952) 400-413.
19. Bessaga C. and A. Pełczyński, *A topological proof that every separable Banach space is homeomorphic to a countable product of lines,* Bull. Acad. Polon. Sci. Sér. sci. math. astr. et phys. 17 (1969) 487-493.
20. Bessaga C. and A. Pełczyński, *The estimated extension theorem, homogeneous collections and skeletons, and their application to the topological classification of linear metric spaces and convex sets,* Fund. Math. 69 (1970) 153-190.
21. Bessaga, C. and A. Pełczyński, *Selected topics in infinite-dimensional topology,* PWN, Warszawa, 1975.

22. Bestvina, M., *Characterizing k-dimensional Menger compacta*, Ph. D. thesis, University of Tennessee, Knoxville, Tennessee.

23. Bing, R.H., *A homeomorphism between the 3-sphere and the sum of two solid horned spheres*, Annals of Math. 56 (1952) 354-362.

24. Borsuk, K., *Sur les rétractes*, Fund. Math. 17 (1931) 152-170.

25. Borsuk, K., *Über eine klasse von lokal zusammenhängende Räumen*, Fund. Math. 19 (1932) 220-242.

26. Borsuk, K., *Drei Sätze über die n-dimensionale euclidische Sphäre*, Fund. Math. 20 (1933) 177-190.

27. Borsuk, K., *Sur les prolongements des transformations continus*, Fund. Math. 28 (1936) 99-110.

28. Borsuk, K., *Un théorème sur les prolongements des transformations*, Fund. Math. 29 (1937) 161-166.

29. Borsuk, K., *Theory of retracts*, PWN Warszawa, 1967.

30. Bourbaki, N., *Topologie Générale*, 2nd ed., Chap. 9, Hermann, Paris, 1958.

31. Bowers, P., M. Bestvina, J. Mogilski and J.J. Walsh, *Characterizing Hilbert space manifolds revisited*, Top. Appl. 24 (1986) 53-69.

32. Brouwer, L.E.J., *On the structure of perfect sets of points*, Proc. Akad. Amsterdam 12 (1910) 785-794.

33. Brouwer, L.E.J., *Über Abbildung von Mannigfaltigkeiten*, Math. Ann. 71 (1912) 97-115.

34. Brouwer, L.E.J., *Invariantz des n-dimensionalen Gebiets*, Math. Ann. 71 (1912)305-313; 72 (1913) 55-56.

35. Brouwer, L.E.J.,*Über den natürlichen Dimensionsbegriff*, Journ. für die reine und angew. Math., 142 (1913) 146-152.

36. Brouwer, L.E.J., *Bemerkungen zum natürlichen Dimensionsbegriff*, Proc. Akad. Amsterdam 27 (1924) 635-638.

37. Brown, A.L. and A. Page, *Elements of functional analysis*, Van Nostrand Reinhold, London, 1970.

38. Brown, M.G., *Some applications of an approximation theorem for inverse limits*, Proc. Am. Math. Soc. 11 (1960) 478-483.

39. Brown, M.G., *Locally flat embeddings of topological manifolds*, Topology of 3-manifolds and Related Topics, Prentice-Hall, Englewood Cliffs, N.J., 1962, pp. 83-91.

40. Burckel, R.B., *Inessential maps and classical euclidean topology*, Jahrbuch überblicke Mathematik, 1981, 119-137.

41. Cantor, G., *Über unendliche, lineare Punktmannichfaltigkeiten*, Math. Ann. 21 (1883) 545-591.

42. Čech, E., *Sur la théorie de la dimension*, C.R. Acad. Paris 193 (1931) 976-977.

43. Čech, E., *Prispevek k theorii dimense*, Casopis Pest. Mat. Fys. 62 (1933) 277-291 (French translation: E. Čech, *Topological papers*, Prague 1968, 129-142).

44. Chapman, T.A., *On the structure of Hilbert cube manifolds*, Comp. Math. 24 (1972) 329-353.

45. Chapman, T.A., *Lectures on Hilbert cube manifolds*, CBMS 28, Providence, 1975.

46. Chapman, T.A., *All Hilbert cube manifolds are triangulable*, preprint.

47. Connely, R., *A new proof of Brown's collaring theorem*, Proc. Am. Math. Soc. 27 (1971) 180-182.

48. Curtis, D.W. and N.T. Nhu, *Hyperspaces of finite subsets which are homeomorphic to $\aleph_0$-dimensional linear metric space*, Top. Appl. 19 (1985) 251-260.

49. Curtis, D.W. and R.M. Schori, *Hyperspaces of Peano continua are Hilbert cubes,* Fund. Math. 101 (1978) 19-38.

50. Daverman, R.J., *Decompositions of manifolds,* Academic Press, New York, 1986.

51. Dijkstra, J.J., *A generalization of the Sierpiński Theorem,* Proc. Am. Math Soc. 91 (1984) 143-146.

52. Dowker, C.H., *Mapping theorems for non-compact spaces,* Amer. Journ. of Math. 69 (1947) 200-242.

53. Dranishnikov, A.N. and E.V. Shchepin, *Cell-like maps. The problem of raising dimension,* Russ. Math. Surveys 41 (1986) 59-111.

54. Dugundji, J., *An extension of Tietze's theorem,* Pac. J. Math. 1 (1951) 353-367.

55. Dugundji, J., *Absolute neighborhoods retracts and local connectedness in arbitrary metric spaces,* Comp. Math. 13 (1958) 229-246.

56. Edwards, R.D., *A theorem and a question related to cohomological dimension and cell-like maps,* Notices Amer. Math. Soc. 25 (1978) A-259 - A-260.

57. Edwards, R.D., *Characterizing infinite dimensional manifolds topologically,* Séminaire Bourbaki 540, Lecture Notes in Mathematics 842 (1979) 278-302.

58. Eilenberg, S. and E. Otto, *Quelques propriétés caractéristiques de la théorie de la dimension,* Fund. Math. 31 (1938) 149-153.

59. Engelking, R., *General Topology,* PWN Warszawa, 1977.

60. Engelking, R., *Dimension Theory,* PWN Warszawa 1978.

61. Engelking, R. and R. Pol, *Compactifications of countable-dimensional and strongly countable-dimensional spaces,* preprint.

62. Erdös, P., *The dimension of the rational points in Hilbert space,* Ann. of Math. 41 (1940) 734-736.

63. Flores, A., *Über n-dimensionale Komplexe die im* R_{2n+1} *absolut selbstverschlungen sind,* Ergebnisse eines math. Koll. 6 (1935) 4-6.

64. Fort, M., *Homogeneity of infinite products of manifolds with boundary,* Pacific J. Math. 12 (1962) 879-884.

65. Freudenthal, H., *Entwicklungen von Räumen und ihren Gruppen,* Comp. Math. 4 (1937) 145-234.

66. Fréchet, M., *Les espaces abstraits,* Paris, 1928.

67. Geoghegan, R., *Open problems in infinite-dimensional topology,* Topology Proc. 4 (1979) 287-338.

68. Groot, J. de, *Non-archimedean metrics in topology,* Proc. Amer. Math. Soc. 7 (1956) 948-953.

69. Hanner, O., *Some theorems on absolute neighborhood retracts,* Arkiv Mat., Svenska Vetens. Akad., 1 (1951) 389-408.

70. Hausdorff, F., *Über halbstetige Fuktionen und deren Verallgemeinerung,* Math. Zeitschr. 5 (1919) 292-309.

71. Hausdorff, F., *Über innere Abbildungen,* Fund. Math. 23 (1934) 279-291.

72. Haver, W.E., *Mappings between* **ANR**'s *that are fine homotopy equivalences,* Pacific J. Math. 58 (1975) 457-461.

73. Henderson, D.W., *An infinite-dimensional compactum with no positive-dimensional compact subsets - a simpler construction,* Amer. J. Math. 89 (1967) 105-121.

74. Hu, S.T., *Theory of retracts,* Wayne State University Press, Detroit, 1965.

75. Hurewicz, W., *Über stetige Bilder von Punktmengen,* Proc. Akad. Amsterdam 29 (1926) 1014-1017.

76. Hurewicz, W., *Normalbereiche und Dimensionstheorie,* Math. Ann. 96 (1927) 736-764.

77. Hurewicz, W., *Über das Verhältnis separabler Räume zu kompakten Räume,* Proc. Akad. Amsterdam 30 (1927) 425-430.

78. Hurewicz, W., *Sur la dimension des produits Cartésiens,* Ann. of Math. 36 (1935) 194-197.

79. Hurewicz, W., *Über Abbildungen topologischer Räume auf die n-dimensionale Sphäre,* Fund. Math. 24 (1935) 144-150.

80. Hurewicz, W. and H. Wallman, *Dimension theory,* Van Nostrand, Princeton, N.J., 1948.

81. Keller, O.H., *Die Homoiomorphie der kompakten konvexen Mengen in Hilbertschen Raum,* Math. Ann. 105 (1931) 748-758.

82. Klee, V.L., *Some topological properties of convex sets,* Trans. Am. Math. Soc. 78 (1955) 30-45.

83. Knaster, B., K. Kuratowski and S. Mazurkiewicz, *Ein Beweis des Fixpunktsatzes für n-dimensionale Simplexe,* Fund. Math. 14 (1929) 132 -137.

84. Kodama, Y., *Cohomological dimension theory,* Appendix to: K. Nagami, *Dimension theory* (New York).

85. Kozlowski, G., *Images of* **ANR**'s, unpublished manuscript.

86. Kuratowski, K., *Sur l'application des espaces fonctionnels à la théorie de la dimension,* Fund. Math. 18 (1932) 285-292.

87. Kuratowski, K., *Sur un theorème fondamental concernant le nerf d'un système d'ensembles,* Fund. Math. 20 (1933) 191-196.

88. Kuratowski, K., *Quelques problèmes concernant les espaces métriques non-séparables,* Fund. Math. 25 (1935) 534-545.

89. Kuratowski, K., *Sur les espaces localement connexes et péaniens en dimension n,* Fund. Math. 24 (1935) 269-287.

90. Kuratowski, K., *Topology II,* New-York, 1968.

91. Lavrentiev, M., *Contribution à la théorie des ensembles homéomorphes,* Fund. Math. 6 (1924) 149-160.

92. Lay, T.L. and J.J. Walsh, ANR's admitting an interval factor are Q-manifolds, Proc. Am. Math. Soc. 73 (1979) 279.

93. Lefschetz, S., *On compact spaces,* Ann. of Math. 32 (1931) 521-538.

94. Lefschetz, S., *Topics in Topology,* Annals of Math. Studies, No. 10, 1942.

95. Lelek, A., *On the dimension of remainders in compact extensions,* Dokl. Akad. Nauk SSSR 160 (1965) 534-537; English translation in Soviet Math. Dokl. 6 (1965) 136-140.

96. Lusternik, L. and L. Schnirelman, *Topological methods in variational calculus* (Russian), Moscow, 1930.

97. Maehara, R., *The Jordan curve theorem via the Brouwer fixed point theorem,* Am. Math. Monthly 91 (1984) 641-643.

98. Mardešić, S. and J. Segal, *Shape Theory,* North-Holland, Amsterdam, 1982.

99. Mazurkiewicz, S., *O arytmetyzacji continuów,* C.R. Varsovie 6 (1913) 305-311.

100. Mazurkiewicz, S., *Sur les problème* κ et λ de Urysohn, Fund. Math. 10 (1927) 311-319.

101. Menger, K., *Über die Dimension von Punktmengen I,* Monatsh. für Math. und Phys., 33 (1923) 148-160.

102. Menger, K., *Über die Dimension von Punktmengen II,* Monatsh. für Math. und Phys., 34 (1926) 137-161.

103. Menger, K., *Dimensionstheorie,* Leipzig, 1928.

104. Michael, E., *Some extension theorems for continuous functions,* Pac. J. Math. 3 (1953) 789-806.

105. Michael, E., *Continuous selections* I, Ann. of Math. 63 (1956) 361-382.

106. Mill, J. van, *Domain invariance in infinite-dimensional linear spaces,* Proc. Am. Math. Soc. 101 (1987) 173-180.

107. Miller, R.J., *Mapping cylinder neighborhoods of some* **ANR**'s, Annals of Math. 106 (1977) 1-18.

108. Mogilski, J., *Characterizing the topology of infinite-dimensional σ-compact manifolds,* Proc. Am. Math. Soc. 92 (1984) 111- 118.

109. Moise, E., *Geometric topology in dimensions 2 and 3,* Springer, Berlin, 1977.

110. Morita, K., *Star-finite coverings and star-finite property,* Math. Japonicae, 1 (1948) 60-68.

111. Munkres, J.R., *Elements of Algebraic Topology,* Addison-Wesley Publishing Company, Menlo Park, 1984.

112. Nadler, S.B., *Hyperspaces of sets,* Marcel Dekker, New York and Basel, 1978.

113. Nöbeling, G., *Über eine n-dimensionale Universalmenge in* $\mathbb{R}^{2n+1}$, Math. Ann. 104 (1931) 71-80.

114. Parthasarathy, K.R., *Probability measures on metric spaces,* Academic Press, New York, 1967.

115. Poincaré, H., *Sur les courbes définies par les équations différentielles,* Jour. de math. pures et appliq. (4) 1 (1885) 167-244.

116. Pol, R., *A weakly infinite-dimensional compactum which is not countable-dimensional,* Proc. Am. Math. Soc. 82 (1981) 634-636.

117. Pol, R., *Countable dimensional universal sets,* Trans. Am. Math. Soc. 297 (1986) 255-268.

118. Pontrjagin, L.S., *Sur une hypothèse fondamentale de la théorie de la dimension,* C.R. Acad. Paris 190 (1930) 1105-1107.

119. Pontrjagin, L.S. and G. Tolstowa, *Beweis des Mengerschen Einbettungssatzes,* Math. Ann. 105 (1931) 734-747.

120. Rubin, L., R.M. Schori and J.J. Walsh, *New dimension-theory techniques for constructing infinite-dimensional examples,* Gen. Top. Appl. 10 (1979) 93 - 102.

121. Rudin, W., *Real and complex analysis,* Mc-Graw-Hill, New York, 1970.

122. Schauder, J., *Der Fixpunktsatz in Funktionalräumen,* Studia Math. 2 (1930) 171-180.

123. Schori, R., *Hyperspaces and symmetric products of topological spaces,* Fund. Math. 63 (1968) 77-88.

124. Schori R.M. and J.E. West, *Hyperspaces of graphs are Hilbert cubes,* Pacific J. Math. 53 (1974) 239-251.

125. Schori R.M. and J.E. West, *The hyperspace of the closed interval is a Hilbert cube,* Trans. Am. Math. Soc. 213 (1975) 217-235.

126. Semadeni, Z., *Banach spaces of continuous functions,* PWN, Warszawa, 1971.

127. Sierpiński, W., *Un théorème sur les continus,* Tôhoku Math. Journal 13 (1918) 300-303.

128. Sierpiński, W., *Sur une propriété topologique des ensembles dénombrables denses en soi,* Fund. Math. 1 (1920) 11-16.

129. Sierpiński, W., *Sur les ensembles connexes et non connexes,* Fund. Math. 2 (1921) 81-95.

130. Spanier, E., *Algebraic Topology*, Springer-Verlag Berlin, 1982.

131. Sperner, E., *Neuer Beweis für die Invarianz des Dimensionszahl und des Gebietes*, Abh. Math. Semin. Hamburg. Univ. 6 (1928) 265-272.

132. Taylor, A.E., *Introduction to functional analysis*, John Wiley and Sons, London, 1964.

133. Taylor, J.L., *A counterexample in shape theory*, Bull. Am. Math. Soc. 81 (1975) 629-632.

134. Toruńczyk, H., *Concerning locally homotopy negligible sets and characterizations of l_2-manifolds*, Fund. Math. 101 (1978) 93-110.

135. Toruńczyk, H., *Characterization of infinite-dimensional manifolds*, Proc. Int. Conf. Geometric Top., PWN Warszawa, 1980, 431-437.

136. Toruńczyk, H., *On CE-images of the Hilbert cube and characterizations of Q-manifolds*, Fund. Math. 106 (1980) 31-40.

137. Toruńczyk, H., *Characterizing Hilbert space topology*, Fund. Math. 111 (1981) 247-262.

138. Tumarkin, L.A., *Beitrag zur allgemeinen Dimensionstheorie*, Mam. Cb. 33 (1926) 57-86.

139. Urysohn, P., *Les multiplicités Cantoriennes*, C.R. Acad. Paris 175 (1922) 440-442.

140. Urysohn, P., *Mémoire sur les multiplicités Cantoriennes (suite)*, Fund. Math. 8 (1926) 225-359.

141. Veblen, O., *Theory of plane curves in nonmetrical analysis situs*, Trans. Am. Math. Soc. 6 (1905) 83-98.

142. Wall, C.T.C., *A geometric introduction to topology*, Addison-Weshley, 1972.

143. Walsh, J.J., *Infinite-dimensional compacta containing no n-dimensional ($n \geq 1$) subsets*, Topology 18 (1979) 91-95.

144. Walsh, J.J., *Dimension, cohomological dimension and cell-like mappings*, Lecture Notes in Math. 870 (1981) 105-118.

145. Walsh, J.J., *Characterization of Hilbert cube manifolds: an alternate proof*, Geometric and Algebraic Topology, Banach Center Publications 18, PWN, Warszawa, 1986, 153-160.

146. West, J.E., *Infinite products which are Hilbert cubes*, Trans. Am. Math. Soc. 150 (1970) 1-25.

147. West, J.E., *The ambient homeomorphy of an incomplete subspace of an infinite-dimensional Hilbert space*, Pacific J. Math. 34 (1970) 257-267.

148. West, J.E., *Mapping Hilbert cube manifolds to* **ANR**'s: *a solution to a conjecture of Borsuk*, Annals of Math., 106 (1977) 1-18.

149. Whitehead, J.H.C., *Simplicial spaces, nuclei, and m-groups*, Proc. London Mathematical Soc. 45 (1939) 243-327.

150. Whitehead, J.H.C., *Note on a theorem of Borsuk*, Bull. Am. Math. Soc. 54 (1948) 1133-1145.

151. Wojdysławski, M., *Rétractes absolus et hyperespaces des continus*, Fund. Math. 32 (1939) 184-192.

152. Wong, R.Y.T., *On products of* **ANR**'s *and characterization of Q-manifolds*, Proc. Topology Conference, Budapest 1978.

153. Wong, R.Y.T and N. Kroonenberg, *Unions of Hilbert cubes*, Trans. Am. Math. Soc. 211 (1975) 289-294.

A

B

C

D

E

F

Z